Canopy Arthropods

Canopy Arthropods

Edited by

N.E. Stork

Head of Biodiversity Division in the Department of Entomology, The Natural History Museum, London, UK; Now Director of the Cooperative Research Centre for Tropical Rainforest Ecology and Management, James Cook University, Cairns, Australia

J. Adis

Senior Scientist, Tropical Ecology Working Group, Max-Planck-Institute for Limnology, Plön, Germany

and

R. K. Didham

Commonwealth Scholar, The Natural History Museum, London and the NERC Centre for Population Biology, Ascot, UK

CHAPMAN & HALL

London · Weinheim · New York · Tokyo · Melbourne · Madras

Published by Chapman & Hall, 2–6 Boundary Row, London SE1 8HN, UK

Chapman & Hall, 2–6 Boundary Row, London SE1 8HN, UK

Chapman & Hall GmbH, Pappelallee 3, 69469 Weinheim, Germany

Chapman & Hall USA, 115 Fifth Avenue, New York, NY 10003, USA

Chapman & Hall Japan, ITP-Japan, Kyowa Building, 3F, 2–2–2 Hirakawacho, Chiyoda-ku, Tokyo 102, Japan

Chapman & Hall Australia, 102 Dodds Street, South Melbourne, Victoria 3205, Australia

Chapman & Hall India, R. Seshadri, 32 Second Main Road, CIT East, Madras 600 035, India

First edition 1997

Typset in 10/12pt Palatino by Florencetype Ltd, Stoodleigh, Devon
Printed in Great Britain by St Edmundsbury Press Ltd, Bury St Edmunds, Suffolk

ISBN 0 412 74900 9

A catalogue record for this book is available from the British Library

Library of Congress Catalog Number: 96–85876

∞ Printed on acid-free text paper, manufactured in accordance with ANSI/NISO 239.48-1992 and ANSI/NISO Z39.48–1984 (Permanence of Paper).

Contents

List of contributors

Aberlenc, H.P.
Laboratoire de Faunistique et de Taxonomie
CIRAD, B.P. 5035
34032 Montpellier
France

Adis, J.
Max-Planck-Institute for Limnology
Tropical Ecology Working Group
Postfach 165, 24302 Plön
Germany

Allison, A.
Bishop Museum
Box 19000-A
Honolulu
Hawaii 96817
USA

Amedegnato, C.
Laboratoire d'Entomologie
Museum National d'Histoire Naturelle
45 Rue de Buffon
75005 Paris
France

Azarbayjani, F.F.
School of Science
University of Western Sydney
Hawkesbury, Richmond, N.S.W.
Australia 2753

Basset, Y.
International Institute of Entomology
56 Queen's Gate
London SW7 3JR
UK

Brendell, M.J.D.
Department of Entomology
The Natural History Museum
Cromwell Road
London SW7 5BD
UK

Burgin, S.
School of Science
University of Western Sydney
Hawkesbury, Richmond, N.S.W.
Australia 2753

Davies, J.G.
Department of Entomology
The Natural History Museum
Cromwell Road
London SW7 5BD
UK
Now at: Cresswell Associates
Environmental Consultants
The Court House
High Street
Bisley
Gloucestershire
GL6 7AA
UK

Davis, A.J.
Department of Pure and Applied Biology
University of Leeds
Leeds LS2 9JT
UK
Now at:
Danum Valley Field Centre
Rakyat Berjaya, Sdn. Bhd.
PO Box 282
91108 Lahad Datu
Sabah, Malaysia

Delvare, G.
Laboratoire de Faunistique et de Taxonomie
CIRAD, B.P. 5035
34032 Montpellier
France

Didham, R.K.
Zoology Department
University of Canterbury
P.O. Box 4800
Christchurch
New Zealand
Now at:
Department of Entomology
The Natural History Museum
Cromwell Road
London SW7 5BD
UK

Floren, A.
Theodor-Boveri-Institut
Zoologie III
Biozentrum
Am Hubland
D-97074 Würzburg
Germany

Foggo, A.
Department of Zoology
University of Oxford
South Parks Road
Oxford OX1 3PX
UK

Fonseca, C.R.V. da
Departamento de Entomologia
Instituto Nacional de Pesquisas da Amazônia
C.P. 478
69011–970, Manaus, AM
Brazil

Guilbert, E.
Laboratoire d'Entomologie
Museum National d'Histoire Naturelle
45 Rue de Buffon
75005 Paris
France
Now at:
Departamento de Biologia
Universidad de Antioquia
AA 1226 Medellin
Colombia

Hambler, C.
Department of Zoology
University of Oxford
South Parks Road
Oxford OX1 3PS
UK

Hammond, P.M.
Department of Entomology
The Natural History Museum
Cromwell Road
London SW7 5BD
UK

Harada, A.Y.
Departamento de Entomologia
Instituto Nacional de Pesquisas da Amazônia
C.P. 478
69011–970, Manaus, AM
Brazil

Now at:
Muséu Paraense Emílio Goeldi
C.P. 399

66040–170 Belém-Pará
Brazil

Hine, S.J.
Department of Entomology
The Natural History Museum
Cromwell Road
London SW7 5BD
UK

Huijbregts, J.
National Museum of Natural History
Postbus 9517
2300 RA Leiden
The Netherlands

Kerck, K.
Fachhochschule Hildesheim/Holzminden
Fachbereich Forstwirtschaft
Büsgenweg 4
37077 Göttingen
Germany

Kirk-Spriggs, A.H.
Department of Entomology
National Museum of Wales
Cathays Park
Cardiff CF1 3NP
UK

Kitching, R.L.
Cooperative Research Centre for Tropical Ecology and Management
Faculty of Environmental Science
Griffith University
Nathan, Brisbane
Australia 4111

Krikken, J.
National Museum of Natural History
Postbus 9517
2300 RA Leiden
The Netherlands

Linsenmair, K.E.
Theodor-Boveri-Institut
Zoologisches Institut III
Biozentrum
Am Hubland
D-97074 Würzburg
Germany

Lutubula, S.
School of Science
University of Western Sydney
Hawkesbury, Richmond, N.S.W.
Australia 2753

Malcolm, J.R.
Florida Museum of Natural History
University of Florida
Gainesville, Florida 32611
USA
Now at:
Faculty of Forestry
University of Toronto
Toronto
Ontario M5S 3B3
Canada

Mawdsley, N.A.
Department of Entomology
The Natural History Museum
Cromwell Road
London SW7 5BD
UK
Now at:
Department of Pure and Applied Biology
University of Leeds
Leeds LS2 9JT
UK

Miller, S.E.
Bishop Museum
Box 19000-A
Honolulu
Hawaii 96817
USA

Mitchell, H.
Cooperative Research Centre for Tropical Ecology and Management
Faculty of Environmental Science
Griffith University
Nathan, Brisbane
Australia 4111

Morse, G.
Cooperative Research Centre for Tropical Ecology and Management
Faculty of Environmental Science
Griffith University
Nathan, Brisbane
Australia 4111

Ozanne, C.M.
School of Life Sciences
Roehampton Institute London
Whitelands College
West Hill
London SW15 3SN
UK

Paarmann, W. and D.
Fachhochschule Hildesheim/Holzminden
Fachbereich Forstwirtschaft
Büsgenweg 4
37077 Göttingen
Germany

Prinzing, A.
Universität Kiel
Zoologisches Institut
LS Ökologie
Olshausenstr. 40
24098 Kiel
Germany
Now at:
Staatl. Museum f. Naturkunde
Zoologie
Pf 6209
76042 Karlsruhe
Germany

Rafael, J.A.
Departamento de Entomologia
Instituto Nacional de Pesquisas da Amazônia
C.P. 478
69011–970, Manaus, AM
Brazil

Richardson, B.J.
School of Science
University of Western Sydney
Hawkesbury, Richmond, N.S.W.
Australia 2753

Riede, K.
Institut für Biologie I (Zoologie)
Albert-Ludwigs-Universität
Albertstr. 21a
D-79104 Freiburg
Germany

Samuelson, G.A.
Bishop Museum
Box 19000-A
Honolulu
Hawaii 96817
USA

Speight, M.R.
Department of Zoology
University of Oxford
South Parks Road
Oxford OX1 3PX
UK

Springate, N.D.
Département d'Entomologie
Muséum d'Histoire Naturelle
C.P. 6434
1211 Genève 6
Switzerland
Now at:
Department of Entomology
The Natural History Museum

Cromwell Road
London SW7 5BD
UK

Stork, N.E.
Department of Entomology
The Natural History Museum
Cromwell Road
London SW7 5BD
UK
Now at:
Cooperative Research Centre for Tropical Rainforest Ecology and Management
James Cook University
P.O. Box 6811
Cairns
Queensland 4870
Australia

Sutton, S.L.
Department of Geography
University of Manchester
Manchester M13 9PL
UK

Thebaud, C.
Department of Environmental Sciences
Griffith University
Nathan
Brisbane
Australia 4111
Now at:
Department of Biology
Imperial College at Silwood Park
Ascot, Berks SL5 7PY
UK

Wagner, T.
Zoologisches Forschungsinstitut und Museum Alexander Koenig
Adenauerallee 160
D-53113 Bonn
Germany

Watanabe, H.
Faculty of Agriculture
Kyoto University
Kyoto 606–01
Japan

Winchester, N.N.
Biology Department
University of Victoria
PO Box 1700, Victoria
British Columbia
Canada V8W 2Y2

Wirtz, H.P.
Universität Kiel
Abt. für marine Systematik und Ökologie
Olshausenstr. 40
24098 Kiel
Germany
Now at:
Sonnenbergstr. 11
72336 Balingen
Germany

Preface

The past 20 years have seen a revolution in the study and understanding of the diversity and structure of arthropod communities in the forest canopy. This has been enhanced by the development of new techniques to access the canopy and to sample arthropods. This volume brings together, for the first time, a wide range of the most recent studies on canopy arthropods.

Why has there been such a dramatic increase in studies of the canopy and, in particular, of arthropods? In recent years, the role of forests, especially tropical forests, in carbon and water cycles, and hence global climate, has been widely recognized. As forests are cut down, altered and fragmented, those organisms that are associated with them are also affected. Predictions of global species extinction rates for animals and plants based on forest loss range from 1–10% of all species per decade. Comprising, as they do, the major part of animal species richness, this inevitably means that exceptionally large numbers of arthropods may become extinct. Studies of canopy arthropods have also been critical in achieving a new appreciation of the total number of animal species on Earth. Estimates range as high as 100 million species, with the crucial assumption that the greater part of global species richness is represented by canopy arthropods in tropical forests. Controversial issues such as this have been important in placing biodiversity firmly on the political agenda, and have highlighted the importance of studying arthropods in rainforest canopies.

There are a variety of means of gaining access to the tops of trees. Ladders, towers, aerial walkways between trees, rope techniques, cranes, and rafts suspended from large balloons are all used by scientists to sample canopy arthropods. Sampling has involved the use of knockdown insecticides, various flight interception traps, light-traps, and hand-collecting. The full range of these often dangerous and costly methods of access and sampling have been used by the authors of this book.

Many canopy studies in recent years have focused on the remarkable diversity of arthropods in the canopy of temperate and tropical trees.

Some researchers, particularly those based in taxonomic institutions, have been able to overcome the taxonomic problems faced when dealing with large and species-rich samples. Typical samples from lowland tropical trees have been found to contain thousands of species, most of which have never been seen before, providing a wealth of new material for systematic studies. It has also been possible to examine a wide range of aspects of community structure including taxonomic composition and species-abundance, body size and guild relationships. The patterns revealed by these studies – and the hypotheses generated from them – have broad ecological implications that extend beyond canopy arthropod assemblages. Herbivory, host-specificity and the problematic issue of the ratio of canopy- to ground-dwelling arthropod species are other themes that have been closely examined by a number of authors in this book.

The book is arranged in five parts. The first part has two introductory chapters reviewing the principal methods used for sampling arthropods from trees, followed by chapters examining the effectiveness of the insecticide fogging method, including length of 'drop time', concentration of insecticide, and faunal recovery. The second part has chapters reviewing aspects of beetle community structure and species richness at different sites around the world, including North and South America, tropical Asia, New Caledonia, Australia, and, for the first time, Africa. Part three focuses on the diversity and community structure of other arthropod groups, such as ants, acridid Orthoptera, and Diptera. Part four has chapters which examine aspects of the biology of canopy arthropods, including arboreal dung beetle behaviour, bioacoustic monitoring of canopy insect communities, and the spatio-temporal habitat utilization of arboreal Collembola. Finally, part five addresses the impact of disturbance and forest fragmentation on canopy arthropods. In a concluding chapter we discuss the role that future canopy arthropod studies can play in providing invaluable information to guide environmental and conservation management decisions.

Interest in the canopy continues to increase at a remarkable rate. Aerial walkways, once a rare research novelty, are now relatively cheap and common-place, and fixed cranes of the kind used in the construction industry are now beginning to be used by researchers. The first of these at the Smithsonian Tropical Research Institute, in Panama, has been in use for several years. Other research groups in Europe and the USA are also erecting cranes in various parts of the world. Similarly, several groups have funding to build 'canopy research stations' in the trees. Networking between canopy researchers has improved dramatically through the electronic mail network funded by the National Science Foundation, and the European Science Foundation has launched a Tropical Rainforest Canopy Research programme. The results of two surveys of canopy researchers, one in the USA (Nadkarni and Parker

(1994) *Selbyana*, **15**, 38-50) and the other in Europe (Stork and Best (1994) *Selbyana*, **15**, 51-62) have been published recently, the latter indicating that entomologists are the dominant group of canopy researchers.

For those interested in broader aspects of tropical forest canopies there are several very general and well illustrated books such as Andrew Mitchell's *The Enchanted Canopy: Secrets from the Rainforest Roof* (1986, published by Collins, Glasgow) and Mark Moffett's *The High Frontier – Exploring the Tropical Rainforest Canopy* (1994, Harvard University Press, Harvard). More recently, Margaret Lowman and Nalini Nadkarni have published a book on *Forest Canopies* (1995, Academic Press, San Diego) which takes a look at some of the scientific issues involved. However, only two of the 24 chapters in their book are concerned with arthropods in the canopy. We hope that our book will serve to provide an in-depth coverage of many of the current research issues involving canopy arthropods and will stimulate research into new areas. Some of the papers included in this book were presented at a 2-day international symposium held at Manchester, UK, as part of the INTECOL Congress of Ecology in August 1994, but others have been added to ensure that a full range of topical issues is covered. All of the papers presented here describe previously unpublished work and have been subject to external review.

In preparing this volume we thank the many contributors for replying so promptly to our requests. We thank Trudy Brannan of The Natural History Museum and Ward Cooper and Kim Worham of Chapman & Hall for their guidance and support. We also thank the many anonymous reviewers who helped to make sure that the chapters in this volume were well prepared, accurate and relevant.

NES, JA & RKD
21 November, 1995

Part One

Methods of Studying Arthropods in Trees

1

Sampling arthropods from tree-crowns by fogging with knockdown insecticides: lessons from studies of oak tree beetle assemblages in Richmond Park (UK)

N.E. Stork and P.M. Hammond

ABSTRACT

1. The knockdown insecticide fogging regime used to study the seasonal occurrence and distribution of arthropods in the canopy of oak (*Quercus robur* L.) trees in Richmond Park, UK, is described.
2. Field trials supported the use of a 0.5% concentration of active ingredient in the insecticide and a 2-hour 'drop-time', and showed that fewer 'tourist' species are collected by fogging in the early morning than in the evening.
3. The composition of fog samples with respect to the total arthropod fauna of trees is discussed and areas of discrepancy are highlighted.
4. The advantages and disadvantages of using knockdown insecticides compared with other sampling methods, and the rate of faunal recovery of fogged trees, are discussed.
5. The past uses and potential uses of fogging for studying a range of biological questions are considered.

Canopy Arthropods. Edited by N.E. Stork, J. Adis and R.K. Didham. Published in 1997 by Chapman & Hall, London. ISBN 0 412 74900 9

INTRODUCTION

As noted by Moran and Southwood (1982) and Crawley (1983), trees provide model habitats for a range of studies in community ecology. In terms of species richness, the largest component of arboreal communities consists of arthropods. However, until recently, difficulties of access have limited the amount and range of studies carried out on the rich arthropod assemblages associated with tree-crowns. In general, direct observation of arthropods in the high canopy of trees has required the use of fixed platforms and aerial walkways (e.g. Wint, 1983). More recent developments, such as balloon rafts (Lowman *et al.*, 1993; Basset *et al.*, 1992) and cranes (Parker *et al.*, 1992), have made a greater area of canopy accessible to systematic investigation. However, the technique that has been responsible for the most dramatic recent growth in appreciation and understanding of the diversity and community structure of arthropods in trees, is the use of knockdown insecticides, especially as applied using fogging machines.

Perhaps the most important advantage of knockdown insecticides for the quantitative study of arboreal arthropods is that samples more or less accurately reflect the standing crop of arthropods present on surfaces (branches, leaves, fruits, etc.) in the canopy (see Discussion). Furthermore, since sampling of the insects does not depend on their activity, the contents of samples from different trees or parts of trees, or from different times of the year, different elevations or latitudes, are directly comparable, irrespective of differences in prevailing weather conditions. Also, although obviously intrusive, the method need not be destructive of habitat.

Following preliminary trials in the previous year, a study of the spatial and seasonal distribution of arthropods in oak trees in Richmond Park (UK) was carried out in 1984, using knockdown insecticides. The sampling regime used, and the protocol on which it was based, are described in the present paper. The results of this investigation are presented elsewhere (Barnard *et al.*, 1986; Palmer, 1986; Stork, 1988; Hammond and Owen, in press; P.M. Hammond and N.E. Stork, unpublished data). The programme of sampling through a full field season also served to test the reliability and robustness of equipment and methods used, before embarking on investigations in the tropics (Stork and Brendell, 1990), and to monitor the effects of a number of variables, such as weather, tree age, etc., on samples. The contribution made by this fogging programme to an inventory of the oak-associated and other beetles of Richmond Park as a whole is considered elsewhere (Hammond and Owen, in press). Based largely on results from the Richmond Park study, the usefulness of insecticide fogging for sampling arthropods indicative of woodland quality is discussed by Hammond and Harding (1991) and Hammond and Owen (in press).

The history and development of the use of knockdown insecticides for canopy studies are not described in detail, as this is discussed by Erwin (1983a,b, 1990). Also, the primary focus of this chapter is on the use of fogging, rather than other methods of releasing or applying insecticides, such as spraying or insecticide 'smoking' canisters (Watanabe, 1997, Chapter 18, this volume). Although often lumped together or confounded with fogging proper, these other methods differ substantially in their advantages and limitations as means of obtaining representative and reliable samples from tree-crowns. Other methods of sampling arthropods from trees (not involving the use of insecticides) are reviewed by Basset *et al.* (1997, Chapter 2, this volume).

METHODS OF COLLECTING CANOPY ARTHROPODS

Before the use of insecticides, information on the occurrence and distribution of insects within trees was obtained primarily by other direct sampling methods such as hand-collecting (Nielsen and Ejlersen, 1977) and beating (New, 1970; Elton, 1973). Collecting methods which rely on the activity of insects, such as light trapping (Wolda, 1980, 1983; Rees, 1983; Sutton, 1983), suction and Malaise trapping (Basset, 1985a,b; Hammond, 1990; Hammond and Harding, 1991; Mawdsley 1994) in trees, provide useful data of a more qualitative nature. Although a number of knockdown insecticide studies have investigated arboreal arthropod communities, few have made any reference to the reliability of the procedures used or provided any measure of the effectiveness of the insecticides employed. The present paper discusses the results of field trials conducted before the main sampling programme mentioned above, to determine the effectiveness of various insecticide concentrations, 'drop times' and the most appropriate time of day to carry out fogging. The advantages and disadvantages of the experimental design chosen for the study, the efficiency of sampling and the rate of faunal recovery of fogged trees are discussed. The more general advantages and disadvantages of fogging as a sampling method and the potential uses of the technique in biological studies are also considered.

Study site

Richmond Park (Figure 1.1) is an area of ancient parkland, approximately 9.5 km^2 in area, located 12 km to the south west of central London, that was enclosed as a Royal Deer Park in the 1630s by King Charles I (Brown, 1985; Hammond and Owen, in press). In 1992, largely on the basis of the rare invertebrates shown to be associated with its old trees (Hammond, 1983), it was designated a Site of Special Scientific Interest.

Although most of the park is open pasture woodland (Harding and Rose, 1986), a number of pockets of woodland are enclosed by high fencing to prevent the entry of deer and the public. One of these enclosures, Sidmouth Wood, occupying 26 ha towards the north-west corner of the park, was selected as a suitable site for the knockdown insecticide sampling programme. Sidmouth Wood is an area of mature, mixed deciduous, closed-canopy woodland on gravelly soil, and is dominated by oak, mostly standards less than 200 years old. The area also contains a few older oaks, some pollarded, as well as good numbers of sweet chestnut, (*Castanea sativa* Mill.), occasional beech (*Fagus sylvatica* L.) and a very few Scots pine (*Pinus sylvestris* L.), horse chestnut (*Aesculus hippocastanum* L.) and sycamore (*Acer pseudoplatanus* L.). In some parts of the wood, particularly along the ride which runs north-east to south-west across the north of the wood, there is an understorey comprised mostly of small birch trees (*Betula* sp.), but with occasional elder (*Sambucus nigra* L.) and patches of dense *Rhododendron* sp. From June to October there is a dense ground cover of bracken (*Pteridium aquilinum* L.) in most parts of the wood. The oaks are almost entirely pedunculate oak (*Quercus robur* L.) with a few sessile oak (*Quercus petraea* L.) towards the south-west part of Sidmouth Wood. Unlike much of Richmond Park, Sidmouth Wood is relatively little managed and much dead wood, including entire fallen trees, is left *in situ*. Twelve groups of three *Q. robur* of similar age (about 100 years old) and size (Appendix 1A) were selected for sampling. These groups of three trees were spaced out through the study area (Figure 1.1) so that the application of insecticide to one group of trees would have no effect on trees that were to be sampled subsequently.

Fogging equipment

For each tree selected for sampling, a rope was thrown over a high branch using a line-throwing gun ('E-Z Liner', Forestry Suppliers Inc., Jackson, Mississippi, USA). A pulley with a second rope running through it, was attached to one end of the first rope, pulled up into the canopy of the tree and the ropes tied off. Insecticide (a 1% solution of 'Reslin E' dissolved in diesel) was applied to the crowns of oak trees using an insecticide fogging machine ('fogger'), a SN11 Swing-Fog (Jaydon Engineering, Sutton, Surrey, UK). Reslin E is a non-residual synthetic pyrethroid which breaks down in minutes in direct sunlight leaving no toxic residues; it is harmless to vertebrates.

The fogger was started and hoisted into the canopy using the rope and pulley system. Release of the insecticide was controlled from the ground by a radio transmitter operating a specially fitted servo unit on the fogger to open and close the insecticide-release valve. The direction

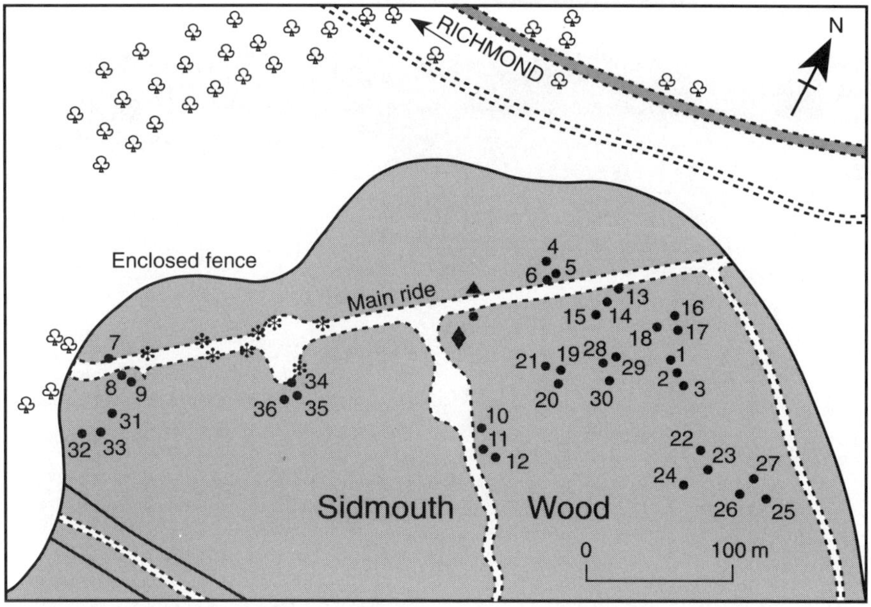

Figure 1.1 Map of the area studied in Sidmouth Wood, Richmond Park, UK. Symbols for fogged trees: numbered •, oaks in Fogs 1–12; unnumbered •, extra oak; ▲ , horse chestnut; ◆, beech; *, birch.

of the insecticide cloud produced by the fogger was controlled by a hand-held line attached to the exhaust end of the machine. Where parts of the tree were inadequately fogged using the pulley system, or it was impossible to erect a pulley in the tree, supplementary fogging was carried out from the ground. On later sampling dates, it was found that most of the trees could be fogged well from the ground and that pulleys were not always necessary. Approximately 2 litres of insecticide mixture were used on the three trees sampled on each occasion.

Sample collection

Samples of arthropods falling from each fogged tree were collected on 20 or more collecting trays, each 1 m^2 in area. The trays were suspended by detachable clips under the canopy of the selected tree, using a 'spiders web' arrangement of ropes strung at head height (Figure 1.2A). The conical-shaped trays were made of a smooth, tough, silicone-coated nylon fabric suspended from a ring of thin aluminium tubing. A hollow plastic bottle top, from which the centre disc had been removed, was tightly fixed to the open pointed end of the cone so that a plastic bottle

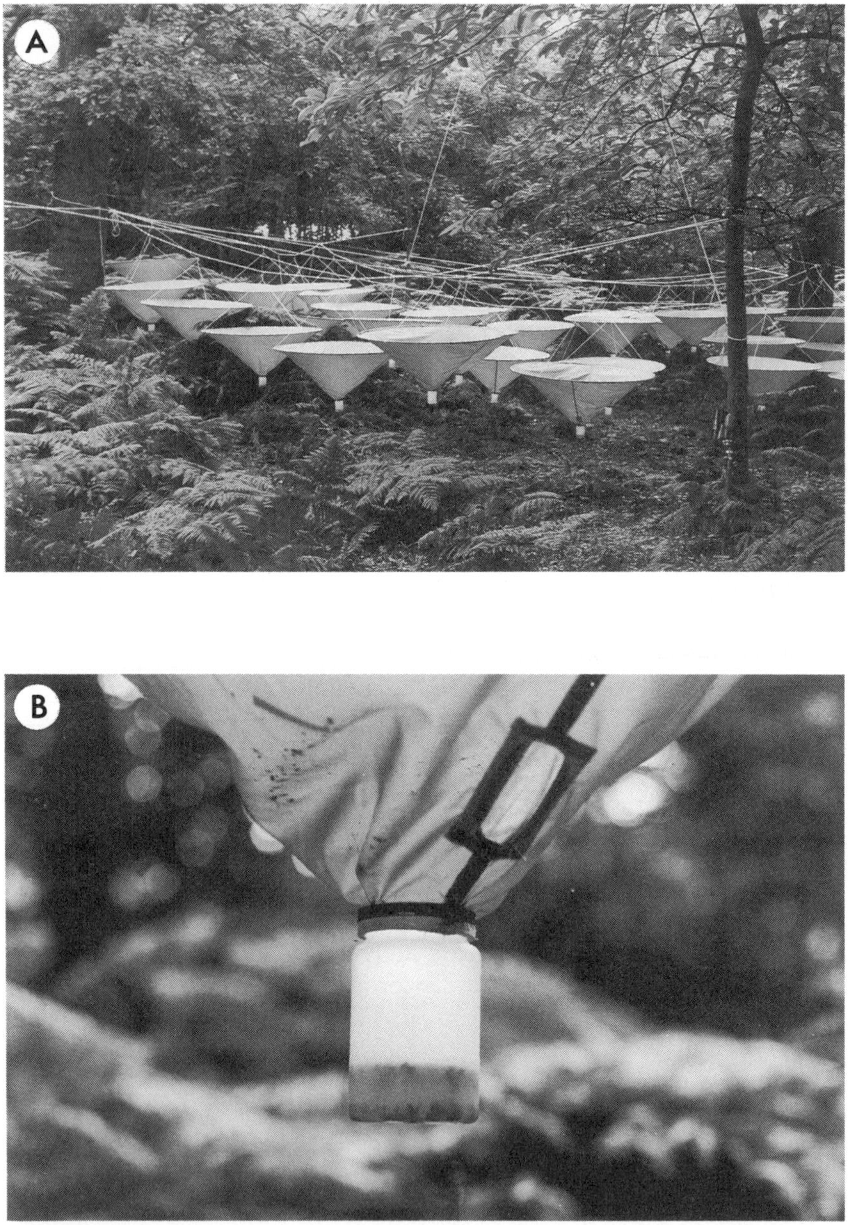

Figure 1.2 (A) Knockdown insecticide sampling trays suspended from ropes. (B) Base of one tray showing attached bottle containing the catch from fogging an oak tree.

containing 70% alcohol, could be screwed on and suspended beneath the tray (Figure 1.2B; Erwin, 1990). A scale drawing was made of each tree, recording the positions of tray-hanging ropes, trays, pulley ropes and the lateral extent of the tree's canopy (see example in Stork and Brendell, 1990). The diameter at breast height (DBH) of the main trunk (bole) of the tree was recorded, and an estimate of tree height made using a Sunnto clinometer. The distance from the tree trunk to the attachment clip (corresponding to the centre of the tray) was recorded for each tray (Appendix 1A). At 2 hours after the application of the insecticide the sides of the trays were gently tapped to aid the drop of the insects. Those stuck to the sides of the trays were washed down with 70% alcohol using a garden sprayer. The bottles were removed, numbered and, on return to the laboratory, their arthropod contents were transferred to fresh alcohol to await sorting.

Fogging regime

Sampling was carried out early in the morning after a rainless night, usually before 06:30 h. At this time of the morning the air was generally very still and the cloud of insecticide fog rose slowly through the canopy. Sampling was not carried out when there was a breeze sufficient to move the leaves on the trees, as the fog fails to reach the tops of the trees and spreads laterally through the woods. Similarly, fogging was not carried out when the leaves were wet from rain during the night. Samples were collected at intervals of two to three weeks (see Appendix 1B).

Sample analysis

The contents of each tray were kept separate and sorted to major Orders. The beetles were identified by P.M. Hammond with reference to the BM(NH) British and World collections. Further details on the distribution of the trays with respect to the oak trees will be presented elsewhere (N.E. Stork and P.M. Hammond, unpublished data).

Time of fogging

Variation in the arthropod fauna at different times of the day was examined by fogging six oak trees in an area of open, grazed woodland between the south end of Sidmouth Wood and the neighbouring Queen Elizabeth's Plantation. Samples were collected on plastic sheets with a 2-hour drop time. Trees A and B were fogged at 17:45 h (10.8.83), trees C and D at 18:45 h (11.8.83), and trees E and F at 07:00 h (12.8.83). Trees A, C and E were all mature oaks (more than 200 years old, < 1.3 m DBH) with some dead branches. Trees B, D and F were all young oaks (less than 100 years old, ca. 0.6 m DBH).

Drop time and insecticide concentration

To test the effect of different concentrations of insecticide on the total insect drop and the rate of fall of insects, eight patches of birch trees (*Betula pendula*) were fogged from the ground (21.6.84). The patches were widely spaced along the edge of the ride running through Sidmouth Wood (Figure 1.1) and each contained birch trees approximately 3–5 m in height. Insecticide concentrations of 0.1, 0.5, 1.0 and 3.0% active ingredient were tested at two sites each, the samples being collected on five trays per site. The samples were collected in the normal manner after 2 hours except for one tray per site where the sample bottles were replaced every 20 minutes for the first hour and then after two 40-minute periods. The sample bottles were also replaced every 20 minutes for two trays placed beneath one of the oaks fogged on 26.4.84 (Fog 1).

Other sampling

Reference is made in the Discussion to data from other insecticide samples (all except some of the Burnham Beech samples collected on plastic sheets laid on the ground):

- three trees (*Nothofagus procerus, Quercus robur, Pinus sylvestris*) in Thetford Chase, approximately 4 km west of Thetford, 27.5.82, 17:00–18:30 h; 2-hour drop time.
- one Turkey oak (*Q.* × *'hispanica'*) and one lime tree (*Tilia* sp.), Woodchester Park, Gloucestershire, 30.6.84, ca. 07:00 h; 2-hour drop time.
- ten even-aged oaks (*Q. robur*), Ham Cross Wood, Richmond Park, 18.8.83, ca. 07:00 h; 1.5-hour drop time (Stork, 1988, unpublished results).
- eight trees (*Q. robur, Fagus sylvatica*), Burnham Beeches, Bucks, 17.7.90, 07:00–09:00 h; 1.5-hour drop time. Five trees (*Q. robur, F. sylvatica, Betula pendula, Populus alba, Alnus glutinosa*), 09:00–11:00 h; 1.5-hour drop time (Purvis and Hammond, 1990; Hammond, 1991).

RESULTS OF FIELD TRIALS

Time of fogging

Fourteen out of 16 oak trees sampled in 1983 were fogged in the evening, while all 34 oaks sampled in 1984 were fogged in the morning. Of the

Table 1.1 Comparison of beetle catches from (a) two trees fogged in the evening and two fogged in the morning in the same area near Queen Elizabeth's Plantation in August 1983, and (b) six trees fogged in the evening at Ham Cross Wood in August 1983 and six trees fogged in the morning in Sidmouth Wood between 14.8.84 and 1.9.84

	Totals		*Tourists*	
	Individuals	*Species*	*Individuals*	*Species*
(a) Evening	173	28	15	8
Morning	110	24	4	4
(b) Evening	427	55	26	17
Morning	490	44	11	7

29 beetle species found in 1983, but not in 1984, 72% were 'tourists' (*sensu* Moran and Southwood, 1982; see also Gaston *et al.*, 1993 and Discussion below), whereas of the 51 species found in 1984, but not in 1983, 19.6% were tourists. This suggests that there were more tourist species in the evening fog samples than in the morning samples. More precise comparisons may be made: (i) two trees fogged in the evening (10.8.83) versus two trees fogged in the morning (12.8.83); and (ii) six trees fogged in the evening in August 1983 (trees selected at random) and six trees fogged in the morning in August/early September 1984 (Fogs 8 and 9) (Table 1.1). In both comparisons, although the overall number of individual beetles collected is similar, the proportion of tourist individuals and species in the evening fogs is two to three times greater than in the morning fogs. Three of the larger oak-associated species (see Discussion) were present only in evening fog samples.

Drop-time and insecticide concentration

The effect of insecticide concentration on the mean number of arthropods collected per tray varies (Table 1.2). No consistent change in the proportional representation of any arthropod group, nor of winged versus non-winged arthropods, with increasing insecticide concentration was observed (Table 1.3). It appears that stronger concentrations of insecticide, at least above 0.5%, do not necessarily increase the overall catch, or the catch of any particular insect group.

The cumulative percentages of the total catch obtained after different drop times are generally similar for the 0.5, 1.0 and 3.0% concentrations, but those for the 0.1% concentration are much lower, at least for the first hour (Figure 1.3). This suggests that at concentrations below 0.5%,

Table 1.2 Results of one-way Anova tests on the mean number of arthropods collected in 1 m^2 trays under young birch trees fogged with various concentrations of insecticide

Insecticide concentration (%)	*Insecticide concentration (%)*		
	0.5	*1.0*	*3.0*
0.1	2.30*	1.37	3.14**
0.5	–	1.44	0.08
1.0	–	–	2.05*

Significantly different means * $P < 0.05$, ** $P < 0.01$.

Table 1.3 The percentage contribution of different insect groups to samples collected using various concentrations of knock-down insecticide in 10 m^2 trays under young birch trees

Arthropod group	*Insecticide concentration (%)*			
	0.1	*0.5*	*1.0*	*3.0*
Hymenoptera	8.3	5.2	5.4	12.9
Lepidoptera	3.6	1.1	1.2	1.8
Coleoptera	9.1	6.1	4.5	6.5
Diptera	19.9	11.4	11.1	15.8
Neuroptera	3.2	1.5	1.0	1.9
Psocoptera	5.2	4.5	9.0	1.9
Hemiptera	36.5	51.2	46.3	42.3
Arachnida	8.3	9.5	7.3	6.3
Collembola	2.9	8.2	12.9	4.8
Others	3.0	1.2	1.2	2.1
% Non-winged arthropods	54.2	67.6	67.6	51.5
Total no. of arthropods	1008	1949	1379	1986

arthropods take longer to succumb to the insecticide. At concentrations of 0.5, 1.0 and 3.0%, half of the projected total is obtained in about 30 minutes compared with 50 minutes at 0.1%. However, 80% of the sample had fallen after 75 minutes at the higher concentrations compared with 85 minutes at the lower concentration.

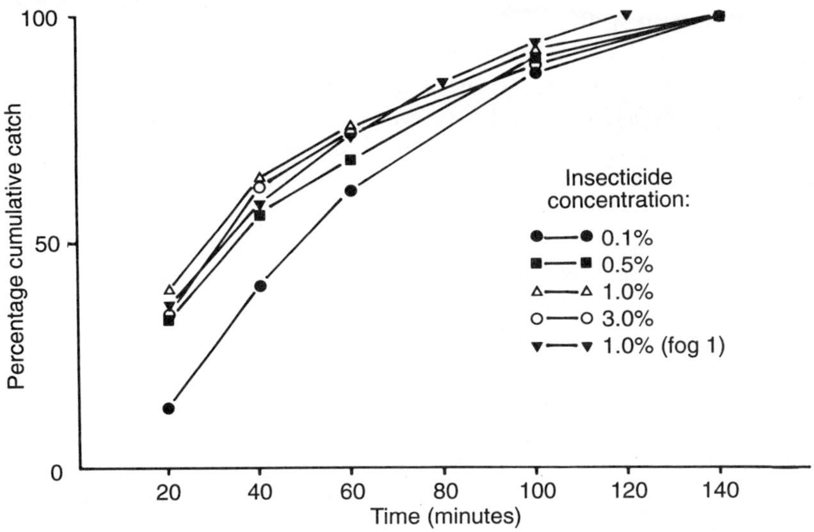

Figure 1.3 Plots of the cumulative catch, expressed as a percentage of the total catch, against time for different concentrations of insecticide.

DISCUSSION

What concentration of insecticide, and how to apply it?

A number of different insecticides have been used in previous studies, but to obviate long-term side effects a synthetic pyrethroid was used in the present study. Reslin E breaks down rapidly in direct sunlight leaving no harmful residues. Therefore, recolonization of the trees, as discussed later, is not hindered. The two methods of application of insecticide most used in ecological studies of arthropod communities are spraying and fogging. In spraying, the air blast from a motor-driven fan breaks up the insecticide into a fine mist. The blast throws the insecticide mist up to 5–10 m. This technique is suitable for sampling from the lower canopy of trees even under slightly breezy conditions, and has been used to investigate the community structure of arthropods in South African and British deciduous trees (Moran and Southwood, 1982; Southwood *et al.*, 1982a,b). In fogging, the insecticide is injected into the exhaust of a small motor, the force of the exhaust breaking it up into considerably smaller droplets than those produced by spraying. Since the exhaust is hot, the warm fog rises up to 30–40 m before dispersing, given still air conditions. Because of the more mobile nature of the fog, this technique can only be used when there is little or no wind. The problem of sampling from trees over 30 m in height was overcome by Montgomery and others (Erwin, 1983a,b; Adis *et al.*, 1984; Stork, 1988; Stork and Brendell, 1990; G. Montgomery, unpublished data) who used a rope and pulley system similar to that described in the present paper. Using this technique, samples have been collected from trees over 70 m in height (Stork, 1987a,b, 1988, 1991). The fitting of a radiocontrolled servo unit to the release valve of the fogger, introduced by Erwin (1983b), prevented unnecessary loss of insecticide and unwanted fogging of the understorey as the machine was hoisted into the tree or when the exhaust pipe was pointing in the wrong direction.

The experiments described above indicate that a 0.5% concentration of insecticide is adequate for sampling by insecticide fogging. In addition, it is clear that the time of day for fogging influences the catch. Many species from the 'background fauna' – especially those characteristic of ephemeral habitats, such as dung or temporary pools of water – fly at dusk (Lewis and Taylor, 1965) and are well represented in samples taken by evening fogging. Early morning fogging samples, on the other hand, contain few such species, or other 'tourists'. Although the number of individuals of non-resident species differs relatively little between samples taken at different times of the day (Table 1.1), large numbers of tourist species, if present only as a few individuals, can influence results considerably with regard to species number and sample composition. On the other hand, the comparison of samples taken by fogging various trees at Burnham Beeches and Thetford (see 'Other sampling', p. 10), as well as the comparison of morning and evening fogging at Richmond Park, suggests that some tree-crown species may be poorly represented in morning samples. Nevertheless, morning sampling is to be recommended since samples are generally more representative of the canopy fauna and this is the calmest time of day.

How to collect and how long to wait?

In earlier studies (Gagné and Martin, 1968; Gagné, 1979; Adis *et al.*, 1984) samples were collected on bed sheets and the insects removed by hand. This was both very time-consuming (sometimes taking days) and inaccurate, with some small insects being missed. Although the square plastic trays used by Erwin (1983a,b) were a great improvement on the bed sheets, their flat design still required the brushing of all insects into the central bottle. The funnel-shape and smooth fabric of the trays used here largely overcome the problems of insects sticking to the trays and damage due to brushing. Most falling insects slide down the smooth sides of the trays into the bottles, while the few remaining insects are washed into the bottles by use of a garden sprayer so that they do not contaminate the next sample taken with the tray. Few insects crawl out of the trays and, because the trays are suspended on ropes, few crawl in. Both of these problems are encountered when using collecting sheets placed on the ground.

Current methods of studying variation in insect populations in trees that involve clipping branches, cutting down whole trees, or a new technique, 'restricted canopy fogging' (Basset, 1990; Basset *et al.*, 1997, Chapter 12, this volume), produce questionable results since these methods disturb the insect population before and during sampling. Restricted canopy fogging involves enclosing branches with large plastic bags and then gassing with carbon dioxide. Some insects, particularly certain beetles and bugs, will fly or drop off as an escape reaction simply at the wave of an arm or similar sudden movement. It is therefore reasonable to assume that many insects may be lost in the process of encapsulating the branch being examined. Although canopy fogging involves some disturbance, it is probably minimal compared with these other techniques. Other methods of examining arthropod abundance in trees, such as flight interception traps (Basset, 1985a; Hammond, 1990; Mawdsley, 1994) are not directly comparable and are not discussed here.

The drop-time experiments described above indicate that the numbers of insects falling declines rapidly after about 1 hour and that a 2-hour drop time is acceptable. With a longer drop time than this there is an increased risk of contamination.

What do the samples represent?

Comparison with results of hand-collecting and sampling by other means in Richmond Park and elsewhere (Hammond and Owen, in press; P.M. Hammond, unpublished data) enables fairly confident assertions to be made about the nature of the fogging samples. Individuals of most species active on foliage, flowers, twigs and on the surface of branches and trunks (including epiphytes) appear to be knocked down in a relatively unselective manner. There are some exceptions to this. Some semi-permanently or permanently attached phloem feeders, such as scale insects and the nymphs of some psyllid species, are not sampled. This may also apply to some aphids and other sucking insects that fail to remove their stylets before succumbing to the insecticide, although visual examination suggests that in practice few of these insects remain after fogging. Other surface-dwellers that may remain attached to the tree include those that inhabit leaf-rolls, especially larvae. For instance, some free-living adults of *Attelabus nitens* (Scopoli) (Coleoptera: Attelabidae) were present in the samples, but not their larvae that are found in leaf-rolls. Some Lepidoptera larvae and small spiders may remain attached by silken threads, but again few of these were found on visual examination of the fogged birch trees described above. Examination of inflorescences of fogged oaks showed that a few small arthropods, especially Thysanoptera, remain deep in the inflorescences. The low numbers of adult bark beetles (Scolytidae) and low numbers or absence of adults, and absence of larvae, of other wood-inhabiting groups, indicate that the technique is not successful in sampling individuals that inhabit burrows in bark, beneath bark, or deep in the wood. In contrast, Malaise traps in trees catch a wider range of such species and in larger numbers. However, adults of some wood-boring beetle species are collected in fair numbers by fogging, but these appear to be mostly those (especially certain Anobiidae) which inhabit shallow burrows at right angles to the surface and/or spend time walking over, or 'resting' on, the surface of the trunk and main branches. The larvae of leaf- and stem-mining species are also not collected, although adults (e.g. of *Rhynchaenus* spp., Curculionidae) are sampled successfully. Some species normally found under loose bark during daytime are well sampled, such as Raphidioptera larvae (Barnard *et al.*, 1986) and the tenebrionid beetle *Cylindronotus laevioctostriatus*, but others are completely absent (e.g. larvae of the dermestid beetle, *Ctesias serra*). The presence in the samples of species normally found under loose bark is not surprising since, in its industrial application, the insecticide is used to flush out cockroaches from inaccessible parts of buildings such as behind skirting boards. Fogging does not appear to be a good method of sampling insects normally found in rot holes in trees or in nests.

Comparison of fogging samples with samples collected from tree-crowns by other techniques may be instructive, but can also be misleading. For instance, Malaise trap samples appear to contain a much higher proportion of large insects. However, direct evidence to suggest that the larger and more mobile insects generally escape from a fogged tree during fogging is lacking. Indeed, large individuals of butterflies, moths, flies and bees, although not plentiful, were present in the samples from Richmond Park (N.E. Stork, unpublished data). Still, the relatively low numbers of large Hymenoptera and Coleoptera in fogging samples does require some explanation. Partially at least, it may be that some of these large insects are inactive and in concealed situations (burrows, etc.) in the early morning when most fogging is done, and are only caught in canopy-level traps during a short period of the day (Hammond *et al.*, 1997, Chapter 10, this volume). Similarly, there is no evidence that very small insects such as chalcid wasps are drawn upwards by the up-draft of the fog. In fact, large numbers of very small chalcids were also found in the samples (Noyes, 1984; Askew, 1985).

Further investigations are required to determine the numbers of insects remaining on trees after the application of insecticide.

The mobile nature of the insecticide fog and the use of a large amount of insecticide generally ensured that the trees were well fogged.

However, the insecticide fog on some occasions failed to penetrate some parts of the trees. For instance, if there was a slight breeze then parts of the trees up-wind of the pulley point, or the top few metres of a tree, failed to be well fogged. In the present study, the insecticide used was one that has a high knockdown component which affects the nervous system of insects. This means that the insects are not killed immediately but still move in an uncontrolled manner for some time after application of the insecticide. The uncontrolled movements of these insects means that any insect dropping onto leaves is likely to fall off again and thus has a greater chance of being collected in the trays below.

What is the recovery time for the fauna of a sampled tree?

Two factors are important in determining the recovery time of the fauna of a fogged tree: (i) the proportion of the tree's fauna not killed by the insecticide; and (ii) the rate of recolonization. Clearly, determining the influence of each of these factors in the re-establishment of a tree's fauna after fogging is extremely difficult. However, evidence from activity-based collecting methods such as aerial Malaise and other flight interception traps and from light trapping in trees, provides information on the movements of arboreal insects and some indication of potential rates of recolonization. An additional complication is that the proportion of a tree's fauna that is 'non-residential' or, in Moran and Southwood's (1982) terminology, forms the tourist guild, remains debatable (Stork, 1987a; Didham, 1997, Chapter 15, this volume), but may be substantial (Hammond *et al.*, 1997, Chapter 10, this volume).

Preliminary results from a simple experiment in Richmond Park where five trees, one from each of Fogs 5–9 (see Appendix 1B), were fogged in early June and again in early September, suggest that the tree-crown faunas had recovered (at least in terms of numbers of individuals) within 3 months (N.E. Stork, unpublished data). A re-fog of a Bornean tree 10 days after the initial fogging produced samples representing roughly 20% of the individuals and species in the initial fog (Stork, 1991). In contrast, a similar experiment carried out in Peru produced samples containing approximately the same numbers of individuals in a re-fog after 10 days as in the first fog (T.L. Erwin, personal communication). Conflicting results such as these are typical of re-fogging experiments and show that further work is required on the faunal recovery rates of fogged trees.

Collecting trays typically cover only some 10% or so of the area under each fogged tree. As the kill component of the insecticide used is low, many insects falling to the ground recover completely. Canopy fogging is now being used to collect living arboreal arthropods for life-history studies (Paarmann and Kerck, 1997, Chapter 3, this volume). For example, Carabidae collected by canopy fogging in Sulawesi were successfully reared and their life-history strategies studied by Paarmann and Stork (1987) and Paarmann and Paarmann (1997, Chapter 20, this volume). W. Paarmann (unpublished data) has since found a 90% recovery rate for a range of arthropods fogged from oaks using natural pyrethrum. Although return to the tree is easy for winged insects, the only path to the canopy for wingless insects is the tree trunk and therefore many will not return. Some, particularly Lepidoptera larvae, may fall prey to birds and other predators while on the ground.

How does fogging compare with other sampling methods?

The intensive survey of Coleoptera in Richmond Park carried out by Hammond and Owen (in press) has revealed the presence there of some 1100 species (approximately 28% of the British beetle fauna). The 1983/1984 fogging samples contained 7596 adult beetles belonging to 202 species. Thirty-six of these species were not represented among the approximately 40 000 adult beetles contained in samples of other types taken during a 6-year survey of the Park. Fogging is an excellent method of sampling canopy insects (see below), but samples only one component of the forest fauna. For general insect surveys it must be seen as a useful tool in combination with a range of other sampling techniques. In assessing the value of knockdown insecticide fogging as a method of sampling insects from trees, comparisons must be drawn between this method and other sampling methods. Although this is not the place for detailed discussions, a few simple comparisons with data collated by one of the present authors (P.M.H.) for one such method, Malaise trapping, illustrate the value of fogging (Hammond and Harding, 1991; Hammond and Owen, in press). One of a number of small Malaise/interception traps used in a study at Awbridge, Hampshire (1980–1982), was placed at a height of 8.5 m in the canopy of an oak tree. Data on beetles collected during 700 trap–days in the months April–December are compared with those for beetles collected by fogging in 1983 and 1984 in the same months (Table 1.4). Almost twice as many species associated with oaks were collected by fogging, but, more importantly, some of the most abundant oak-associated beetle species, including the larvae of some species, collected by fogging were not collected at all in the Malaise trap, e.g. *Dromius quadrinotatus* (Carabidae), *Dalopius marginatus* (Elateridae), *Hemicoelus fulvicornis* (Anobiidae), *Scymnus auritus* (Coccinellidae), *Cis vestitus* (Cisidae), *Cylindronotus laevioctostriatus*, *Cryptocephalus pusillus* (Chrysomelidae), *Phyllobius pyri* and *Strophosoma melanogrammum* (Curculionidae). The absence of some of these species is no doubt due to site differences and low vagility (e.g. *C. laevioctostriatus* and *S. melanogrammum* are flightless).

Table 1.4 Comparison of the numbers of oak-associated and other beetle species collected over 700 trap–days (April–December, 1980–1982) in a Malaise trap at 8.5 m height in the crown of an oak tree at Awbridge, Hampshire, with those from fogging oak trees in Richmond Park in 1983–1984. 'Exclusive' indicates species collected only by that method; 'shared' indicates collected by both methods

Collection method	*Total no. collected*	*All beetle species*		*Oak-associated species*	
		Exclusive	*Shared*	*Exclusive*	*Shared*
Fogging	186	111	75	69	5
Malaise	160	85		35	

The main advantages and disadvantages of knockdown insecticide sampling by fogging over other activity-based sampling methods can be summarized as follows:

Advantages

1. Relatively unselective – selectivity can be estimated by visual searches.
2. Not dependent on activity of arthropods.
3. No 'attractants' involved.
4. Not influenced by 'trap behaviour', unlike activity-based traps.
5. Sample composition not directly influenced by weather.
6. Collects some species difficult to sample by other means.
7. Fairly precise origin of specimens known.
8. Understorey species can be avoided if required.
9. No need to climb or disturb trees, particularly during sampling.
10. No servicing or maintenance problems on long-term sites.
11. Largely non-destructive to trees and surrounding habitat.
12. Samples are usually exceptionally clean and easy to sort.
13. Insecticide is non-residual and of limited toxicity.
14. Fogging (as opposed to spraying) can sample the high canopy.

Disadvantages

1. Labour-intensive.
2. Equipment costly and bulky.
3. Limited by weather conditions, i.e. cannot be done when windy.
4. Not effective in sampling some externally feeding groups, e.g. phloem feeders and leaf-miners, and some internal feeders (e.g. wood borers).

5. The insecticide fog may be difficult to control, causing unnecessary fogging of other trees.

The value of fogging for different types of study

Fogging is of considerable use for a wide range of biological studies, largely because of the comparable nature of samples collected and because of its advantages over other sampling methods, as summarized earlier (see also Erwin, 1990; Hammond, 1992). Studies of fogging and other canopy samples from many parts of the world have examined arboreal arthropod communities with respect to a wide range of factors including taxonomic group membership, guild, biomass, body size, and species-abundance patterns (summarized in Stork, 1988; see also Morse *et al.*, 1988; Lawton, 1989, 1991; Blackburn *et al.*, 1990; Stork and Brendell, 1990, 1993; Stork, 1991; Basset, 1990, 1992; and chapters in this volume). Because of their reliability and comparability, fogging samples have also been used to look at seasonal changes in the ratios of different morphs of the ladybird *Adalia decempunctata* (P.M. Hammond, unpublished data). The use of literature citations to obtain lists of tree-associated species has come under recent strong criticism (Owen, 1987). Fogging samples provide a novel and direct method of examining the insect species complement of trees (Southwood *et al.*, 1982b; P.M. Hammond and N.E. Stork, unpublished data) and the similarity of the faunas of different trees (Stork, 1987b) that obviates many of these criticisms. Overall, fogging probably has its greatest potential use in ecological studies where the reliability and comparability of samples is of paramount importance. Such problems as remain in this area derive largely from the interpretation of results, rather than the sampling methods themselves.

The potential has been recognized for using fogging to collect comparable samples of beetles from different parklands/pasture woodlands in order to assess their status as 'ancient woodland' sites (Harding and Rose, 1986; Hammond and Harding, 1991; Hammond and Hine, 1994). The Richmond Park study produced a large number of beetles of Red Data Book status and others of rarity (Hammond and Owen, in press).

To some, the most significant contribution of fogging is the ease with which samples of insects can be collected from trees for taxonomic purposes (Barnard *et al.*, 1986; Palmer, 1986). Of importance to some taxonomists studying phytophagous insects is the fact that these insects can be associated with identified trees.

The ability to collect living insects from the canopy using insecticides is of potential value to those interested in the life histories of arboreal insects (Paarmann and Paarmann, 1997, Chapter 20, this volume). For instance, an understanding of the reproductive and life history strategy of one species of arboreal Carabidae was gained by rearing larvae and subsequently adults from one pair of adults collected by canopy fogging in Sulawesi (Paarmann and Stork, 1987; Paarmann and Paarmann, 1997, Chapter 20, this volume).

Acknowledgements

We are grateful to M.B. Brown, Parks Superintendent, for giving us permission to work in Richmond Park and for use of storage facilities, and to the late P. Chadwick of Wellcome Research Laboratories for provision of the insecticide. We thank T. Smith and our many colleagues, particularly M.J.D. Brendell and C.R. Smith, for their assistance and support in fieldwork and sample sorting. We thank C.R. Vardy for making available the Malaise trap samples from Awbridge.

REFERENCES

Adis, J., Lubin, Y.D. and Montgomery, G.G. (1984) Arthropods from the canopy of inundated and terra firme forests near Manaus, Brazil, with critical considerations on the pyrethrum-fogging technique. *Studies on the Neotropical Fauna and Environment*, **19**, 223–36.

Askew, R.R. (1985) A London fog. *Chalcid Forum*, **4**, 17.

Barnard, P.C., Brooks, S.J. and Stork, N.E. (1986) The seasonality and distribution of Neuroptera, Raphidioptera and Mecoptera on oaks in Richmond Park, Surrey, as revealed by insecticide knock-down sampling. *Journal of Natural History*, **20**, 1321–31.

Basset, Y. (1985a) Comparaison de quelques méthodes de piégeage de la faune dendrobie. *Bulletin Romand d'Entomologie*, **3**, 1–14.

Basset, Y. (1985b) Les peuplements d'arthropodes sur *Pinus mugo* Turra dans les tourbieres du Haut-Jura neuchâtelois. *Bulletin de la Société Neuchâteloise des Sciences Naturelles*, **108**, 63–76.

Basset, Y. (1990) The arboreal fauna of the rainforest tree *Argyrodendron actinophyllum* as sampled with restricted canopy fogging: composition of the fauna. *The Entomologist*, **109**, 173–83.

Basset, Y. (1992) Influence of leaf traits on the spatial distribution of arboreal arthropods within an overstory tree. *Ecological Entomology*, **17**, 8–16.

Basset, Y., Aberlenc, H.-P. and Delvare, G. (1992) Abundance, diversite et stratification verticale de l'entomofaune d'une forêt tropicale humide africaine, in *Biologie d'une Canopée de Forêt Equatoriale - II. Rapport de Mission: Radeau des cimes Octobre–Novembre 1991, Réserve de Campo, Cameroun* (eds F. Hallé and O. Pascal), Fondation Elf, Paris, pp. 45–52.

Basset, Y., Springate, N.D., Aberlenc, H.P. and Delvare, G. (1997) A review of methods for sampling arthropods in tree canopies, in *Canopy Arthropods* (eds N.E. Stork, J. Adis and R.K. Didham), Chapman & Hall, London, pp. 27–52 (this volume).

Blackburn, T.M., Harvey, P.H. and Pagel, M.D. (1990) Species number, population density and body size relationships in natural communities. *Journal of Animal Ecology*, **59**, 335–46.

Brown, M.B. (1985) *Richmond Park: the History of a Royal Deer Park*, Robert Hale, London.

Crawley, M.J. (1983) *Herbivory. Studies in Ecology*, Vol. 10, Blackwell Scientific Publications, Oxford.

Didham, R.K. (1997) Dipteran tree-crown assemblages in a diverse southern temperate rainforest, in *Canopy Arthropods* (eds N.E. Stork, J. Adis and R.K. Didham), Chapman & Hall, London, pp. 320–43 (this volume).

Elton, C.S. (1973) The structure of invertebrate populations inside neotropical rain forest. *Journal of Animal Ecology*, **42**, 55–104.

Erwin, T.L. (1983a) Beetles and other insects of tropical forest canopies at Manaus, Brazil, sampled by insecticidal fogging, in *Tropical Rain Forest: Ecology and Management* (eds S.L. Sutton, T.C. Whitmore and A.C. Chadwick), Blackwell Scientific Publications, Oxford, pp. 59–75.

Erwin, T.L. (1983b) Tropical forest canopies: the last biotic frontier. *Bulletin of the Entomological Society of America*, **29**(1), 14–19.

Erwin, T.L. (1990) Canopy arthropod biodiversity: a chronology of sampling techniques and results. *Revista Peruana de Entomologia*, **32**, 71–7.

Gagné, W.C. (1979) Canopy-associated arthropods in *Acacia koa* and *Metrosideros* tree communities along an altitudinal transect on Hawaii Island. *Pacific Insects*, **21**, 56–82.

Gagné, W.C. and Martin, J.L. (1968) The insect ecology of red pine plantations in Central Ontario. V. The Coccinellidae (Coleoptera). *Canadian Entomologist*, **100**, 835–46.

Gaston, K.J., Blackburn, T.M., Hammond, P.M. and Stork, N.E. (1993) Relationships between abundance and body size: where do tourists fit? *Ecological Entomology*, **18**, 310–14.

Hammond, P.M. (1983) *Beetles in Richmond Park, with Special Reference to Species Associated with Old Oak Trees*. Unpublished Report to Nature Conservancy Council.

Hammond, P.M. (1990) Insect abundance and diversity in the Dumoga-Bone National Park, N. Sulawesi, with special reference to the beetle fauna of lowland rain forest in the Toraut region, in *Insects and the Rain Forests of South East Asia (Wallacea)* (eds W.J. Knight and J.D. Holloway), Royal Entomological Society of London, London, pp. 197–254.

Hammond, P.M. (1991) *An Annotated Checklist of the Coleoptera of Burnham Beeches SSSI*. The Natural History Museum, unpublished report to the City of London Corporation.

Hammond, P.M. (1992) Species inventory, in *Global Biodiversity. Status of the Earth's Living Resources* (ed. B. Groombridge), Chapman & Hall, London, pp. 17–39.

Hammond, P.M., Stork, N.E. and Brendell, M.J.B. (1997) Arboreal beetle assemblages in context: comparison of canopy and other ecotones of lowland rainforest in Sulawesi, in *Canopy Arthropods* (eds N.E. Stork, J. Adis and R.K. Didham), Chapman & Hall, London, pp. 184–223 (this volume).

Hammond, P.M. and Harding, P.T. (1991) Saproxylic invertebrate assemblages in British woodlands: their conservation significance and its evaluation, in *Pollard and Veteran Tree Management* (ed. H.J. Read), Richmond Publications Co. Ltd., Slough, pp. 29–37.

Hammond, P.M. and Hine, S.J. (1994) *Welsh Parklands and their Saproxylic Invertebrates*. The Natural History Museum, unpublished report to the Countryside Council for Wales.

Hammond, P.M. and Owen, J.A. *The Saproxylic and Other Beetles of Richmond Park SSSI. A Case Study*. English Nature Science (in press).

Harding, P.T. and Rose, F. (1986) *Pasture Woodlands in Lowland Britain: A Review of their Importance for Wildlife Conservation*. Institute of Terrestrial Ecology, Dorset Press.

Lawton, J.H. (1989) What is the relationship between population density and body size in animals? *Oikos*, **55**, 429–34.

Lawton, J.H. (1991) Species richness, population abundances, and body sizes in insect communities: tropical versus temperate comparisons, in *Plant–Animal Interactions: Evolutionary Ecology in Tropical and Temperate Regions* (eds P.W. Price, T.M. Lewinsohn, G.W. Fernandes and W.W. Benson), John Wiley & Sons, pp. 71–89.

Lewis, T. and Taylor, L.R. (1965) Diurnal periodicity of flight by insects. *Transactions of the Royal Entomological Society of London*, **116**, 393–469.

Lowman, M., Moffett, M. and Rinker, H.B. (1993) A new technique for taxonomic and ecological sampling in rain forest canopies. *Selbyana*, **14**, 75–9.

Mawdsley, N.M. (1994) *Community Structure of the Coleoptera Assemblage in a Bornean Tropical Forest*. PhD Thesis, University of London, UK.

Moran, V.C. and Southwood, T.R.E. (1982) The guild composition of arthropod communities in trees. *Journal of Animal Ecology*, **51**, 289–306.

Morse, D.R., Stork, N.E. and Lawton, J.H. (1988) Species number, species abundance and body length relationships of arboreal beetles in Bornean lowland rain forest trees. *Ecological Entomology*, **13**, 25–37.

New, T.R. (1970) The relative abundance of some British Psocoptera on different species of trees. *Journal of Animal Ecology*, **39**, 521–40.

Nielsen, B.O. and Ejlersen, A. (1977) The distribution pattern of herbivory in a beech canopy. *Ecological Entomology*, **2**, 293–9.

Noyes, J.S. (1984) In a fog. *Chalcid Forum*, **3**, 4–5.

Owen, D.F. (1987) Insect species richness on the Rosaceae: are the primary data reliable? *Entomologists' Gazette*, **38**, 209–13.

Paarmann, W. and Kerck, K. (1997) Advances in using the canopy fogging technique to collect living arthropods from tree crowns, in *Canopy Arthropods* (eds N.E. Stork, J. Adis and R.K. Didham), Chapman & Hall, London, pp. 53–66 (this volume).

Paarmann, W. and Paarmann, D. (1997) Studies on the biology of a canopy-dwelling carabid beetle collected by canopy fogging, in *Canopy Arthropods* (eds N.E. Stork, J. Adis and R.K. Didham), Chapman & Hall, London, pp. 433–41 (this volume).

Paarmann, W. and Stork, N.E. (1987) Canopy fogging, a method of collecting live insects for investigations of life history strategies. *Journal of Natural History*, **21**, 563–6.

Palmer, J.M. (1986) Thrips in English oak trees. *Entomologists' Gazette*, **37**, 245–52.

Parker, G.G., Smith, A.P. and Hogan, K.P. (1992) Access to the upper forest canopy with a large tower crane. *BioScience*, **42**(9), 664–70.

Purvis, W.O. and Hammond, P.M. (1990) *Report on the Plants and Insects of Burnham Beeches SSSI*. Unpublished report, The Natural History Museum, London, UK.

Rees, C.J.C. (1983) Microclimate and flying Hemiptera fauna of a primary lowland rain forest in Sulawesi, in *Tropical Rain Forest: Ecology and Management* (eds S.L. Sutton, T.C. Whitmore and A.C. Chadwick), Blackwell Scientific Publications, Oxford, pp. 121–36.

Southwood, T.R.E, Moran, V.C. and Kennedy C.E.J. (1982a) The assessment of arboreal insect fauna: comparisons of knockdown sampling and faunal lists. *Ecological Entomology*, **7**, 331–40.

Southwood, T.R.E, Moran, V.C. and Kennedy C.E.J. (1982b) The richness, abundance and biomass of the arthropod communities on trees. *Journal of Animal Ecology*, **51**, 635–49.

Stork, N.E. (1987a) Guild structure of arthropods from Bornean rain forest trees. *Ecological Entomology*, **12**, 69–80.

Stork, N.E. (1987b) Arthropod faunal similarity of Bornean rain forest trees. *Ecological Entomology*, **12**, 219–26.

Stork, N.E. (1988) Insect diversity: facts, fiction and speculation. *Biological Journal of the Linnean Society*, **35**, 321–37.

Stork, N.E. (1991) The composition of the arthropod fauna of Bornean lowland rain forest trees. *Journal of Tropical Ecology*, **7**, 161–80.

Stork, N.E. and Brendell, M.J.D. (1990) Variation in the insect fauna of Sulawesi trees with season, altitude and forest type, in *Insects and the Rain Forests of South East Asia (Wallacea)* (eds W.J. Knight and J.D. Holloway), Royal Entomological Society of London, London, pp. 173–90.

Stork, N.E. and Brendell, M.J.D. (1993) Arthropod diversity studies in lowland rain forest of Seram, in *The Natural History of Seram* (eds I.D. Edwards, A.A. MacDonald and J. Proctor), Intercept Ltd., Hampshire, UK, pp. 115–30.

Sutton, S.L. (1983) The spatial distribution of flying insects in tropical rain forests, in *Tropical Rain Forest: Ecology and Management* (eds S.L. Sutton, T.C. Whitmore and A.C. Chadwick), Blackwell Scientific Publications, Oxford, pp. 77–91.

Watanabe, H. (1997) Estimation of arboreal and terrestrial arthropod densities in the forest canopy as measured by insecticide smoking, in *Canopy Arthropods* (eds N.E. Stork, J. Adis and R.K. Didham), Chapman & Hall, London, pp. 401–12 (this volume).

Wint, G.R.W. (1983) Leaf damage in tropical rain forest canopies, in *Tropical Rain Forest: Ecology and Management* (eds S.L. Sutton, T.C. Whitmore and A.C. Chadwick), Blackwell Scientific Publications, Oxford, pp. 229–39.

Wolda, H. (1980) Seasonality of tropical insects. I. Leafhoppers (Homoptera) in Las Cumbres, Panama. *Journal of Animal Ecology*, **49**, 277–90.

Wolda, H. (1983) Spatial and temporal variation in abundance in tropical animals, in *Tropical Rain Forest: Ecology and Management* (eds S.L. Sutton, T.C. Whitmore and A.C. Chadwick), Blackwell Scientific Publications, Oxford, pp. 93-105.

Appendix 1A Data relating to oak trees investigated in Sidmouth Wood. All measurements are in metres. DBH, diameter of bole at breast height; HT, tree height; Spread, maximum and minimum spread of canopy from trunk of tree; Trays, number of sample trays used; Fog, method of fogging (by rope and pulley, P, or from the ground, G, or both); *, birch understorey present

Tree	*DBH*	*HT*	*Spread*	*Trays*	*Fog*	*Position and condition of tree*
1	0.7	19.2	4.9–7.5	19	P	Edge of open area, buds not open
2	0.6	21.1	4.2–7.0	22	P	Part edge of open area, 50% buds open
3	0.6	21.0	2.2–10.5	22	P	Part edge of open area, buds not open
4	0.7	24.8	7.4–8.3	21	P	Continuous canopy, 100% buds open
5	0.6	20.5	5.6–7.1	21	P	Part edge of ride, 90% buds open
6	0.6	18.2	7.2–8.0	21	P	Edge of ride, 90% buds open
7	0.7	22.1	5.3–9.4	20	P	Continuous canopy, some trunk foliage
8	0.7	17.5	4.0–6.5	23	P	Edge of ride, some trunk foliage
9	0.8	22.1	6.2–8.2	23	P	Continuous canopy
10	0.9	22.8	8.0–10.	23	PG	Part edge of open area
11	0.8	21.8	5.4–6.4	21	PG	Part edge of open area*
12	0.8	25.1	6.0–9.5	21	PG	Continuous canopy, large dead branch
13	0.6	20.1	4.2–7.8	21	PG	Edge of ride, some low foliage*
14	0.6	19.6	4.2–8.7	22	PG	Semi-continuous canopy
15	0.7	19.3	6.2–6.5	23	PG	Semi-continuous canopy
16	0.6	23.0	5.6–8.4	21	PG	Semi-continuous canopy*
17	0.6	24.1	3.7–6.3	21	PG	Semi-continuous canopy*
18	0.6	18.0	5.5–6.5	22	PG	Continuous canopy
19	0.7	21.7	3.3–8.2	21	PG	Continuous canopy
20	0.8	21.3	4.5–7.0	21	PG	Semi-continuous canopy, dead branch*
21	0.7	22.5	4.5–7.0	22	PG	Part edge of open area
22	0.7	18.0	3.0–9.0	21	PG	Semi-continuous canopy
23	0.7	20.1	2.0–6.5	20	PG	Semi-continuous canopy
24	0.7	22.6	4.0–8.7	22	PG	Semi-continuous canopy
25	0.7	26.5	4.0–8.3	21	G	Semi-continuous canopy*
26	0.6	30.2	4.0–7.3	21	G	Part edge of open area
27	0.7	28.1	3.5–6.7	21	G	Part edge of open area
28	0.7	20.4	4.6–5.0	14	G	Continuous canopy, low trunk foliage
29	0.5	22.6	4.0–5.0	13	G	Continuous canopy, low trunk foliage
30	0.6	22.1	5.0–6.0	13	G	Continuous canopy, some trunk foliage
31	0.8	21.0	4.4–8.7	13	G	Edge of ride
32	0.7	18.4	2.5–7.5	14	G	Semi-continuous canopy

Appendix 1A continued

Tree	*DBH*	*HT*	*Spread*	*Trays*	*Fog*	*Position and condition of tree*
33	0.7	21.2	2.7–4.6	13	G	Semi-continuous canopy*
34	0.7	18.3	4.2–5.3	13	G	Edge of open area
35	0.7	15.8	3.1–5.4	12	G	Semi-continuous canopy*
36	0.7	21.0	2.8–5.3	15	G	Semi-continuous canopy*

Appendix 1B Dates and wind conditions for the 12 sampling periods (Fog No.) carried out in Sidmouth Wood. For information on the individual trees see Appendix 1A and Figure 1.1

Fog No.	*Date*	*Trees*	*Wind condition*
1	26.04.84	1–3	Very light wind
2	15.05.84	4–6	Very light wind
3	30.05.84	7–9	Occasionally breezy
4	12.06.84	10–12	Very slight wind
5	26.06.84	13–15	Still
6	11.07.84	16–18	Almost still
7	25.07.84	19–21	Still
8	14.08.84	22–24	Still
9	01.09.84	25–27	Slight wind
10	11.09.84	28–30	Light wind
11	26.09.84	31–33	Light wind
12	16.10.84	34–36	Occasionally breezy

2

A review of methods for sampling arthropods in tree canopies

Y. Basset, N.D. Springate, H.P. Aberlenc and G. Delvare

ABSTRACT

A review of most of the 'non-fogging' methods which have been used to sample arthropods in tree canopies, particularly in tropical forests, is presented, emphasizing the strengths and limitations of each method, as well as the rationale for its use. In particular, methods used with the canopy raft are examined. The review is supplemented with a comparison of the selectivity of four methods used in Papua New Guinea for collecting adult leaf-feeding beetles: pyrethrum knockdown, composite flight-interception traps, branch clipping and hand-collecting/beating. The total number of species collected was highest with composite flight-interception traps, whereas the number of species collected and known to feed on the tree species sampled was highest with hand-collecting/beating. The results emphasize the need for spatial and seasonal replicates in faunal surveys and the abundance of transient species in these replicates. None of the sampling methods examined can be considered as a panacea for investigating a wide range of topics. It is imperative that several, complementary methods should be used for general arthropod surveys. A key is provided to assist ecologists in selecting sampling methods appropriate for their research.

INTRODUCTION

Early studies on canopy arthropods were more concerned with taxonomic inventories than with investigation of specific ecological topics. With improvements in gaining access to the canopy and the development and refinement of several sampling methods, the study of canopy

Canopy Arthropods. Edited by N.E. Stork, J. Adis and R.K. Didham. Published in 1997 by Chapman & Hall, London. ISBN 0 412 74900 9

arthropods has matured and now embraces a wide range of ecological issues, as this book testifies. In this chapter, the term 'canopy' is used in its broader sense, meaning the crown of trees, whereas its stricter sense, 'canopy ecotone', refers to the interface between the uppermost layer of leaves and the atmosphere (Hallé and Blanc, 1990).

The growing scientific interest in canopy arthropods emphasizes the need for appropriate sampling methods. Obviously, it is important to standardize techniques whenever studies attempt to answer similar questions at different locations. However, it would be misleading to assume that one or a few sampling methods would be appropriate to overcome the numerous challenges that the ecologist faces when studying canopy arthropods. Each method has its inherent advantages and biases and some may be more appropriate than others to investigate specific topics (Southwood, 1978). For example, pyrethrum knockdown is a productive method which has been used widely both in temperate and tropical forests (Stork and Hammond, 1997, Chapter 1, this volume). However, this method would be unsuitable to study arthropod diel activity in tree canopies, since the same tree cannot be re-sampled within a few hours (since full recolonization cannot be expected during that period of time).

Instead, this review examines these concerns and acknowledges the strengths and the limitations of each method, as well as the rationale for its use. It is intended primarily for the ecologist, is focused on sampling tropical arthropods, and provides representative examples rather than an exhaustive list of references. It is based on field experience and is supplemented with a comparison of four methods used for sampling leaf-feeding beetles at one tropical location. Students interested in sampling canopy arthropods may wish to consult, in addition to textbooks on statistics and multivariate analyses, papers covering important topics which are not discussed here. These include protocols for quantitative studies of assemblages, sample dimensions and complementarity, and the extrapolation of species richness (Hutcheson, 1990; Coddington *et al.*, 1991; Eberhardt and Thomas, 1991; Hammond and Harding, 1991; Colwell and Coddington, 1994; Hammond, 1994; Longino, 1994).

Before commencing the review of sampling methods, it should be mentioned that there are a number of methods for gaining access to the canopy or for establishment of sampling equipment in the canopy, such as the spikes-and-belt method, single-rope technique, towers, cranes, walkways, dirigible and canopy raft (reviewed in Mitchell, 1982; Lowman *et al.*, 1993a; Moffett, 1993). Since it is possible to use a wide range of sampling methods with the recent innovation of the canopy raft, a separate section is devoted to the latter.

A REVIEW OF METHODS

Insecticide knockdown

Stork and Hammond (1997, Chapter 1, this volume) review the use of insecticide for sampling canopy arthropods. The main advantage of this method includes relatively quick implementation (making it suitable for short-term studies), high productivity (high numbers of arthropods collected) and 'clean' samples, which may be processed easily. The method appears ideal for general surveys of forest tracts and large-scale taxonomic work (Erwin and Scott, 1980; Erwin, 1983; Stork, 1987a,b). However, sampling may be highly dependent on weather conditions and, usually, needs to be performed when air conditions are calm, e.g. at daybreak. Where an emphasis is on the determination of arthropod densities (expressed by the number of arthropods collected per surface area of tray) (Greenwood, 1990; Stork and Brendell, 1990, 1991; Stork, 1991; Russell-Smith and Stork, 1994), the sample size represented by each fogging tray is known imprecisely, since the amount of foliage above it is difficult to quantify. Usually, arthropods are collected dead (but see Paarmann and Stork, 1987; Adis *et al.*, 1997, Chapter 4, this volume; Paarmann and Kerck, 1997, Chapter 3, this volume) and their origin from a specific habitat within the tree sampled may be difficult to trace with confidence. Another restriction, as emphasized previously, is that it is not possible to re-sample the same tree before allowing for recolonization. The cost of equipment and chemicals and the time required to both clear the area beneath the target tree and emplace the collecting trays must be taken into consideration.

Foliage samples

Most methods targeting foliage arthropods, including those in flowers and seeds, are grouped here. Limitations of these methods include: (i) the disturbance of foliage causing active insects to fly or jump off, although this is less of a problem when using cranes; (ii) depending on which method is used to gain access to the canopy (e.g. single-rope technique), it is often impossible to sample in the periphery of the crown, unless access is gained from adjacent trees (but see discussion on the canopy raft); and (iii) it is difficult to sample from moist foliage. Most of these methods are inexpensive, with the possible exception of 'gassing', but often time consuming.

Hand-collecting

The most direct technique is the inspection of foliage and subsequent collection of arthropods in tubes or with entomological nets or aspirators/pooters (Morris, 1955; Moran *et al.*, 1994; Basset, 1997, Chapter 12, this volume). The origin of specimens is known and they may be collected alive. However, the method is not productive and its results are considerably dependent on the experience of the investigator. Direct examination of leaves has provided estimates of densities of minute arthropods, such as mites (Walter *et al.*, 1994).

Extraction

Minute arthropods associated with leaves, flowers and seeds may be extracted using Berlese–Tullgren apparatus or similar devices (Harris, 1971; Basset, 1985). The method is destructive, but sample size can be determined easily so that both numbers of individuals and species may be compared among samples. Small arthropods may be extracted from leaves by exposing them to the vapour of certain chemicals or washed from leaves using various solutions (Southwood, 1978).

Branch clipping

A few branches are cut and enclosed in a large plastic bag, whose content is later examined in the laboratory (Ohmart *et al.*, 1983; Majer and Recher, 1988; Costa and Crossley, 1991; Basset *et al.*, 1992a) (Figure 2.1). Blanton (1990) described a convenient collapsible-bag sampler and compared samples taken with this technique with those obtained with pyrethrum knockdown. Arthropods can be anaesthetized by dropping a small ball of cotton-wool, saturated with ethyl acetate, inside the bag. Sample size (leaf area) can be determined, usually with precalculated regressions of dry weight against leaf area. It is an inexpensive method to estimate the actual density of foliage arthropods, despite being destructive, biased towards sedentary taxa and relatively non-productive. Further, individual bags often sample such a small portion of habitat that they must be pooled to be sufficiently representative (Blanton, 1990).

'Gassing'

This is a small-scale variation of the fogging method, also known as the 'selecteur' or 'restricted canopy fogging' (Lepointe, 1956; Dempster, 1961; Basset, 1985, 1990). A few branches are enclosed in a container or a plastic bag, which is then gassed with carbon dioxide, for example, and the anaesthetized arthropods retrieved. Leaves may be cut or counted to estimate sample size and to provide estimates of arthropod densities. This method has been used for studying arthropod stratification within tree crowns (Basset, 1992). However, the method is inadequate for sampling arthropods from the trunk and large limbs and the foliage is disturbed when the container or bag is positioned.

Beating

A beating tray is held under a few branches, which are then struck with a stick. Usually, fallen arthropods are collected with aspirators (Lepointe, 1956; Harris *et al.*, 1972; Turner, 1974; New, 1979; Mouna *et al.*, 1985). However, quantitative samples may be derived by collecting most arthropods into a large plastic funnel, fitted with a collecting jar filled with fluid (Wilson, 1962; Basset, 1985; unpublished data). Insects and falling debris may be pushed gently into the collecting jar with a brush. Sample size may vary considerably, depending on the type of foliage. This method is particularly effective for dislodging free-living caterpillars, but less so for active or small arthropods.

Sweeping

This is a popular method for sampling arthropods in the field layer, in which the vegetation is swept with a net. Since sweeping requires perambulation, this method has been used rarely in tree crowns (Dowdy, 1950; Lepointe, 1956; Lowman *et al.*, 1993a,b). Like beating, sample size (measured here as one or a few sweeps) can vary considerably, depending on the nature of the foliage and, therefore, samples are difficult to compare. Sweeping is less effective in dense vegetation and depends on the experience of the investigator (Lamotte *et al.*, 1969). Active arthropods tend to be better sampled than sedentary ones (Noyes, 1989) and small specimens tend to be overlooked (Hespenheide, 1979). To remedy the latter problem and to process large numbers of arthropods, LeSage (1991) proposed a sweeping technique in which the entire contents of the net are placed in a killing jar. A study comparing the relative efficiency of pyrethrum knockdown, beating and sweeping for sampling arboreal arthropods is in progress (M.D. Lowman, personal communication).

Non-attractive traps

The methods reviewed in this section do not provide a measurement of density, but relative measurement of activity. Flying insects are targeted, but adult Lepidoptera are difficult to identify after being immersed in or entrapped by the collecting agent (although killing-jars may be used). Non-attractive traps may provide less biased general surveys than attractive traps. The former are often inexpensive, but the investment in time for cleaning the samples must be considered. Several authors have used various criteria to compare the effectiveness of non-attractive and attractive traps (see p. 33; Juillet, 1963; Southwood, 1978; Hosking, 1979; Osmelak, 1987; Noyes, 1989; Hammond, 1990; Muirhead-Thomson, 1991).

Malaise traps

These refer to a class of tent-like traps of different designs. They have been used extensively in the field layer, but less so within tree-crowns (Crossley *et al.*, 1973; Basset, 1985; Hammond, 1990) (Figure 2.1). In the field layer they are supported by ropes and pegs, but require suspension within a rigid frame for use in the canopy. They target insects whose tendency is to fly upwards when encountering a vertical surface, and are particularly effective for collecting Diptera and Hymenoptera. Dufour (1980) described a trap combining features of Malaise and light trap, suitable for sampling arthropods both during day- and night-time.

Flight-interception traps

These consist of vertical panels and collecting trays, the latter filled with water or other collecting fluid, which have been used in tree canopies (Merrill and Skelly, 1968; Crossley *et al.*, 1973; Hosking and Knight, 1975). They are more effective for collecting arthropods, such as Coleoptera, which fall when encountering vertical surfaces. Masner and Goulet (1981) described a model on which contact insecticide is applied, thus increasing the effectiveness of collection of minute and slow-flying taxa. Wilkening *et al.* (1981) illustrated an inexpensive omnidirectional flight trap suitable for sampling in tree-crowns.

Composite flight-interception traps

These traps combine features of both Malaise and flight-interception traps, thus resulting in less bias toward specific taxa (Basset, 1988; Basset *et al.*, 1992b) (Figure 2.1). One recent model (Springate and Basset, 1996) was particularly well-adapted for selective sampling of tree-crown faunas. The main body of the trap consists of a rectangular cross-panel of black netting with a roof of white netting, connected to a collecting jar via a clear plastic tube. A clear plastic funnel is attached below the main body of the trap and connected to a large collecting jar. In the lower collecting jar, a solution of water saturated with salt is used as collecting fluid, which remains effective even during heavy rainfall. The width (80 cm) and height (250 cm) of the trap allow convenient emplacement and re-positioning after survey within tree-crowns. Similarly, Robert (1992) described a model combining features of Malaise, window, water and pitfall traps ('piège entomologique composite') and illustrated its use within the tree layer in Madagascar. A recent study showed that a trapping period of 24 hours with this model was sufficient to characterize entomological samples obtained at different sites, but further trapping was required to estimate total species richness at these sites (J.C. Robert, personal communication).

Sticky traps

Sticky traps (wood, plastic or stout cardboard coated with glue) have been used in tropical tree-crowns relatively rarely (Sutton and Hudson, 1980; House, 1989). Recently, these traps were used on a large scale in Papua New Guinea, for collecting arboreal weevils (O. Missa, personal communication), using a small roof in order to protect the glue from rain and falling debris. This method is inexpensive, hence enabling a large number of replicates to be taken, and is highly suitable for the study of arthropod spatial distribution and stratification. However, the glue is difficult to apply and removal and identification of trapped specimens, particularly fragile insects, are difficult. The traditional removal of insects using non-polar (and usually carcinogenic) solvents is being superseded by the use of citrus oil (Miller *et al.*, 1993). Sticky traps may be modified with different coloured surfaces and chemical baits to become attractive (Knodel and Agnello, 1990; Muirhead-Thomson, 1991).

Photo-eclectors

Usually, arboreal photo-eclectors consist of black funnels surmounted by clear collecting containers, affixed to tree trunks. These traps target arthropods foraging on tree trunks (Funke, 1971; Adis, 1981; Nicolai, 1986; New *et al.*, 1991) and may be modified to study upward or downward migration of arthropods (Moeed and Meads, 1983; Adis and Schubart, 1984).

Attractive traps

As in the previous section, the emphasis is on recording arthropod activity. Since the attractiveness of the trap may vary from one taxon to another, the strength of a certain trap model for particular taxa may become a weakness when used in general surveys. The distance from which insects are attracted is often difficult to evaluate, so that selective sampling of the fauna associated with particular tree species may be less effective than that with non-attractive traps, since non-resident arthropods may well be included in any sample.

Light traps

Nocturnal insects which may be attracted to light, such as many species of moths, are often trapped with this technique in tree canopies (Sutton, 1979; Wolda, 1979; Sutton and Hudson, 1980; Smythe, 1982; Rees, 1983). Several models of light traps are available, combining different designs

(a)

(b)

Figure 2.1 Some entomological methods used with the canopy raft. (a) The canopy raft emplaced (Cameroon, Campo, October 1991; photograph P. Grard). (b) A Malaise trap on the canopy raft (French Guiana, Petit-Saut, October, 1989; photograph G. Delvare). (c) A composite flight-interception trap lowered to ground-level (Cameroon, Campo, October 1991; photograph.

(c)

(d)

R. Gaillarde, Gamma). (d) The first author examining the content of a branch-clipping sample in the laboratory with six Berlese funnels, of Orousset's (in press) design, in the background (Cameroon, Campo, October 1991; photograph H. Setsumasa).

and light sources, the latter including acetylene lamps, tungsten filament electric lights, mercury-quartz lamps, 'black' lamps, etc. (reviewed by Southwood, 1978). Gerber *et al.* (1992) described a portable, solar-powered charging system for blacklight traps. Robert (1983) devised a model of directional light trap used at different heights to study arthropod stratification in Madagascan forests. Light trapping often provides high numbers of specimens. However, for comparison between traps, correction factors must be computed for the effects of temperature, moonlight, cloud cover and background illumination (Bowden, 1982). The cost of equipment and its operation is often high, particularly if a high number of replicates are required, although this may be remedied by a recent and inexpensive version of the Robinson-pattern mercury-vapour lamp moth-trap, costing about $US 30.00 (G. Robinson, personal communication).

Baited traps

This category includes traps which are designed specifically to catch a narrow spectrum of taxa. Various trap models and baits have been employed, the former ranging from small plastic bottles to large buckets (Togashi, 1990; Allemand and Aberlenc, 1991) and the latter from food-matter to pheromones or their mimics (Hammond, 1990; Togashi, 1990; Muirhead-Thomson, 1991). Austin and Riley (1995) describe two models of a portable bait trap for butterflies, discuss their use (usually hung between 5–10 m) with fruit- and 'stink'-baits in the Neotropical region and provide a useful set of references about trap design. They noted that 'stink'-baits proved effective in trapping Orthoptera, Hemiptera, Diptera and Hymenoptera, in addition to Lepidoptera. Sourakov and Emmel (1995) discuss briefly the use of three types of bait trap, at several heights (up to 20 m) and with various lures, in Kenya. Allemand and Aberlenc (1991) described an inexpensive trap made from a plastic water bottle, which proved efficient and less selective. A liquid bait based on red wine was often used, but other successful baits included beer, fermented fruit, fish, shrimps, cheese, meat and excrement. It is possible to collect live insects with this method, depending on the bait used. Usually, specimens require cleaning and rinsing before storage or mounting.

Water pan traps

Usually, these consist of shallow card, plastic or aluminium food containers, painted yellow and filled with water and detergent (although other colours may be used with success; see Kirk, 1984). These traps have been used extensively in the field layer but less so in tree-crowns

(Krizelj, 1971; Couturier, 1973; Basset, 1985). They are particularly attractive for Diptera, Hymenoptera and Thysanoptera and high numbers of replicates are reasonably inexpensive. However, their ultimate effectiveness is likely to depend upon trap specification and siting. A disadvantage of these traps is their sensitivity to rainfall, wind and desiccation, which may easily ruin catches, thus requiring frequent servicing.

Other methods

D-Vac sampler

Some authors have used suction apparatus fixed on towers or suspended within trees to vacuum arthropods from the surrounding foliage (Lepointe, 1956; Rees, 1983). A more promising approach is the use of portable D-Vac samplers with long, flexible pipes (Dietrick, 1961). These devices have been used frequently to vacuum arthropods from the field layer but, to date, do not appear to have been used on a large scale in tree canopies. The apparatus may be carried conveniently and models relying on both electrical and combustion engines are available, the latter being more powerful and of higher autonomy. It is possible to modify inexpensive commercial leaf-blowers (Wilson *et al.*, 1993). Arthropods could also be sampled from other habitats than foliage (e.g. trunk, branches). However, disadvantages include cost, weight, exhaust gases, clogging with debris, possible damage to specimens and the definition of sample size.

Extraction of epiphytes

Inhabitants of epiphytes and of 'suspended soils' in the canopy have been sampled using Berlese–Tullgren apparatus or Winkler/Moczarksi eclectors (Delamare-Debouteville, 1951; Nadkarni and Longino, 1990; Paoletti *et al.*, 1991). Depending on the robustness of the taxa, large volumes of samples can be sifted and processed with Winkler/Moczarksi eclectors, which are independent from power source and light and permit the extraction of live arthropods (Besuchet *et al.*, 1987). A convenient, light and collapsible Berlese–Tullgren apparatus has been devised by Orousset (in press).

Rearing of branch and other samples

Arthropods from specific arboreal habitats may be obtained by rearing galls, leaf mines, stem-borers and samples of flowers, fruits, seeds, dead branches, etc. collected in the canopy. Living or dead branches may be cut, left for a few weeks or months and then placed in rearing cages

(Owen, 1989, 1992), plastic bags (C.R. Vardy, personal communication) or Tullgren extraction apparatus (Paviour-Smith and Elbourn, 1993). In French Guiana, this technique is being used on a large scale for assessing the host-specificity of longicorn beetles, by felling many trees which are later cut up and placed in rearing cages (G. Tavakilian, personal communication). Emergence traps for bark-dwelling arthropods (Glen, 1976) are an alternative method, but their effectiveness and results depend on the state of decay of the habitat sampled (Basset, 1985).

SAMPLING METHODS USED WITH THE CANOPY RAFT

The 'canopy raft' ('radeau des cimes') represents a recent development for investigating selectively the canopy ecotone of tropical forests. This is a large-scale operation in which an air-inflated dirigible is used to transport and emplace a hexagonal platform of 580 m^2 on the canopy (Cleyet-Marrel, 1990; Ebersolt, 1990) (Figure 2.1). Investigators use single-rope techniques to gain access to the platform (the raft), which consists of air-inflated beams and Aramide netting. Descriptions of the operation, as well as examples of scientific projects performed with the canopy raft, are reported in Hallé and Blanc (1990) and in Hallé and Pascal (1992).

A wide range of methods for sampling arthropods has been used from the canopy raft. During the first scientific expedition in French Guiana, Delvare and Aberlenc (1990) used hand-collecting, sweeping, Malaise traps, light traps, baited traps and rearing. In addition, they used a large net towed by the dirigible at night and illuminated by 500-W lights, thus creating a mobile light trap above the canopy. They concluded that Malaise and light traps were effective for entomological survey of the canopy, but some moths flying upward to the light trap were unlikely to be canopy residents. Nancé (1990) used beating and sweeping to collect spiders in the canopy.

During the second expedition in Cameroon, Basset *et al.* (1992a,b) used branch clipping, Malaise traps and composite flight-interception traps. They concluded that branch clipping was appropriate to estimate densities of sedentary arthropods and that composite flight-interception traps provided a wider spectrum of taxa than Malaise traps for general surveys. Lowman *et al.* (1992, 1993a) used sweeping for general surveys, while McKey (1992) and Dejean (1992) used hand-collecting and direct observation to study ants and Yumoto (1992) netted insect pollinators.

Both expeditions provided the opportunity to test methods for obtaining botanical and entomological samples from a variety of locations in the canopy. A small triangular platform of 16 m^2 ('sledge' or 'luge') was suspended 10 m below the dirigible, as the latter glides over the canopy at low speed (Ebersolt, 1990; Lowman *et al.*, 1993a). The sledge is suitable for three investigators, who may command its precise placement. Two teams of investigators took entomological samples using different methods: Basset *et al.* (1992a,b) used branch clipping, whereas Lowman *et al.* (1992, 1993a) used sweeping. Both teams concluded that the sledge is effective, since it allowed rapid access to many otherwise inaccessible locations. The use of D-Vac samplers, both on the raft and with the sledge, appears to be particularly promising for future expeditions.

A COMPARISON OF FOUR METHODS IN PAPUA NEW GUINEAN TREES

Materials and methods

From the above discussion it can be seen that while certain methods may be appropriate for sampling a particular target group of arthropods, or a specific habitat within the canopy, no single method exists as a sampling panacea for general surveys. For the future it may be useful to select several, complementary methods for use in a 'sampling package' (Stork, 1994). These need not be expensive and may provide much useful and comparable data from a variety of habitats (Gadagkar *et al.*, 1990). To illustrate such a review and the use of a sampling package, data are presented comparing four methods used in a survey of leaf-feeding beetles (i.e. Chrysomelidae, Curculionidae, Lagriidae, etc.). Beetles associated with 10 tree species were sampled during one year of field work on the slopes of Mount Kaindi, near Wau, Papua New Guinea (details in Basset, 1997, Chapter 12, this volume). Here, the question whether certain sampling methods provided a better general survey of beetle species and of specialists than others is investigated.

Beetles were collected using four methods:

1. Hand-collecting/beating: these two methods were considered jointly, since the foliage was struck immediately after its visual inspection. These samples represented, for each tree species, about 50 hours of hand-collecting activity and 300 beating samples, distributed among different trees.
2. Branch clipping represented, for each tree species, 55 samples of about 33 m^2 of leaf surface, obtained from different trees.
3. One composite flight-interception trap was established in the crown of one individual of each tree species. The trap collected insects continuously during an entire year and was surveyed approximately every 11 days.
4. One individual of each tree species was sampled using pyrethrum knockdown (5% Pyranone® and kerosene), using 12–20 trays (1 m^2

Table 2.1 Number of individuals and morphospecies of adult leaf-feeding beetles and mean body length (and standard error) of morphospecies collected with each sampling method and within each feeding category. HC/B, hand-collecting/beating; BC, branch clipping; FIT, composite flight-interception traps; FOG, pyrethrum knockdown

Category	*HC/B*		*BC*		*FIT*		*FOG*		*All methods*	
	No. of individuals	*No. of morpho-species*	*No. of individuals*	*No. of morpho-species*	*No. of individuals*	*No. of morpho-species*	*No. of individuals*	*No. of morpho-species*	*No. of individuals*	*No. of morpho-species*
Specialists	413	40	94	20	319	29	177	20	1003	40
Generalists	370	32	98	11	457	13	272	14	1197	33
Uncertains	32	13	1	1	137	4	13	4	183	13
Incidentals	230	93	100	21	302	44	693	23	1325	99
Additionals	1	1	19	13	647	146	263	66	930	197
Proven feeders	815	85	193	32	913	46	462	38	2383	86
Leaf-feeding beetles	1046	179	312	66	1862	236	1418	127	4638	382
Mean body length (mm) (standard error)	6.32 (0.399)		5.10 (0.883)		5.64 (0.278)		4.62 (0.268)		5.56 (0.195)	

surface area), depending on tree size (total 159 trays used for all tree species).

The first three of these methods were used during both day and night, whereas pyrethrum knockdown was performed at daybreak only.

Live beetles were stored in plastic vials, at room temperature and in conditions of near-saturated relative humidity. They were provided with fresh foliage and tested for feeding. These tests (details in Basset, 1997, Chapter 12, this volume) allowed the assignment of beetles into the following categories: (i) 'specialists', i.e. feeding only on one tree species; (ii) 'generalists', feeding on two or more tree species; (iii) 'uncertains', feeding, but not enough information to assign either to specialists or generalists; (iv) 'incidentals', not feeding; and (v) 'additionals', collected dead, by various methods and, therefore, not tested. Categories (i), (ii) and (iii) were referred to as 'proven feeders', i.e. species known to feed on the foliage of tree species sampled.

Since sampling effort, as well as the number of habitats (trees) sampled, varied for each method, it is difficult to compare the effectiveness of the different methods for surveying foliage beetles. In particular, no attempt was made to use rarefaction techniques to estimate the number of species for a common sample size since the results of these computations would be heavily dependent upon the arbitrary definition of a 'sample' for each method (e.g. 1 hour or 1 day of hand-collecting, one week or one month of trap-collecting, one or several fogging trays, etc.). Further, the accuracy and precision of the jack-knife estimate is highly dependent upon the number of replicates (samples) available (Coddington *et al.*, 1991). As an alternative, for each sampling method, the rate of species accumulation within the cumulative number of individuals collected was considered. Thus, to some extent, a comparison could be made between the rates of discovery of 'new' species within the entire material sampled and this within the different feeding categories defined.

Results and discussion

A total of 4638 leaf-beetles, representing 382 morphospecies in the families Chrysomelidae, Curculionidae and Lagriidae, were collected with the four sampling methods. The total numbers of individuals and species collected with each method and within each feeding category of beetles are indicated in Table 2.1. While the total number of morphospecies collected was greatest with the traps, the number of species of specialists and proven feeders was particularly high using hand-collecting/beating. These totals were lower for the trap, fogging and branch clipping samples. This was not unexpected since hand-collecting/beating

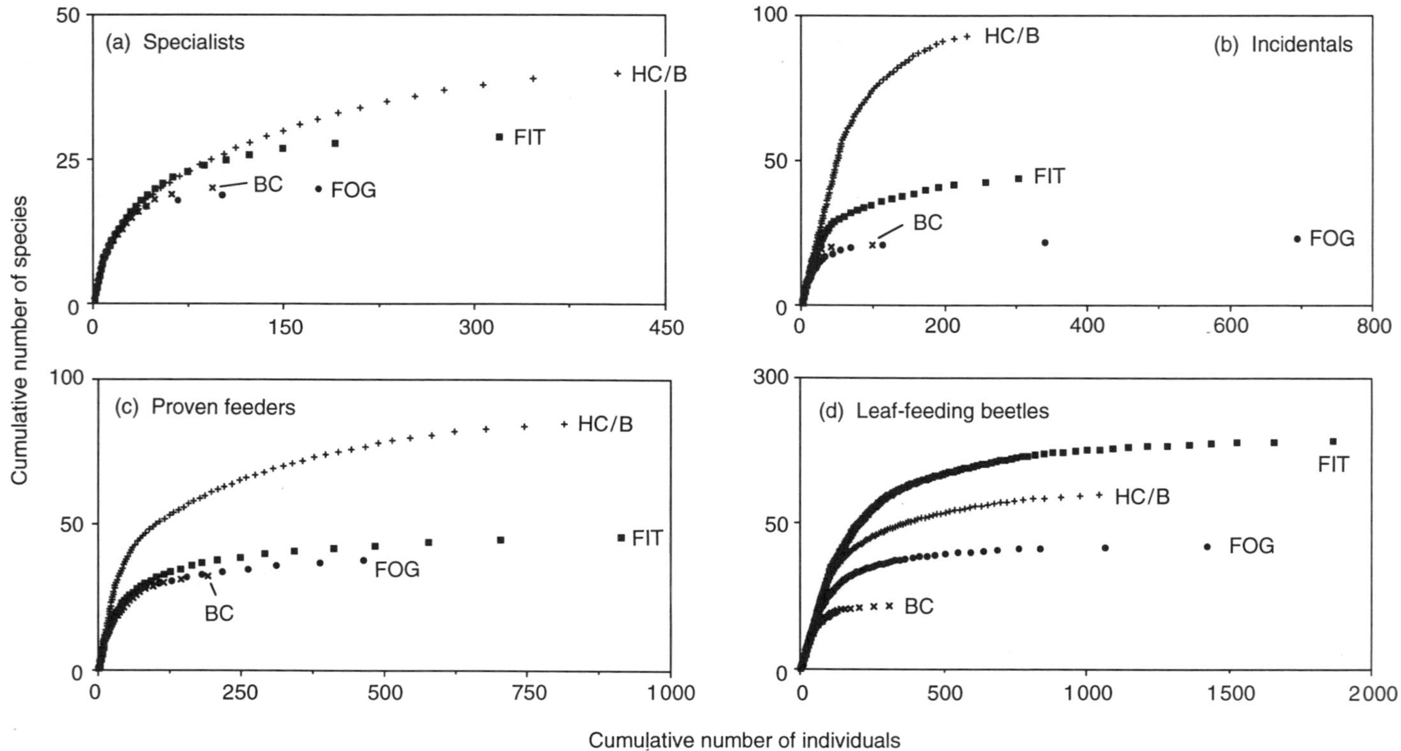

Figure 2.2 Cumulative number of individuals plotted against cumulative number of species for (a) specialists, (b) incidentals, (c) proven feeders and (d) all leaf-feeding beetles detailed by sampling method. HC/B, hand-collecting/beating (+); BC, branch clipping (x); FIT, composite flight-interception traps (■); FOG, pyrethrum knockdown (●).

was performed on several trees of the same species, at different periods of the year (as was branch clipping), while the traps sampled arthropod populations obtained from one tree, but at different periods of the year, and fogging focused on one individual at one period of time.

Despite high proportions of incidental species being collected with hand-collecting/beating and trapping, the number of species collected with each sampling method was distributed uniformly when the proven feeder and incidental categories were compared (G-test, G = 5.38, P = 0.146). However, a similar comparison of the distribution of the number of individuals collected was non-uniform, with a high proportion of incidentals collected by fogging (G = 439.4, P <0.001).

Figure 2.2 shows the cumulative plots of the number of individuals and species for each sampling method and for selected feeding categories. If a common sample size among the different sampling methods of, for example, 300 individuals, is considered, the above observations remain valid. It is probable that trapped material will be more speciose than that derived from other methods, when all leaf-feeding beetles are considered. However, it is probable that when proven feeders or specialists are considered, material obtained by hand-collecting/beating will be more speciose. However, this effect is only likely to be noteworthy for material containing more than 300 individuals.

Several interesting observations may be inferred from these plots:

1. Spatial as well as seasonal replicates ensure that samples are representative of the total species richness present. Here, the apparent poor performance of pyrethrum knockdown is explained by the lack of such replicates. For proven feeders, spatial replicates appear more important than seasonal replicates, as few differences existed between trap and fogging curves (Figure 2.2c). For the purpose of this study, hand-collecting/beating was the superior method, since many more trees were visited during the time needed to sample different trees with pyrethrum knockdown or with composite flight-interception traps. However, it is evident that the applicability of the hand-collecting/beating method is highly dependent on the ease of gaining access to the canopy.
2. Sampling other habitats will result in collecting more incidentals and, therefore, more transient species. It is probable that 'incidentals' included both transients and species genuinely associated with tree species studied, but which were feeding on parts other than the foliage. The relation between transient species and the diversity of tropical vegetation is discussed elsewhere (Basset, 1997, Chapter 12, this volume). Since the arthropod fauna associated with vegetation surrounding the trees sampled will change throughout the year, seasonal replicates may yield more transient species. Similarly,

diurnal/nocturnal replicates may also influence species richness (Blanton, 1990)

3. Branch clipping was the least effective of all methods, but was unique in providing a precise estimate of the amount of habitat sampled. This technique should be used to compare relative densities of arthropods, rather than estimating species richness.

Some 120 beetle species were collected exclusively with the traps, 68 by hand-collecting/beating, 53 by fogging and only 10 by branch clipping. To some extent, this reflects the effectiveness of the different methods with the present sampling protocol. Not unexpectedly, a cluster analysis with the 382 species of leaf-feeding beetles (data not presented) showed that hand-collecting/beating and branch clipping were the closest of the four methods, with trapping and fogging more distant. This reflects that the first two methods target foliage arthropods, whereas the others sample indiscriminately the fauna foraging on the foliage, trunk and branches (some weevil species in the 'incidental' and 'additional' categories may be wood-borers).

Further analysis indicated that the average body size of morphospecies collected varied significantly between sampling methods (Kruskal–Wallis $W = 17.113$, $P < 0.01$). In particular, the average body size of morphospecies collected with hand-collecting/beating was higher than that collected with pyrethrum knockdown (Table 2.1). Either this reflects the poor performance of hand-collecting/beating at collecting small species, or the poor performance of pyrethrum knockdown at collecting large species, or possibly both.

CONCLUSION: CHOOSING AN APPROPRIATE COLLECTING METHOD

Both the literature review and the comparison of sampling methods for surveying leaf-feeding beetles in Papua New Guinea support the contention that none of the methods examined can be considered as the panacea for investigating a wide range of ecological topics. Rather, and in particular for general surveys, the implementation of a range of methods, used in conjunction and providing spatial, seasonal and diurnal replicates, will provide larger and more diverse samples. For example, a combination of hand-collecting/beating, composite flight-interception traps and pyrethrum knockdown may be one such strategy. Depending on the research goals of the investigator, other techniques, used singly or in conjunction, may be more suitable.

The few studies of the stratification of arthropods in tropical forests and of the entomofauna of the canopy ecotone suggest that the faunal composition of the canopy is very different from that found in lower

Table 2.2 Key to assist the ecologist in the choice of a method for sampling arthropods in tree canopies (numbers in parentheses refer to pages in the text)

1.	No emphasis on sampling particular habitats within trees	2
–	Emphasis on sampling particular habitats within trees	9
2.	Emphasis on estimating relative activities of arthropods, regular sampling of the same individual tree; flying arthropods targeted	3
–	Emphasis on estimating actual densities of arthropods and not based upon regular sampling from the same individual tree; both flying and flightless arthropods targeted	**insecticide knockdown** (29)
3.	Emphasis on nocturnal arthropods	**light traps** (33)
–	No emphasis on nocturnal arthropods	4
4.	Non- or relatively non-specific sampling of arthropod faunas	5
–	Specific taxon or taxa targeted	8
5.	Emphasis on studying spatial distribution, particularly with a high number of replicates	**sticky traps** (33)
–	Emphasis on studying seasonal distribution	6
6.	Traps not particularly biased towards either light or heavy arthropods	**composite flight-interception traps** (32)
–	Trapping method with known tendency to selectivity	7
7.	Traps biased towards Diptera and light arthropods	**Malaise traps** (32)
–	Traps biased towards Coleoptera and heavy arthropods	**flight-interception traps** (32)
8.	Visual attractants	**water traps** (36)
–	Olfactory attractants	**baited traps** (36)
9.	Emphasis on foliage arthropods	10
–	No emphasis on foliage arthropods	16
10.	No emphasis on comparing samples of known size	11
–	Emphasis on comparing samples of known (or relatively known) size	12
11.	Electrical/solar power source or fuel available for sampling over long duration	**D-Vac sampler** (37)
–	No power source or fuel available	**hand-collecting** (29)
12.	Emphasis on estimation of actual densities of arthropods	13
–	Emphasis on estimation of relative densities of arthropods	15
13.	Minute arthropods targeted	**extraction of leaves** (30)
–	Minute arthropods not targeted	14
14.	Volume sampled relatively small (destructive method)	**branch clipping** (30)
–	Volume sampled higher (method not necessarily destructive)	**gassing** (30)
15.	Perambulation possible; active arthropods targeted	**sweeping** (31)
–	Perambulation restricted; sedentary arthropods targeted	**beating** (31)
16.	Emphasis on concealed fauna	**rearing** (37)
–	Emphasis on 'exposed' fauna	17
17.	Emphasis on epiphytic fauna	**extraction of epiphytes** (37)
–	Emphasis on trunk-foraging fauna	**photo-eclectors** (33)

strata (Sutton and Hudson, 1980; Delvare and Aberlenc, 1990; Basset *et al.*, 1992a,b; J.C. Robert, personal communication). This difference emphasizes the need for selective sampling of the canopy ecotone, which, to date, can be achieved with the canopy raft and sledge alone, and, in a more restricted way, by cranes. Furthermore, the variety of sampling methods used on the canopy raft emphasizes that different methods are needed to pursue specific research goals.

Based on field experience and a literature review, a key (Table 2.2) is provided to assist (rather than direct) the ecologist in the selection of suitable methods for sampling arthropods in tree canopies. It should be noted that certain methods could be placed in different sections or may appear twice, particularly where they have multiple functions. Additional factors the ecologist should consider are the efficiency of methods in relation to both cost and investment of time, and the suitability and sensitivity of a particular method to climatic conditions in tropical forests.

Acknowledgements

The assistance in French Guiana and Cameroon of Francis Hallé, Gilles Ebersolt, Dany Cleyet-Marrel and the staff of Opération Canopée is gratefully acknowledged. In Papua New Guinea, the first author acknowledges the assistance of Nathan Daniel and Martin Kasbal and the support of Harry Sakulas and the staff of the Wau Ecology Institute. G.A. Samuelson checked the assignment of beetles to morphospecies. Many colleagues commented on earlier drafts of the manuscript and provided us with unpublished information, particularly Nigel Stork, Peter Hammond, Daniel Burckhardt, Willy Matthey, Scott Miller, Jeremy Holloway, Gaden Robinson, Barry Bolton, Paul Williams, Steve Brooks, Tom Huddleston and Andrew Polaszek. Studies with the canopy raft were funded by the Elf Foundation, Opération Canopée and other sponsors indicated in Hallé and Blanc (1990). Studies in Papua New Guinea were funded in part by a grant from the Swiss National Science Foundation to the first author and were made possible by Bishop Museum technical staff. The manuscript was prepared while the first author was in receipt of a Christensen Research Institute Fellowship and National Science Foundation grant DEB-94-07927 (with Scott Miller and Allen Allison). This is contribution no. 150 of the Christensen Research Institute.

REFERENCES

Adis, J. (1981) Comparative ecological studies of the terrestrial arthropod fauna in Central Amazonian inundation forests. *Amazoniana*, **7**, 87–173.

Adis, J. and Schubart, H.O.R. (1985) Ecological research on arthropods in Central Amazonian forest ecosystems with recommendation for study procedures, in *Trends in Ecological Research for the 1980s* (eds J. Cooley and F.B. Golley), Plenum Press, New York, pp. 111–44.

Adis, J., Paarmann, W., Fonseca, C.R.V. da and Rafael, J.A. (1997) Knockdown efficiency of natural pyrethrum and survival rate of arthropods obtained by canopy fogging in Central Amazonia, in *Canopy Arthropods* (eds N.E. Stork, J. Adis and R.K. Didham), Chapman & Hall, London, pp. 67–79 (this volume).

Allemand, R. and Aberlenc, H.-P. (1991) Une méthode efficace d'échantillonnage de l'entomofaune des frondaisons: le piège attractif aérien. *Mitteilungen der Schweizerischen Entomologischen Gesellschaft*, **64**, 293–305.

Austin, G.T. and Riley, T.J. (1995) Portable bait traps for the study of butterflies. *Tropical Lepidoptera*, **6**, 5–9.

Basset, Y. (1985) Comparaison de quelques méthodes de piégeage de la faune dendrobie. *Bulletin Romand d'Entomologie*, **3**, 1–14.

Basset, Y. (1988) A composite interception trap for sampling arthropods in tree canopies. *Journal of the Australian Entomological Society*, **27**, 213–19.

Basset, Y. (1990) The arboreal fauna of the rainforest tree *Argyrodendron actinophyllum* as sampled with restricted canopy fogging: composition of the fauna. *The Entomologist*, **109**, 173–83.

Basset, Y. (1992) Influence of leaf traits on the spatial distribution of arboreal arthropods within an overstorey rainforest tree. *Ecological Entomology*, **17**, 8–16.

Basset, Y. (1997) Species abundance and body size relationships in insect herbivores associated with New Guinea forest trees, with particular reference to insect host specificity, in *Canopy Arthropods* (eds N.E. Stork, J. Adis and R.K. Didham), Chapman & Hall, London, pp. 237–64 (this volume).

Basset, Y., Aberlenc, H.-P. and Delvare, G. (1992a) Abundance and stratification of foliage arthropods in a lowland rain forest of Cameroon. *Ecological Entomology*, **17**, 310–18.

Basset, Y., Aberlenc, H.-P. and Delvare, G. (1992b) Abundance, diversité et stratification verticale de l'entomofaune d'une forêt tropicale humide africaine, in *Biologie d'une Canopée de Forêt Equatoriale – II. Rapport de Mission: Radeau des Cimes Octobre-Novembre 1991, Réserve de Campo, Cameroun* (eds F. Hallé and O. Pascal), Fondation Elf, Paris, pp. 45–52.

Besuchet, C., Burckhardt, D. and Löbl, I. (1987) The 'Winkler/Moczarski' eclector as an efficient extractor for fungus and litter Coleoptera. *The Coleopterists' Bulletin*, **41**, 392–4.

Blanton, C.M. (1990) Canopy arthropod sampling: a comparison of collapsible bag and fogging methods. *Journal of Agricultural Entomology*, **7**, 41–50.

Bowden, J. (1982) An analysis of the factors affecting catches of insects in light traps. *Bulletin of Entomological Research*, **72**, 535–56.

Cleyet-Marrel, D. (1990) Bilan technique partie dirigeable, in *Biologie d'une Canopée de Forêt Equatoriale. Rapport de Mission. Radeau des Cimes Octobre-Novembre 1989, Guyane Française* (eds F. Hallé and P. Blanc), Montpellier II et CNRS-Paris VI, Montpellier/Paris, pp. 11–14.

Coddington, J.A., Griswold, C.E., Dávila, D.S., Peñaranda, E. and Larcher, S.F. (1991) Designing and testing sampling protocols to estimate biodiversity in tropical ecosystems, in *The Unity of Evolutionary Biology* (ed. E.C. Dudley), Proceedings of the Fourth International Congress of Systematics and Evolutionary Biology. Dioscorides Press, Portland, pp. 44–60.

Colwell, R.K. and Coddington, J.A. (1994) Estimating terrestrial biodiversity through extrapolation. *Philosophical Transactions of the Royal Society of London, Series B*, **345**, 101–18.

Costa, J.T. III. and Crossley, D.A.J. (1991) Diel patterns of canopy arthropods associated with three tree species. *Environmental Entomology*, **20**, 1542–8.

Couturier, G. (1973) Etude éthologique et biocénotique du peuplement d'insectes dans un verger naturel. *Travaux et Documents O.R.S.T.O.M.*, **22**, 1–96.

Crossley, D.A.J., Coulson, R.N. and Gist, C.S. (1973) Trophic level effects on species diversity in arthropod communities of forest canopies. *Environmental Entomology*, **2**, 1097–100.

Dejean, A. (1992) Les relations plantes-fourmis, in *Biologie d'une Canopée de Forêt Equatoriale – II. Rapport de Mission: Radeau des Cimes Octobre-Novembre 1991, Réserve de Campo, Cameroun* (eds F. Hallé and O. Pascal), Fondation Elf, Paris, pp. 74–75.

Delamare-Debouteville, C. (1951) *Microfaune du Sol dans les Pays Tempérés et Tropicaux*, Hermann, Paris.

Delvare, G. and Aberlenc, H.P. (1990) Des entomologistes sur la canopée, in *Biologie d'une Canopée de Forêt Equatoriale. Rapport de Mission. Radeau des Cimes Octobre-Novembre 1989, Guyane Française* (eds F. Hallé and P. Blanc), Montpellier II et CNRS-Paris VI, Montpellier/Paris, pp. 211–21.

Dempster, J.P. (1961) A sampler for estimating populations of active insects upon vegetation. *Journal of Animal Ecology*, **30**, 425–7.

Dietrick, E.J. (1961) An improved back pack motor fan for suction sampling of insect populations. *Journal of Economic Entomology*, **54**, 394–5.

Dowdy, W.W. (1950) A community study of an Oak–Hickory association with special reference to invertebrates. *American Midland Naturalist*, **43**, 667–95.

Dufour, C. (1980) Un nouveau piège lumineux pour la capture des Tipulidae et autres Diptères Nématocères: une tente 'Malaise' lumineuse. *Mitteilungen der Schweizerischen Entomologischen Gesellschaft*, **53**, 313–20.

Eberhardt, L.L. and Thomas, J.M. (1991) Designing environmental field studies. *Ecological Monographs*, **61**, 53–73.

Ebersolt, G. (1990) Bilan technique partie radeau, in *Biologie d'une Canopée de Forêt Equatoriale. Rapport de Mission. Radeau des Cimes Octobre-Novembre 1989, Guyane Française* (eds F. Hallé and P. Blanc), Montpellier II et CNRS-Paris VI, Montpellier/Paris, pp. 15–26.

Erwin, T.L. (1983) Beetles and other insects of tropical forest canopies at Manaus, Brazil, sampled by insecticidal fogging, in *Tropical Rain Forest: Ecology and Management* (eds S.L. Sutton, T.C. Whitmore and A.C. Chadwick), British Ecological Society Special Publication 2, Blackwell Scientific Publications, Oxford, pp. 59–75.

Erwin, T.L. and Scott, J.C. (1980) Seasonal size and patterns, trophic structure, and richness of Coleoptera in the tropical arboreal ecosystem: the fauna of the tree *Luehea seemannii* Triana and Planch in the Canal Zone of Panama. *The Coleopterists' Bulletin*, **34**, 305–22.

Funke, W. (1971) Food and energy turnover of leaf-eating insects and their influence on primary production. *Ecological Studies*, **2**, 81–93.

Gadagkar, R., Chandrashekara, K. and Nair, P. (1990) Insect species diversity in the tropics: sampling methods and a case study. *Journal of the Bombay Natural History Society*, **87**, 337–53.

Gerber, G.H., Walkof, J. and Juskiw, D. (1992) Portable, solar-powered charging system for blacklight traps. *The Canadian Entomologist*, **124**, 553–4.

Glen, D.M. (1976) An emergence trap for bark-dwelling insects, its efficiency and effects of temperature. *Ecological Entomology*, **1**, 91–4.

Greenwood, S.R. (1990) Patterns of herbivory on the edge of the tropical forest in North Sulawesi, in *Insects and the Rain Forests of South East Asia (Wallacea)* (eds W.J. Knight and J.D. Holloway), The Royal Entomological Society of London, London, pp. 309–12.

Hallé, F. and Blanc, P. (eds) (1990) *Biologie d'une Canopée de Forêt Equatoriale. Rapport de Mission. Radeau des Cimes Octobre-Novembre 1989, Guyane Française.* Montpellier II et CNRS-Paris VI, Montpellier/Paris.

Hallé, F. and Pascal, O. (eds) (1992) *Biologie d'une Canopée de Forêt Equatoriale – II. Rapport de Mission: Radeau des Cimes Octobre-Novembre 1991, Réserve de Campo, Cameroun.* Fondation Elf, Paris.

Hammond, P.M. (1990) Insect abundance and diversity in the Dumoga-Bone National Park, N. Sulawesi, with special reference to the beetle fauna of lowland rain forest in the Toraut region, in *Insects and the Rain Forests of South East Asia (Wallacea)* (eds W.J. Knight and J.D. Holloway), The Royal Entomological Society of London, London, pp. 197–254.

Hammond, P.M. (1994) Practical approaches to the estimation of the extent of biodiversity in speciose groups. *Philosophical Transactions of the Royal Society of London, Series B*, **345**, 119–36.

Hammond, P.M. and Harding, P.T. (1991) Saproxylic invertebrate assemblages in British woodlands: their conservation significance and its evaluation, in *Pollard and Veteran Tree Management* (ed. H.J. Read), Richmond Publishing, Slough, pp. 30–7.

Harris, J.W.E., Collis, D.G. and Magar, K.M. (1972) Evaluation of the tree-beating method for sampling defoliating insects. *The Canadian Entomologist*, **104**, 723–9.

Harris, M. (1971) Sampling pear foliage for nymphs of the pear *Psylla*, using the Berlese–Tullgren funnel. *Journal of Economic Entomology*, **64**, 1317–18.

Hespenheide, H.A. (1979) Are there fewer parasitoids in the tropics? *American Naturalist*, **113**, 766–9.

Hosking, G.P. (1979) Trap comparison in the capture of flying Coleoptera. *New Zealand Entomologist*, **7**, 87–92.

Hosking, G.P. and Knight, F.B. (1975) Flight habits of some Scolytidae in the spruce-fir type of northern Maine. *Annals of the Entomological Society of America*, **68**, 917–21.

House, S.M. (1989) Pollen movement to flowering canopies of pistillate individuals of three rain forest tree species in tropical Australia. *Australian Journal of Ecology*, **14**, 77–94.

Hutcheson, J. (1990) Characterization of terrestrial insect communities using quantified malaise-trapped Coleoptera. *Ecological Entomology*, **15**, 143–51.

Juillet, J.A. (1963) A comparison of four types of traps used for capturing flying insects. *Canadian Journal of Zoology*, **41**, 219–23.

Kirk, W.D. (1984) Ecologically selective coloured traps. *Ecological Entomology*, **9**, 35–41.

Knodel, J.J. and Agnello, A.M. (1990) Field comparison of nonsticky and sticky pheromone traps for monitoring fruit pests in western New York. *Journal of Economic Entomology*, **83**, 197–204.

Krizelj, S. (1971) L'entomofaune circulante dans une chênaie à charmes de haute Belgique, in *Productivité des Ecosystèmes Forestiers* (ed. UNESCO), UNESCO, Paris, pp. 397–402.

Lamotte, M., Gillon, D., Gillon, Y. and Ricou, G. (1969) L'échantillonnage quantitatif des peuplements d'invertebrés en milieux herbacés, in *Problèmes d'Ecologie: l'échantillonnage des Peuplements Animaux des Milieux Terrestres* (eds M. Lamotte and F. Bourlière), Masson, Paris, pp. 7–54.

Lepointe, J. (1956) Méthodes de capture dans l'écologie des arbres. *Vie et Milieu*, **7**, 233–41.

LeSage, L. (1991) An improved technique for collecting large samples of arthropods. *Entomological News*, **102**, 97–104.

Longino, J.T. (1994) How to measure arthropod diversity in a tropical rainforest. *Biology International*, **28**, 3–13.

Lowman, M.D., Moffett, M.W. and Rinker, B. (1992) Insect sampling in forest canopies: a new method, in *Biologie d'une Canopée de Forêt Equatoriale – II. Rapport de Mission: Radeau des Cimes Octobre-Novembre 1991, Réserve de Campo, Cameroun* (eds F. Hallé and O. Pascal), Fondation Elf, Paris, pp. 41–44.

Lowman, M., Moffett, M. and Rinker, H.B. (1993a) A new technique for taxonomic and ecological sampling in rain forest canopies. *Selbyana*, **14**, 75–9.

Lowman, M., Taylor, P. and Block, N. (1993b) Vertical stratification of small mammals and insects in the canopy of a temperate deciduous forest: a reversal of tropical forest distribution? *Selbyana*, **14**, 25.

Majer, J.D. and Recher, H.F. (1988) Invertebrate communities on western Australian eucalyptus, a comparison of branch clipping and chemical knockdown procedures. *Australian Journal of Ecology*, **13**, 269–78.

Masner, L. and Goulet, H. (1981) A new model of flight-interception trap for some hymenopterous insects. *Entomological News*, **92**, 199–202.

McKey, D. (1992) Interactions between ants and plants: comparison of canopy, understorey and clearing environments, in *Biologie d'une Canopée de Forêt Equatoriale – II. Rapport de Mission: Radeau des Cimes Octobre-Novembre 1991, Réserve de Campo, Cameroun* (eds F. Hallé and O. Pascal), Fondation Elf, Paris, pp. 66–73.

Merrill, W. and Skelly, J.M. (1968) A window trap for collection of insects above the forest canopy. *Journal of Economic Entomology*, **61**, 1461–2.

Miller, R.S., Passoa, S., Waltz, R.D. and Mastro, V. (1993) Insect removal from sticky traps using a citrus oil solvent. *Entomological News*, **104**, 209–13.

Mitchell, A.W. (1982) *Reaching the Rain Forest Roof*, Leeds Philosophical and Literary Society, Leeds.

Moeed, A. and Meads, M.J. (1983) Invertebrate fauna of four tree species in Orongorongo Valley, New Zealand, as revealed by trunk traps. *New Zealand Journal of Ecology*, **6**, 39–53.

Moffett, M.W. (1993) *The High Frontier. Exploring the Tropical Rainforest Canopy*. Harvard University Press, Cambridge, MA.

Moran, V.C., Hoffmann, J.H., Impson, F.A.C. and Jenkins, J.F.G. (1994) Herbivorous insect species in the tree canopy of a relict South African forest. *Ecological Entomology*, **19**, 147–54.

Morris, R.F. (1955) The development of sampling techniques for forest insect defoliators, with particular reference to the spruce budworm. *Canadian Journal of Zoology*, **33**, 225–94.

Mouna, M., Bigot, L. and Fabre, J.P. (1985) Comparaison des communautés frondicoles des Coléoptères du Cèdre (*Cedrus atlantica* Massetti) en France (Provence) et au Maroc (Moyen-Atlas). *Vie et Milieu*, **35**, 99–106.

Muirhead-Thomson, R.C. (1991) *Trap Responses of Flying Insects*, Academic Press, London.

Nadkarni, N.M. and Longino, J.T. (1990) Invertebrates in canopy and ground organic matter in a Neotropical montane forest, Costa Rica. *Biotropica*, **22**, 286–9.

Nancé, E. (1990) Recherches arachnologiques sur la canopée. Biologie d'une canopée de forêt équatoriale, in *Biologie d'une Canopée de Forêt Equatoriale. Rapport de Mission. Radeau des Cimes Octobre-Novembre 1989, Guyane Française* (eds F. Hallé and P. Blanc), Montpellier II et CNRS-Paris VI, Montpellier/Paris, pp. 205–7.

New, T.R. (1979) Phenology and relative abundance of Coleoptera on some Australian Acacias. *Australian Journal of Zoology*, **27**, 9–16.

New, T.R., Adis, J.U., de Morais, J.W. and Gomes Rodrigues, J.M. (1991) Notes on phenology and abundance of Psocoptera from primary and secondary dryland forest in Central Amazonia, Brazil. *Studies on Neotropical Fauna and Environment*, **26**, 243–7.

Nicolai, V. (1986) The bark of trees: thermal properties, microclimate and fauna. *Oecologia*, **69**, 148–60.

Noyes, J.S. (1989) A study of five methods of sampling Hymenoptera (Insecta) in a tropical rainforest, with special reference to the Parasitica. *Journal of Natural History*, **23**, 285–98.

Ohmart, C.P., Stewart, L.G. and Thomas, J.R. (1983) Phytophagous insect communities in the canopies of three *Eucalyptus* forest types in south-eastern Australia. *Australian Journal of Ecology*, **8**, 395–403.

Orousset, J. (1997) Les techniques de récolte des coléoptères hypogés (4e partie). *Supplément au Bulletin de liaison de l'ACOREP*, Paris (in press).

Osmelak, J.A. (1987) A comparison of five different traps for monitoring leafhopper activity (Homoptera). *General and Applied Entomology*, **19**, 49–56.

Owen, J.A. (1989) An emergence trap for insects breeding in dead wood. *British Journal of Entomology and Natural History*, **2**, 65–7.

Owen, J.A. (1992) Experience with an emergence trap for insects breeding in dead wood. *British Journal of Entomology and Natural History*, **5**, 17–20.

Paarmann, W. and Kerck, K. (1997) Advances in using the canopy fogging technique to collect living arthropods from tree crowns, in *Canopy Arthropods* (eds N.E. Stork, J. Adis and R.K. Didham), Chapman & Hall, London, pp. 53–66 (this volume).

Paarmann, W. and Stork, N.E. (1987) Canopy fogging, a method of collecting living insects for investigations of life history strategies. *Journal of Natural History*, **21**, 563–6.

Paoletti, M.G., Taylor, R.A.J., Stinner, B.R., Stinner, D.H. and Benzing, D.H. (1991) Diversity of soil fauna in the canopy and forest floor of a Venezuelan cloud forest. *Journal of Tropical Ecology*, **7**, 373–83.

Paviour-Smith, K. and Elbourn, C.A. (1993) A quantitative study of the fauna of small dead and dying wood in living trees in Wytham Woods, near Oxford. *English Nature Science*, **7**, 33–57.

Rees, C.J.C. (1983) Microclimate and the flying Hemiptera fauna of a primary lowland rain forest in Sulawesi, in *Tropical Rain Forest: Ecology and Management* (eds S.L. Sutton, T.C. Whitmore and A.C. Chadwick), British Ecological Society Special Publication 2, Blackwell Scientific Publications, Oxford, pp. 121–38.

Robert, J.C. (1983) Le piégeage lumineux directionnel. Ses utilisations. *Cahiers de Liaisons O.P.I.E.*, **17**, 21–6.

Robert, J.C. (1992) Le piège entomologique composite (P.E.C): une technique d'échantillonnage à large spectre de l'entomofaune terrestre circulante. *Mitteilungen der Schweizerischen Entomologischen Gesellschaft*, **65**, 395–411.

Russell-Smith, A. and Stork, N.E. (1994) Abundance and diversity of spiders from the canopy of tropical rainforests with particular reference to Sulawesi, Indonesia. *Journal of Tropical Ecology*, **10**, 545–58.

Smythe, N. (1982) The seasonal abundance of night-flying insects in a Neotropical forest, in *The Ecology of a Tropical Forest: Seasonal Rhythms and Long-Term Changes* (eds), E.G. Leigh Jr, A.S. Rand and D.M. Windsor), Smithsonian Institution Press, Washington, pp. 309–18.

Sourakov, A. and Emmel, T.C. (1995) Bait trapping for butterflies in Kenya. *Tropical Lepidoptera*, **6**, 1–2.

Southwood, T.R.E. (1978) *Ecological Methods, with Particular Reference to the Study of Insect Populations*, 2nd edn, Chapman & Hall, London.

Springate, N.D. and Basset, Y. (1996) Diel activity of arboreal arthropods associated with Papua New Guinean trees. *Journal of Natural History*, **30**, 101–12.

Stork, N.E. (1987a) Guild structure of arthropods from Bornean rain forest trees. *Ecological Entomology*, **12**, 69–80.

Stork, N.E. (1987b) Arthropod faunal similarity of Bornean rain forest trees. *Ecological Entomology*, **12**, 219–26.

Stork, N.E. (1991) The composition of the arthropod fauna of Bornean lowland rainforest trees. *Journal of Tropical Ecology*, **7**, 161–80.

Stork, N.E. (1994) Inventories of biodiversity: more than a question of numbers, in *Systematics and Conservation Evaluation* (eds P.L. Forey, C.J. Humphries and R.I. Vane-Wright), Systematics Association Special Volume No. 50, Clarendon Press, Oxford, pp. 81–100.

Stork, N.E. and Brendell, M.J.D. (1990) Variation in the insect fauna of Sulawesi trees with season, altitude and forest type, in *Insects and the Rain Forests of South East Asia (Wallacea)* (eds W.J. Knight and J.D. Holloway), The Royal Entomological Society of London, London, pp. 173–190.

Stork, N.E. and Brendell, M.J.D. (1991) Arthropod abundance in lowland forest of Seram, in *Natural History of Seram* (eds I.D. Edwards, A.A. Macdonald and J. Proctor), Intercept, Andover, pp. 115–30.

Stork, N.E. and Hammond, P.M. (1997) Sampling arthropods from tree-crowns by fogging with knockdown insecticides: lessons from studies of oak tree beetle assemblages in Richmond Park, UK, in *Canopy Arthropods* (eds N.E. Stork, J. Adis and R.K. Didham), Chapman & Hall, London, pp. 3–26 (this volume).

Sutton, S.L. (1979) A portable light trap for studying insects of the upper canopy. *The Brunei Museum Journal*, **4**, 156–60.

Sutton, S.L. and Hudson, P.J. (1980) The vertical distribution of small flying insects in the lowland rain forest Zaire. *Zoological Journal of the Linnean Society*, **68**, 111–23.

Togashi, I. (1990) Symphyta (Hymenoptera) lured to benzyl acetate and alphapyrene in traps. *Transactions of the Shikoku Entomological Society*, **19**, 117–25.

Turner, B.D. (1974) The population dynamics of tropical arboreal Psocoptera (Insecta) on two species of conifers in the Blue Mountains, Jamaica. *Journal of Animal Ecology*, **43**, 323–37.

Walter, D.E., O'Dowd, D. and Barnes, V. (1994) The forgotten arthropods: foliar mites in the forest canopy. *Memoirs of the Queensland Museum*, **36**, 221–6.

Wilkening, A.J., Foltz, J.L., Atkinson, T.H. and Connor, M.D. (1981) An omnidirectional flight trap for ascending and descending insects. *The Canadian Entomologist*, **113**, 453–5.

Wilson, L.F. (1962) A portable device for mass-collecting or sampling foliage-inhabiting arthropods. *Journal of Economic Entomology*, **55**, 807–8.

Wilson, S.W., Smith, J.L. and Purcell, A.H. (1993) An inexpensive vacuum collector for insect sampling. *Entomological News*, **104**, 203–8.

Wolda, H. (1979) Abundance and diversity of Homoptera in the canopy of a tropical forest. *Ecological Entomology*, **4**, 181–90.

Yumoto, T. (1992) Pollination ecology in a tropical rain forest canopy in Cameroon, in *Biologie d'une Canopée de Forêt Equatoriale – II. Rapport de Mission: Radeau des Cimes Octobre-Novembre 1991, Réserve de Campo, Cameroun* (eds F. Hallé and O. Pascal), Fondation Elf, Paris, pp. 110–114.

3

Advances in using the canopy fogging technique to collect living arthropods from tree-crowns

W. Paarmann and K. Kerck

ABSTRACT

The insecticide fogging technique was tested as a method for collecting canopy-dwelling arthropods alive in two oak forests close to Göttingen, Germany, during 1990–1991. If natural pyrethrum without a killing agent is used a high percentage of the knocked-down animals recover. The recovery rate was raised by a direct transfer of each animal into a single container at the fogging site. No differences in the knockdown efficiency or recovery rate were found with natural pyrethrum concentrations ranging from 0.15–0.90% and exposure times of 15–60 seconds. Most arthropods fall down from the canopy in the first hour after fogging. Different carrier oils were also tested. Alternatives to diesel, such as highly raffinized white oils (for example Ondina and Risella, Shell) and Biodiesel can be used. Thus, the insecticide fogging technique is a highly effective method for collecting living arthropods from tree-crowns. The only problem found in connection with the method is that some long-legged arthropods lose their legs. Suggestions to deal with this problem are made.

INTRODUCTION

The insecticide fogging method has been used mainly in pest control. The first attempt to use this method in studies of forestry ecology was made by Altenkirch (1965, 1968). The insecticide he used was DDT, which has a very low knockdown efficiency. Recently introduced pyrethroids have a very high knockdown efficiency, raising new interest in the method for collecting canopy-dwelling insects (Erwin, 1982; Adis *et al.*, 1984; Stork 1987). Erwin (1982) and Stork (1988) developed new

Canopy Arthropods. Edited by N.E. Stork, J. Adis and R.K. Didham. Published in 1997 by Chapman & Hall, London. ISBN 0 412 74900 9

estimates of the number of living arthropod species on Earth, based on data collected by insecticide fogging of tropical trees. Its use on oak trees in Richmond Park (London, UK) gave surprising results concerning the species richness and seasonality of Neuroptera, Rhaphidioptera and Mecoptera (Barnard *et al.*, 1986).

Studying the biology of canopy-dwelling arthropods directly, by looking at them in the canopy, can be very difficult because of the low population densities of most rainforest species. In 1985 a first attempt was made to collect living insects from trees by fogging (Paarmann and Stork, 1987). Two individuals of the beetle species *Colpodes buchanani* Hope (Carabidae) were collected by fogging with 'Reslin E'. The offspring of this pair of beetles are still being bred after 10 years in the laboratory in Göttingen and many interesting aspects of the biology of this species have been revealed (Paarmann and Bolte, 1990; Paarmann and Paarmann, 1997, Chapter 20, this volume).

In the present study an attempt has been made to improve the method of collecting live insects by using natural pyrethrum without the killing additive piperonyl butoxide and by varying the carrier oils (Paarmann, 1994). Attempts have also been made to improve the survival rate of collected arthropods.

METHODS AND STUDY SITES

A Swingfog SN 50, manufactured by Motan Swingtec in Insny (Germany) was used. Fogging was always conducted at dawn. Table 3.1 summarizes the dates, insecticide concentrations and types of carrier oils for all foggings. Each run of the fogger lasted about 5 minutes.

Arthropods were collected on plastic sheets spread on the ground (5.8×7.9 m). During 1990 arthropods were transferred into large plastic containers and separated later in the laboratory. During 1991 arthropods were isolated directly from the sheet to avoid harmful interactions between individuals and kept singly in plastic vials on moist peat. Live arthropods were fed regularly according to their trophic level, with oak or beech leaves, honey, *Drosophila* flies, aphids, pieces of mealworms or dried fish-food.

To obtain more detailed information about the influence of pyrethrum concentration and time of exposure on knockdown efficiency and recovery rate, caged oak crickets (*Meconema thalassinum* de Geer) were fogged from a distance of 5–6 m.

Fogging was carried out in oak forests south of Göttingen (Mollenfelde), except fogs 27 and 28, which were carried out north of Göttingen (Hagenberg bei Moringen).

Table 3.1 Date, pyrethrum concentration and type of carrier oil for different fogs in the fogging experiments

Fog no.	*Date*	*Carrier oil*	*Insecticide concentration (%)*
1	16.05.90	Diesel	0.30
2	16.05.90	Diesel	0.45
3	23.05.90	Diesel	0.15
4	23.05.90	Diesel	0.60
5	01.06.90	Diesel	0.70
6	01.06.90	Diesel	0.15
7	13.06.90	Diesel	0.45
8	20.06.90	Diesel	0.75
9	20.06.90	Diesel	0.00
10	27.06.90	Diesel	0.00
11	27.06.90	No fog*	
12	30.07.90	Diesel	0.90
13	30.07.90	Diesel	0.90
14	07.08.90	Diesel	0.60
15	07.08.90	Diesel	0.60
16	27.08.90	Diesel	0.20
17	27.08.90	Diesel	0.45
18	13.09.90	Diesel	0.75
19	13.09.90	Diesel	0.75
20	13.09.90	Olive oil+	0.00
21	20.09.90	Salat oil	0.45
22	20.09.90	Penatenöl‡	0.45
23	10.10.90	Penatenöl	0.15
24	30.11.90	Diesel	0.45
25	29.05.91	Diesel	0.15
26	29.05.91	Shell Ondina§	0.15
27	12.06.91	Shell Risella§	0.45
28	12.06.91	Shell Risella§	0.15

* Only sheet deposited; +, test of fog quality; ‡, baby oil; §, Shell Ondina oil G17 and Shell Risella oil G18 are basic oils for cosmetic products.

RESULTS

Time-scale of the arthropod rain

Most arthropods drop down during the first hour after the end of fogging (Figure 3.1; Table 3.2). In the longest lasting experiment (135 min; Table 3.2) there was 48% knockdown during the first 30 minutes and a total of 67% during the first hour. Not all arthropod groups respond as quickly; for example, the drop-time required for Collembola seems to be greater than that required for Hymenoptera or Diptera (Table 3.2).

Table 3.2 Drop rate of Hymenoptera, Diptera, Collembola and total arthropods in fogs 17 and 18 at 15-minute intervals after fogging

	Fog number							
	17				*18*			
Time (min)	*Total*	*Hymenoptera*	*Diptera*	*Collembola*	*Total*	*Hymenoptera*	*Diptera*	*Collembola*
15	228	109	40	6	228	63	83	7
30	146	70	48	0	63	19	23	7
45	82	31	0	6	186	62	56	18
60	67	14	17	2	132	57	30	16
75	68	12	14	18	78	17	22	14
90	46	13	14	6	54	7	10	16
105	77	20	17	13	77	17	16	25
120	52	22	10	1				
135	19	0	12	0				

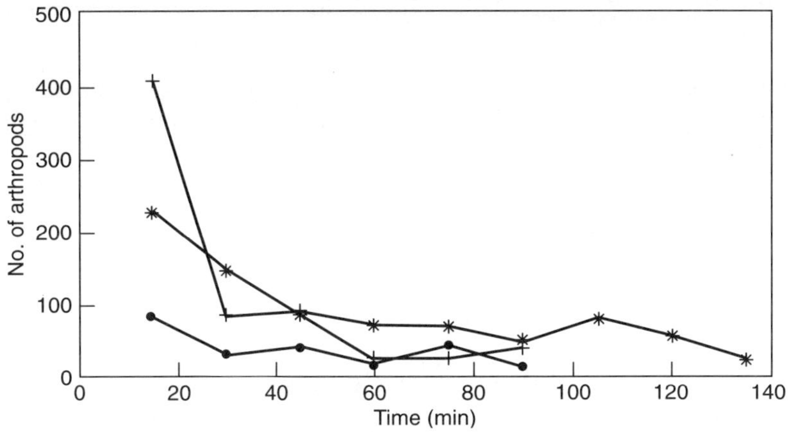

Figure 3.1 Knockdown rates for arthropods in three different fogs. Pyrethrum concentrations: •, 0.45%; +, 0.75%; ∗, 0.45%.

Three-day survival rate

Table 3.3 summarizes 126 series of 3-day survival experiments. A 3-day survival rate of more than 80% was found in 46 series (37%), 60% or more in 67 series (53%), and 50% or more in 80 series (63%). There is no clear correlation between insecticide concentration and survival rate. Often the insects die directly after treatment. In these experiments, death may also be caused by interactions between the animals during transportation to the laboratory. If insects recover, but subsequently die in the laboratory, death is mainly caused by inadequate culture conditions.

Interactions between specimens were avoided by separating them directly in the forest. Survival rates in such experiments are summarized in Table 3.4. The 3-day survival rate was over 80% in 10 of the 13 series and over 50% in 12 of 13 series. Survival was only below 50% for Bibionidae (Diptera), although these are known to have a short life-span. All three female bibionids deposited eggs before death. Hence, in general, survival rate is higher if the specimens are separated and transferred into culture conditions directly after knockdown from the canopy.

Survival for more than 3 days

Survival rates at 10 and 30 days after insecticide treatment are shown in Table 3.4. There is only a very small drop in survival rate between 3 and 10 days, with the exception of Bibionidae (the last specimen of which died 9 days after capture). Also, in six out of 10 experiments the 30-day survival rate was higher than 50%. In some series the animals were kept for more than 30 days. For example, four nymphs of Blattodea

Table 3.3 Percentage of arthropods still alive 3 days after insecticide treatment (1990 fogs). Note that fog 11 was a control and no insecticide was used, only a sheet was deposited

		Lepidoptera larvae			*Araneae*			*Meconema thalassinum oak cricket*		
Pyrethrum concentration (%)	*Fog. no.*	*No. of specimens*	*No. dying immediately after fogging*	*Survival rate (%)*	*No. of specimens*	*No. dying immediately after fogging*	*Survival rate (%)*	*No. of specimens*	*No. dying immediately after fogging*	*Survival rate (%)*
–	11				15	2	87	1		0
0.00	10				10		80			
0.00	9	3	0	66	15	0	100			
0.00	20				28	4	75			
0.10	19	11	1	73						
0.15	3	21	0	57	39	5	85	7	1	29
0.15	6	2	0	100						
0.15	22	15	0	15	31	2	68	1	0	0
0.15	23	3	0	33	25	0	96			
0.20	16	38	20	3	19	0	100			
0.30	1	26	3	61	5	0	100	6	0	50
0.45	2	14	0	21	26	8	58			
0.45	7	1	1	0	20	0	95	1	0	100
0.45	17	15	8	0	45	9	80	8	0	88
0.45	21	24	2	91						
0.60	4	43	5	37	11	6	45	24	9	50
0.60	14	3	0	0	7	2	71	1	0	100
0.60	15	6	0	0	8	3	63			
0.70	5	23	4	48	8	8	0			
0.75	8	1	0	0	11	0	100	3	0	100
0.75	18	15	1	13				1	0	0
0.90	12	50	15	10	86	54	37			
0.90	13				26	9	58	8	2	25

Pyrethrum concentration (%)	Fog. no.	Coleoptera*			Heteroptera			Diptera		
		No. of specimens	No. dying immediately after fogging	Survival rate (%)	No. of specimens	No. dying immediately after fogging	Survival rate (%)	No. of specimens	No. dying immediately after fogging	Survival rate (%)
–	11	7	0	100						
		7	0	57C						
0.00	10	6	0	100						
		4	0	50C						
0.00	9	15	0	67C	1	0	100			
0.10	19	26	1	77						
0.15	3	18	0	78	109	3	32	32	2	34
0.15	6	6	2	67C				4	0	100
0.15	22	8	0	100				14		0
		13	0	100Co						
0.30	1	10	0	70	50	1	22	11	0	0
0.45	2	8	0	100	165	9	40	92	20	51
0.45	7	15	0	100	100	2	36	10	1	60
0.45	26	9	0	100				3	0	66
0.60	4	22	4	55	214	17	30	177	68	12
		96	10	63C						
0.60	14	11	0	100Co						
0.60	15	12	0	91Co	3	0	100	13	0	69
0.70	5	52	9	61	903	295	11			
0.75	8	7	0	86	148	0	3	116	22	22
		57	0	39C						
0.75	18	6	0	83						
0.90	12	17	0	53C	13	0	84	120		18
0.90	13				6	0	100			

Table 3.3 continued

		Homoptera			*Small Hymenoptera*			*Psocoptera*		
Pyrethrum concentration (%)	*Fog. no.*	*No. of specimens*	*No. dying immediately after fogging*	*Survival rate (%)*	*No. of specimens*	*No. dying immediately after fogging*	*Survival rate (%)*	*No. of specimens*	*No. dying immediately after fogging*	*Survival rate (%)*
–	11	3	0	100						
0.15	3	5	1	80	30	6	57	2	2	0
0.30	1	3	0	33	3	0	33	4	0	50
0.45	2	3	0	66	11	3	36	12	2	50
0.45	7	6	0	33				9	1	11
0.60	4	19	8	47	29	15	21			
0.75	8	16	0	0						

		Dermaptera			*Collembola*			*Acari*		
Pyrethrum concentration (%)	*Fog. no.*	*No. of specimens*	*No. dying immediately after fogging*	*Survival rate (%)*	*No. of specimens*	*No. dying immediately after fogging*	*Survival rate (%)*	*No. of specimens*	*No. dying immediately after fogging*	*Survival rate (%)*
–	11				4	0	100	1	0	100
0.00	10							1	0	100
0.00	9	3	0	100						
0.15	22	6	0	83	8	2	63			
0.45	7	1	0	100	3	0	100	2	0	100
0.45	21				22	0	82			
0.75	18	3	1	66	16^{+}	5				

Pyrethrum concentration (%)	Fog. no.	Formicidae			Mecoptera			Blattodea		
		No. of specimens	*No. dying immediately after fogging*	*Survival rate (%)*	*No. of specimens*	*No. dying immediately after fogging*	*Survival rate (%)*	*No. of specimens*	*No. dying immediately after fogging*	*Survival rate (%)*
0.15	3	1	0	100						
0.45	2	9	0	33						
0.45	21							2	0	100
0.60	4				5	1	60			
0.70	5	1	0	100						
0.75	8				7	0	43			

* For Coleoptera, figures are for all beetles, except where stated: C, Curculionidae; Co, Coccinellidae.
⁺ Eight kept in tubes for survival experiment; all were alive after 3 days.

Table 3.4 Percentage of arthropods alive 3, 10 and 30 days after being collected by fogging. 1991 fogs – data are pooled for fogs 25/26 and 27/28

Fog no.	*No. of arthropods*	*Days* 3	*Days* 10	*Days* 30		
Lepidoptera larvae						
25/26	7	86	86	71		
27/28	66	85	73	55		
Araneae						
25/26	9	66	66	44		
27/28	7	86	86	43		
Meconema thalassinum						
25/26	20	55	45	5		
27/28	11	100	91	64		
Coleoptera*						
25/26	56	100	96	34	(62 days: 25%)	C
	33	100	91	70		O
27/28	75	97	83			C
20	100	90	60			O
Heteroptera						
27/28	13	100	85	0		
Diptera (Bibionidae)						
25/26	41	49	0			
Blattodea						
25/26	4	100	75	75	(53 days: 75%)	

* C, Curculionidae; O, other

were collected, three of which reached the imaginal stage and were still alive after 53 days. In curculionid beetles, 25% were still alive 62 days after capture. Egg deposition was recorded in eight beetles and five spiders (four of the spider egg sacs hatching). The nymphs of *Meconema thalassinum* (Orthoptera) developed during 30 days captivity to adults (seven) or to the last nymphal instar (one). Also, all surviving caterpillars either pupated or the moth hatched. In six cases parasites hatched within 30 days.

A particular problem arose with long-legged arthropods such as certain Nematocera, Opiliones and Araneae. Pyrethrum seems to trigger the autotomy reflex, so that the animals lost their legs.

Table 3.5 Influence of different pyrethrum concentrations on the 10-day survival rate of Curculionidae (Coleoptera, *Phyllobius* and *Polydrusus* spp.) and the 30-day survival rate of caterpillars. Fogs 27 and 28

Pyrethrum concentration (%)	*No. of arthropods*	*Proportion alive (%)*	
Curculionidae			
0.15	40	75	
0.45	35	80	
			$(\chi^2_{(1,0.05)} = 0.266)$
Caterpillars			
0.15	14	50	
0.45	52	55	
			$(\chi^2_{(1,0.05)} = 0.066)$

There is no clear evidence of a connection between pyrethrum concentration and survival rate (Table 3.3). In some of the experiments (for example, fogs 27 and 28, summarized in Table 3.4) it is possible to compare the influence of two different concentrations (Table 3.5). Higher concentrations of pyrethrum do not appear to cause a significantly higher mortality ($\chi^2_{(1,0.05)} = 3.841$, n.s., Table 3.5), although in both fogs 27 and 28 the survival rate is marginally higher with a higher dose.

One problem with insecticide fogging is the difficulty of producing a homogeneous distribution of fog within the canopy. Some insects encounter smaller amounts of insecticide than others during the same fog. Similarly, the degree of exposure of the insect is also important (e.g. exposed on a leaf or hidden in a leaf roll). Drifting of the fog can change the exposure time considerably. In experiments with caged oak crickets exposed to different pyrethrum concentrations for different lengths of time, no significant differences in knock-down efficiency were seen (Table 3.6).

Influence of carrier oils on survival rate

To produce a thick and visible cloud the carrier oil must have a similar viscosity to diesel. This was found only with the highly raffinized Shell oils, Ondina and Risella, and with baby oil, which is based on a similar raffinized oil. The plant oils used produced a thin fog and therefore were less suitable. Comparison between different carrier oils with the same pyrethrum concentration (0.15%) was only possible for curculionid beetles in fogs 25 and 26. No difference was evident ($\chi^2_{(1,0.05)} = 0.198$, n.s., Table 3.7).

Table 3.6 Results of five experiments testing the fogging of caged oak crickets (*Meconema thalassinum*) with different pyrethrum concentrations and different fogging duration. Carrier oil: diesel

Experiment no.	*Date/time*		*Pyrethrum concentration (%)*	*Time of exposure (s)*					
				15		*30*		*60*	
				n	*%*	*n*	*%*	*n*	*%*
1	13.9.90	07.15 h	0.10						
	Dropped-down after 1 h			–	–	2	100	1	100
	Alive 3 days later			–	–	2	100	1	100
2	13.5.90	05.10 h	0.45						
	Dropped-down after 1 h			8	100	8	100	8	100
	Alive 3 days later			8	100	6	75	8	100
3	7.8.90	06.55 h	0.60						
	Dropped-down after 1 h			2	100	4	100	3	100
	Alive 3 days later			2	100	4	100	3	100
4	21.6.90	05.50 h	0.75						
	Dropped-down after 1 h			3	43	7	100	8	100
	Alive 3 days later			6	86	7	100	8	100
5	30.7.90	05.30 h	0.90						
	Dropped-down after 1 h			3	100	0	0	3	100
	Alive 3 days later			3	100	3	100	3	100

DISCUSSION

The survival rate of arthropods collected by canopy fogging does not appear to differ much in a concentration range from 0.15–0.9% pyrethrum. Similar results were also found for concentrations of 0.5, 1.0 and 1.5% in Amazonia, Brazil (Adis *et al.*, 1997, Chapter 4, this volume). In addition, different times of exposure did not appear to affect knock-down rate or survival rate of arthropods. This would suggest that most arthropods are extremely sensitive to very low pyrethrum concentrations and short exposure times, making this technique highly suitable for collecting living arthropods from the canopy.

The survival rate of arthropods can be raised by separating them directly at the fogging site, thus avoiding harmful interactions between individuals. Death after recovery from pyrethrum knockdown is mainly caused by inadequate culture conditions. Better knowledge of the life habits of the species collected may lead to increased survival rates in the future. No long-term effects of pyrethrum on development

Table 3.7 Survival rate of Curculionidae (Coleoptera, *Phyllobius* and *Polydrusus* spp.) after 10 days. Pyrethrum concentration 0.15%, carrier oils Diesel and Ondina

	No. of arthropods	*Proportion alive (%)*
Diesel	16	94
Ondina	40	90
		($\chi^2_{(1,0.05)}$ = 0.198)

or reproduction were observed. Also, in a 10-year laboratory study of the biology of one species collected with the synthetic pyrethroid 'Reslin E', no long-term effects of the insecticide have been seen (Paarmann and Stork, 1987; Paarmann and Bolte, 1990; Paarmann and Paarmann, 1997, Chapter 20, this volume).

One limitation of fogging is leg loss in some long-legged arthropods, although this problem may be solved by using very low insecticide doses and rapid transfer from the plastic sheet to a natural substrate.

Instead of the less environmentally friendly carrier oil diesel, the highly raffinized white oils Ondina and Risella (Shell) can be used. Recently it was found that Biodiesel (Rapemethylester) can also be used as a carrier oil (W. Paarmann, unpublished data), the fog produced in this case containing only natural substances.

In conclusion, insecticide fogging with natural pyrethrum without a killing additive is a highly effective method for collecting living arthropods from tree-crowns.

Acknowledgements

The study was supported by a grant from the German Research Foundation (DFG: Pa 99/14–1). The authors are very much indebted to Dr W. Altenkirch of the Niedersächsische Forstliche Versuchsanstalt for his scientific advice; they also thank the Forstamtsleiter O.A. Beck and Dr B. von Lüpke, and the Revierleiter H.W. Hänel and W. Hildebrandt for permission to carry out these studies in their forest, and former students S. Greeb, J. Happe, C. Hilbert. C. Hinkelmann, B. Staritz and J. Seegert for their assistance. The white oils Risella and Ondina, used in the study, were generously provided by Shell.

REFERENCES

Adis, J., Lubin, Y.D. and Montgomery, G.G. (1984) Arthropods from the canopy of inundated and terra firme forests near Manaus, Brazil, with critical considerations on the pyrethrum-fogging technique. *Studies on Neotropical Fauna and Environment*, **19**, 223–36.

Adis, J, Paarmann, W., Fonseca, C.R.V. da, and Rafael, J.A. (1997) Knockdown efficiency of natural pyrethrum and survival rate of arthropods obtained during canopy fogging in Central Amazonia, in *Canopy Arthropods* (eds N.E. Stork, J. Adis and R.K. Didham), Chapman & Hall, London, pp. 67–79 (this volume).

Altenkirch W. (1965) Synökologische Untersuchungen der Kronenfauna eines Eichen-Hainbuchenwaldes. Teil I. *Zeitschrift für Angewandte Entomologie*, **55**, 326–47.

Altenkirch W. (1968) Synökologische Untersuchungen der Kronenfauna eines Eichen-Hainbuchenwaldes. Teil II. *Zeitschrift für Angewandte Entomologie*, **58**, 1–69.

Barnard, P.C., Brooks, S.J., and Stork, N.E. (1986) The seasonality and distribution of Neuroptera, Raphidioptera and Mecoptera on oaks in Richmond Park, Surrey, as revealed by insecticide sampling. *Journal of Natural History*, **20**, 1321–31.

Erwin, T.L. (1982) Tropical forests: their richness in Coleoptera and other arthropod species. *The Coleopterists' Bulletin*, **36**, 74–5.

Paarmann, W. (1994) Von der Tötung zum Lebendfang. *Entwicklung der Insektizidnebeltechnik. Allgemeine Forstzeitschrift*, **49**, 1457–9.

Paarmann, W. and Bolte, H. (1990) Studies on the biology of *Colpodes buchanani* (Hope) (Coleoptera, Carabidae), collected in Sulawesi rain forest (Indonesia) by application of knock-down insecticides with a canopy fogger, in *Insects and the Rain Forests of South East Asia (Wallacea)* (eds W.J. Knight and J.D. Holloway), The Royal Entomological Society of London, London, pp. 191–5.

Paarmann, W. and Paarmann, D. (1997) Studies on the biology of a canopy-dwelling carabid beetle collected by canopy fogging, in *Canopy Arthropods* (eds N.E. Stork, J. Adis and R.K. Didham), Chapman & Hall, London, pp. 433–41 (this volume).

Paarmann, W. and Stork N.E. (1987) Canopy fogging, a method of collecting living insects for investigations of life history strategies. *Journal of Natural History*, **21**, 563–6.

Stork, N.E. (1987) Guild structure of arthropods from Bornean rain forest trees. *Ecological Entomology*, **12**, 69–80.

Stork, N.E. (1988) Insect diversity: facts, fiction and speculation. *Biological Journal of the Linnean Society*, **35**, 321–37.

4

Knockdown efficiency of natural pyrethrum and survival rate of living arthropods obtained by canopy fogging in Central Amazonia

J. Adis, W. Paarmann, C.R.V. da Fonseca and J.A. Rafael

ABSTRACT

1. One individual of the widely distributed Amazonian tree species *Goupia glabra* Aubl. (Celastraceae, height 45 m) was fogged with knockdown insecticide at both 06:00 h and 08:00 h on two successive days in a primary upland forest ('Reserva Ducke') near Manaus. Natural pyrethrum (without synergist, 1.0% diluted in diesel oil), a non-killing fogging agent, was used for the first three fogging events. Approximately 57% of total arthropod abundance (n = 3685) were caught within 2 hours of the first fogging, 23% after the second and 13% after the third fogging. Only 7% of total arthropod abundance was obtained during the fourth fogging, where the synthetic pyrethrum 'Baythroid' (0.15% diluted in diesel oil) was used as a killing agent. On each fogging occasion about 70% of the arthropods collected fell within the first hour of a 2-hour drop time.
2. Canopies of the more locally occurring Amazonian tree, *Calophyllum brasiliense* Camb. (Guttiferae, height 10 m), were fogged after dawn on the same day at five adjacent localities in a 20-year-old plantation in the reserve, using natural pyrethrum (0.5%, 1.0%, 1.5%), Baythroid (0.3%) and diesel oil, respectively. About 75% of all arthropods were obtained within the first hour (independent of concentration and agent used) and the remaining 25% after trees had been shaken throughout the next hour.

Canopy Arthropods. Edited by N.E. Stork, J. Adis and R.K. Didham. Published in 1997 by Chapman & Hall, London. ISBN 0 412 74900 9

3. The average survival rate of arthropods, sampled separately on the plantation with 1.0% and 0.5% active ingredient of natural pyrethrum and kept alive under field conditions, was 60% and 75%, respectively, after 7 days. These concentrations seem to be suitable for collecting living arthropods from neotropical tree canopies for further experimentation.

INTRODUCTION

In 1991, Paarmann and co-workers reported that a natural pyrethrum extracted from chrysanthemum flowers without synergist ('pyrethrum pale') could be used as a non-killing knockdown agent to collect and breed living arthropods from tree canopies in Germany (Paarmann *et al.*, 1991). We wanted to test if this agent could be used to collect live arthropods from Neotropical trees for habitat manipulation experiments. In this paper we present the results of two field experiments.

In the first experiment we tested the knockdown rate of arthropods from a single tree 45 m in height. In the second experiment we sampled arthropods from the canopy of a 20-year-old plantation to look at: (i) knockdown rates; (ii) differences in concentrations of natural pyrethrum and differences between natural pyrethrum and synthetic pyrethroid; and (iii) differences in the survival rates of arthropods using different knockdown methods.

METHODS

Study site

Our study area, the Adolpho Ducke Forest Reserve ('Reserva Ducke'), is covered by 90 km^2 of undisturbed rainforest on terra firme latosol and is located 26 km north-east of Manaus (02°55'S, 59°59'W). It belongs to the National Institute for Amazonian Research (INPA) and represents one of the most intensively studied upland forest sites in Central Amazonia (Willis, 1977; Penny and Arias, 1982; Adis and Schubart, 1984; Adis, 1988; Hero, 1990; Prance, 1990; Höfer *et al.*, 1994).

Experiment one

The tree species selected for fogging, *Goupia glabra* Aubl. (Celastraceae, common name 'Cupiuba'), has a high local abundance and is widely distributed in Amazonian upland forests (Brazil, Colombia, Venezuela and the Guianas; Loureiro and Silva, 1968). The canopy of a single tree (crown diameter about 15 m) was fogged for 5 minutes during the dry season in August 1991 (Ribeiro and Adis, 1994) at 06:00 h and again at 08:00 h on two successive days, with a Swingfog SN50 (Montan Swingtec). The fogging machine was hoisted into the lower canopy on a rope and pulley system (Adis *et al.*, 1984; Erwin, 1989; Stork, 1991). Release of the insecticide was controlled from the ground by radio-control. Subsequently, the fog was directed to all parts of the tree crown by rotating the fogger 180° with the rope from which it was suspended. A 1.0% solution of natural pyrethrum, diluted in diesel oil, was used for the first three fogging events and 0.15% solution of the synthetic pyrethrum, Baythroid, for the fourth fogging. The knockdown effect of the killing agent Baythroid was stated by the manufacturer (Bayer, Leverkusen, Germany, personal communication) to be 10 times higher than that of natural pyrethrum.

After each fogging event, knocked-down arthropods were intercepted by 26 funnel-shaped trays, each of 1 m^2 in area, for 2 h. We used the same trays as Stork and co-workers (Stork and Brendell, 1990) which are made of a fine, smooth nylon fabric on a metal frame. Eighteen trays were placed directly below the canopy and hung about 1 m above the ground and up to 5 m from the tree trunk on a web of ropes tied about head-height on available tree trunks (Figure 4.1). Two additional trays were placed at each of the four compass points, the first at 10 m and the second at 20 m from the tree trunk, to monitor possible wind drift of falling arthropods. At the end of each drop period, specimens were washed down the funnel walls with a garden sprayer filled with 70% ethanol into plastic bottles attached to the funnel outlet (Erwin, 1989).

Experiment two

Calophyllum brasiliense Camb. (Guttiferae, common name 'Jacareuba', maximum height 30 m) occurs in Amazonian lowland forests, particularly in inundation-forests along black- and clear-water rivers, and in Central Brazil as well (Loureiro and Silva, 1968). The canopies of trees in five plots about 50 m apart in a 20-year-old plantation in Reserva Ducke were fogged from the ground at the same time at dawn during the rainy season (February, 1992). Each of the five plots included six to eight trees, each 2 m apart. The trees were about 10 m in height, the diameter of the tree trunks was about 16 cm and canopy diameter was approximately 4 m. The five plots were fogged for 3 minutes with 0.5%, 1.0% and 1.5% natural pyrethrum, Baythroid (0.3%) and diesel oil, respectively. A sixth plot served as control and was not fogged. For each plot the falling arthropods were collected in five funnel-shaped trays as described above. The trays were randomly distributed within the plots. Arthropods were collected for 1 hour after fogging. The trees on all plots were then shaken and arthropods were collected for a further 1 hour.

To test the survival rates of arthropods following different insecticide treatments of different concentrations, additional specimens were collected on plastic sheets laid out on the forest floor between the trays in each

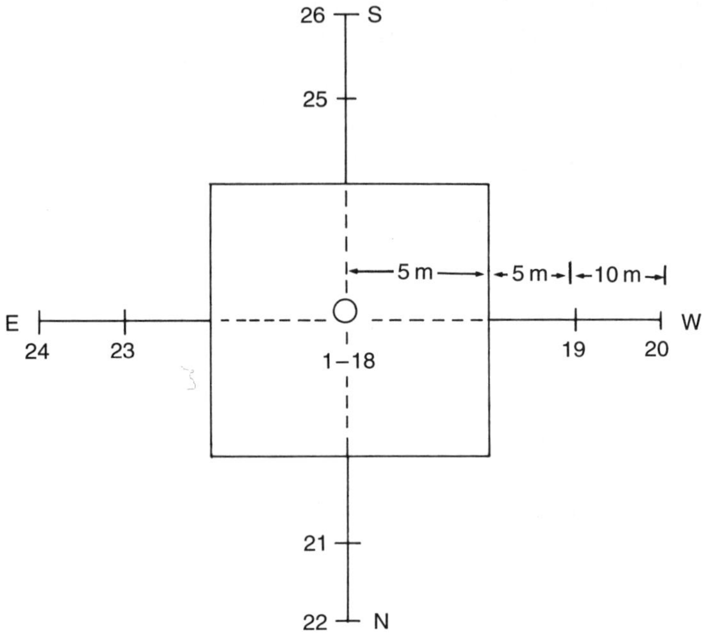

Figure 4.1 Arrangement of 26 trays under the canopy of *Goupia glabra* Aubl. (Celastraceae), fogged in August, 1991 in a primary upland forest near Manaus, Brazil.

of the plots. Each arthropod collected in this way was kept separately on moist peat in a glass vial under natural field conditions. Survival rate was monitored every second day for 15 days and the animals fed with either banana, sugar, dried fish-food, dead soft-bodied insects or 'Jacareuba' leaves, depending on their assumed feeding preference.

RESULTS

Knockdown rates in *Goupia glabra*

In total, 2107 arthropods (117.1 ± 61.5 individuals/m^2) were collected during the first fogging event in the 18 trays installed directly below the canopy (Figure 4.2). Hymenoptera (51.0%), mostly Formicidae (45.0% of all arthropods), and Diptera (20.6%) dominated. Patchiness of most taxa was high (Table 4.1). Only 1.1% of the total arthropods collected in all 26 trays (n = 2375) were found in the trays 20 m from the tree trunk. Catches in these trays were represented mostly by small winged insects

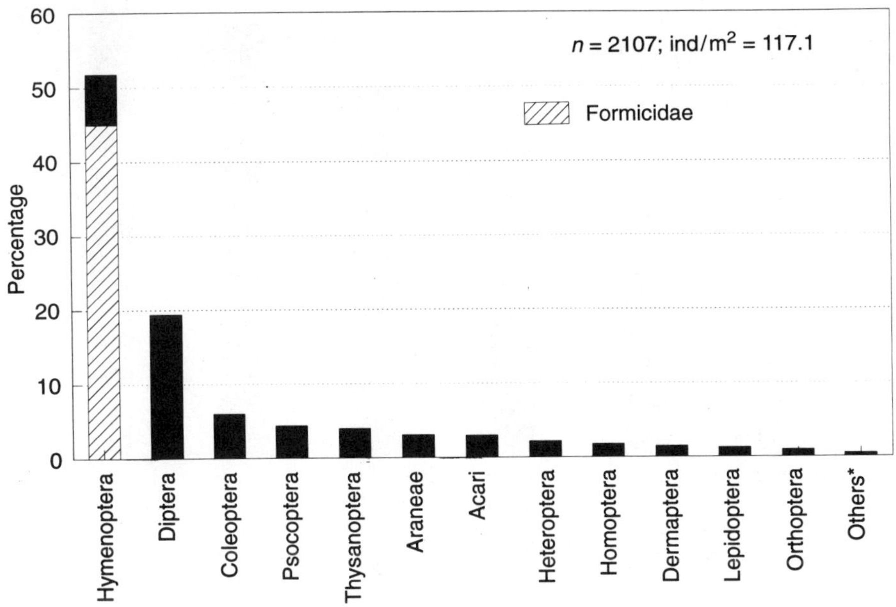

Figure 4.2 Percentages of arthropods obtained from *Goupia glabra* Aubl. after the first fog with 1.0% natural pyrethrum (without synergist) on 21 August, 1991 (06:00 h) near Manaus (drop time 120 min, 18 trays of 1 m^2 each). Others* represents Thysanura, Diplopoda, Opiliones, Isoptera, Dermaptera, Embioptera and Neuroptera.

(Thysanoptera, 31%; Diptera, 29%; micro-Hymenoptera, 14%; and Homoptera, 12%) which are known to disperse widely on the wind.

Approximately 72% of all the arthropods collected over the 2-hour drop period of the first fog fell within the first hour (Figure 4.3). The drop rates of different taxa varied.

With respect to the total number of trays required for canopy fogging, no statistical difference was found in the relative abundance of the seven most frequent orders caught in five or 18 trays after the first fogging (χ^2-test; Figure 4.4). However, the number of trays required for species-level accumulation was higher (e.g. 14 in Formicidae; cf. Harada and Adis, 1997, Chapter 17, this volume). This is especially true for those orders where more than half of all species collected were represented by a single individual (e.g. Araneae; cf. Höfer *et al.*, 1994).

A total of 3685 arthropods were collected from all four fogging events in the 18 trays directly installed under the canopy. Of these, 57.2% were

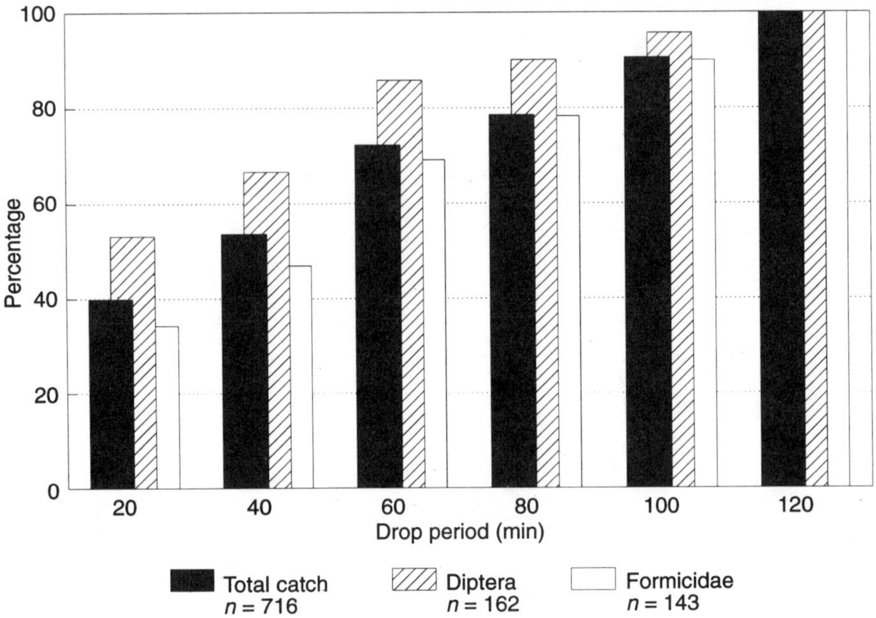

Figure 4.3 Percentages of total arthropod abundance and the two most abundant taxa, Diptera and Formicidae, against drop time after canopy fogging of *Goupia glabra* Aubl. with 1.0% natural pyrethrum (without synergist) near Manaus (fogging at 06 : 00 h on 21 August, 1991, total drop time 120 min, five selected trays of 1 m^2 each).

caught after the first fog, 22.8% after the second, 12.7% after the third and only 7.3% after the fourth (Figure 4.5). Again, Hymenoptera (mainly Formicidae) and Diptera were the most abundant orders. After the first and fourth fogs the six most abundant orders were the same, but their rank had changed considerably.

Knockdown rates in *Calophyllum brasiliense*

About 75% of the arthropods were obtained in the first hour (independent of concentration and agent used) and the remaining 25% after the trees had been shaken (Figure 4.6a). Exclusion of Formicidae from this analysis produced similar results (Figure 4.6b; χ^2-test: both data, $P < 0.001$).

Table 4.1 Collection of arthropod taxa (in rank order) obtained after canopy fogging *Goupia glabra* Aubl. (Celastraceae) with 1.0% natural pyrethrum (without synergist) near Manaus (fogging at 06:00 h on 21 August 1991; drop time 120 min, 18 traps of 1 m^2 each)

Taxa	*No. collected*	*Proportion (%)*	*Individuals/m² (mean ± S.D.)*		
Hymenoptera	1075	51.0	59.7	±	36.4
Formicidae*	(949)	(45.0)	(52.7)	±	(39.0)
Others*	(126)	(6.0)	(7.0)	±	(10.2)
Diptera	434	20.6	24.1	±	16.5
Homoptera	124	5.9	6.9	±	5.8
Orthoptera	92	4.4	5.1	±	11.3
Coleoptera	82	3.9	4.6	±	5.5
Thysanoptera	64	3.0	3.6	±	3.7
Araneae	63	3.0	3.5	±	2.9
Acari	45	2.2	2.5	±	3.6
Collembola	38	1.8	2.1	±	2.7
Psocoptera	31	1.5	1.7	±	2.7
Heteroptera	26	1.2	1.8	±	1.4
Lepidoptera	20	0.9	1.1	±	1.6
Thysanura	6	0.3	0.3	±	0.7
Diplopoda	2	0.1	0.1	±	0.3
Opiliones	1	<0.1	<0.1	±	0.2
Isoptera	1	<0.1	<0.1	±	0.2
Dermaptera	1	<0.1	<0.1	±	0.2
Embioptera	1	<0.1	<0.1	±	0.2
Neuroptera	1	<0.1	<0.1	±	0.2
Total	2107	100.0	117.1	±	61.5

* Formicidae + Other (non-Formicidae) = Hymenoptera

Knockdown rates versus pyrethrum concentration and type of agent in *Calophyllum brasiliense*

No statistical difference between 0.5% and 1.0% natural pyrethrum and 0.3% Baythroid was found for arthropod abundance in the first hour after fogging (ANOVA: Tukey-test, $P < 0.01$). Similar results were found following the exclusion of Formicidae from the analysis (Figure 4.6c). In both analyses the abundance of arthropods was significantly lower from the plots where diesel and no insecticide (control) was used. The relatively low drop rates obtained with natural pyrethrum 1.0% are attributable to technical problems during fogging.

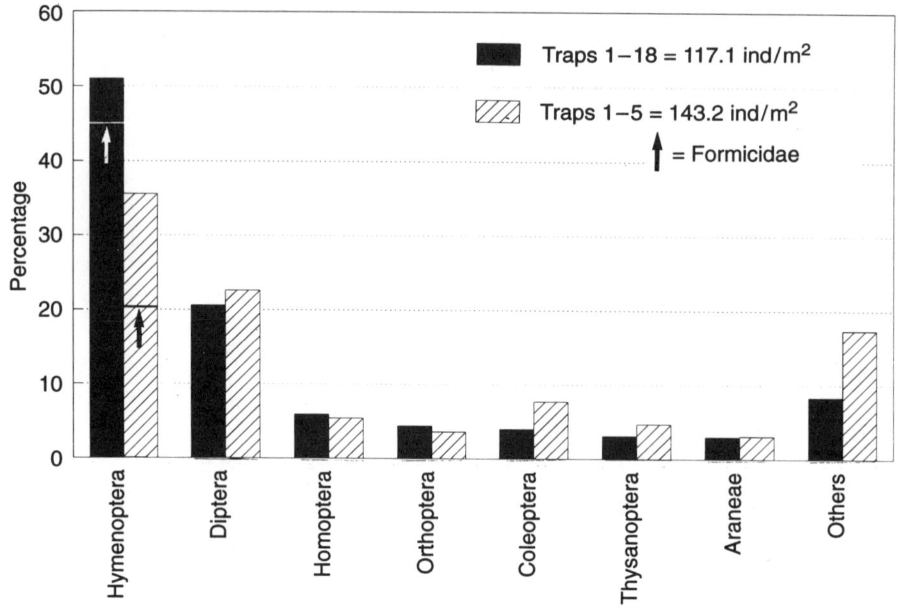

Figure 4.4 Percentages of different taxa and of total arthropods in five and 18 trays, respectively, obtained 120 min after fogging the canopy of *Goupia glabra* Aubl. with 1.0% natural pyrethrum (without synergist) near Manaus (fogging procedure at 06:00 h on 21 August, 1991).

Survival rates of fogged specimens

The average survival rate of insects obtained with 1.0% natural pyrethrum (see Experiment one) was 60% after 7 days (Figure 4.7a) (note that the decrease to 50% after 15 days was attributed to problems in culturing the insects). The highest survival rate observed (75%) was obtained with the lowest pyrethrum concentration of 0.5%. The lowest survival rate (50%) was for insects fogged with the killing agent Baythroid. Survival rates varied considerably with order and family (Figure 4.7). Some 79% of Orthoptera, 74% of Formicidae and 50% of Coleoptera fogged with 0.5% natural pyrethrum survived 7 days. Similarly, different concentrations and different agents resulted in different survival rates within these groups. This was attributed mainly to the individual condition of each animal, which varied within and between species. Most surviving immature insects completed their life-cycle. Adult Carabidae were successfully bred. Most of the Araneae

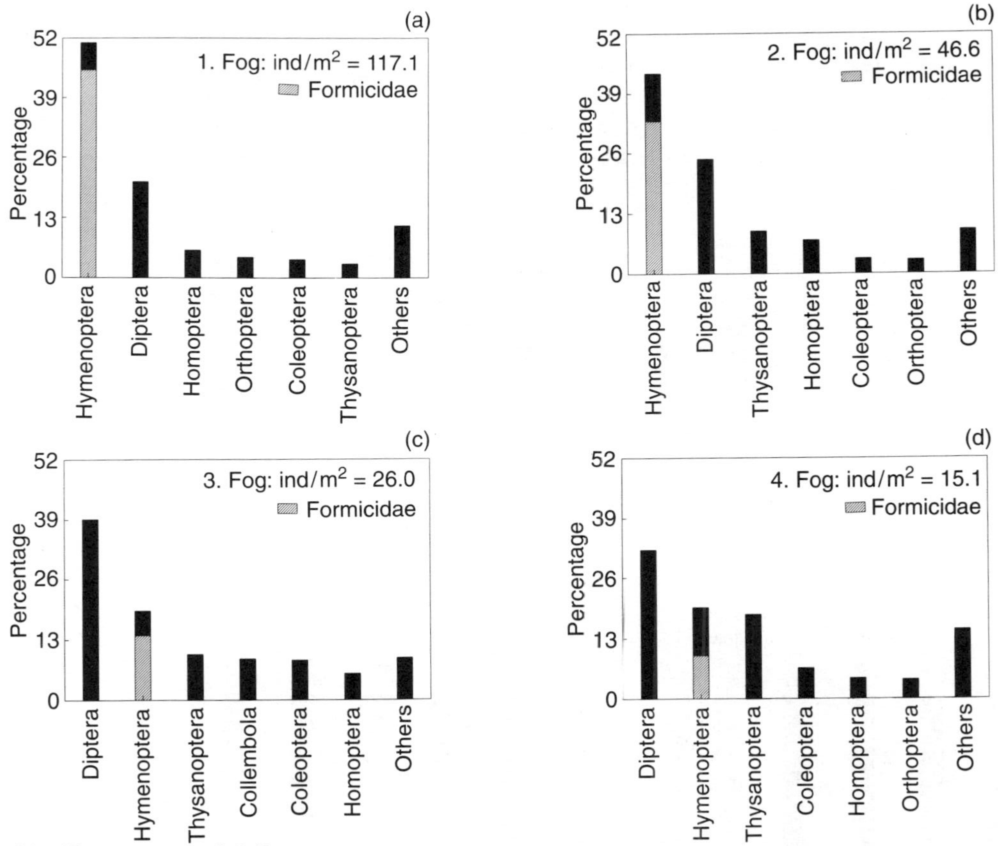

Figure 4.5 Percentages of different taxa and of total arthropods in 18 trays, 120 min after four consecutive fogs (FOG) of *Goupia glabra* Aubl.: (a), 21 August, 1991 at 06:00 h, 1.0% natural pyrethrum; (b), 21 August, 1991 at 08:00 h, 1.0% natural pyrethrum; (c), 22 August, 1991 at 06:00 h, 1.0% natural pyrethrum; (d), 22 August, 1991 at 08:00 h, 0.15% synthetic pyrethrum.

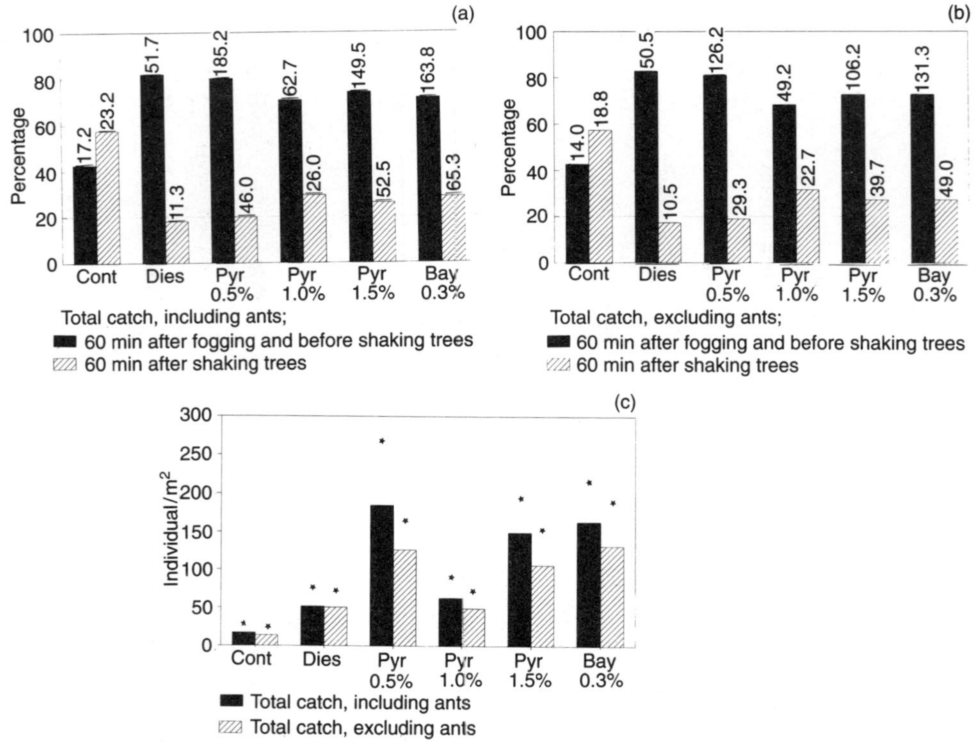

Figure 4.6 Mean number of arthropods and percentage of total catch in five trays from each of six plots in a forest plantation after fogging *Calophyllum brasiliense* Camb. (Guttiferae) near Manaus on 21 February, 1992.
(a), Total catch (%) including ants, 60 min after fogging (before shaking trees) and 60 min after shaking trees; (b), Total catch (%) excluding ants, 60 min after fogging (before shaking trees) and 60 min after shaking trees; (c), Relative abundance (individuals/m^2 with sample standard deviation, *) of total arthropods, with and without ants, 60 min after fogging (before shaking trees). PYR, natural pyrethrum without synergist; BAY, synthetic pyrethrum, Baythroid; DIE, diesel; CONT, control plot, no agent used; see text for details. Figures above bars give the mean number of arthropods sampled per m^2.

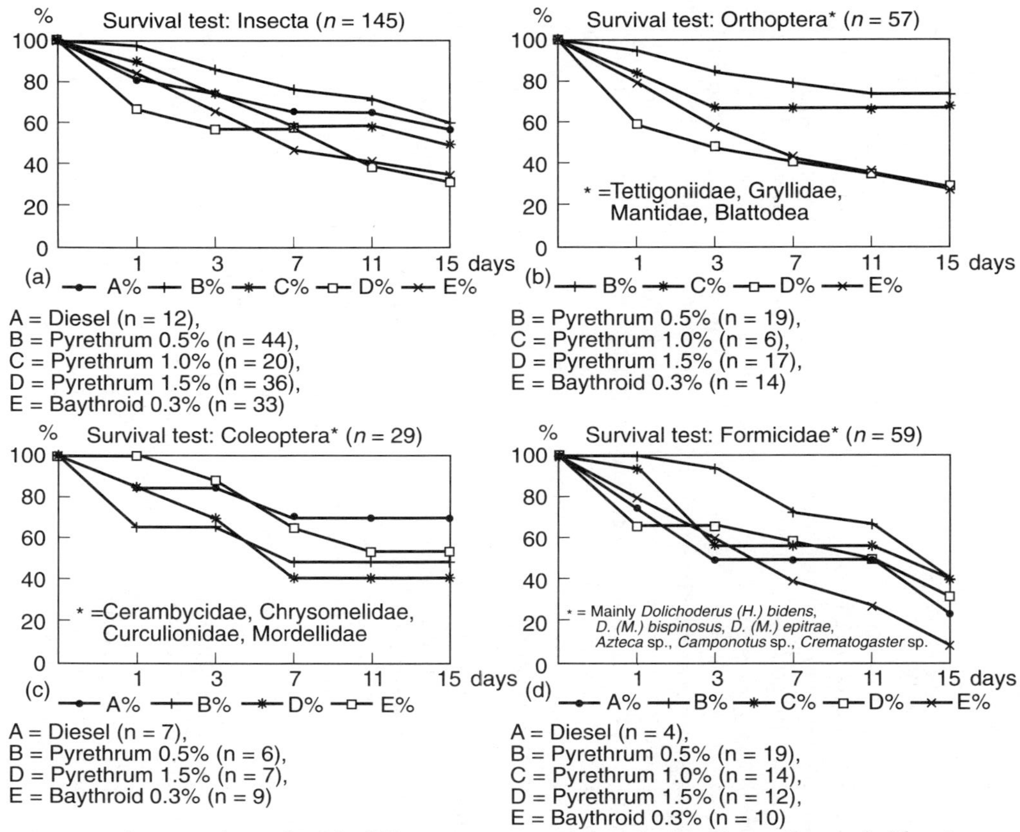

Figure 4.7 Survival rates of insects fogged with different concentrations of natural pyrethrum (without synergist) and the synthetic pyrethrum Baythroid from *Calophyllum brasiliense* Camb. (Guttiferae) on 21 February 1992 in a plantation near Manaus. (a), Total Insecta; (b), Orthoptera; (c), Coleoptera; (d), Formicidae. (Specimens kept separately for 15 days from 21 February, 1992 until 6 March, 1992; see text for details.)

which survived the first day after fogging continued to live for the next 14 days (H. Höfer and A. Brescovit, personal communication). Their average survival rate was 46% for 0.5% pyrethrum (n = 11), 50% for 1.0% pyrethrum (n = 4), 67% for 1.5% pyrethrum (n = 9) and 54% for 0.3% Baythroid (n = 15).

DISCUSSION

Although pyrethroid-based chemicals with synergists have been used for more than 20 years to fog arthropods from tropical trees 45 m or more in height in the tropics (Erwin, 1983, 1989), few data are available on the knockdown efficiency of these agents and statements based solely on anecdotal visual observations differ. Similarly, the survival rates of arthropods fogged with different insecticides are unknown. The widely used but expensive 'pure' or 'natural' pyrethrum with piperonyl butoxide as a synergist has been replaced in many recent studies by cheaper synthetic pyrethroid chemicals which are now available.

Stork and co-workers have used 'Reslin E' in Indonesia, which they state is a "non-residual insecticide with high knockdown and low kill components" (Stork, 1987, 1988, 1991; Stork and Brendell, 1990). Recovery and breeding of fogged Carabidae has been successful (Paarmann and Stork, 1987; Paarmann and Paarmann, 1997, Chapter 20, this volume). Most specimens were observed to fall within the first 20 minutes of fogging, but a 2-hour drop-time was allowed before collecting. Brief visual tests after fogging experiments in Sulawesi and Great Britain suggested "that few insects were left on the leaves" (Stork, 1991). The drop-time of insects after fogging has only been tested for British oak trees (N. Stork and P. Hammond, unpublished results), not in the tropics.

Erwin and co-workers have used the synthetic pyrethroid chemical 'Resmythrin' (also called 'Respond') in northern South America, which they state is "highly effective and similar to 'Reslin E'" (Erwin, 1983, 1989; Farrell and Erwin, 1988). Erwin suggested that the drop-time for beetles was 30–45 minutes after fogging, but that Orthoptera "took longer" to fall (Erwin, 1989). After 2 hours there were "virtually no arthropods left" in the fogged area and almost nothing was collected by various follow-up tests (branch shaking, re-fogging, second 2-hour drop collections) between 1982 and 1985 in Tambopata, Peru (Erwin, 1989). Recovery and breeding experiments were not made.

Our decision to test natural pyrethrum without synergist under tropical conditions was mainly based on the promising results obtained by Paarmann and co-workers, who sampled live arthropods from Oak trees in Germany for life-history studies (Paarmann *et al.*, 1991; Paarmann, 1994; Paarmann and Kerck, 1997, Chapter 3, this volume). They obtained

67% of the arthropods in the first 60 minutes after fogging, using a 0.45% solution of natural pyrethrum without synergist. Survival rate of seven taxa (170 arthropods) 10 days after fogging with 0.15% natural pyrethrum ranged from 49–100%. One of our goals is to provide those involved in biodiversity studies on arthropods in the central Amazon (Höfer *et al.*, 1994; Harada and Adis, 1997, Chapter 17, this volume) and in Borneo (Floren and Linsenmair, 1994, 1997, Chapter 16, this volume) with living specimens from the canopy. In conclusion, a survival rate of 60–75% (depending on the concentration of natural pyrethrum applied) seems to be adequate for collecting living arthropods from neotropical tree canopies for habitat manipulation experiments. Use of the synthetic pyrethroid, Baythroid, even in small concentrations, cannot be recommended for this purpose.

Acknowledgements

This study is part of a 6-year project on 'Mechanisms which maintain tropical diversity', funded by the German Research Foundation (DFG: project PA 99/15–1,2,3), the German Agency of Technical Cooperation (GTZ: project 85.2522.2–06100) and the Brazilian Research Foundation (CNPq: project CNPq/MPG 91.0304/90–4) since 1991. Additional funding was received from the Tropical Ecology Working Group of the Max-Planck-Institute for Limnology in Plön, Germany and Manaus, Brazil, the Fachhochschule Hildesheim, Holzminden in Göttingen, Germany, the German Academic Exchange Service (DAAD) and the National Amazon Research Institute (INPA) in Manaus, Brazil. We thank Nigel Stork of The Natural History Museum in London, Great Britain, for his participation and for help received during our first canopy fogging in August 1991. We are especially grateful to Wolfgang J. Junk, Maria Teresa Fernandez Piedade and Jörg Ohly for logistical support in Manaus via the 'Projeto INPA/Max-Planck'. We heartily thank all scientists, the technical staff of INPA as well as the participants of the post-graduate course 'Entomological Field Ecology' of INPA/Univ. Amazonas (February 1992) who joined the canopy fogging studies. Elizabeth Franklin, 'Projeto INPA/ Max-Planck', kindly helped with the statistical analyses.

REFERENCES

Adis, J. (1988) On the abundance and density of terrestrial arthropods in Central Amazonian dryland forests. *Journal of Tropical Ecology*, **4**, 19–24.

Adis, J. and Schubart, H.O.R. (1985) Ecological research on arthropods in Central Amazonian forest ecosystems with recommendations for study procedures, in *Trends in Ecological Research for the 1980s. NATO Conference Series I: Ecology* (eds J.H. Cooley and F.B. Golley), Plenum Press, New York, pp. 111–44.

Adis, J., Lubin, Y.D. and Montgomery, G.G. (1984) Arthropods from the canopy of inundated and terra firme forests near Manaus, Brazil, with critical considerations on the pyrethrum-fogging technique. *Studies on Neotropical Fauna and Environment*, **19**, 223–36.

Erwin, T.L. (1983) Tropical forest canopies: the last biotic frontier. *Bulletin of the Entomological Society of America*, **29**, 14–19.

Erwin, T.L. (1989) Canopy arthropod biodiversity: a chronology of sampling techniques and results. *Revista Peruana de Entomologia*, **32**, 71–7.

Farrell, B.D. and Erwin, T.L. (1988) Leaf-beetles (Chrysomelidae) of a forest canopy in Amazonian Peru: synoptic list of taxa, seasonality and host-affiliations, in *The Biology of the Chrysomelidae* (eds P. Jolivet, E. Petitpierre and T. Hsiao), W. Junk Publishers, The Hague, pp. 73–90.

Floren, A. and Linsenmair, K.E. (1994) Zur Diversität und Wiederbesiedelungsdynamik von Arthropoden auf drei Baumarten in einem Regenwald in Sabah, Malaysia. *Andrias*, **13**, 23–8.

Floren, A. and Linsenmair, K.E. (1997) Diversity and recolonization dynamics of selected arthropod groups on different tree species in a lowland rainforest in Sabah, Malaysia, in *Canopy Arthropods* (eds N.E. Stork, J. Adis and R.K. Didham), Chapman & Hall, London, pp. 344–81 (this volume).

Harada, A.Y. and Adis, J. (1997) The ant fauna of tree canopies in central Amazonia: a first assessment, in *Canopy Arthropods* (eds N.E. Stork, J. Adis and R.K. Didham), Chapman & Hall, London, pp. 382–400 (this volume).

Hero, J.M. (1990) An illustrated key to tadpoles occurring in the Central Amazonian rainforest, Manaus, Amazonas, Brazil. *Amazoniana*, **11**, 201–62.

Höfer, H., Brescovit, A.D., Adis, J. and Paarmann, W. (1994) The spider fauna of Neotropical tree canopies in Central Amazonia: first results. *Studies on Neotropical Fauna and Environment*, **29**, 23–32.

Loureiro, A.A. and Silva, M.F. da (1968) *Catálogo das madeiras da Amazônia*, Vol. 1, SUDAM, Belém.

Paarmann, W. (1994) Von der Tötung zum Lebendfang – Entwicklung der Insektizidnebeltechnik. *Allgemeine Forstzeitschrift für Waldwirtscaft und Umweltfürsorge (AFZ)*, **49**, 1457–9.

Paarmann, W. and Kerck, K. (1997) Advances in using the canopy fogging technique to collect living arthropods from tree crowns, in *Canopy Arthropods* (eds N.E. Stork, J. Adis and R.K. Didham), Chapman & Hall, London, pp. 53–66 (this volume).

Paarmann, W and Paarmann, P. (1997) Studies on the biology of a canopy-dwelling carabid beetle collected by canopy fogging in the rainforest of Sulawesi (Indonesia), in *Canopy Arthropods* (eds N.E. Stork, J. Adis and R.K. Didham), Chapman & Hall, London, pp. 433–41 (this volume).

Paarmann, W. and Stork, N.E. (1987) Canopy fogging, a method of collecting living insects for investigations of life history strategies. *Journal of Natural History*, **21**, 563–6.

Paarmann, W., Kerck, K., Altenkirch, W. and Hilbert, C. (1991) *Nebeltechnik als Methode zum Studium der Ökologie kronenbewohnender Arthropoden*. Kurzfassungen Entomologen-Tagung DGaae/SEG/ÖEG (2.-6.4.1991), Wien, Austria.

Penny, N.D. and Arias, J.R. (1982) *Insects of an Amazon Forest*, Columbia University Press, New York.

Prance, G.T. (1990) The floristic composition of the forests of central Amazonian Brazil, in *Four Neotropical Rainforests* (ed. A.H. Gentry), Yale University Press, New Haven, pp. 159–74.

Ribeiro, M. de N.G. and Adis, J. (1984) Local rainfall variability – a potential bias for ecological studies in the Central Amazon. *Acta Amazonica*, **14**, 159–74.

Stork, N.E. (1987) Arthropod faunal similarity of Bornean rain forest trees. *Ecological Entomology*, **12**, 219–26.

Stork, N.E. (1988) Insect diversity: facts, fiction and speculation. *Biological Journal of the Linnean Society*, **35**, 321–37.

Stork, N.E. (1991) The composition of the arthropod fauna of Bornean lowland rain forest trees. *Journal of Tropical Ecology*, **7**, 161–80.

Stork, N.E. and Brendell, M.J.D. (1990) Variation in the insect fauna of Sulawesi trees with season, altitude and forest type, in *Insects and the Rain Forests of South East Asia (Wallacea)* (eds W.J. Knight and J.D. Holloway), Royal Entomological Society of London, London, pp. 173–90.

Willis, E.O. (1977) Lista preliminar das aves da parte noroeste e áreas vizinhas da Reserva Ducke, Amazonas, Brasil. *Revista Brasileira de Biologia*, **37**, 585–601.

Part Two

Community Structure of Coleoptera Assemblages

5

Beetle species diversity and faunal similarity in Venezuelan rainforest tree canopies

J.G. Davies, N.E. Stork, M.J.D. Brendell and S.J. Hine

ABSTRACT

Insecticide fogging was used to study in detail the beetle communities associated with the canopies of six rainforest trees in Parque Nacional Henri Pittier, in Northern Venezuela. A total of 6132 individual beetles were collected and sorted to 978 species in 65 families. Diversity indices for the six trees were very high. One of the trees, *Mangifera indica*, had a significantly higher beetle species diversity than the other five trees. The species composition of the tree-crown faunas was strongly site-dependent, with very little faunal overlap between the two trees at one site and the four trees at the second site located 20 km away. This may be due to differences in forest type between the two sites, or simply a function of the distance between them. In contrast, there was relatively high faunal similarity between trees at the same site. Faunal similarity at the family level was also high, with the ranking of the 15 most species-rich beetle families found to be highly constant over the six trees. Comparisons with other studies also revealed constancy at the family level for beetle faunas from different continents. Trophic guild structure of the beetle fauna was surprisingly consistent across trees, despite differences in tree taxonomic status and phenology.

INTRODUCTION

Knockdown insecticide studies of canopy arthropod richness and community structure in tropical America have, to date, focused on four locations: Erwin's widely quoted and classic study of beetle species richness on *Luehea seemanii* (Erwin and Scott, 1980; Erwin, 1982) was based

Canopy Arthropods. Edited by N.E. Stork, J. Adis and R.K. Didham. Published in 1997 by Chapman & Hall, London. ISBN 0 412 74900 9

in the Canal Zone of Panama (see also Wolda, 1979), and his later studies were based around Manaus in Brazil (Erwin, 1983a,b) and at Tambopata (Erwin, 1988, 1990) and Pakitza (Erwin, 1990) in the Madre de Dios district of southern Peru. More detailed aspects of the canopy arthropod fauna in the forests around Manaus have been discussed by Adis and co-authors (Adis *et al.*, 1984; Adis and Schubart, 1985; Adis *et al.*, 1997, Chapter 4, this volume; Harada and Adis, 1997, Chapter 17, this volume; Paarmann and Kerck, 1997, Chapter 3, this volume; Paarmann and Paarmann, 1997, Chapter 20, this volume).

These studies have concentrated in particular on several aspects of the canopy arthropod community. First, they have provided data on the abundance of canopy arthropod Orders (Erwin, 1983a,b, 1990; Adis *et al.*, 1984; Adis and Schubart, 1985), clearly demonstrating that ants are the dominant group at these Amazonian sites (see also Harada and Adis, 1997, Chapter 17, this volume). Second, they have shown the immense richness of the beetle fauna (Erwin and Scott, 1980; Erwin, 1982, 1983a,b). Finally, the most detailed of these studies (Erwin and Scott, 1980) examined how beetle diversity, guild structure and biomass varied between individuals of the same tree species and at different times of the year.

In a new study of five sites in the Parque Nacional Henri Pittier in Venezuela, a total of 18 identified trees were sampled by insecticide fogging (J.G. Davies and N.E. Stork, unpublished data). The species richness and guild structure of beetles from six trees at two of these sites were analysed in detail. We analyse the composition of the beetle fauna with respect to site, tree species and tree phenology, and compare the beetle fauna with that from sites on other continents.

METHODS

Study area

The study was conducted in Parque Nacional Henri Pittier, Estado Aragua, in northern Venezuela between March and May 1990. The national park (Figure 5.1) covers 110 000 ha of the Coastal Cordillera mountain range and supports a wide variety of forest types, ranging from dry deciduous woodland on the lower slopes at sea level, to dwarf cloud forest on the peaks (1800 m). Plant diversity in the park has been shown to be unusually high (Huber, 1986). In the north, the mountains drop to arid scrubland bounded by the Caribbean coast, and to the south are fertile valleys and Lake Valencia. The six trees selected for the beetle analysis were located at two different sites: four tree species, *Talisia* sp., *Brownea grandiflora*, Polygonaceae sp. and *Chrysophyllum lucentifolium*, in dry deciduous forest at La Trilla, km 36 between Maracay and Ocumare (300 m above sea level), and the other two tree species, *Cassia grandis*

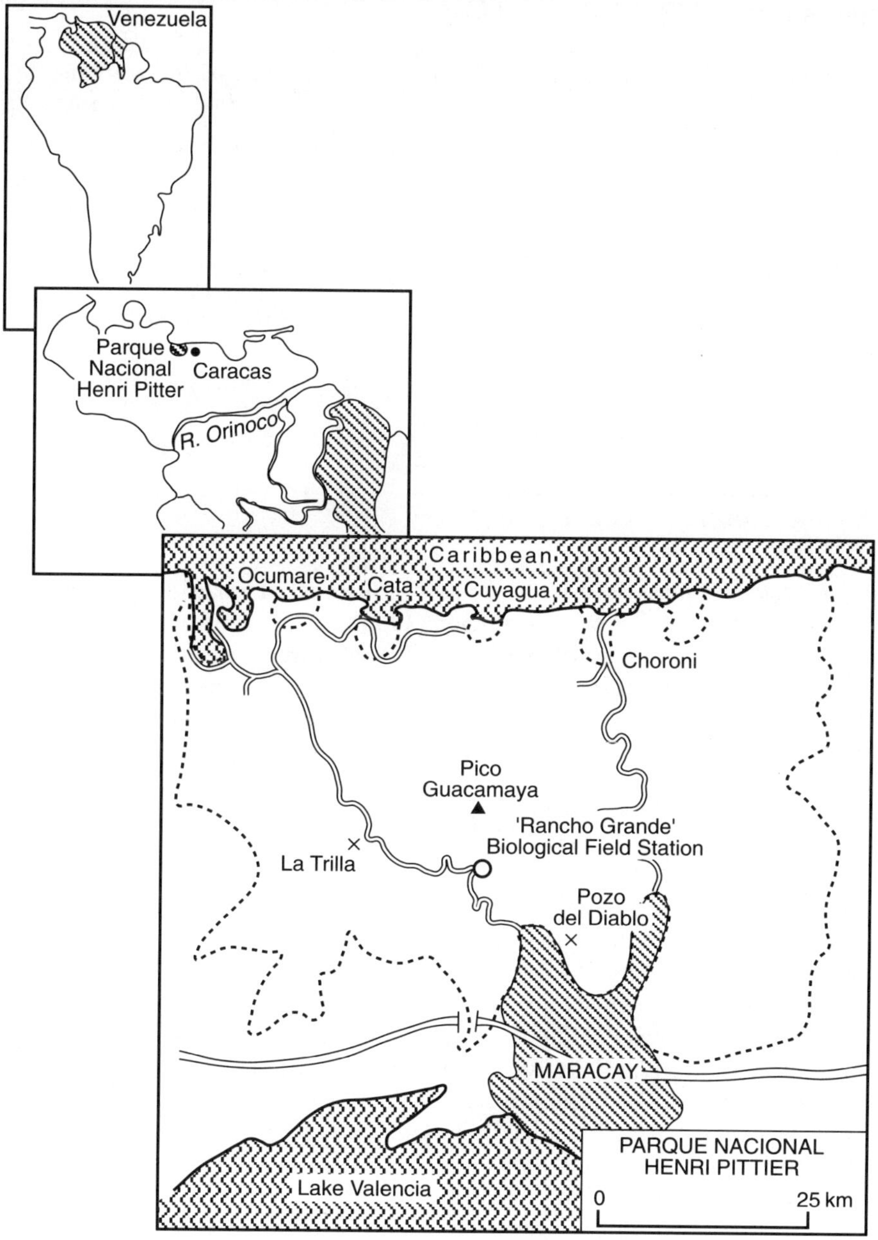

Figure 5.1 Location of the two study sites within Henri Pittier National Park.

Table 5.1 Details of the six fogged trees

Tree species	*Family*	*Trays*	*Altitude* (m)*	*Forest type*	*Canopy type*	*Volume (m^3)*	*Life stage*
Talisia sp.	Sapindaceae	24	300	Deciduous	Medium/ open	900	Sterile
Brownea grandiflora	Leguminosae	29	300	Deciduous	Small/ open	110	Flowering
Unidentified sp.	Polygonaceae	30	300	Deciduous	Medium/ dense	195	Sterile
Chrysophyllum lucentifolium	Sapotaceae	30	300	Deciduous	Small/ dense	195	Sterile
Cassia grandis	Leguminosae	30	400	Gallery	Large/ open	4575	Fruiting
Mangifera indica	Anacardiaceae	30	400	Gallery	Large/ dense	2145	Fruiting

* Above sea level

and *Mangifera indica*, in a patch of riparian forest at Pozo del Diablo, in the southern foothills near Maracay (400 m a.s.l.). The two sites were approximately 20 km apart. Details of the six trees are given in Table 5.1. The main criterion for tree selection was that the canopy should be isolated from other canopies to ensure that the arthropods collected could reasonably be expected to have been associated with the target tree.

Sampling procedure

Canopy fogging of each tree was carried out using an SN11 Swingfog insecticide fogger, which was hoisted high into the tree canopy on a rope and pulley system suspended from a large branch. The direction of release of the fog was controlled by a hand line attached to the nozzle of the fogger. The insecticide used was K-Othrin, a non-residual synthetic pyrethroid with a 2.5% active ingredient of deltamethrin. In order to make the fog visible, diesel was used as a carrier for the insecticide.

In all cases, fogging was carried out early in the morning on a day when the wind conditions were still (windless conditions are required to allow the fog to rise up through the canopy without drifting), and only after a dry night (to prevent the risk of knocked-down insects sticking to wet leaves in the canopy). Samples were collected in 24–30 (depending on the size of the canopy) 1 m^2, funnel-shaped trays made of smooth nylon, beneath which was suspended a collecting jar containing 70% ethanol. The trays were suspended from a network of ropes at head height. Sufficient insecticide was released to envelop the entire canopy (usually taking 15–20 minutes), and a drop time of 2 hours was allowed for the arthropods to fall from the tree. Trays were then gently tapped or brushed so that insects trapped on the sides dropped into the collecting pots.

Trees were identified from leaf, flower, fruit, bark and other samples at the Royal Botanical Gardens, Kew (UK), and at the herbarium of the Universidad Central de Venezuela in Maracay. Canopy area was mapped from the ground by taking bearings and measuring distances at several points around the perceived perimeter of the canopy. Canopy volumes were calculated using the formula for a sphere, with the radius being estimated from the perimeter map of the canopy (Table 5.1). Relative density of foliage is taken from the botanical description of each tree (K. Edwards, personal communication).

Sample sorting and data analysis

Beetles from the six trees were extracted, mounted, labelled and sorted to recognizable taxonomic units (RTUs), or morphospecies, at the Natural History Museum in London. A suite of five different measures was then used to quantify species diversity: Margalef's index of species richness, the Shannon–Wiener index, Simpson's index, evenness, and rarefaction (Magurran, 1988). K-dominance curves (Lambshead *et al.*, 1983) were also plotted to compare the relative beetle diversities of the six trees. Faunal similarity was investigated using Principle Components Analysis (PCA) and the Normalized Expected Species Shared (NESS) similarity index (Grassle and Smith, 1976). Friedman's method for randomized blocks and Kendall's coefficient of concordance were used to investigate familial constancy, that is the degree to which the ranking of the main beetle families (in terms of the number of species represented) varied between trees.

Allocation of feeding guilds

When little or nothing is known about the feeding ecology of individual beetle species, patterns of trophic guild structure can be investigated through the assignment of families (or subfamilies) to different trophic groups, based on the known biology of related taxa. This is particularly appropriate when studying a diverse and poorly known fauna. However, such a method is highly susceptible to error, both because of the dearth of autecological knowledge about individual species, and as a result of complications as to how to assign a species when its adult and larval stages belong to different guilds (Stork, 1987a; Hammond, 1990). Some of these problems can be minimized by allocating guilds on a proportional basis (e.g. regarding the Buprestidae as 75% herbivorous

Table 5.2 Diversity measures for the beetles sampled from the six fogged trees

Tree species	*Family*	*No. of species*	*No. of individuals*	*Shannon index*	*Margalef index*	*Simpson index*	*Evenness*	*Rarefaction*
Talisia sp.	Sapindaceae	292	975	4.90	42.3	0.0148	0.863	169
Brownea grandiflora	Leguminosae	168	393	4.43	28.0	0.0316	0.865	169
Unidentified sp.	Polygonaceae	170	517	4.18	27.0	0.0388	0.815	142
C. lucentifolium	Sapotaceae	211	827	4.02	31.3	0.0786	0.751	134
C. grandis	Leguminosae	253	1620	4.23	34.1	0.0416	0.764	132
M. indica	Anacardiaceae	313	1800	4.62	41.6	0.0239	0.804	138

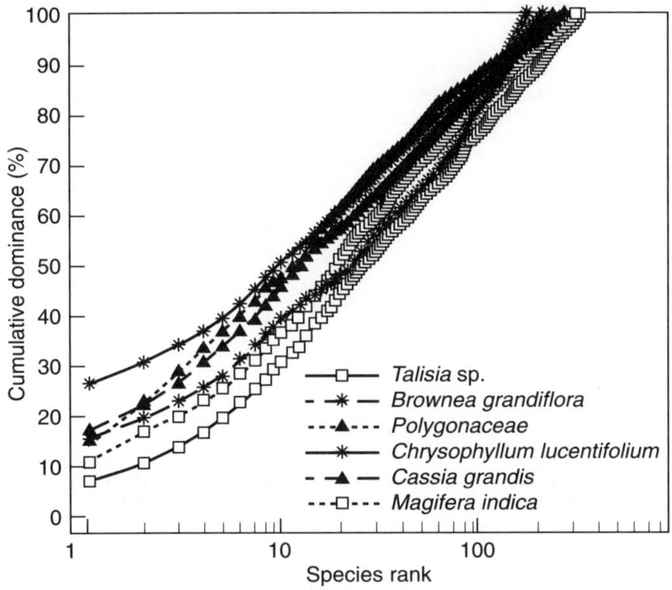

Figure 5.2 K-dominance curves illustrating beetle diversity for the six trees.

and 25% xylophagous) using whatever limited information is available for each family, but this method is somewhat arbitrary.

Trophic guild structure in this study was analysed by allocating each beetle family an estimated percentage guild affiliation based on Stork (1987a), Hammond (1990) and P.M. Hammond (personal communication). The estimated proportions of each guild represented on each of the six trees were then calculated from the number of individuals collected per beetle family.

RESULTS AND DISCUSSION

Beetle diversity

A total of 6132 beetles from the six trees were sorted to 65 families and 978 morphospecies. The numbers of species and individuals per tree are very high, ranging from 168 to 313 species, and from 393 to 1800 individuals (Table 5.2). Diversity measures for each of the six fogged trees are given in Table 5.2. Figure 5.2 illustrates the diversities of the six trees in graphic form; higher diversity is represented by a steeper curve and an origin closer to zero. One-way analysis of variance (ANOVA) indicated that there were significant differences in arthropod diversity on different trees ($F_{5,84}$ = 52.85, P <0.001). *Post hoc* tests revealed

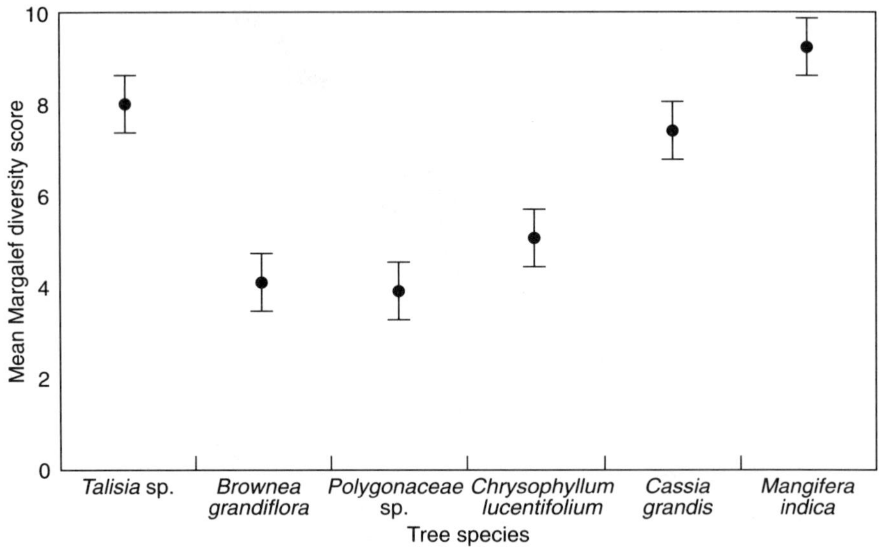

Figure 5.3 Mean (± 95% confidence limits) beetle species diversity (Margalef index) of individual 1 m^2 tray samples (n = 15, selected at random) from each of the six tree species sampled. Means whose intervals do not overlap are significantly different.

that the diversity of beetles collected from *Mangifera indica* was significantly higher than for all the other five trees (Tukey's HSD, P <0.05) (Figure 5.3). The diversity of beetles on both *Cassia grandis* and *Talisia* sp. was also significantly higher than on *Chrysophyllum lucentifolium, Brownea grandiflora* and Polygonaceae sp. The high species diversity on the non-fruiting *Talisia* sp. could be due to a combination of a high species-abundance ratio (Margalef score, Table 5.2) and a high degree of evenness. The low score for *C. lucentifolium* is largely due to the high dominance of its most common beetle species (as indicated by the high Simpson score, Table 5.2) and the consequent low evenness. The relatively low score for *C. grandis*, considering its high number of species, was also due to low evenness (Table 5.2), possibly caused by a dominance of a small number of species of Bruchidae and Apionidae.

A similar study in which 10 Bornean rainforest trees were fogged (Stork, 1991) revealed a lower total beetle species richness (859 species) than that recorded here, even though the total area sampled (i.e. tray area) was 17% greater in the Bornean study (Table 5.3). However, the higher number of species in the Venezuelan samples may be because there were 33% more specimens collected than in the Bornean samples.

Table 5.3 Comparison of beetle species richness from fogging studies in Brunei (Stork, 1987a,b, 1991) and Venezuela

Region	*Total area sampled (m^2)*	*No. of species*	*No. of individuals*	*Species richness (Margalef)*
Brunei	200	859	4043	103.32
Venezuela	173	978	6132	112.03

Similarity of the canopy beetle fauna between trees and sites

Faunal similarity at the species level was investigated both within and between the two sites. NESS and PCA faunal similarity analyses of the six trees (Figures 5.4 and 5.5, respectively) demonstrate a clear difference between the beetle communities of the La Trilla site and the Pozo del Diablo site. All but one between-site comparison scored lower than 10% similarity. In contrast, all but two within-site comparisons scored above 30% similarity (Figure 5.4). Statistical comparison of these similarity index values reveals within-site similarity to be significantly higher than between-site similarity (t = 16.92, d.f. = 12, P <0.001). Although the beetle faunas of *Cassia grandis* and *Mangifera indica* are much more similar to each other (30.7%) than either is to the four trees from the other site (Figure 5.4), PCA ordination nevertheless revealed a clear separation of beetle species composition on these two tree species (Figure 5.5).

The dense, dark canopy of *M. indica* contrasts starkly with the open, sparse canopy of *C. grandis*. The most likely explanation for the observed difference in beetle species composition between the two tree species is the difference in fruit form. *M. indica* has fleshy fruit, whereas *C. grandis* has dry and probably less accessible seed pods. While the former, an exotic, provides a simple meal for a wide variety of opportunistic frugivorous beetles, the latter is likely to support a more specialized fauna adapted to penetrating hard seed pods.

The faunal similarity data are corroborated by analysis of the site- and tree-fidelity of different beetle species. Some 92.5% of beetle species were restricted to one site, emphasizing the importance of forest type and/or inter-site distance in determining community composition. This may also be due in part to the small sample size and the clear difference in forest type. In a similar study, Erwin (1983a,b) fogged trees in four different habitat types near Manaus, Brazil, and found that 83% of the 1080 beetle species he collected were restricted to only one forest type, and a further 14% were shared between just two of the sites.

Tree fidelity of the Venezuelan beetles was also found to be high, with 72% of species restricted to only one of the trees. Similarly, Erwin and

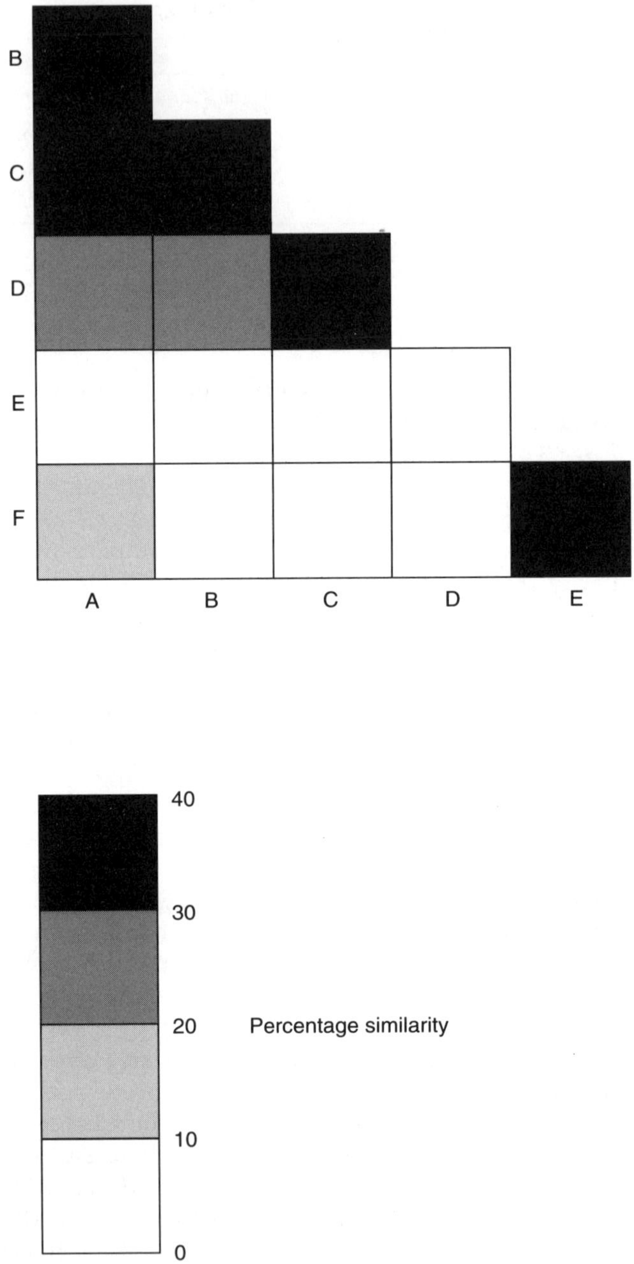

Figure 5.4 NESS faunal similarities for beetle communities on six Venezuelan rainforest trees. A, *Talisia* sp.; B, *Brownea grandiflora*; C, Polygonaceae; D, *Chrysophyllum lucentifolium*; E, *Cassia grandis*; F, *Mangifera indica*.

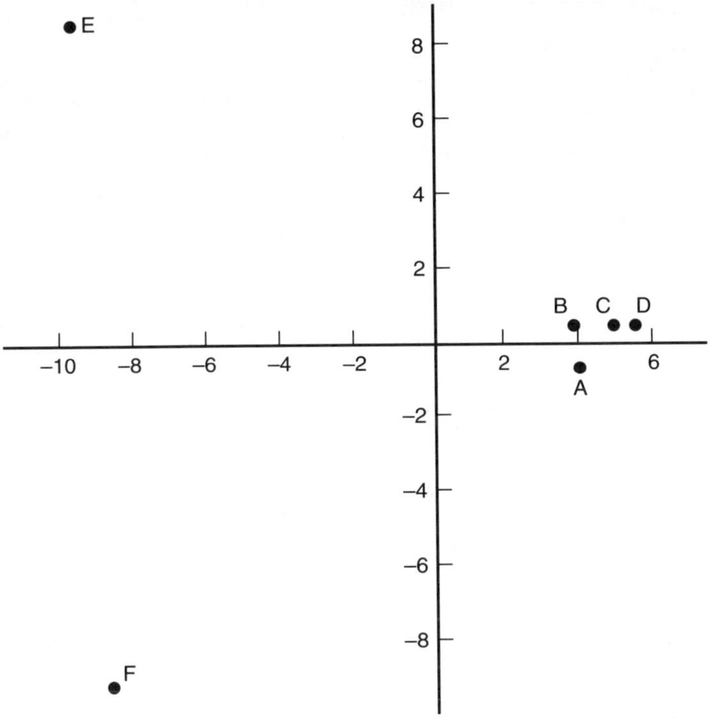

Figure 5.5 Principal components analysis (PCA) of the beetle faunas from six Venezuelan rainforest trees. A, *Talisia* sp.; B, *Brownea grandiflora;* C, Polygonaceae; D, *Chrysophyllum lucentifolium;* E, *Cassia grandis;* F, *Mangifera indica.*

Scott (1980) found that of more than 1200 beetle species, each herbivore species was found, on average, on only 2.36 of the 19 trees, while predator species were found, on average, on only 1.76 of 19 trees. Considering this, it is perhaps surprising that even beetle faunas from the same site exhibit such a high degree of parity. These data suggest that there is a very high proportion of 'tourists' (*sensu* Moran and Southwood, 1982; Stork, 1987b) in the samples.

Familial constancy

In contrast to the low species fidelity between sites, there was a much greater degree of family constancy (in both number of species and individuals) across the six trees (Table 5.4). The most speciose families were Curculionidae, Staphylinidae, Coccinellidae and Anobiidae, although the highest number of individuals belonged to Bruchidae. Almost 50% of the families (30 out of 65) were represented by fewer

Table 5.4 The number of beetle species and individuals per tree belonging to each of the 65 families

	A		*B*		*C*		*D*		*E*		*F*		*Total*	
Family	*Sp.*	*No.*	*Sp.*	*No.*	*Sp.*	*No.*	*Sp.*	*No.*	*Sp.*	*No.*	*Sp.*	*No.*	*Sp.*	*No.*
Acanthoceridae	0	0	1	1	0	0	0	0	0	0	1	1	2	2
Aderidae	9	11	3	5	7	9	8	10	0	0	7	17	24	52
Alleculidae	2	5	0	0	0	0	1	2	0	0	0	0	3	7
Anobiidae	15	107	8	18	12	19	19	79	23	141	26	101	72	465
Anthicidae	2	28	2	2	1	2	1	6	3	17	4	28	8	83
Anthribidae	6	7	3	3	6	7	7	11	1	1	4	21	20	50
Apionidae	21	114	9	84	10	139	11	60	12	31	22	66	48	494
Attelabidae	1	2	0	0	0	0	0	0	0	0	0	0	1	2
Biphyllidae	0	0	1	2	0	0	2	3	0	0	0	0	2	5
Brentidae	0	0	0	0	0	0	1	1	0	0	1	1	2	2
Bruchidae	8	36	9	10	1	1	5	7	32	284	42	564	60	902
Buprestidae	0	0	0	0	0	0	0	0	2	6	2	2	4	8
Cantharidae	0	0	0	0	0	0	0	0	1	2	4	11	5	13
Carabidae	1	2	2	4	2	2	3	6	4	7	5	124	9	145
Cerambycidae	5	6	0	0	1	1	3	4	9	12	4	6	22	29
Cerylonidae	1	1	0	0	1	1	3	6	2	6	0	0	6	14
Chrysomelidae	20	41	5	15	2	6	10	14	15	54	22	79	52	209
Ciidae	8	32	4	9	11	65	3	4	3	11	3	12	18	133
Cleridae	0	0	0	0	0	0	0	0	2	5	3	3	5	8
Coccinellidae	24	135	6	23	10	12	12	26	18	68	30	112	72	376
Colydiidae	3	4	0	0	2	3	2	5	3	5	0	0	9	17
Corylophidae	10	46	10	17	6	41	10	271	8	16	5	19	21	410
Cucujidae	0	0	3	4	1	1	3	7	1	4	0	0	6	16
Curculionidae	46	125	25	48	21	52	20	86	12	433	18	122	100	866
Dermestidae	1	1	1	1	1	2	1	3	12	195	7	67	16	269
Elateridae	3	5	1	1	2	6	0	0	6	9	18	63	28	84
Endomychidae	1	3	1	2	1	1	1	1	0	0	1	1	3	8
Erotylidae	0	0	0	0	1	1	0	0	0	0	1	1	2	2
Eucnemidae	0	0	1	1	0	0	0	0	0	0	1	1	2	2
Histeridae	1	1	0	0	0	0	0	0	2	2	0	0	3	3
Lagriidae	0	0	0	0	0	0	0	0	0	0	2	22	2	22
Languriidae	4	6	0	0	1	1	3	3	0	0	0	0	7	10
Lathridiidae	6	56	5	21	4	24	9	46	2	2	6	23	13	172
Leiododae	0	0	0	0	0	0	0	0	1	1	0	0	1	1
Lycidae	2	7	0	0	0	0	0	0	0	0	0	0	2	7
Melandryidae	1	2	0	0	0	0	0	0	0	0	0	0	1	2
Melyridae	1	1	1	1	2	2	3	4	1	5	1	2	7	15
Merophysiidae	2	2	0	0	1	3	0	0	0	0	0	0	2	5
Monommidae	1	1	0	0	0	0	0	0	1	1	1	1	2	3
Mordellidae	2	2	1	1	2	2	3	4	1	1	4	4	9	14
Mycetophagidae	1	1	0	0	0	0	1	10	1	1	0	0	1	12
Mycteridae	0	0	0	0	1	1	1	1	0	0	1	2	2	4
Nitidulidae	5	20	3	5	0	0	13	29	8	15	4	6	26	75

Table 5.4 continued

	A		*B*		*C*		*D*		*E*		*F*		*Total*	
Family	*Sp.*	*No.*	*Sp.*	*No.*	*Sp.*	*No.*	*Sp.*	*No.*	*Sp.*	*No.*	*Sp.*	*No.*	*Sp.*	*No.*
Nosodendridae	0	0	0	0	0	0	0	0	1	11	0	0	1	11
Oedemeridae	0	0	0	0	0	0	0	0	0	0	3	3	3	3
Peltidae	0	0	0	0	0	0	0	0	1	3	1	16	1	19
Phalacridae	12	32	3	3	1	4	2	23	5	34	14	127	23	223
Platypodidae	2	2	1	1	1	2	2	3	1	1	0	0	4	9
Pselaphidae	1	1	1	1	1	1	0	0	1	2	0	0	3	5
Ptiliidae	2	3	4	6	2	6	1	1	3	3	1	1	7	20
Ptinidae	2	2	0	0	1	3	0	0	6	32	5	19	11	56
Rhizophagidae	1	1	1	1	1	1	1	3	2	6	1	1	4	13
Salpingidae	1	10	0	0	1	1	3	3	0	0	0	0	3	14
Scarabaeidae	0	0	0	0	0	0	0	0	0	0	2	2	2	2
Scaphidiidae	0	0	0	0	1	1	0	0	0	0	0	0	1	1
Scirtidae	0	0	0	0	0	0	0	0	0	0	2	3	2	3
Scolytidae	15	27	16	26	15	28	11	24	11	15	6	30	53	150
Scraptiidae	0	0	0	0	1	1	0	0	1	41	3	53	4	95
Scydmaenidae	5	8	9	15	6	8	4	9	0	0	3	17	19	57
Silvanidae	3	17	2	4	0	0	2	6	7	56	0	0	10	83
Smicripidae	1	1	0	0	2	2	1	1	1	3	0	0	3	7
Staphylinidae	27	41	23	52	23	48	17	23	16	39	11	14	90	217
Tenebrionidae	5	18	2	5	5	8	5	18	8	30	5	17	21	96
Throscidae	0	0	0	0	0	0	2	2	0	0	4	8	6	10
Trogossitidae	2	2	1	1	0	0	1	2	4	8	1	2	8	15

(A, *Talisia* sp.; B, *Brownea grandiflora*; C, Polygonaceae sp.; D, *Chrysophyllum lucentifolium*; E, *Cassia grandis*; F, *Mangifera indica*)

than five species. The ranking of the 15 most species-rich families in each of the six trees, and the numbers of species within each family, was highly consistent across the six trees, both within and between sites. This constancy is highly significant (Friedman's method for randomized blocks: $\chi^2 = 47.1$, d.f. = 14, $P < 0.001$; Kendall's coefficient of concordance, W = 0.561, $P < 0.001$).

Using rank correlation, Stork (1993) analysed the similarity of beetle faunas of two very different regions (central America, Erwin and Scott, 1980; south-east Asia, Stork, 1991) at the family level and found that even across continents there was a high degree of constancy in the number of species in each family ($r_s = 0.624$, $n = 79$, $P < 0.001$). Here we have again used rank correlation analysis to compare the beetle faunas from the present study (Table 5.4) with those of the other two studies described above. Surprisingly, the results indicate that there is less familial similarity between the trees at the two relatively close sites of

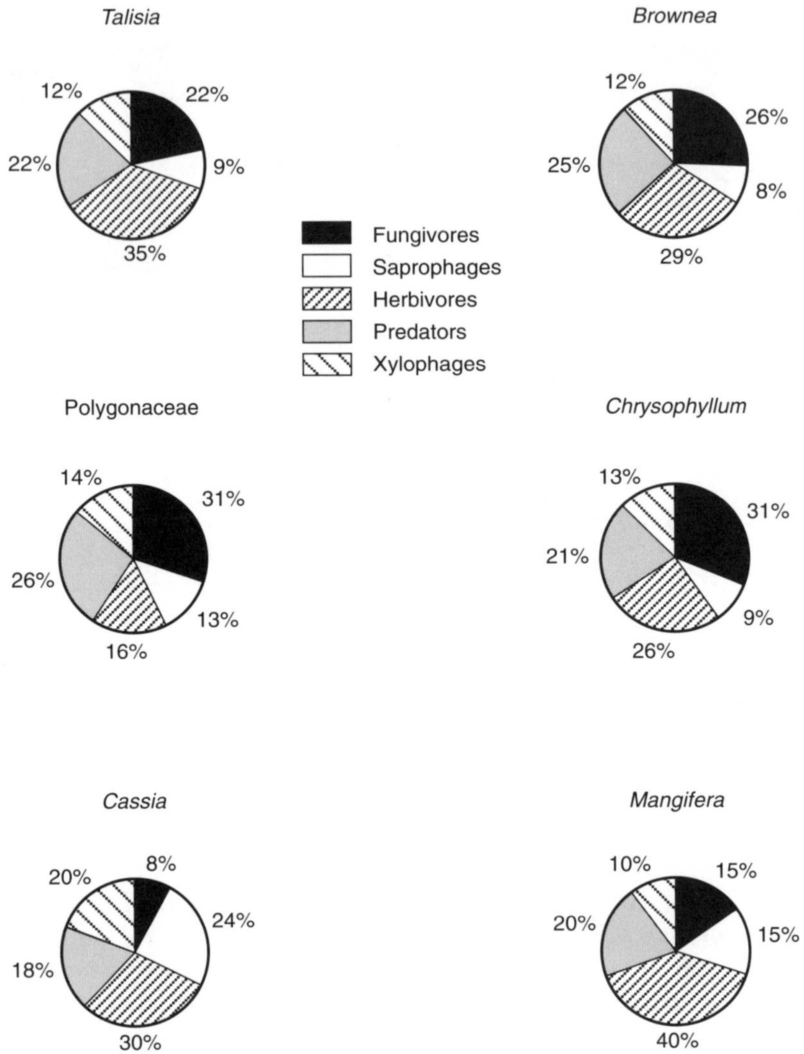

Figure 5.6 The proportion of beetle species belonging to different feeding guilds in the six fogged trees.

Panama and Venezuela than between either of these sites and the Borneo trees, although all comparisons are highly significant (Venezuela–Panama: $r_s = 0.491$, $n = 92$, $P < 0.001$; Venezuela–Borneo: $r_s = 0.558$, $n = 92$, $P < 0.001$). Overall, these comparisons between the faunas of different continents indicate that, at least at this level of organization, there is a taxonomic constancy on a global scale.

Trophic guild structure and constancy

Moran and Southwood (1982) found that there was a remarkable and statistically significant constancy in the proportional representation of some guilds of arthropod species in knockdown samples from different tree species in temperate (Britain) and sub-tropical (South Africa) sites. Stork (1987a) found similar results in arthropod samples from different tropical tree species from Borneo. By contrast, the proportional representation of numbers of individuals of beetles in different guilds varied considerably in both studies.

Similar patterns of guild constancy were observed in the present study (Figures 5.6 and 5.7). The proportional species richness of the feeding guilds remain relatively constant across tree species. In particular, predatory species constitute a relatively constant proportion of about 20% of the fauna in all six trees. Even in fruiting trees the only major variation is an increase in the proportions of herbivores and saprophages at the expense of the relative proportion of fungivores. A much greater variability in the proportional abundance of different guilds was found, with the abundance of herbivores and fungivores (notably Bruchidae, Curculionidae, Apionidae, and Corylophidae) increasing markedly in those trees with fruit or fungi available as a plentiful resource. Thus, in contrast to their poor representation on the two fruiting trees *M. indica* and *C. grandis* (12% and 13% of individuals, respectively), fungivores exhibited markedly higher proportional representation on the two non-fruiting trees Polygonaceae sp. and *C. lucentifolium* (46% and 53% of individuals, respectively). Conversely, herbivores dominated the faunas of the fruiting and flowering trees, but occurred in much lower numbers on the two sterile trees. The high abundance of herbivores on the sterile *Talisia* sp. suggests that this species may have palatable leaves, or is in some other way attractive to phytophages.

Stork (1987a), in a similar study in Borneo, found a constant trophic representation of arthropod species, even between distantly related tree species. However, comparison between the guild structure from the Borneo study with that from this study is precluded by the use of contrasting methods and different trophic categorizations. A tentative comparison with guild data from Sulawesi, Indonesia (Hammond, 1990), for which the same methodology was used, reveals a remarkable level of constancy (Table 5.5). However, statistical analysis reveals that they are not significantly correlated, largely because of the small sample size (Pearson's $r = 0.823$, d.f. = 3, $P = 0.0872$).

The high degree of inter-tree variation in herbivore and fungivore, and to a lesser extent saprophage and xylophage, guild proportions is perhaps because their relative richness is strongly dependent on whether the tree is fruiting, flowering or sterile. Other sources of variation

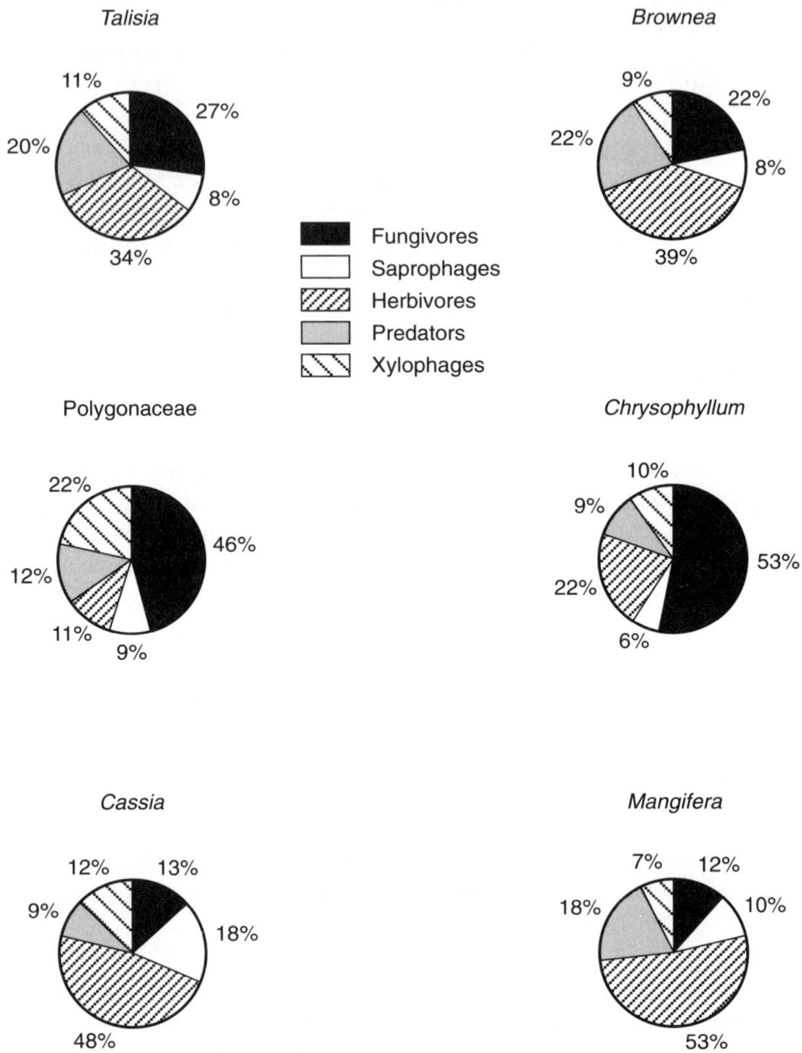

Figure 5.7 The proportion of individual beetles belonging to different feeding guilds in the six fogged trees.

between trees, such as differences in leaf chemistry, presence of deadwood, or presence of fungi, will also affect the proportions of these guilds much more than they affect predators (hence the relative constancy of the latter). This may be because the majority of predators are relatively opportunistic and will feed on most suitable prey, independent of trophic affiliation. Differences in the abundance of fungivores

Table 5.5 Comparison of the trophic guild structure of beetle faunas from the canopies of rainforest trees in Sulawesi (Hammond, 1990) and Venezuela (numbers are percentage species representation)

Region	*Fungivores*	*Saprophages*	*Herbivores*	*Predators*	*Xylophages*
Venezuela	22.2	12.9	29.7	21.9	13.3
Sulawesi	23.4	12.7	24.9	29.0	8.5

may be entirely due to climatic differences between the sites, with higher abundances being found on those trees at the moister closed deciduous forest at La Trilla than on those trees in the drier gallery forest of Pozo del Diablo.

CONCLUSIONS

Studies such as this help us to understand the extent and nature of biodiversity in different habitats. The sheer number of species of beetles collected from the six trees gives us some indication of the magnitude of alpha diversity in rainforest canopies. The low level of similarity between the faunas of trees from different sites suggests that beta diversity is also very high. Further investigations of the relationship between faunal similarity and the distance between trees using individuals of the same species of tree may provide a better understanding of the patterns of spatial distribution of beetle species. Over what distance, for example, does the beetle fauna of a tree species begin to show marked changes? And what are the implications for beta and gamma diversity if there is indeed large-scale variation in species composition between trees of the same species located in different areas? Stork 1988, Richardson *et al.* (1997, Chapter 24, this volume) and N.E. Stork and P.M. Hammond (unpublished results) have made some preliminary investigations of the variation in arthropod faunal complement within and between conspecific trees.

As well as being of great scientific interest, information on diversity and faunal composition, such as that provided in this study, allows us to make prescriptions and priorities for conservation management based on ecological principles. It enables us to identify areas of high biodiversity for priority protection and also provides base-line data for decision-making in the face of human development. For example, an assumption that two areas of apparently similar forest share matching faunas could be used to justify the logging of one of the areas, under the pretext that the loss of biodiversity would be minimal. A better knowledge of the composition of the arthropod faunas of the two sites would help determine whether or not this were indeed the case, and thus help identify the level of threat to regional biodiversity should the development be sanctioned. Although arthropods have largely been overlooked as an important issue in conservation, recent interest in biodiversity is likely to result in an increased emphasis on this enormously diverse group.

Acknowledgements

The authors thank Kate Edwards, Dr Terry Pennington and Dr Alfonso Cardozo for providing identifications of the trees. Jon Davies gratefully acknowledges the support of the University of Bristol/Universidad Central de Venezuela 'Henri Pittier' Expedition (Simon Garrett, Kate Edwards, Tania Roe, Robert Thompson and Catherine Duckett), and the assistance provided by Professor Alberto Fernandez-Badillo and Dr Fransisco Cerda of UCV in Maracay. The financial support provided by the various donors to the expedition is also gratefully acknowledged: in particular, we thank Tioxide Group plc for an additional grant to the Natural History Museum's Development Trust which enabled Jon Davies to complete the study. Finally, considerable thanks are also due to Dr Clive Moncrieff for assistance with the NESS and PCA analyses, to Dr Tim Ferrero and to Harriet Eeley and Raphael Didham for comments made on this manuscript.

REFERENCES

Adis, J. and Schubart, H.O.R. (1985) Ecological research on arthropods in central Amazonian forest ecosystems with recommendations for study procedures, in *Trends in Ecological Research for the 1980s. Nato Conference Series, Series 1: Ecology* (eds J.H. Cooley and F.B. Golley), Plenum Press, New York, London, pp. 111–14.

Adis, J., Lubin, Y.D. and Montgomery, G.G. (1984) Arthropods from the canopy of inundated and terra firme forests near Manaus, Brazil, with critical considerations on the pyrethrum-fogging technique. *Studies on Neotropical Fauna and Environment*, **19**, 223–36.

Adis, J., Paarmann, W., Fonseca C.R.V. da and Rafael, J.A. (1997) Knockdown efficiency of natural pyrethrum and survival rate of arthropods obtained by canopy fogging in Central Amazonia, in *Canopy Arthropods* (eds N.E. Stork, J. Adis and R.K. Didham), Chapman & Hall, London, pp. 67–79 (this volume).

Erwin, T.L. (1982) Tropical forests: their richness in Coleoptera and other arthropod species. *The Coleopterists' Bulletin*, **36**, 74–5.

Erwin, T.L. (1983a) Beetles and other insects of tropical forest canopies at Manaus, Brazil, sampled by insecticidal fogging, in *Tropical Rain Forest: Ecology and Management* (eds S.L Sutton, T.C. Whitmore and A.C. Chadwick), Blackwell Scientific Publications, Oxford, pp. 59–75

Erwin, T.L. (1983b) Tropical forest canopies: the last biotic frontier. *Bulletin of the Entomological Society of America*, **30**, 14–19.

Erwin, T.L. (1988) The tropical forest canopy: the heart of biotic diversity, in *Biodiversity* (eds E.O. Wilson and F.M. Peter), National Academy Press, Washington DC, pp. 123–9.

Erwin, T.L. (1990) Canopy arthropod biodiversity: a chronology of sampling techniques and results. *Revista Peruana de Entomologia*, **32**, 71–7.

Erwin, T.L. and Scott, J.C. (1980) Seasonal and size patterns, trophic structure, and richness of Coleoptera in the tropical arboreal ecosystem: the fauna of the tree *Luehea seemannii* Triana and Planch, in the Canal Zone of Panama. *The Coleopterist's Bulletin*, **34**, 305–22.

Grassle, J.F. and Smith, W. (1976) A similarity measure sensitive to the contribution of rare species, and its use in investigations of variation in marine benthic communities. *Oecologia (Berlin)*, **25**, 643–59.

Hammond, P.M. (1990) Insect abundance and diversity in the Dumoga-Bone National Park, N. Sulawesi, with special reference to the beetle fauna of lowland rain forest in the Toraut region, in *Insects and the Rain Forests of South East Asia (Wallacea)* (eds W.J. Knight and J.D. Holloway), Royal Entomological Society of London, pp. 197–254.

Harada, A.Y. and Adis, J. (1997) The ant fauna of tree canopies in central Amazonia: a first assessment, in *Canopy Arthropods* (eds N.E. Stork, J. Adis and R.K. Didham), Chapman & Hall, London, pp. 382–400 (this volume).

Huber, O. (1986) Las selvas nubladas de Rancho Grande: observaciones sobre su fisionomia, estructura y fenologia, in *La Selva Nublada de Rancho Grande, Parque Nacional 'Henri Pittier'* (ed. O. Huber), Fondo Editorio Acta Cientifica Venezolana, Caracas, Venezuela, pp. 131–70.

Lambshead, P.J.D., Platt, H.M. and Shaw, K.M. (1983) The detection of differences among assemblages of marine benthic species based on an assessment of dominance and diversity. *Journal of Natural History*, **17**, 859–74.

Magurran, A.E. (1988) *Ecological Diversity and its Measurement*, Croom Helm, London.

Moran, V.C. and Southwood, T.R.E. (1982) The guild composition of arthropod communities in trees. *Journal of Animal Ecology*, **59**, 296–306.

Paarmann, W. and Kerck, K. (1997) Advances in using the canopy fogging technique to collect living arthropods from tree crowns, in *Canopy Arthropods* (eds N.E. Stork, J. Adis and R.K. Didham), Chapman & Hall, London, pp. 53–66 (this volume).

Paarmann, W. and Paarmann, P. (1997) Studies on the biology of a canopy-dwelling carabid beetle collected by canopy fogging in the rainforest of Sulawesi (Indonesia), in *Canopy Arthropods* (eds N.E. Stork, J. Adis and R.K. Didham), Chapman & Hall, London, pp. 433–41 (this volume).

Richardson, B.J., Burgin, S., Azarbayjani, F.F. and Lutubula, S. (1997) Distinguishing the woods from the trees, in *Canopy Arthropods* (eds N.E. Stork, J. Adis and R.K. Didham), Chapman & Hall, London, pp. 497–509 (this volume).

Stork, N.E. (1987a) Guild structure of arthropods from Bornean rain forest trees. *Ecological Entomology*, **12**, 69–80.

Stork, N.E. (1987b) Arthropod faunal similarity of Bornean rainforest trees. *Ecological Entomology*, **12**, 219–26.

Stork, N.E. (1988) Insect diversity: fact, fiction or speculation. *Biological Journal of the Linnean Society*, **35**, 321–37.

Stork, N.E. (1991) The composition of the arthropod fauna of Bornean lowland rain forest trees. *Journal of Tropical Ecology*, **7**, 161–80.

Stork, N.E. (1993) How many species are there? *Biodiversity and Conservation*, **2**, 215–32.

Wolda, H. (1979) Abundance and diversity of Homoptera in the canopy of a tropical forest. *Ecological Entomology*, **4**, 181–90.

6

Host-specificity and the effective specialization of tropical canopy beetles

N.A. Mawdsley and N.E. Stork

ABSTRACT

In this chapter we estimate the 'effective specialization' (May, 1990) of the beetle assemblages from a series of four tropical forest tree species. We define host-specificity as the distribution of insects among different host species and discuss the application of this to canopy insect faunas. Our method of estimating effective specialization overcomes the problem of incomplete sampling by comparing the distribution patterns within a series of conspecific trees to that within a series of different tree species. The results indicate that although sampling effects dominate our data, the signal of host-specificity is still present and gives support for the notion of relatively low host-specificity in adult tropical canopy Coleoptera faunas. We also find evidence that the effect of host relatedness on faunal overlap of the beetle herbivore assemblages is primarily at the family level, suggesting that the functional diversity of the tropical tree flora is also at the family taxonomic level.

INTRODUCTION

Several recent attempts to understand the magnitude of global species richness have focused on the association of insects with plants (Erwin, 1982, 1983a; Stork, 1988a; May 1990; Thomas, 1990; Gaston, 1992; for fungi–plant associations, see Hawksworth, 1991). At local and regional scales, insect species richness appears to scale positively with plant species richness (Murdoch *et al.*, 1972; Abbott, 1974; Szentkiralyi and Kozar, 1991; Gaston, 1992). One mechanism that may generate such a

Canopy Arthropods. Edited by N.E. Stork, J. Adis and R.K. Didham. Published in 1997 by Chapman & Hall, London. ISBN 0 412 74900 9

Table 6.1 The taxonomic identities and beetle species richness of 10 Bornean trees fogged near Bukit Sulang, Brunei. (Determinations by Dr Meg Huby)

Tree number (k)	*Family*	*Genus*	*Abbreviation*	*No. of beetle species per tree* (N_k)
1	Dipterocarpaceae	*Shorea johorensis* Foxwood	sj1	130
2	Dipterocarpaceae	*Shorea johorensis* Foxwood	sj2	270
3	Dipterocarpaceae	*Shorea johorensis* Foxwood	sj3	166
4	Dipterocarpaceae	*Shorea johorensis* Foxwood	sj4	141
5	Dipterocarpaceae	*Shorea macrophylla* (De Vriese)	sm1	145
6	Dipterocarpaceae	*Shorea macrophylla* (De Vriese)	sm2	142
7	Anacardaceae	*Pentaspadon motleyi* Hook	pm1	45
8	Anacardaceae	*Pentaspadon motleyi* Hook	pm2	100
9	Fagaceae	*Castanopsis* sp.	cast	103
10	Unidentified*	Unidentified	unid	119

* Unidentified tree, but of different family to other trees

pattern – and one that is widely assumed to be important – is that different plant species have specialized insects (particularly phytophagous insects) not found on other plant species. Consequently, the total insect species richness of an area may be largely dependent upon plant species richness, the number of insect species per host plant species and the level of host-specificity of the insect fauna (Erwin, 1982; May, 1990; Thomas, 1990; Table 6.1).

The most widely cited use of this host-specificity function is that of Erwin (1982) who calculated that there are 30 million species of insects in the world's tropical forests. He suggested that, on average, 13.5% of the 1200 beetle species he had collected from a single tropical tree species (*Luehea seemannii* (Tiliaceae)) may be specific to that particular tree species. Briefly, Erwin classified the beetles into four trophic groups – herbivores, predators, fungivores and scavengers – and assumed that the proportion of species specific to a single tree was 20, 5, 10 and 5% respectively. This gave a weighted average of 13.5% of beetle species being host-specific to this tree species, which Erwin (1982) then used to scale up to his widely cited global estimate of 30 million insect species in tropical rainforests.

The highly generalized and subjective assumption about the degree of host-specificity in tropical insect faunas forms one of the four key 'areas of uncertainty' (May, 1990) associated with Erwin's estimate (see also Stork, 1988a, 1993, 1995; Hodkinson and Casson, 1991). Clearly, there is marked overlap in the insect faunas of different plant species (Futuyma and Gould, 1979; Leather, 1985; see also Strong *et al.*, 1984), but despite these studies Erwin's estimated level of host-specificity has yet to be empirically tested for insects associated with tropical rainforest trees (Basset, 1992). Furthermore, there are several hypotheses accounting for both lower and higher host-specificity in the tropics (Janzen, 1973; Marquis and Braker, 1993), but the empirical evidence, in general, points towards lower host-specificity in tropical herbivorous insects (Beaver, 1979; Gauld, 1986; Basset, 1992; Marquis and Braker, 1993; see Discussion). Here we provide an initial attempt at assessing the host-specificity of a tropical canopy fauna using the notion of 'effective specialization' (May, 1990) in the Coleoptera associated with five tree species in an area of primary Bornean rainforest.

In this chapter we present findings that generally support the notion that tropical insect species are less specific to their host-trees than Erwin (1982) assumed. First, we discuss the phenomenon of host-specificity and detail means of estimating its magnitude for insects sampled from a number of tree species. There are many problems associated with estimating the host-specificity of insects in forests (see also Futuyma and Gould, 1979) and we discuss these in the next section. Secondly, we present the frequency distribution of the number of host-trees utilized by beetle species sampled from series of different rainforest tree species and compare it with that for beetles from conspecifics of the same rainforest tree species. Thirdly, we use these data to calculate the effective specialization (May, 1990) for the beetle fauna of one tree species, correcting for sampling effects, and compare this with Erwin's hypothesis. Finally, we analyse the effects of taxonomic relatedness on the faunal overlap between pairs of trees (Ehrlich and Raven, 1964), and ask at which level in the floral taxonomic hierarchy (i.e. species, genera or family) are the effects of host-specificity most prevalent.

HOST-SPECIFICITY, EFFECTIVE SPECIALIZATION AND THE DISTRIBUTION OF INSECTS ON PLANTS

Definitions and concepts

Definitions of host-specificity in herbivorous insects usually emphasize the taxonomic range of plants upon which a herbivore feeds and focus on the fitness benefits of being associated with particular host species (Janzen, 1973; Ward, 1992). Host-specificity is a property of both individuals and species, such that the host range of a species will be greater than that of any individual. Analysis of species lists of insects on host plant species inevitably concerns host-specificity as a population attribute.

The host-specificity of herbivorous insects is theoretically dependent upon features of the plant, of the plant community and its habitat (Courtney and Chew, 1987), of the insects themselves and of the degree of intimacy of the plant–insect association (see reviews by Strong *et al.*, 1984; Bernays and Graham, 1988; Basset, 1992). The mechanisms determining host-specificity in insects, such as the relative roles of plant chemistry and ecological factors have been the subject of controversy (Bernays and Graham, 1988; Basset, 1992). Given the multiplicity of factors, both intrinsic and extrinsic to the insect–plant interaction, that may potentially act to restrict the breadth of insect diets and distribution (Gilbert and Smiley, 1978; Strong *et al.*, 1984), it is not surprising that the actual host range of a herbivore is a subset of the plant species upon which it can potentially feed, as indicated by feeding trials (Gilbert and Smiley, 1978; Janzen, 1985a). In essence this is the same distinction as Hutchinson's (1957) fundamental- and realized-niches (Futuyma and Moreno, 1988).

In addition to herbivorous insects that use a particular tree species in some way for successful growth and reproduction, Erwin (1982) extended the idea of specificity to include species in other trophic groups (e.g. fungivores, scavengers, predators) that are tied in some way to the host-plant and/or its herbivores. Some insects, for example, use plants for mating purposes (Corbet, 1961), avoidance of unfavourable ground conditions, and avoidance of predators (see Stork, 1988b). In the extreme, these species may come under the term of 'tourist', for during their transient association with a 'host' plant they obtain little nutritional benefits (Gaston *et al.*, 1993). Tourist species are likely to be less host-specific than other insects, especially phytophagous ones, and their inclusion in the concept of host-specificity will lead to lower levels of host-specificity in whole faunas.

We take a generalized view that defines host-specificity as the *distribution* of insects among different hosts in their natural habitats, as opposed to feeding-specificity. For phytophagous insects, such a definition indicates a more restricted distribution than that derived from analysis of feeding preferences, and therefore greater host-specificity. However, feeding preferences can only apply to herbivorous insects, whereas our distributional notion of host-specificity can in theory be applied to all plant-associated taxa, from the most intimate leaf-miner through to the most ephemeral of associations such as the perching of a dragonfly. Thus, level of host-specificity of an entire insect fauna is predicted to be less than that of the herbivores.

Current understanding of host-specificity, which is inevitably biased to temperate regions, suggests that most insects appear to feed on a restricted range of plants, thereby supporting a view of relatively high host-specificity (Strong *et al.*, 1984; Claridge, 1987; Ward and Spalding, 1993; but see Janzen, 1988; Marquis, 1991; Marquis and Braker, 1993). Care must be taken not to overgeneralize, because patterns of host-specificity are variable in space (see Thomas, 1990; Leather, 1991), time and taxonomic group (Gilbert and Smiley, 1978; Marquis and Braker, 1993; Ward and Spalding, 1993). However, almost nothing is known about the host-specificity of the insect faunas of tropical rainforest tree species, in part due to the methodological difficulties in studying their faunas (Basset, 1992).

Estimation and measurement: effective specialization

Measuring the host-specificity of an insect fauna has similar inherent problems to measuring species diversity (Magurran, 1988; Hayek, 1993) in that an insect that utilizes two host plants may be equally distributed on each, or have predominantly more individuals on one or other species (Janzen, 1973). Furthermore, some insects are specific to certain plant species, a number are found on two plant species, others on three species, and so on. This means that any measure of host-specificity must inversely weight species in accordance with the number of hosts on which they are found. A final problem is that of different levels of host-specificity associated with different taxonomic levels of plants; some species of insect are restricted to a single plant species, others to a plant genus and others to a plant family (Leather, 1985). These levels of host-specificity (specific, generic and familial) reflect the role of plant taxonomy and co-evolution between insects and plants (Ehrlich and Raven, 1964). In this chapter we use presence–absence data which assume that insect abundance is equally distributed among species, and which may underestimate the true level of host-specificity.

May (1990) describes a simple method of calculating the host-specificity or effective specialization, f_k, of the insect fauna of a particular tree species. The central element of this method is the frequency distribution of the proportion of insect species ($p_k(i)$) of a particular tree species k (where k = 1, 2....M) that are also found on i other tree species (where i = 1,2...M). An insect that is shared between two tree species (i.e. i = 2) is deemed to be only half as specialized as an insect restricted to a single host species, and so its contribution to the effective specialization is weighted by 0.5 (i.e. $1/i$). At the limits, all the insect species of a tree may be unique to it, and so the effective specialization is 1.0. At the other extreme, each tree species may support identical insect faunas, and the effective specialization becomes $1/M$. The effective specialization of a given tree species, f_k, that is, the proportion of insect species on a host that are on average host-specific, is therefore given by (May, 1990):

$$f_k = \sum_{i=1}^{M} (1/i)p_k(i). \tag{1}$$

If there are N_k insect species on each tree, then the number of species effectively specialized to each tree species, N_f, is given by:

$$N_f = N_k \cdot f_k \tag{2}$$

Given complete faunal lists (see Ward, 1988) for a series of different tree species, we can calculate an average effective specialization, F, as:

$$F = \frac{\sum N_f}{\sum N_k} \tag{3}$$

Erwin's theory may then be interpreted in terms of this average effective specialization, F. An empirically derived value of F should be compared with the 13.5% suggested by Erwin's hypothesis.

Erwin's assumption that 13.5% of species are 'effectively specialized' to a single species of tree is actually extremely difficult to directly test (Gaston, 1993; Moran *et al.*, 1994). In addition to the problems of measurement outlined above, consideration of host-specificity requires attention to taxonomic (i.e. host-plant variety and phylogenetic relationships), temporal and spatial scale (Fox and Morrow, 1981; May, 1994). If monitored over a long time period, a given insect species would be recorded from an increasing number of plant species; that is, it would tend to be viewed as less host-specific over longer periods of time. Spatial scale, in terms of the number and range of plants and habitats used for comparative purposes would also affect the measurement of host-specificity (Thomas, 1990; Leather, 1991). If the range of plant species were from a single family, then their associated insects might reasonably be expected to be found on a greater proportion of species than if each plant species was a representative of a different family (Leather, 1991). Thus, any estimate of a 'global' level of host-specificity will be a large and somewhat tenuous extrapolation from a necessarily restricted local situation.

SAMPLING PROBLEMS AND MEASUREMENT OF HOST-SPECIFICITY

Generally, medium-term studies (5–10 years) may be necessary to characterize properly the insect faunas of different tree species in a tropical forest (see Janzen, 1988). The logistics and the resources required for such studies mean that they are rarely feasible. Relatively cheap, short-term studies may also be able to give initial indications of host-specificity, especially if calibrated with longer-term projects. Insecticide fogging is a common and convenient way of sampling local canopy insect faunas, it may be replicated across a relatively large number of trees species in a short period of time, and it is thought to give adequate data for comparing the relative species richness of arboreal faunas (Southwood *et al.*, 1982).

Short–term sampling introduces at least two major problems for estimating the absolute level of host-specificity in an assemblage. First, differences between successive samples may be generated as a function of sample size even when sampling the same statistical population (Koch, 1987; McArdle, 1990). These sampling-dependent differences in community composition, known as pseudoturnover, are a common problem in community ecology (Nilsson and Nilsson, 1985), and their importance scales inversely with sampling effort. The effect of sampling on f_k and $p_k(i)$ is likely to be profound, especially in tropical communities where many species are rare and large samples are required to fully inventory a particular (micro)habitat, although nothing is known about this important aspect of assessing effective specialization (May, 1990, 1994).

Second, most insect sampling techniques commonly used in tropical biology (e.g. canopy fogging, flight interception trapping and light-trapping) usually only sample adult insects. Host-plant associations for other life stages (egg, larva, pupa) are not recorded. It is not possible to tell whether a species collected by these mass-collecting techniques is actually associated with a particular host or is simply a 'tourist' (*sensu* Moran and Southwood, 1982). For this reason it is essential to state explicitly that such samples derive information essentially about host-effects on the local distribution of adults rather than host utilization *per se*. This accords with our notion of host-specificity and does not affect the value of our analysis because the distribution of insects among different host-plants is the phenomenon of interest. Consideration of larval stages will simply uncover a different picture which may be interpreted in its own right. In the following sections we investigate the distribution and effective specialization of adult Coleoptera on five species of tropical rainforest trees in Borneo.

METHODS

Sampling design

The sampling design for the present study requires a series of samples from trees of the same species to estimate the effect of sampling pseudoturnover (PS) coupled with inter–tree variability (V) in faunal composition that is independent of host identity (see, for example, McGarvin *et al.*, 1986; Thomas, 1987), and a series of samples from trees of different species which would be affected by the additional factor of host-specificity. The ideal situation would be a randomized block design with tree species as treatments within blocks. For each tree species we could calculate true host-specificity (HS) using the following scheme:

conspecific turnover	=	PS + V
between-host turnover	=	PS + V + HS
host-specificity	=	HS

Such data are presently unavailable. Instead, this study uses samples taken in Brunei Darussalam, north Borneo, from 10 trees representing five tree species in four genera and four different families (Table 6.1), as described in detail by Stork (1987a,b, 1991). Constraints on time in the field meant that our data-set is in fact akin to an incomplete block design, with four replicate samples taken from only one species of tree, *Shorea johorensis* Foxwood, and two replicate samples being taken from two other tree species, *Shorea macrophylla* and *Pentaspadon motleyi*.

In our data-set, we have a maximum of four conspecific trees sampled. This is the maximum number of trees among which we can therefore compare the effects of host-specificity and sampling pseudoturnover. This is not entirely satisfactory because it constrains values of proportional occurrence, that is the proportion of the four trees on which a species is found ($1/i$), to 0.25, 0.5, 0.75 and 1. Moreover, the nature of the data further constrains the analysis because for a sample of four trees of different species, we cannot select four trees from the same genus, nor four trees from different genera in the same plant family, but can only have species from different families (see Tables 6.1 and 6.2). Therefore for the estimation of effective specialization we excluded the two *S. macrophylla* trees. Thus, our analysis primarily considers the effect of host-specificity at the host-plant family level. The influence of taxonomic relatedness (generic–familial–suprafamilial) on the relative strengths of host-specific effects is analysed in the final part of this chapter.

Location and collecting methodology

Tree-crowns were sampled within an area of alluvial lowland forest at Bukit Sulang in the Ladang Hills Forest Reserve of Brunei, north Borneo, between 29 August and 9 September 1982. The 10 trees were separated by no more than 1 km and were sampled using synthetic pyrethroid knockdown insecticide (Reslin E) released from a Swingfog fogging machine suspended from rope and pulley systems within the canopy of

Table 6.2 Schematic diagram showing the combinations of trees used in the determination of beetle species occurrences and effective specialization on sets of four trees. Only one combination of conspecific trees was possible (combination 1), with eight permutations of tree species in different families being used (combinations 2–9). The mean specialization of beetle species among trees of different families is used in this paper. (For tree abbreviations see Table 6.1)

Plant taxonomic relatedness	*Combination*	*Tree*							
		sj1	*sj2*	*sj3*	*sj4*	*pm1*	*pm2*	*cast*	*unid*
Conspecific	1	+	+	+	+				
Different families	2	+				+		+	+
	3		+			+		+	+
	4			+		+		+	+
	5				+	+		+	+
	6	+					+	+	+
	7		+				+	+	+
	8			+			+	+	+
	9				+		+	+	+

each tree. Sampling was carried out at dawn when there was almost no wind and the insecticidal fog produced from the exhaust of the fogger could rise through the canopy of the tree. The falling arthropods were collected on two plastic sheets totalling 20 m^2 in area placed under each tree being sampled. The numbers of arthropods falling declines considerably 1 hour after fogging and a 2-hour drop-time was allowed before the arthropods were gently brushed from the sheets into alcohol. The arthropods were sorted to order and species by specialists at The Natural History Museum in London (see Stork 1987a,b, 1991 for details of sampling and sorting). A total of 859 species of Coleoptera were represented in these samples and these data are used in the analyses presented here.

Data analysis

The patterns of species distribution (number of host-trees per insect species) analysed for the four *S. johorensis* trees were termed the 'conspecific' data (Table 6.2). The same analyses were carried out for series of four trees in each of eight possible permutations using data from the four non-dipterocarp trees in addition to the *S. johorensis* data. These eight four-tree series, with each tree representing a different taxonomic plant family, were termed the 'different family' combinations. This provided a total of nine combinations of four-tree series which are detailed in Table 6.2; eight four-tree combinations of trees in different plant families, and one combination of the four conspecific trees.

Species occurrence among hosts

The simplest measure of host-specificity is the number of host-trees used by a beetle species. Using series of four hosts, as described above, we calculated the number of host trees used by the 859 species of beetles, both for the series of four conspecific trees (Table 6.2, combination 1) and for each of the eight series of host-trees representing different families (Table 6.2, combinations 2–9). We compared their frequency distributions, using the average for combinations 2–9, and tested the null hypothesis of no difference between the two distributions using log-linear analysis and the Mann–Whitney *U*-test.

Estimating effective specialization: overcoming sampling artefacts

The notion of effective specialization refers to the number or proportion of species on a given individual tree or species of tree that are, on average, host-specific. Here we are looking specifically at the level of the individual tree because we are trying to separate the effects of within-tree species sampling from the effects of host-specificity.

For each tree, we calculated its effective specialization using the formula in Equation (6.1) (May, 1990). We did this for each of the nine four-tree combinations in Table 6.2. Full details of these calculations are given in Appendix 6A.

For each of the *S. johorensis* trees we compared the distributions of $p_k(i)$ values for the conspecific data-series and for an average of the eight different-family series. Differences between the distributions of the different-family and conspecific series were tested using Mann–Whitney *U*-tests. The null hypothesis was no difference between different-family and conspecific series, which would indicate that the effect of host-specificity in these samples was relatively negligible. Consequently, differences in the magnitude of effective specialization for a series of different tree species relative to that between a series of conspecific trees is the result of host-specificity. Such an approach is in the spirit of analysis of variance (ANOVA) where it is necessary to compare the variance within groups with the variance between groups.

Plant taxonomy and faunal similarity between trees

Finally, the effect of taxonomic relatedness on pair-wise faunal similarity between all 10 trees was analysed using the Morisita–Horn index (see

Wolda, 1981; Stork, 1987b). This was calculated separately for the whole Coleoptera fauna and for herbivorous Coleoptera. These pair-wise similarity measures were grouped into three levels of taxonomic relatedness: (i) those between conspecific trees; (ii) those between species of the same genus and family; and (iii) those between trees in different families. It was expected that host-specific effects would be greater for herbivores than for all trophic groups combined.

These data suffer from a lack of statistical independence because the species lists from the 10 trees are used to generate the 45 measures of pair-wise faunal similarity. This is likely to cause a type II statistical error; that is, differences in faunal similarity between the three levels of tree taxonomic relatedness will be reduced. We elected to use a parametric test with reduced power using only N–2 degrees of freedom, where N is the number of trees used to calculate faunal overlap in each test (Harrison *et al.*, 1992; Mawdsley, 1994). We used *t*-tests on log-transformed data to assess the effect of taxonomic relatedness on faunal overlap. Although appropriate, no attempt was made to incorporate the effect of tree phylogeny on this analysis (Felsenstein, 1985; Harvey and Pagel, 1991).

RESULTS

Frequency distribution of species occurrence between host trees

The frequency distribution of the occurrence of Coleoptera species between different tree species (Figure 6.1; Table 6.3) shows that, on average, 82% of species are restricted to a single tree species, 12% are found on two species, and only 6% occur on three or all four tree species in each combination. This type of distribution, showing few widespread species and many rare species, is typical of canopy fogging samples and indeed of samples from many other collecting methods. In the absence of any information on the distribution of insect species on conspecific trees, this result would lead to the erroneous conclusion that faunal overlap between tree species is low and host-specificity high (Erwin, 1983b, 1988, 1991; Farrell and Erwin, 1988; Casson and Hodkinson, 1991).

In fact, the within-tree species data show a similarly high proportion of species, 78%, restricted to just a single tree (Figure 6.1; Table 6.3). If the assumption that conspecific trees at the same site actually support the same fauna is valid, then we can conclude that studies of the distribution of insect species on trees, using the fogging methods described here, are dominated by sampling effects.

Results of a log–linear analysis for independence between species occurrence and the taxonomic relatedness of trees (conspecific versus different-families) show that there is no significant difference between

Table 6.3 Comparison of the frequency distribution of number of host trees for beetle species sampled from a series of four conspecific trees and four trees from different plant families. The latter is an average for the eight 'different-family' combinations detailed in Table 6.2

Species occurrence class (no. of trees)	*Conspecific trees (Shorea johorensis) Combination 1*	*Between-family trees (different families) Mean of combinations 2–9*
1	421	311
2	76	46
3	30	17
4	11	5
Total	538	379

conspecific trees and trees of different families in the number of host-trees on which beetle species occur (independent model: taxonomic status and species occurrence, $\chi^2 = 2.28$, 3 d.f., $P = 0.515$). Taxonomic status and species occurrence can independently account for most of the variance in the data, such that there is no reason to believe there is any difference between the frequency distributions of the occurrence of beetle

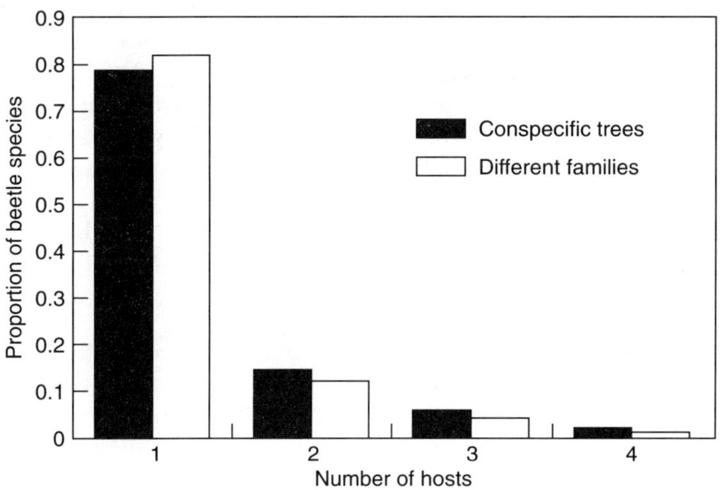

Figure 6.1 Histogram showing the frequency distribution of the number of host trees used by beetle species in fogging samples from trees in Brunei. Trees are divided into four-tree series representing conspecific trees and trees of different taxonomic families (see text for details). Frequency is given as the proportion of beetle species in each class.

species on trees of the same species and on trees of different families. A Mann–Whitney *U*-test of the number of hosts used by species in the conspecific series and the different-family series also shows no significant difference (U = 97 946, W = 169 956, Z = –1.45, P = 0.147, two–tailed test). This gives an initial indication that the effect of tree taxonomy on patterns of beetle species distribution between trees is relatively low in comparison with sampling effects.

The effective specialization of tropical forest Coleoptera

Full calculation of effective specialization for individual trees within each combination is given in Appendix 6A. The four *S. johorensis* trees in the conspecific series (combination 1) have estimates of effective specialization for each tree ranging from 0.71–0.81. When combined with trees from different Families, the effective specialization of the four *S. johorensis* trees has a range of 0.84–0.87. Again, we see that the absolute magnitude of these estimates is high, but that much of this is likely to be sampling effects as indicated by the high value for series of conspecific trees.

Figure 6.2 compares the frequency distribution of beetle species occurrence on host trees measured proportionally (i.e. $p_k(i)$) for the four *S. johorensis* trees. For the faunas of each tree, most species occur on a single host tree (Figure 6.2). However, unlike Figure 6.1, there is a significant trend for beetle species on conspecific trees to be more widely distributed than between trees of different families (see Figure 6.2 and Table 6.4). In short, we can see that host-plant identity and host-specificity are having a significant effect on the distribution of canopy beetle species. The absolute difference between conspecific and different-family combinations defines the magnitude of host-specificity for the fauna of these trees.

We have used the data depicted in Figure 6.2 to calculate the effective specialization of each tree using Equation (6.1) (see Appendix 6A).

Table 6.4 Results of the Mann-Whitney *U*-tests for differences between the proportional occurrence among host trees of beetle species sampled from the canopies of each individual *S. johorensis* tree as seen in Figure 6.2

Tree	*No. of beetle species sampled* (N_k)	*U*	*W*	*Z*	*P (two tails)*
S. johorensis 1	130	6 798	18 617	–3.17	0.002
S. johorensis 2	270	32 890.5	76 594.5	–2.44	0.015
S. johorensis 3	166	11 700.5	29 716.5	–2.87	0.004
S. johorensis 4	141	7 751	22 141	–3.81	<0.001

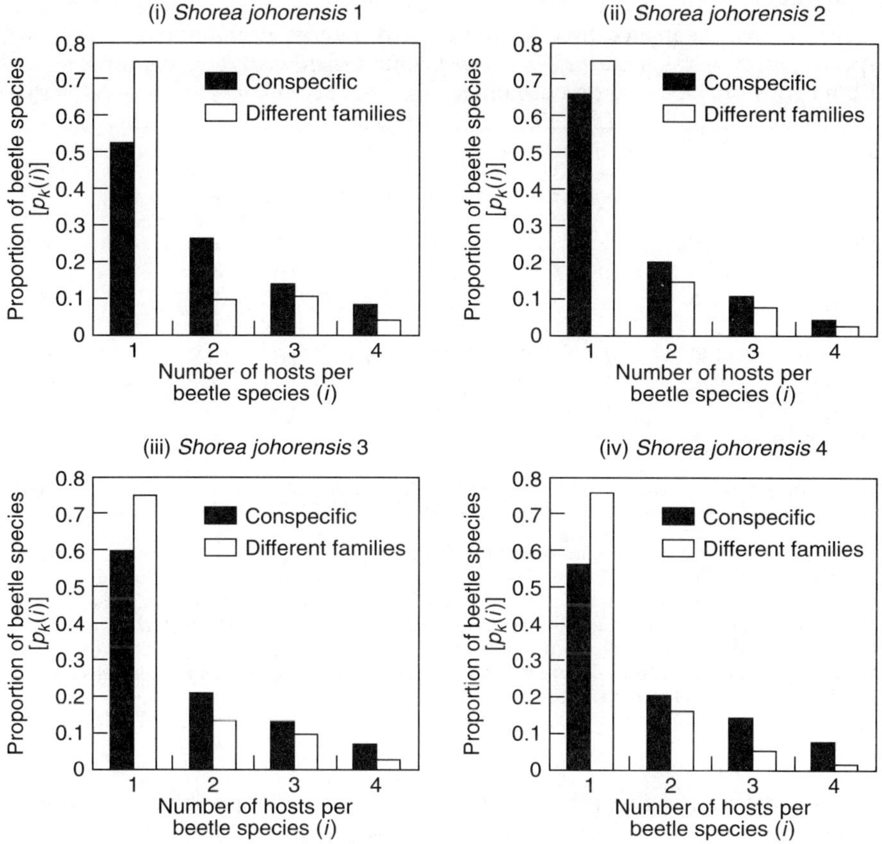

Figure 6.2 Histograms showing the frequency distribution of the parameter $p_k(i)$ which defines the effective specialization (f_k) of the fauna of a given tree. Each of the four plots represents one of four different *Shorea johorensis* trees, and shows results for a four-tree series representing conspecific trees and trees of different taxonomic families (see text for details). Frequency is given as the proportion of beetle species in each class.

For the four *S. johorensis* trees we present estimates of the proportion (f_k) of the fauna and number (N_f) of beetle species that are effectively specialized to each tree (Table 6.5). We have calculated this for each tree when in a conspecific series and for each tree in a series of trees of different families. Each *S. johorensis* tree is represented twice because there are two 'different-family' combinations for each tree (see Tables 6.1 and 6.2).

The first row in Table 6.5 shows the estimate for *S. johorensis* 1. Of the 130 beetle species sampled from this tree, 93 of these are effectively

Table 6.5 Effective specialization of the beetle faunas of each of four *Shorea johorensis* trees estimated by calculating the difference in the number of species effectively specialized in a conspecific series relative to the number from series of trees of different families

Σ		*Conspecific series*			*Different families*				
	spp (N_k)	*comb*	f_k*con*	N_f*con*	*comb*	f_k*dif*	N_f*dif*	ΔN_f	f_k*real*
shorjoh 1	130	1	0.71	93	2	0.86	112	19	0.149
shorjoh 2	270	1	0.81	216	3	0.87	234	18	0.068
shorjoh 3	166	1	0.76	126	4	0.85	142	16	0.096
shorjoh 4	141	1	0.74	104	5	0.87	123	19	0.137
shorjoh 1	130	1	0.71	93	6	0.84	109	16	0.127
shorjoh 2	270	1	0.81	216	7	0.84	227	11	0.04
shorjoh 3	166	1	0.76	126	8	0.85	141	16	0.094
shorjoh 4	141	1	0.74	104	9	0.86	121	18	0.124
Average effective specialization, $F = \Sigma\,(\Delta N_f)/\Sigma N_k$									0.104

Abbreviations: comb, combination number from Table 6.2; spp (N_k), number of beetle species per tree; $f_k$con, effective specialization of the beetle fauna of a tree within a conspecific series; $N_k$con, number of beetle species effectivly specialized to a tree within a conspecific series; $f_k$dif, effective specialization of fauna of a tree within a series of trees of different families; $N_k$dif, number of beetle species effectively specialized to a tree within a series of trees different families; ΔN_f,($N_f$dif – $N_f$con); $f_k$real, true effective specialization corrected for sampling effects = $\Delta N_f/N_k$.

specialized relative to other conspecifics. These 93 species are assumed to be specific to individual tree-crowns as a result of sampling pseudo-turnover and natural variation in faunal composition within a host tree species. When *S. johorensis* 1 is combined in a series with trees of different families, as in combination 2 (Table 6.2), then there are 112 species effectively specialized to it. The difference between 112 and 93, that is 19, is the actual number of species that are effectively specialized to this tree in combination 2. Dividing by total beetle species richness on that tree (130) gives an effective specialization of 0.149, or approximately 15%.

The number of species effectively specialized to a particular *S. johorensis* tree is significantly greater in the combinations with trees of different families (paired *t*-test on log-transformed data of conspecific N_f and mean of different families N_f: $t = 5.19$, 3 d.f., $P = 0.014$). The values of f_k span a range from 4% to 15%, and the mean effective specialization for this data-set is calculated at approximately 10%. We can conclude that relative to the four tree species analysed, about 10% of the species on *S. johorensis* are effectively specialized to it.

The effect of taxonomic relatedness on faunal similarity between trees

We have shown that there is a detectable effect of host-specificity on the distribution of insects on tropical trees, but that its magnitude is relatively small. In this section we ask at what level in the floral taxonomic hierarchy, that is from species to genus to family, is this effect most pronounced. We test the null hypothesis that the faunal overlap of pair-wise combinations of trees is no different for pairs of trees of the same species, trees of different genera and trees of different families. The results for all species of Coleoptera (Table 6.6) show a non-significant (Table 6.7) decline in pair-wise similarity of trees of decreasing taxonomic relatedness. The statistical problem of non-independence means that we can only cautiously conclude that taxonomic relatedness has no effect on the overlap of the total Coleoptera fauna between these tree species. Similarly, Futuyma and Gould (1979) found that host taxonomic

Table 6.6 Values of average percent faunal similarity (Morisita–Horn index × 100) of total Coleoptera and herbivorous Coleoptera faunas, respectively, between Bornean rainforest trees of varying taxonomic relatedness

Trees of:	*All Coleoptera*	*Herbivorous Coleoptera*
Same species	47.3	30.8
Different species, same genus	44.9	33.5
Different families	40.4	21.7

Table 6.7 Results of t-tests for differences in the average faunal similarity (Morisita–Horn index) of total Coleoptera and herbivorous Coleoptera faunas (data in Table 6.6) of rainforest trees of varying taxonomic relatedness. Faunal similarity is significantly different only when trees from different families are included. As a result of non-independence of similarity data (see text), the number of degrees of freedom for each test are given by N–2, where N = the number of independent samples (i.e. trees) that make up the dataset for each test

Taxonomic relatedness		*All Coleoptera*			*Herbivorous Coleoptera*		
Treatment 1	*Treatment 2*	*t-value*	*d.f.*	*P*	*t-value*	*d.f.*	*P*
Same species	Different families	1.52	8	n.s.	2.72	8	0.05
Same species	Different species	0.39	6	n.s.	0.14	6	n.s.
Different species	Different families	1.17	8	n.s.	2.59	8	0.05

n.s., $P > 0.05$.

relatedness had little effect on the overlap of the leaf-chewing guild of insect larvae between tree species in a temperate broadleaf forest.

For the herbivorous beetles in these samples, the picture is very different (Tables 6.6 and 6.7). Herbivorous beetles have a considerably lower faunal similarity between individual tree-crowns than the total beetle fauna. Conspecific and congeneric trees have a faunal overlap of about 30% (Table 6.6), whereas for trees of different families this figure falls to 21.7%. *t*-tests in Table 6.7 show that the difference in faunal overlap is only significant for trees of different families. This suggests that for these tropical forest trees, host-specific effects in the distribution of herbivorous beetles are seen predominantly between trees of different families. For these herbivorous beetles at least, the host diversity of their tropical forest habitat may be best characterized by tree familial diversity rather than species diversity.

DISCUSSION

Host-specificity in tropical canopy insect faunas

We have shown that it is possible to estimate the extent and magnitude of host-specificity in canopy-associated tropical insects using fogging samples that give a snap-shot picture of the distribution of plant-associated insects. We estimate that between trees of different families, about 10% of adult Coleoptera species in canopy fogging samples are effectively specialized to a single host-tree species. These data are limited, however, in that they document adult distribution, are qualitative and do not take account of relative abundance, and are based on comparisons between just four trees of unknown status. This estimate of effective specialization may well reflect these influences, and so be low relative to that of the quantitative estimates using larvae and adults for whole insect faunas.

Nevertheless, this result is in accordance with a growing number of studies that suggest host-specificity in tropical forests faunas is lower than previously thought (Beaver, 1979; Wood and Olmstead, 1984; Gauld, 1986; Holloway, 1989; Thomas, 1990; Marquis, 1991; Price, 1991; Basset, 1992; Marquis and Braker, 1993; Wood, 1993). In comparison with Erwin's hypothesis of an average of 13.5% of the fauna specialized between host tree *species*, our estimate of 10% between host tree *families* does indicate much lower levels of host-specificity than Erwin suggested. Moreover, we only found host-specificity to have an effect at the family level in our data-set (Table 6.7).

Absolute levels of host-specificity vary between insect taxa, and several studies have shown that Coleoptera tend to be less host-specific than many other groups such as the Hemiptera (Hodkinson and Casson,

1991; Ward and Spalding, 1993; Fensham, 1994). Thus, our low estimate of host-specificity may in turn reflect the taxonomic restriction of our study. Further analysis across all taxa is required to better understand the host-specificity of the entire fauna.

The generalization that tropical insects are less host-specific than temperate species (Futuyma and Gould, 1979; Basset, 1992) is also unlikely to hold across all taxa, for there is limited evidence that some groups, such as butterflies, which use primarily non-woody vegetation tend to be more host-specific in tropical localities (Marquis and Braker, 1993). The nature of a plant's defences and their effects on herbivore community structure depend upon host apparancy (Feeny, 1976) and will differ between trees and herbs (see also Strong *et al.*, 1984; Basset, 1992). Herbivore pressure on herbs (including vines) may select for qualitative chemical defences, whereas tree foliage may be better defended by quantitative defences such as leaf toughness or polyphenols. Consequently, we would expect canopy insects to have low host-specificity as compared with insects on herbaceous plants (Janzen, 1985b; Basset, 1992).

Explaining tropical insect host-specificity: the role of plant relatedness

Potential explanatory factors for host-specificity in tropical arboreal insect assemblages have been reviewed by Basset (1992), who concluded that plant chemistry, phenology and apparancy, coupled with nutrient availability and the abundance of predators, may all affect the host-specificity of rainforest canopy insects in complex, and as yet undiscovered, ways. Similar hypotheses concerning resource fragmentation, seasonality, high predation rates and the effects of allelochemicals also exist to explain patterns in the host-specificity and species richness of insect parasitoids (Gauld *et al.*, 1992; Hawkins *et al.*, 1992).

Further to these explanations, we suggest that plant relatedness may also produce lower levels of host-specificity in tropical as compared with temperate canopy insects. In this study the effect of host-specificity on faunal similarity of herbivorous beetles is essentially confined to the family level (Table 6.7). In a tropical forest with over 200 tree species per hectare, but only 100–150 genera and 50 families (Whitmore, 1984; Newberry *et al.*, 1992; Campbell, 1994), tree species are on average more closely related than in a temperate forest (Table 6.8). In a temperate forest there may be only two tree species per family, but in a tropical forest, the figure may be closer to nine. The taxonomic range of hosts present in a hectare of tropical forest may not actually be proportional to its higher tree species richness. Instead, the 'gradient' of host-types may be more finely divided. Thus, for trees, host plant diversity in a tropical forest is essentially represented at the family level, whereas in

Table 6.8 Tree taxonomic richness at species, genus and family levels in tropical and temperate forest assemblages

Taxonomic level	*Tropical**	*Temperate*[+]	*Tropical:temperate ratio*
Family	59	10	6
Genus	164	15	11
Species	511	20	26

* From Newberry *et al.* (1992): inventory of two 4-ha plots at Danum Valley, Sabah; [+] data are from a species-rich local area of woodland in southern England (N.A. Mawdsley, unpublished data).

temperate forests host diversity is greatest at the species level. Therefore, we would expect tropical herbivores to, on average, feed on more host species, without necessarily being any more host-specific at higher taxonomic levels.

Effects of sampling on the estimation of host-specificity and ecological distribution

The problems associated with collecting samples from a highly speciose biota whose constituent species are found at very low densities are well known (Bullock, 1971). Typically, insufficient sampling effort is expended for the species accumulation–sampling effort curve to reach an asymptote, leading to difficulties in assessing the total number of species present (Karban and Ricklefs, 1983; Soberon and Llorente, 1993; Colwell and Coddington, 1994; Mawdsley, 1994). Less widely appreciated is the effect that this has on attempts to describe and compare the species composition of faunas. For progress to be made in determining the degree of host-specificity of tropical insects, either full faunal lists need to be compiled, which requires intensive study over a period of time (see, for example, Janzen, 1988), or the effects of sampling effort need to be explicitly incorporated into analyses of distributional data. We believe that our ANOVA style method for overcoming the effects of sampling has some value towards this goal.

One of the limitations of using canopy fogging data for the estimation of host-specificity and effective specialization is that the samples represent just a snap-shot in time. It has been suggested that herbivory is greater at night in tropical forest canopies (Basset and Springate, 1992) so that samples taken at dawn miss this peak in canopy insect activity. Furthermore, by only considering adults we potentially underestimate the levels of host-specificity. For example, in Costa Rican deciduous forest the time of year that is unfavourable for many insect larvae may be passed as adults in reproductive diapause. Consequently, the adult insects may exhibit low host-specificity to cope with the low availability of preferred host plants (Janzen, 1973). The larval stages of some species of insect – particularly those that are on exposed surfaces such as leaves – may also be very short-lived and once they reach adulthood they may disperse rapidly from the host-tree. Such host-specificity may therefore go unrecorded.

Species richness estimation and insect–host plant relations

It is unlikely that estimates of host-specificity will be useful in determining global species richness as envisioned by Erwin (1982), for at least two reasons. One is that patterns vary between taxa, as discussed above. The other reason is that scale-dependence, a characteristic of most ecological phenomena (May, 1994), is critical in the estimation and interpretation of host-specificity (Fox and Morrow, 1981; Strong *et al.*, 1984). The use of local estimates of host-specificity in extrapolating up to a global estimate of species richness assumes scale-independence of host-specificity (May, 1990). Thus, studies of local host-specificity will only be directly relevant to local patterns of species richness, community structure and insect–plant interactions.

Our study also suggests that tropical insect assemblages are not organized into discrete component communities (*sensu* Root, 1973) associated with particular host-tree species (see Figures 6.1 and 6.2; also Futuyma and Gould, 1979; Cyrtinowitz, 1991). In turn, with high variability in host-specificity, as discussed earlier, we would expect the relationship between insect species richness and plant species richness to be rather loose (Farrell and Erwin, 1988). Plant species richness may therefore be a poor indicator of insect species richness in the world's tropical forests.

We can, however, speculate about the species richness of the canopy beetle fauna within a single locality. Consider an area of lowland rainforest with 60 tree families (Table 6.8). Published data suggest the total Coleoptera fauna of single tropical tree species may have an average in the region of 200–400 species (Erwin and Scott, 1980; Stork, 1991; Allison *et al.*, 1993; Mawdsley, 1994). With an effective specialization of 10%, this gives a total host-specific beetle species richness of up to $400 \times 60 \times 0.1 = 2400$ species. The factors affecting our estimate of host-specificity will in turn affect our evaluation of the magnitude of local canopy beetle species richness.

The host-specificity of tropical insect faunas is central in the use of insect–plant associations to scale up estimates of insect species richness (May, 1990; Thomas, 1990; Gaston, 1992). There is a real need to understand variation in the host-specificity of different tree species and different forest types and whether, in general, turnover of insect species on plants occurs at a faster or slower rate as one moves from local up to regional scales. That is, whether the effective specialization of the insect fauna increases or decreases with spatial scale (May, 1994). This is a major unknown because it is dependent upon knowledge of geographical ranges, which is presently lacking for all but a few taxonomically restricted groups of tropical insects (Johnson and Siemens, 1992).

Acknowledgements

We thank Bob May, John Lawton, Yves Basset and Vojtech Novotny who commented on various drafts of this paper. Nick Mawdsley was supported by a Natural History Museum PhD studentship.

REFERENCES

Abbott, I. (1974) Numbers of plant, insect and land bird species on 19 remote islands in the Southern hemisphere. *Biological Journal of the Linnean Society*, **6**, 143–52.

Allison, A., Samuelson, G.A. and Miller, S.E. (1993) Patterns of beetle species diversity in New Guinea rain forest as revealed by canopy fogging: preliminary findings. *Selbyana*, **14**, 16–20.

Basset, Y. (1992) Host-specificity of arboreal and free-living insect herbivores in rain forests. *Biological Journal of the Linnean Society*, **47**, 115–33.

Basset, Y. and Springate, N.D. (1992) Diel activity of arboreal arthropods associated with a rainforest tree. *Journal of Natural History*, **26**, 947–52.

Beaver, R.A. (1979) Host-specificity of temperate and tropical animals. *Nature*, **281**, 139–41.

Bernays, E. and Graham, M. (1988) On the evolution of host-specificity in phytophagous arthropods. *Ecology*, **69**, 886–92.

Bullock, J.A. (1971) The investigation of samples containing many species. II. Sample comparison. *Biological Journal of the Linnean Society*, **3**, 23–56.

Campbell, D.G. (1994) Scale and patterns of community structure in Amazonian forests, in *Large-Scale Ecology and Conservation Biology* (eds P.J. Edwards, R.M. May and N.R. Webb), Blackwell Scientific Publications, Oxford, pp. 179–97.

Casson, D.S. and Hodkinson, I.D. (1991) The Hemiptera (Insecta) communities of tropical rain forest in Sulawesi. *Zoological Journal of the Linnean Society*, **102**, 253–75.

Claridge, M.F. (1987) Insect assemblages — diversity, organization, and evolution, in *Organization of Communities: Past and Present* (eds J.H.R. Gee and P.S. Giller), Blackwell Scientific Publications, Oxford, pp. 141–62.

Colwell, R.K. and Coddington, J.A. (1994) Estimating terrestrial biodiversity through extrapolation. *Philosophical Transactions of the Royal Society of London, Series B*, **345**, 101–18.

Corbet, P.S. (1961) Entomological studies from a high tower in Mpanga forest, Uganda. VIII. The age-composition of biting mosquito populations according to time and level. *Transactions of the Royal Entomological Society of London*, **113**, 336–45, references 362–8.

Courtney, S.P. and Chew, F.S. (1987) Coexistence and host-use by a large community of pierid butterflies: Habitat is the templet. *Oecologia*, **71**, 210–20.

Cyrtinowitz, M. (1991) Resource size and predictability and local herbivore richness in a subtropical Brazilian cerrado community, in *Plant–Animal Interactions: Evolutionary Ecology in Tropical and Temperate Regions* (eds P.W. Price, T.M. Lewinsohn, G.W. Fernandes and W.W. Benson), John Wiley & Sons, New York, pp. 651–89.

Ehrlich, P.R. and Raven, P.H. (1964) Butterflies and plants: a study in coevolution. *Ecology*, **18**, 586–608.

Erwin, T.L. (1982) Tropical forests: their richness in Coleoptera and other arthropod species. *The Coleopterist's Bulletin*, **36**, 74–5.

Erwin, T.L. (1983a) Tropical forest canopies: the last biotic frontier. *Bulletin of the Entomological Society of America*, **29**, 14–19.

Erwin, T.L. (1983b) Beetles and other insects of tropical forest canopies at Manaus, Brazil, sampled by insecticidal fogging, in *Tropical Rain Forest: Ecology and Management* (eds S.L. Sutton, T.C. Whitmore and A.C. Chadwick), Blackwell Scientific Publications, Oxford, pp. 59–75.

Erwin, T.L. (1988) The tropical forest canopy: the heart of biotic diversity, in *Biodiversity* (eds E.O. Wilson and F.M. Peter), National Academic Press, Washington DC, pp. 123–9.

Erwin, T.L. (1991) How many species are there? Revisited. *Conservation Biology*, **5**, 330–3.

Erwin, T.L. and Scott, J.C. (1980) Seasonal and size patterns, trophic structure, and richness of Coleoptera in the tropical arboreal ecosystem: the fauna of the tree *Luehea seemannii* Triana and Planch in the canal zone of Panama. *The Coleopterist's Bulletin*, **34**(3), 305–22.

Farrell, B.D. and Erwin, T.L. (1988) Leaf-beetle community structure in an Amazonian rainforest community, in *Biology of Chrysomelidae* (eds P. Jolivet, E. Petitpierre and T.H. Hsiao), Kluwer Academic Publications, pp. 73–90.

Feeny, P. (1976) Plant apparancy and chemical defence. *Recent Advances in Phytochemistry*, **10**, 1–40.

Felsenstein, J. (1985) Phylogenies and the comparative method. *American Naturalist*, **125**, 1–15.

Fensham, R.J. (1994) Phytophagous insect–woody sprout interactions in tropical eucalypt forest. II. Insect community structure. *Australian Journal of Ecology*, **19**, 189–96.

Fox, L.R. and Morrow, P.A. (1981) Specialization: species property or local phenomenon? *Science*, **211**, 887–93.

Futuyma, D.J. and Gould, F. (1979) Associations of plants and insects in a deciduous forest. *Ecological Monographs*, **49**, 33–50.

Futuyma, D.J. and Moreno, G. (1988) The evolution of ecological specialization. *Annual Review of Ecology and Systematics*, **19**, 207–33.

Gaston, K.J. (1992) Regional numbers of insect and plant species. *Functional Ecology*, **6**, 243–7.

Gaston, K.J. (1993) Herbivory at the limits. *Trends in Ecology and Evolution*, **8**, 193–4.

Gaston, K.J., Blackburn, T.M., Hammond, P.M. and Stork, N.E. (1993) Relationships between abundance and body size: where do tourists fit? *Ecological Entomology*, **18**, 310–14

Gauld, I.D. (1986) Latitudinal gradients in ichneumonid species richness in Australia. *Ecological Entomology*, **11**, 155–61.

Gauld, I.D., Gaston, K.J. and Janzen, D.H. (1992) Plant allelochemicals, tritrophic interactions and the anomalous diversity of tropical parasitoids: the 'nasty' host hypothesis. *Oikos*, **65**, 353–7.

Gilbert, L.E. and Smiley, J.T. (1978) Determinants of local diversity in phytophagous insects: host specialists in tropical environments, in *Diversity of Insect Faunas* (eds L.A. Mound and N. Waloff), Blackwell Scientific Publications, Oxford, pp. 89–104.

Harrison, S., Ross, S.J. and Lawton, J.H. (1992) Beta diversity on geographic gradients in Britain. *Journal of Animal Ecology*, **61**, 151–8.

Harvey, P.H. and Pagel, M.D. (1991) *The Comparative Method in Evolutionary Biology. Oxford Series in Ecology and Evolution*, Oxford University Press, Oxford.

Hawkins, B.A., Shaw, M.A. and Askew, R.R. (1992) Relations among assemblage size, host specialization, and climatic variability in North American parasitoid communities. *American Naturalist*, **139**, 58–79.

Hawksworth, D.L. (1991) The fungal dimension of biodiversity: magnitude, significance, and conservation. *Mycological Research*, **95**, 641–55.

Hayek, L.C. (1993) Analysis of amphibian biodiversity data, in *Measuring and Monitoring Biological Diversity. Standard Methods for Amphibians* (eds W.R. Heyer, M.A. Donnelly, R.W. McDiarmid, L.C. Hayek and M.S. Foster), Smithsonian Institution Press, Washington, pp. 207–69.

Hodkinson, I.D. and Casson, D. (1991) A lesser predilection for bugs: Hemiptera (Insecta) diversity in tropical rain forests. *Biological Journal of the Linnean Society*, **43**, 101–9.

Holloway, J.D. (1989) Moths, in *Tropical Rain Forest Ecosystems. Biogeographical and Ecological Studies* (eds H. Leith and M.J.A. Werger), Elsevier, Amsterdam, pp. 437–53.

Hutchinson, G.E. (1957) Concluding remarks. *Cold Spring Harbour Symposium of Quantitative Biology*, **22**, 415–27.

Janzen, D.H. (1973) Comments on host-specificity of tropical herbivores and its relevance to species richness, in *Taxonomy and Ecology* (ed. V.H. Heywood), The Systematics Association, Special Volume No. 5, Academic Press, London, pp. 201–11.

Janzen, D.H. (1985a) A host plant is more than its chemistry. *Illinois Natural History Survey Bulletin*, **33**, 141–74.

Janzen, D.H. (1985b) Plant defences against animals in the Amazonian rainforest, in *Amazonia* (eds G.T. Prance and T.E. Lovejoy), Pergamon Press, Oxford, pp. 207–17.

Janzen, D.H. (1988) Ecological characterization of a Costa Rica dry forest caterpillar fauna. *Biotropica*, **20**, 120–35.

Johnson, C.D. and Siemens, D.H. (1992) Observations on host plant and biogeographical differences in bruchid beetles in Southern Ecuador. *Biotropica*, **24**, 195–9.

Karban, R. and Ricklefs, R.E. (1983) Host characteristics, sampling intensity and species richness of Lepidoptera larvae on broad–leaved trees in southern Ontario. *Ecology*, **64**, 636–41.

Koch, C.F. (1987) Prediction of sample size effects on the measured temporal and geographical distribution patterns of species. *Paleobiology*, **13**, 100–7.

Leather, S. R. (1985) Does the bird cherry have its 'fair share' of insect pests? An appraisal of the species–area relationships of the phytophagous insects associated with British *Prunus* species. *Ecological Entomology*, **10**, 43–56.

Leather, S.R. (1991) Feeding specialisation and host distribution of British and Finnish *Prunus* feeding macrolepidoptera. *Oikos*, **60**, 40–8.

Magurran, A.E. (1988) *Ecological Diversity and its Measurement*, Chapman & Hall, London.

Marquis, R.J. (1991) Herbivorous fauna of *Piper* (Piperaceae) in a Costa Rica wet forest: diversity, specificity and impact, in *Plant–Animal Interactions:*

Evolutionary Ecology in Tropical and Temperate Regions (eds P.W. Price, T.M. Lewinsohn, G.W. Fernandes and W.W. Benson), John Wiley & Sons, New York, pp. 179–208.

Marquis, R.J. and Braker, H.E. (1993) Plant–herbivore interactions: diversity, specificity, and impact, in *La Selva: Ecology and Natural History of a Neotropical Rain Forest* (eds L.A. McDade, K.S. Bawa, H.A. Hespenheide and G.S. Hartshorn), University of Chicago Press, Chicago, pp. 261–81.

Mawdsley, N.A. (1994) *Community Structure of the Coleoptera Assemblage in a Bornean Tropical Forest*, unpublished PhD Thesis, Imperial College, University of London.

May, R.M. (1990) How many species? *Philosophical Transactions of the Royal Society London, Series B*, **330**, 293–304.

May, R.M. (1994) The effects of spatial scale on ecological questions and answers, in *Large–Scale Ecology and Conservation Biology* (eds P.J. Edwards, R.M. May and N.R. Webb), Blackwell Scientific Publications, Oxford, pp. 1–17.

McArdle, B.H. (1990) When are rare species not there? *Oikos*, **57**, 276–7.

McGarvin, M., Lawton, J.H. and Heads, P.A. (1986) The herbivorous insect communities of open and woodland bracken: observations, experiments and habitat manipulations. *Oikos*, **47**, 135–48.

Moran, V.C. and Southwood, T.R.E. (1982) The guild composition of arthropod communities in trees. *Journal of Animal Ecology*, **51**, 289–306.

Moran, V.C., Hoffman, J.H., Impson, F.A.C. and Jenkins, J.J.G. (1994) Herbivorous insect species in the tree canopy of a relict South African forest. *Ecological Entomology*, **19**, 147–54.

Murdoch, W.M., Evans, F.C. and Peterson, C.H. (1972) Diversity and patterns in plants and insects. *Ecology*, **53**, 819–29.

Newberry, D.M., Campbell, E.J.F., Lee, Y.F., Ridsdale, C.E. and Still, M.J. (1992) Primary lowland dipterocarp forest at Danum valley, Sabah, Malaysia: structure, relative abundance and family composition. *Philosophical Transactions of the Royal Society London, Series B*, **335**, 341–56.

Nilsson, I.N. and Nilsson, S.G. (1985) Experimental estimates of census efficiency and pseudoturnover on islands: error trend and between–observer variation when recording vascular plants. *Journal of Ecology*, **73**, 65–7.

Price, P.W. (1991) Patterns in communities along latitudinal gradients, in *Plant–Animal Interactions: Evolutionary Ecology in Tropical and Temperate Regions* (eds P.W. Price, T.M. Lewinsohn, G.W. Fernandes and W.W. Benson), John Wiley & Sons, New York, pp. 51–69.

Root, R.B. (1973) Organisation of a plant–arthropod association in simple and diverse habitats: the fauna of collards (*Brassica oleracea*). *Ecological Monographs*, **43**, 95–124.

Soberon, M.J. and Llorente, B.J. (1993) The use of species accumulation functions for the prediction of species richness. *Conservation Biology*, **7**, 480–8.

Southwood, T.R.E., Moran, V.C. and Kennedy, C.E.J. (1982) The assessment of arboreal insect fauna: comparisons of knockdown sampling and faunal lists. *Ecological Entomology*, **7**, 331–40.

Stork, N.E. (1987a) Guild structure of arthropods from Bornean rain forest trees. *Ecological Entomology*, **12**, 69–80.

Stork, N.E. (1987b) Arthropod faunal similarity of Bornean rain forest trees. *Ecological Entomology*, **12**, 219–26.

Stork, N.E. (1988a) Insect diversity: facts, fiction and speculation. *Biological Journal of the Linnean Society*, **35**, 321–37.

Stork, N.E. (1988b) Adaptations of arboreal carabids to life in trees. *Acta Phytopathologica Hungarica*, **22**, 273–91.

Stork, N.E. (1991) The composition of the arthropod fauna of Bornean lowland rain forest trees. *Journal of Tropical Ecology*, **7**, 161–80.
Stork, N.E. (1993) How many species are there? *Biodiversity and Conservation*, **2**, 215–32.
Stork, N.E. (1995) Measuring global biodiversity and its decline, in *Biodiversity II. Understanding and Protecting our Natural Resources* (eds M.L. Reaka-Kudla, D.E. Wilson and E.O. Wilson), Joseph Henry Press, Washington DC.
Strong, D.R., Lawton, J.H. and Southwood, T.R.E. (1984) *Insects on Plants: Community Patterns and Mechanisms*, Blackwell Scientific Publications, Oxford.
Szentkiralyi, F. and Kozar, F. (1991) How many species are there in apple insect communities?: testing the resource diversity and intermediate disturbance hypotheses. *Ecological Entomology*, **16**, 491–503.
Thomas, C.D. (1987) Behavioural determination of diet breadth in insect herbivores: the effect of leaf age on choice of host species by beetles feeding on *Passiflora* vines. *Oikos*, **48**, 211–16.
Thomas, C.D. (1990) Fewer species. *Nature*, **347**, 237.
Ward, L.K. (1988) The validity and interpretation of insect food-plant records. *British Journal of Entomology and Natural History*, **1**, 153–62.
Ward, L.K. and Spalding, D.F. (1993) Phytophagous British insects and mites and their food-plant families: total numbers and polyphagy. *Biological Journal of the Linnean Society*, **49**, 257–76.
Ward, S. A. (1992) Assessing functional explanations of host-specificity. *American Naturalist*, **139**, 883–91.
Whitmore, T.C. (1984) *Tropical Rain Forests of the Far East*, Oxford University Press, Oxford.
Wolda, H. (1981) Similarity indices, sample size and diversity. *Oecologia*, **50**, 296–302.
Wood, T.K. (1993) Diversity in the New World Membracidae. *Annual Review of Entomology*, **38**, 409–35.
Wood, T.K. and Olmstead, K.L. (1984) Latitudinal effects on treehopper species richness (Homoptera: Membracidae). *Ecological Entomology*, **9**, 109–15.

Appendix 6A Calculation of effective specialization from the species occurrence data of beetle species on four *Shorea johorensis* trees. See text for details.

Combination 1

		$N_k(i)[p_k(i)]$						
Tree species and no. (k)	*No. spp.* (N_k)	$i=1$	$i=2$	$i=3$	$i=4$	f_k	N_f	$F=\Sigma N_f/\Sigma N_k$
shorjoh 1	130	67 [0.52]	34 [0.26]	18 [0.14]	11 [0.08]	0.71	93	
shorjoh 2	270	177 [0.66]	53 [0.20]	29 [0.11]	11 [0.04]	0.81	216	
shorjoh 3	166	98 [0.59]	35 [0.21]	22 [0.13]	11 [0.07]	0.76	126	
shorjoh 4	141	79 [0.56]	30 [0.21]	21 [0.15]	11 [0.08]	0.74	104	
Totals	707						539	0.76

Combination 2

Tree species and no. (k)	No. spp. (N_k)	$N_k(i)[p_k(i)]$ i=1	i=2	i=3	i=4	f_k	N_f	$F=\Sigma N_f/\Sigma N_k$
shorjoh 1	130	102 [0.78]	8 [0.06]	14 [0.11]	6 [0.05]	0.86	112	
pentmot 1	45	24 [0.53]	9 [0.20]	6 [0.13]	6 [0.13]	0.71	32	
castan	103	65 [0.63]	21 [0.20]	11 [0.11]	6 [0.06]	0.78	81	
unknown	119	119 [0.69]	20 [0.17]	11 [0.09]	6 [0.05]	0.82	97	
	397						322	0.81

Combination 3

Tree species and no. (k)	No. spp. (N_k)	$N_k(i)[p_k(i)]$ i=1	i=2	i=3	i=4	f_k	N_f	$F=\Sigma N_f/\Sigma N_k$
shorjoh 2	270	208 [0.77]	36 [0.13]	20 [0.07]	6 [0.02]	0.87	234	
pentmot 1	45	22 [0.49]	9 [0.20]	8 [0.18]	6 [0.13]	0.68	31	
castan	103	55 [0.53]	25 [0.24]	17 [0.17]	6 [0.06]	0.72	75	
unknown	119	66 [0.55]	32 [0.27]	15 [0.13]	6 [0.05]	0.74	89	
	537						428	0.801

Combination 4

Tree species and no. (k)	No. spp. (N_k)	$N_k(i)[p_k(i)]$ i=1	i=2	i=3	i=4	f_k	N_f	$F=\Sigma N_f/\Sigma N_k$
shorjoh 3	166	124 [0.75]	23 [0.14]	15 [0.09]	4 [0.02]	0.85	142	
pentmot 1	45	22 [0.49]	13 [0.29]	6 [0.13]	4 [0.09]	0.70	32	
castan	103	60 [0.58]	25 [0.24]	14 [0.14]	4 [0.04]	0.76	78	
unknown	119	74 [0.62]	25 [0.21]	16 [0.13]	4 [0.03]	0.78	93	
	433						344	0.79

Combination 5

Tree species and no. (k)	No. spp. (N_k)	$N_k(i)[p_k(i)]$ i=1	i=2	i=3	i=4	f_k	N_f	$F=\Sigma N_f/\Sigma N_k$
shorjoh 4	141	109 [0.77]	22 [0.16]	7 [0.05]	3 [0.02]	0.87	123	
pentmot 1	45	21 [0.47]	15 [0.33]	6 [0.13]	3 [0.07]	0.69	31	
castan	103	61 [0.59]	32 [0.31]	7 [0.07]	3 [0.03]	0.78	80	
unknown	119	75 [0.63]	31 [0.26]	10 [0.08]	3 [0.03]	0.79	95	
	408						329	0.81

Combination 6

Tree species and no. (k)	No. spp. (N_k)	$N_k(i)[p_k(i)]$ i=1	i=2	i=3	i=4	f_k	N_f	$F=\Sigma N_f/\Sigma N_k$
shorjoh 1	130	94 [0.72]	18 [0.14]	13 [0.10]	5 [0.04]	0.84	109	
pentmot 2	100	66 [0.66]	22 [0.22]	7 [0.07]	5 [0.05]	0.81	81	
castan	103	64 [0.62]	21 [0.20]	13 [0.13]	5 [0.05]	0.78	80	
unknown	119	78 [0.66]	21 [0.18]	15 [0.13]	5 [0.04]	0.80	95	
	452						364	0.81

Combination 7

Tree species and no. (k)	No. spp. (N_k)	$N_k(i)[p_k(i)]$ i=1	i=2	i=3	i=4	f_k	N_f	$F=\Sigma N_f/\Sigma N_k$
shorjoh 2	270	197 [0.73]	44 [0.16]	22 [0.08]	7 [0.03]	0.84	228	
pentmot 2	100	61 [0.61]	20 [0.20]	12 [0.12]	7 [0.07]	0.77	77	
castan	103	56 [0.54]	23 [0.22]	17 [0.17]	7 [0.07]	0.73	75	
unknown	119	65 [0.55]	29 [0.24]	18 [0.15]	7 [0.06]	0.73	87	
	592						467	0.79

Combination 8

Tree species and no. (k)	No. spp. (N_k)	$N_k(i)[p_k(i)]$ i=1	i=2	i=3	i=4	f_k	N_f	$F=\Sigma N_f/\Sigma N_k$
shorjoh 3	166	123 [0.74]	21 [0.13]	17 [0.10]	5 [0.03]	0.85	140	
pentmot 2	100	71 [0.71]	14 [0.74]	10 [0.10]	5 [0.05]	0.83	83	
castan	103	63 [0.61]	19 [0.18]	16 [0.16]	5 [0.05]	0.77	79	
unknown	119	71 [0.60]	26 [0.22]	17 [0.14]	5 [0.04]	0.76	91	
	488						393	0.81

Combination 9

Tree species and no. (k)	No. spp. (N_k)	$N_k(i)[p_k(i)]$ i=1	i=2	i=3	i=4	f_k	N_f	$F=\Sigma N_f/\Sigma N_k$
shorjoh 4	141	106 [0.75]	24 [0.17]	8 [0.06]	3 [0.02]	0.86	121	
pentmot 2	100	68 [0.68]	20 [0.20]	9 [0.09]	3 [0.03]	0.82	82	
castan	103	62 [0.60]	29 [0.28]	9 [0.09]	3 [0.03]	0.78	80	
unknown	119	72 [0.61]	31 [0.26]	13 [0.11]	3 [0.03]	0.78	93	
	463						376	0.81

7

Determinants of species richness in assemblages of canopy arthropods in rainforests

R.L. Kitching, H. Mitchell, G. Morse and C. Thebaud

ABSTRACT

Australia offers special opportunities for the study of rainforest ecology, with sites ranging from tropical to cool temperate, littoral to montane. Canopy arthropod assemblages display high levels of species richness, low levels of taxonomic knowledge and multiple opportunities for community ecological studies.

We have spent some years assembling baseline data on faunal patterns in tropical and subtropical rainforest sites. We now address an open-ended set of questions about the underlying mechanisms generating the observed levels of species richness. We address four topics:

1. The role of the tree species. In estimates of global species richness, high levels of species specificity have been assumed for plant–animal interactions. For Australian data these appear to be overestimates. The issue is further clouded by degrees of inter-relatedness among host tree species.
2. The role of epiphyte abundance and richness in determining canopy arthropod richness remains largely uninvestigated. We have begun work in this area through examination of the arthropod fauna of leaf litter contained within epiphytic ferns and that of associated ground litter. Distinct differences have been noted both in terms of overall arthropod densities and the ordinal level composition of the two sets of samples.
3. The role of the parent rock in determining arthropod diversity has been investigated. Clear differences have been observed across basaltic and granitic based forests which occur in otherwise similar situations and have comparable histories of human disturbance.
4. The role of 'edges' – rainforest/non-rainforest, forest type/forest type, light gap/undisturbed – is likely to be important in determining gamma diversity within

Canopy Arthropods. Edited by N.E. Stork, J. Adis and R.K. Didham. Published in 1997 by Chapman & Hall, London. ISBN 0 412 74900 9

a heterogeneous environment. We are examining the ways in which arthropod canopy assemblages vary across and along rainforest/sclerophyll forest edges, focusing firstly on ordinal profiles and secondly on the Coleoptera of these habitats. Overlap between rainforest arthropod diversity and that of adjacent moist sclerophyll forest is small and the fauna of the edge itself has many distinctive features.

INTRODUCTION

The very high levels of species richness of arthropods in the rainforest canopies of the world has been unequivocally established. Our data from subtropical and tropical Australian rainforest canopies (Kitching and Arthur, 1993; Kitching *et al.*, 1992, 1994; Kitching, 1994; Kitching and Zalucki, 1996; Hammond *et al.*, in press) complements available data from Neotropical (Erwin and Scott, 1980; Erwin, 1982) and south-east Asian studies (Stork, 1987a,b; Stork and Brendell, 1990, 1993). What these high levels of species diversity in rainforest canopies can tell us about global species richness is a much more contentious area which we do not propose to explore here (but see Erwin, 1991; Gaston, 1991; Stork, 1993; Kitching *et al.*, 1994, and references therein).

Observations and analyses of the global and local patterns of species richness lead directly to a series of hypotheses which attempt to identify the underlying processes that generate high local and/or regional species richness. These processes will be historical, geographical and ecological in nature. Our programme of research in Australian rainforests examines the ecological determinants of canopy diversity as part of the investigations of rainforest dynamics being undertaken under the aegis of the Cooperative Research Centre for Tropical Rainforest Ecology and Management.

Our studies of rainforest canopies in Australia build on a number of key pieces of earlier work. Basset (1990, 1991a,b, 1992) studied the fauna associated with a single rainforest tree species, *Argyrodendron actinophyllum*. Lowman and co-workers examined levels of herbivory and associated insects across five species of Australian rainforest tree (Lowman, 1985, 1986, 1992; Lowman and Box, 1983). These studies were restricted to subtropical forests of northern New South Wales and south-east Queensland. In addition, key studies on the canopies of sclerophyll forests and woodlands have been made by Majer and Recher (1988) and Majer *et al.* (1990).

In this paper we summarize some of the emergent patterns for which we must now seek explanations. We present a general templet within which we seek explanations for the patterns we see in our results, and present selected results from these process studies. In particular we discuss the role of the tree species, interactions with epiphytes, the relationship between the ground and the canopy fauna, and the impact of edges on arthropod biodiversity.

METHODS AND SITES

We have been sampling selected rainforest canopies in Australia since 1988. Most of our activities have been focused on two sites: a subtropical forest at 1100 m altitude on the Queensland/New South Wales Border ('Green Mountains') and a lowland tropical forest at Cape Tribulation in North Queensland. In addition, very recently we have added a high elevation tropical location, also in North Queensland, to our regular study sites ('Atherton').

At each site we have carried out basic faunistic studies sampling the canopy using insecticide knockdown methods. Additional techniques such as light-trapping, funnel extraction, local insecticide spraying and interception trapping have been employed as required to target particular locations or taxa.

The characteristics of our subtropical and lowland tropical sites plus details of our sampling procedures have been presented by Kitching *et al.* (1992) and McIntyre *et al.* (1994).

For all samples we have employed a sequential sorting protocol partitioning each sample to order in field laboratories, then sorting selected orders to families and morphospecies in the laboratory. For every Order selected for special attention we have worked in collaboration with professional taxonomists who, *inter alia*, have confirmed the 'species' designations we have made or, indeed, have been responsible themselves for the species-level sorting. In this fashion we have focused on the Coleoptera, Lepidoptera, Psocoptera, Collembola, Araneida and oribatid mites, plus selected families within the Diptera.

PATTERNS

At each site we have first made general canopy collections, the first of their kind in Australian rainforests. This has allowed us to become familiar with general patterns of richness, examining ordinal, familial and species levels of complexity, from which a number of overlapping and complementary hypotheses have emerged relating to the determinants of the biodiversity patterns we have observed. Some of the emergent patterns have already been published. These have included comparisons of the ordinal profiles from tropical, subtropical and temperate rainforest canopies (Kitching *et al.*, 1992), the differential impact of canopy drought across arthropod orders (Kitching and Arthur, 1993), the overlap between the faunas of three tropical tree species (Kitching and Zalucki, 1996) and the patterns of size/abundance plots of Coleoptera from subtropical canopies (Kitching *et al.*, 1994). A series of works on the faunistics of selected canopies is well advanced: specifically, on Psocoptera (D. Brown, R.L. Kitching and E. Schmidt, unpublished data), on Coleoptera (Hammond *et al.*, in press), on spiders (R. Raven and R.L. Kitching, unpublished data), on oribatid mites (D. Walters and R.L. Kitching, unpublished data), on tipulid (Kitching and Theischinger, in press), dolichopodid and lauxaniid flies and on moths (D. Harmsen, S. Boulton and R.L. Kitching, unpublished data).

Table 7.1 summarizes some of the hereto unpublished faunistic and taxonomic results available to date. For all the groups so far examined we note high levels of species richness and low levels of previous taxonomic knowledge. An extreme case is presented by the spiders where, out of 69 morphospecies identified by Dr Robert Raven from 2035 specimens from subtropical rainforest canopy in south-east Queensland, only one had been named previously. As elsewhere in the world, access to previously inaccessible canopy components of the rainforest arthropod community is proving to be both a new and rich resource to local taxonomists and a challenge to existing levels of available expertise.

THE EXPLANATORY FRAMEWORK

Students of arthropod biodiversity deal with a growing and open-ended number of emergent properties of the assemblages which they study. A selection of these are summarized in Table 7.2. It is these properties that we seek to explain when we move to more detailed process-oriented studies. The biological factors which produce these properties are many and varied, ranging, on a temporal scale, from the geological to the immediate and, on the spatial scale, from the global to the local. Table 7.3 defines these temporal and spatial axes and attempts to locate some of the emergent properties and the processes that operate to produce them within these axes. Such a model has focused our thoughts when attempting to account for the patterns of diversity that we observe and places particular process studies in a proper context.

Consider, by way of example, the set of arthropods that we obtain from a selected patch of rainforest canopy in northern Australia. The composition of the sample will be influenced, *inter alia*, by:

1. The tectonic history of Australia which has led to the exclusion of some taxa and the presence of others at the continental level (Darlington, 1965; Crook, 1981).
2. Long-term patterns of climate change reflecting the tertiary and quaternary history of the location (Kemp, 1981; Truswell, 1990).
3. The current physicochemical environment present at that particular location – the local climate, topography and pedology (Clunebell *et al.*, 1995).
4. The disturbance history of the site, including disturbance due to natural processes (cyclonic, volcanic, or hydrologic) and anthropogenic processes (clearing, forestry, pollution and so forth), and the

resulting habitat mosaic (Lovejoy *et al.*, 1983; Hopkins, 1990; Olson, 1994; Souza and Brown, 1994; Daily and Ehrlich, 1995).
5. The non-arthropod assemblages, particularly of plants and fungi, present at the sampled location (Balick *et al.*, 1978; Lawton, 1978).
6. The internal composition of the assemblage itself with respect to predator–prey, parasitoid–host and mutualistic interactions (Kitching, 1986; May, 1990; LaSalle and Gauld, 1993; Majer, 1993).

Within this very general model of the determination of biodiversity we have selected a number of specific studies for detailed attention within our present research programme. Each such subprogramme is an on-going one and results presented here represent those of pilot studies and preliminary analyses.

THE ROLE OF THE TREE SPECIES

Specific associations between species of insects and species of plant or fungi are a commonplace result of long-term co-evolutionary processes in all ecosystems. The most obvious 'substrate' species within rainforests are the trees and a reasonable hypothesis to account for arthropod biodiversity in forest canopies is that it will reflect, in some repeatable way, the underlying diversity of trees in the forest.

Erwin (1982) used a figure of 13.5% for the specificity of the association between species of beetles and the species of tree from which they were collected. This percentage represented the first in the sequence of hypotheses he erected in order to arrive at an estimate of global arthropod diversity for rainforests. As the first of the multipliers used in Erwin's calculation, it remains one to which the overall calculation is very sensitive. We have examined the levels of specificity shown in assemblages from a small number of tropical trees from northern Australia. This summary is based on a preliminary analysis of some of the results presented by Kitching and Zalucki (1996). Other results remain under analysis.

We selected a number of understorey trees, the growth forms of which facilitate sampling. Each of these was between 6 and 12 m in height, growing more or less isolated from its neighbours. In each case a set of 0.5 m^2 collecting trays were located around the base of the subject tree and a painters' scaffold erected adjacent to it. The canopy of the tree was then sprayed with a commercially available mixture of pyrethrins (see Kitching *et al.*, 1992 for details of composition, dilution, etc.) for about 6 minutes using a backpack dispenser operated from the scaffold.

Samples were sorted to order at field laboratories and returned to the laboratory in ethanol. Subsequently, particular orders were further curated and sorted to family and morphospecies. We have focused on

Table 7.1 Summary of taxonomic results currently available from Australian rainforest canopies

Taxon	*Location*	*No. of specimens*	*Approximate no. of taxa*	*Novelty level**	*Co-worker(s)*
Psocoptera	Green Mountains	1941	51 spp.	18 species discribed 49 named to genus	D. Brown, E. Schmidt
Coleoptera	Green Mountains	2238	465 spp.	No spp. identification attempted 48 families present	N. Stork, P. Hammond
Diptera	Green Mountains	4812	41 families	2 new families	P. Cranston
Tipulidae	Green Mountains	446	55 spp.	28 species described All named to genus	G. Theischinger
Tipulidae	Cape Tribulation	?	11	1 species described All named to genus	G. Theischinger
Lauxaniidae	Green Mountains	148	26 spp.	4 species named 25 named to genus	S. Kim
Dolichopodidae	Green Mountains	235	19 spp.	5 species named 7 manuscript names All named to genus	D. Bickel
Collembola					
Entomobryidae	Green Mountains	4502	21 spp.	2 species described all named to genera	Sutrisno, P. Greenslade
Acarina					
Orbitida	Green Mountains	1327	86 spp.	4 species described At least four new genera	D. Walters
Araneida	Green Mountains		49 spp.	1 species described 2 new families indicated	R. Raven

* at time of collection.

Table 7.2 Some emergent properties of arthropod canopy assemblages

- Ordinal, familial and species profiles
- Size/abundance relationships
- Guild structures
- Dominance, evenness and richness measures
- Levels of endemism
- Foodweb properties
- Seasonality

the Coleoptera – an order in which we are developing particular expertise and one which allows for direct comparison with the results of workers elsewhere in the world.

In our initial work we examined six trees of each of two species: *Ryparosa javanica* (Blume) Kurz ex Koord. and Valeton (Flacourtiaceae), a widespread species in Australia and elsewhere; and *Noahdendron nicholasii* P.K. Endress, B. Hyland and Tracey (Hamamelidaceae), a very restricted endemic found only in the Noah Creek and adjacent catchments. The results are presented and discussed by Kitching and Zalucki (1996). At the level of the order the profile of insects showed some key differences that warrant further attention: Collembola and Psocoptera are far more abundant on *R. javanica* than *N. nicholasii* whereas the reverse is true for Orthoptera, Homoptera and Hymenoptera. The relative abundance of some orders (e.g. Blattodea, Coleoptera and Thysanoptera) appears unaffected by the tree species. Of Coleoptera, 135 individuals sorted to 59 species, 38 species occurred on *N. nicholasii* and 35 species on *R. javanica* (14 species shared). These 59 species represented 22 families, dominated on both tree species by Curculionidae, Coccinellidae and Corylophidae.

Only a few species were present in any numbers, the greater part of species diversity being contributed by rare species. Of the six most common arthropod species, only one was not shared by both tree species.

Obviously, thinking of any particular species of beetle, or other taxon, as being restricted to a single species of tree may be misleading. Although there are many examples of monophagous associations, we expect there to be more cases in which some degree of oligophagy will be the rule. Tree species exhibit various sorts of phylogenetic relationships and it is entirely reasonable that two species of tree of the same genus, for instance, might be expected to share more arthropod species than two less closely related ones. The continuation and extension of our work in Australian forests will eventually allow quantification of

Table 7.3 A templet summarizing how emergent properties and processes associated with contemporary biodiversity may fit into the temporal and spatial environment at a variety of scales

TEMPORAL SCALE	SPATIAL AND CURRENT FACTORS				
	Processes	*Geological/ Pedological*	*Synoptic Climate*	*Fragmentation/ Patch pattern*	*Co-evolution Co-ecology*
Far Past	Tectonic	Basic biogeographic sets			
				Evolutionary radiation	
	Glaciation-related climatic change		Species 'pumps'		Food-web dynamics • trophic • mutualistic • structural
					Guild structures
	Island biogeographical dynamics	Topographic effects		Gamma-diversity	Establishment facilitation by predecessor species
			Levels of endemism	Assemblage heterogeneity	
Recent	Cyclones Clearing Exploitation	Recovery processes	Seasonality Anthropogenic extinctions		Impacts of exotic biota

the role of such factors in determining this level of overlap in diversity of the canopy faunas of co-occurring trees.

EPIPHYTES

If, as seems likely, the diversity of tree-crowns is insufficient in itself as a predictive or explanatory tool for the observed high levels of arthropod diversity, then we must look for some additional basis by which the very high species richness may be explained. We have concluded that an obvious factor worthy of investigation is the diversity and overall load of epiphytes carried by trees in tropical forest and the impact they may have on canopy arthropods (see also Wayne and Bazzaz, 1991).

Some 10% of all vascular plants are epiphytic (Kress, 1986). This proportion is likely to be much higher in ecosystems such as humid tropical or subtropical forests where epiphytes contribute so significantly to plant diversity (Gentry and Dodson, 1987). The vast majority of epiphytes are obligate (Kress, 1986) and have evolved a complex set of adaptations which fit them for their canopy existence. These include litter-impounding growth forms, foliar trichomes, insectivory, myrmecochory and poikilohydric foliage to acquire and conserve nutrients (Nadkarni and Matelson, 1991). The heterogeneity of the canopy environment has produced high levels of speciation in some groups of canopy epiphytes (e.g. the Orchidaceae, Jones, 1988; bromeliads, Benzing, 1980). It is likely that this high level of epiphytic diversity will have an impact on the arthropod diversity within the canopy.

There are a number of ways in which epiphyte mats might affect arthropod assemblages. First, the presence of epiphytes produces a soil and litter environment in the treetops in addition to the plants themselves (Delamare-Deboutteville, 1948; Nadkarni and Matelson, 1991). As a habitat for invertebrates, such suspended soil and litter may be regarded as ephemeral, although the existence of huge 'container' epiphytes, such as *Asplenium* ferns and some of the larger bromeliads, mitigates against this view. Two contradictory views have been expressed concerning the arthropods associated with this habitat. Nadkarni and Longino (1990) working in Costa Rica found such habitats to be poor in arthropods, with some taxa that are common in the ground zone absent from the canopy. In contrast, the work of Paoletti *et al.* (1991) in Venezuelan rainforest found very rich assemblages in these suspended environments, including many species which differed from their ground-zone counterparts.

Second, there will be a folivorous component associated with the epiphytes themselves. The composition and richness of this component may reflect factors such as the risk of ant predation, nutritional quality of the epiphytes themselves, the proportion of particular arthropod life-cycles that are completed in or around the epiphyte mats, and the set of secondary species dependent upon the herbivore guild.

Last, there will be pollinator assemblages associated with angiospermatous epiphytes and they too may be more or less specific.

We have begun making comparisons of the arthropod assemblages associated with perched and adjacent ground litter at a number of sites in subtropical and tropical forest. As a pilot study, a series of samples was taken from the litter contained within the epiphytic fern *Asplenium australasicum*, together with parallel samples from the adjacent ground layer. These were collected at Wongabel State Forest on the Atherton Tablelands of North Queensland, where the ground zone is made up of basaltic boulders with only a poorly developed litter layer. Accordingly, it is not surprising, in this particular habitat, that the perched litter was significantly richer than that from the ground samples (counts/g dry weight of litter ± s.e.: insects in perched litter 9.94 ± 2.731; ground litter 0.41 ± 0.105; non-insect arthropods in perched litter 20.32 ± 7.998; ground litter 1.07 ± 0.568). Perhaps of more interest are the qualitative differences between the two series of samples (Figure 7.1). There are large and significant differences across the two litter types for Collembola and Acarina. These results clearly indicate that, in some forests at least, the perched litter may represent a major arthropod habitat in the canopy. Litter-containing epiphytes, such as *Asplenium* spp., are generally abundant and have a wide, more or less even, distribution within the forest. These and other aspects of the perched litter are being investigated further by Denis Rodgers, a graduate student within our unit.

SUBSTRATE TYPE

One of the most influential factors affecting the vegetational assemblages in rainforests is substrate type (Webb, 1968, 1969). The most important manifestation of this in the tropical rainforests of North Queensland is the contrast between forests on basaltic or granitic parent rock.

We have recently studied the canopy arthropods from two sites 16 km apart on the Atherton Tablelands, within the same continuous forested area. One of these sites is on a basaltic base rock and the other on granite. Both sites have a more or less comparable disturbance history and occur at the same altitude and aspect. A complete survey of the woody vegetation from our three sampling points at each of the two sites identified 47 species of plant of which only 16 were shared between the two sites. Arthropod numbers differed across these sites (insects at the basaltic site 487.7 ± 170.08, granitic site 341.0 ± 74.50; non-insect arthropods at the basaltic site 55.0 ± 16.70, granitic site 151.3 ± 57.33), but variances are high and further sampling is required. Again, qualitative differences between sites warrant further attention (Figure 7.2). In particular, the relative abundances of Collembola (see Figure 7.2) and Acarina (not illustrated) show large differences between sites. These groups were also those for which clear differentiation between forest types at different latitudes was shown by Kitching *et al.* (1992).

PATTERNS ACROSS EDGES

The regional biodiversity of any group of organisms will reflect the habitat mosaic present over the landscape of interest. Regional diversity will then represent some complex weighted sum of the within-habitat diversity of each of the habitat units. Rainforests in Australia, as elsewhere, exist as patches of various sizes generated either by past human use or past climatic events. These are usually surrounded by eucalypt-dominated sclerophyll formations of one sort or another. It is of considerable interest to know how arthropod diversity changes across these edges, which may be abrupt (in those cases where the sclerophyllous vegetation is directly fire-maintained) or gradual (as where there is a topographic boundary with associated microclimatic changes).

We have examined both sorts of rainforest/sclerophyll boundary using a variety of arthropod trapping techniques. Figure 7.3 illustrates the results obtained from Malaise trapping across a sharp pyric boundary between subtropical rainforest and sclerophyll woodland at Lamington National Park, south-east Queensland. As with all Malaise trap samples, it is Diptera-dominated. Figure 7.4 presents the same data set with the Diptera removed so that differences and similarities reflecting the relative abundances of the other orders become apparent. Some trends are clear: the gradual increase in Collembola numbers across the edge from rainforest to sclerophyll, the rarity of Homoptera in the rainforest compared with the sclerophyll (largely reflecting an absence of psyllids in the rainforest), and the abundance of Psocoptera in the rainforest. Differences become even clearer at the level of the morphospecies. Table 7.4 contains the results from the same Malaise trap catches, but analysed to the family and species level for Coleoptera. A clear decline in numbers of Coleoptera overall is seen from rainforest to sclerophyll, with an associated decline in numbers of families and species. We have calculated the degree of overlap among the three components of the overall catch using Sørensen's index, which is based solely on the presence or absence of a species (Table 7.5). There is a 23% overlap between the rainforest and sclerophyll species. More remarkable is the low level of overlap between the rainforest and the edge sample, and between the edge and the sclerophyll sample (34% and 38% respectively). This suggests, as expected, that there is a large set of species restricted to either the rainforest or the sclerophyll area. It may be also that there is a characteristic set of 'edge' species in addition. Further analyses, following division of the species into appropriate

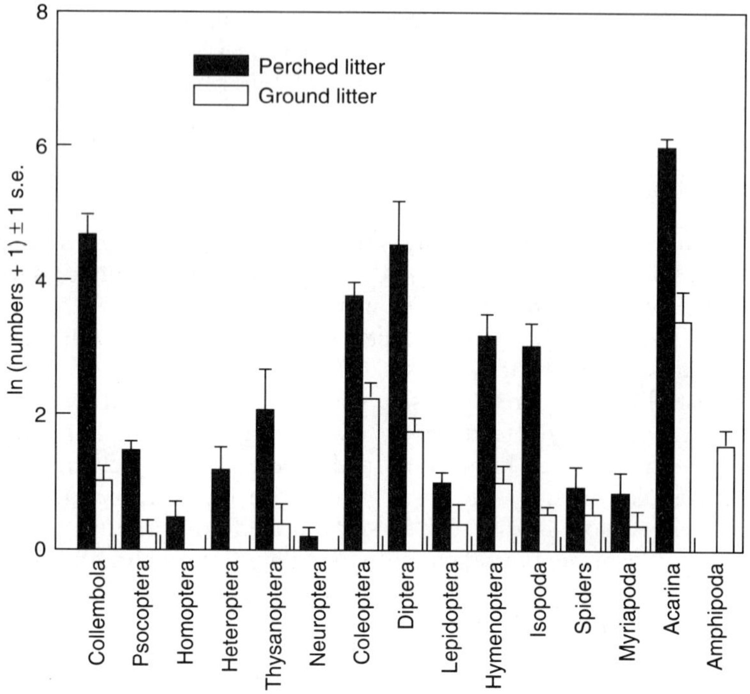

Figure 7.1 Ordinal profiles obtained from arthropod samples extracted from leaf litter held in epiphytic *Asplenium* ferns and obtained from adjacent ground sites.

feeding guilds, will focus on these ideas and the underlying biological reasons for the observed patterns.

GENERAL DISCUSSION

There are several avenues of investigation that we are currently pursuing, some of which have been touched upon in the preceding account. In summary:

1. We continue to accumulate data on the assemblages of organisms that occur in association with particular species of tree. We are currently analysing data based upon our sampling of 10 individuals of each of four tropical tree species, *Lindsayomyrtus racemoides* Greves Craven (Myrtaceae), *Medicosma sessiliflora* (C.T. White) T.G. Hartley (Rutaceae), *Noahdendron nicholasii* and *Ryparosa javanica*.
2. We have sampled sets of sites on basaltic and granitic sites at mid-elevations on the Atherton Tablelands for canopy arthropods (see

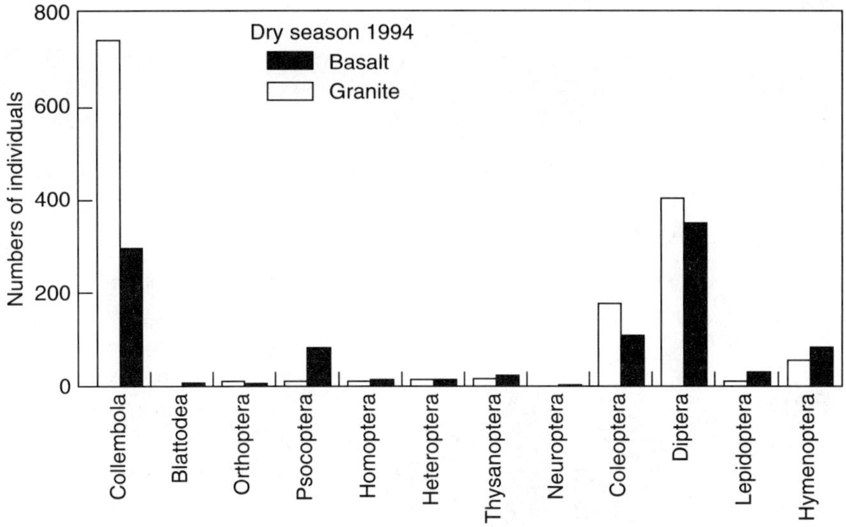

Figure 7.2 Ordinal profiles for insects obtained by insecticide knockdown from mixed canopies of forests on basaltic and granitic parent soils on the Atherton Tablelands, dry season 1994.

above), during both wet and dry seasons. The analysis of this database is in progress.

3. Our pilot work on the fauna of perched litter associated with epiphyte mats is the first part of a project examining the role of epiphytes in general. Other data based on sampling the fibrous base of such mats, on knockdown samples from the epiphytes themselves, and on hand-collecting epiphytophagous insects directly is to hand and will be added to in the near future.
4. Much further taxonomic analysis remains to be carried out based on collections already made. This is limited in rate of progress only by the availability of expert and willing collaborating taxonomists.
5. Data on changes in arthropod diversity across rainforest/non-rainforest edges presented above are part of a larger data-set under analysis. This includes additional catch results based on yellow pan traps, window traps, pitfall traps, litter extractions and bark spraying. Work across a 'tight' pyric edge has been duplicated at both tropical and subtropical sites. A comparable set of samples has been taken along the Mt Haig 'gradsect' (Gillison, 1993) on the Atherton Tablelands, representing a natural gradient between rainforest and sclerophyll forest based on aspect, altitude, soil type and microclimate.

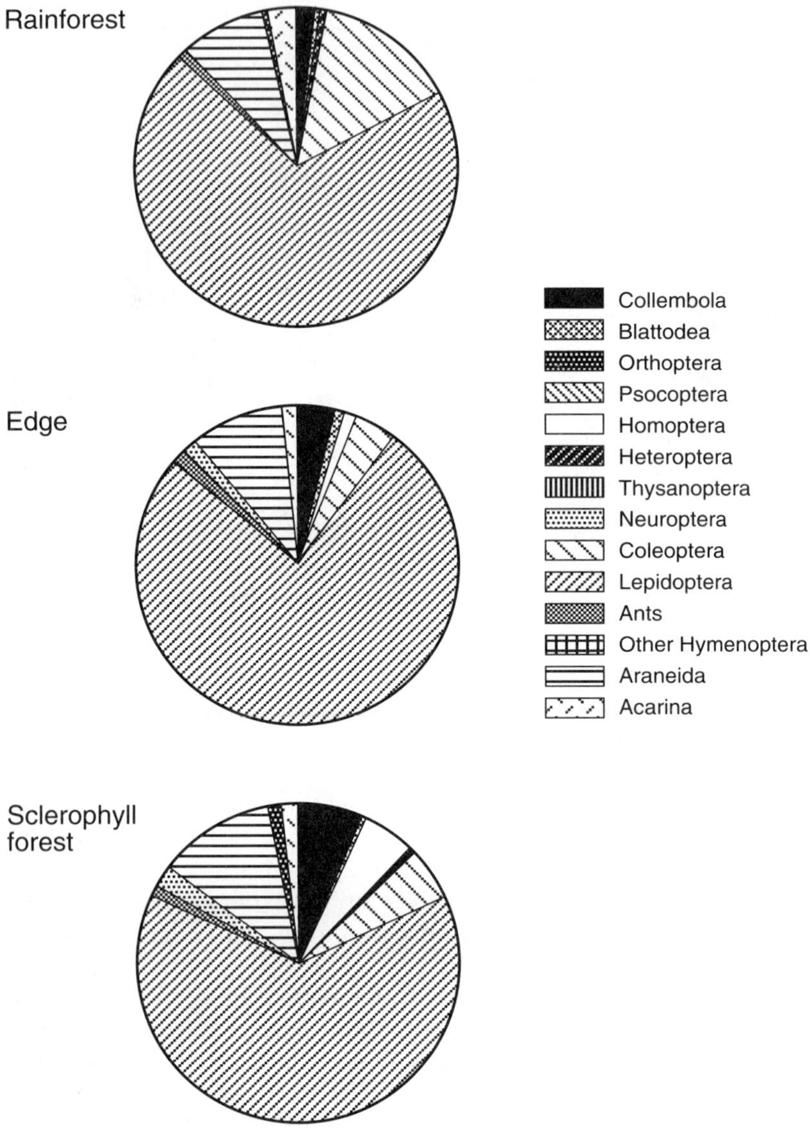

Figure 7.3 Ordinal profiles of arthropods from Malaise trap sampling across a rainforest/sclerophyll edge: all groups.

6. In addition to the results mentioned in this paper, we have also accumulated collections of moths from the canopy and associated ground sites at both subtropical and tropical sites using light-traps. These indicate that certain families of moths, notably the

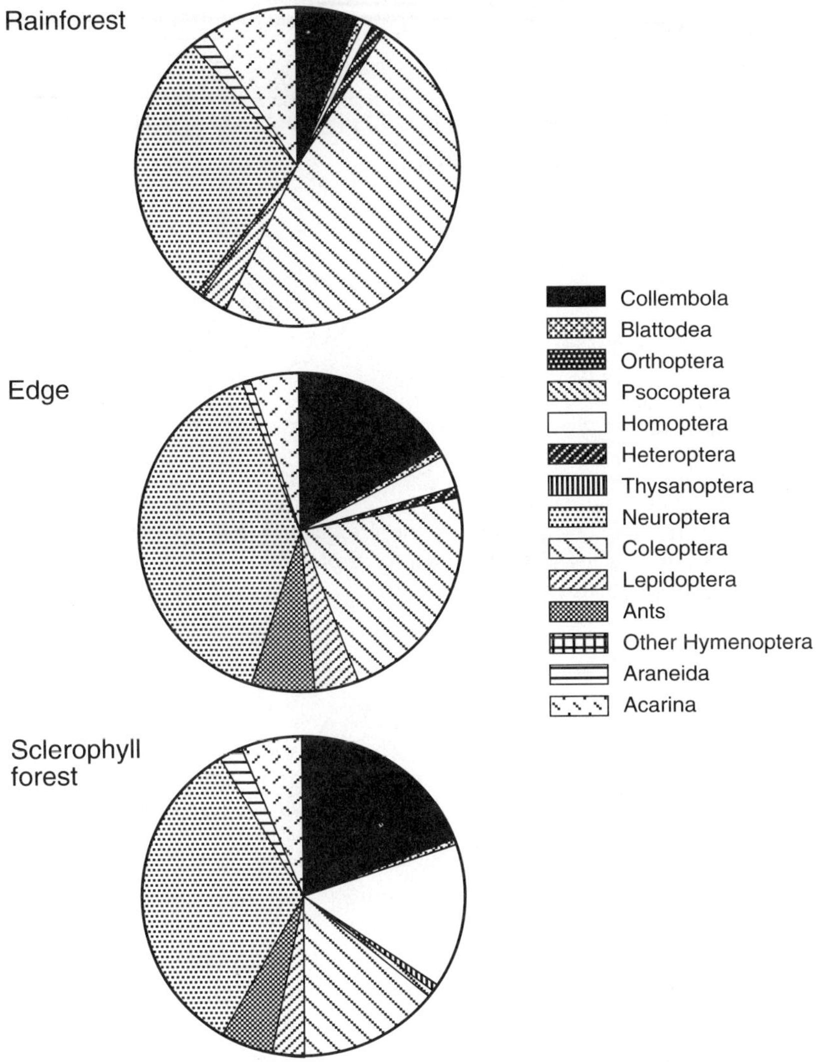

Figure 7.4 Ordinal profiles of arthropods from Malaise trap sampling across a rainforest/sclerophyll edge: all groups but Diptera.

Geometridae, have canopy assemblages distinct from those attracted to ground-based lights. These results will be presented shortly. In addition to these on-going activities we retain a 'wish-list' of additional, related projects which we believe will yield useful and interesting results and for which the general approach we use and the Australian environments to which we have ready access will

Table 7.4 Coleoptera obtained from Malaise trap samples across a rainforest/sclerophyll edge (see text)

	Rainforest	*Ecotone*	*Sclerophyll*
Total individuals caught	377	202	126
Total families	24	23	23
Unique	7	3	3
Shared with edge	15	–	18
Total species	97	87	61
Unique	62	42	29
Shared with edge	31	–	28

Table 7.5 Values of Sörensen's index of similarity based upon presence or absence of beetle species in Malaise trap samples taken across a rainforest/sclerophyll forest edge (see text)

	Rainforest	*Edge*	*Sclerophyll*
Rainforest	1	0.34	0.23
Edge	–	1	0.38
Sclerophyll	–	–	1

form a strong foundation. These will be undertaken as resources become available.

7. The littoral interface between rainforest and mangrove forest, with or without an interposing paperbark forest band, presents a fascinating and important ecotone readily identified and commonplace in tropical Queensland. The arthropod diversity along this gradient awaits study.
8. Edges between rainforests and other ecosystems are important in estimating gamma-diversity at the scale of the landscape. Important edges occur within rainforest ecosystems across geological boundaries and even across light gaps. Intuitively we believe these have impacts on regional diversity within the forest ecosystem. Such situations await study in Australia.
9. We have hereto studied the herbivore guild by analysis of knockdown samples from selected patches of canopy or species of trees. An obvious next stage – and one we intend to start upon within the next year – is to look much more closely at the actual herbivores of

selected tree species, to collect larvae and other insects directly from the foliage of sample trees and to rear these through for identification and comparison with existing host plant records. Such an analysis will be combined with examination of levels of leaf damage at selected levels upon defined leaf age classes within the rainforest.

10. Several forest ecosystems floristically allied to rainforest but lying, in Queensland, inland of the moist coastal forests are of great biogeographic interest. Several of these formations have been substantially cleared to produce cattle and sheep pastures. Most notable of these are the brigalow (*Acacia harpohylla* F. Muell. ex Benth.) shrublands of central Queensland. There are outstanding conservation reasons why the biodiversity of these systems should be examined in the patches that remain.

Acknowledgements

The work summarized in this paper has been supported by the Australian Research Council, Earthwatch (Australia), the Cooperative Research Centre for Tropical Rainforest Ecology and Management, the Wet Tropics Management Authority, the Australian Geographical Society and the Ian Potter Foundation. We are grateful to these bodies for their generous support. In addition, a great many individuals have contributed taxonomic expertise, technical assistance and constructive criticism at various stages of our work. They are acknowledged fully in other works.

REFERENCES

Balick, M.J., Furth, D.G. and Cooper-Driver, G. (1978) Biochemical and evolutionary aspects of arthropod predation on ferns. *Oecologia*, **35**, 55–89.

Basset, Y. (1990) The arboreal fauna of the rainforest tree *Argyrodendron actinophyllum* as sampled with restricted canopy fogging: composition of the fauna. *The Entomologist*, **109**, 173–83.

Basset, Y. (1991a) The seasonality of arboreal arthropods foraging within an Australian rainforest tree. *Ecological Entomology*, **16**, 265–78.

Basset, Y. (1991b) The taxonomic composition of the arthropod fauna associated with an Australian rainforest tree. *Australian Journal of Zoology*, **39**, 171–90.

Basset, Y. (1992) Influence of leaf traits on the spatial distribution of arboreal arthropods within an overstorey rainforest tree. *Ecological Entomology*, **17**, 8–16.

Benzing, D.H. (1980) *The Biology of the Bromeliads*, Mad River Press, Eureka, California.

Clunebell, R.R., Phillips, O.L., Gentry, A.H., Stark, N. and Zuuring, H. (1995) Prediction of neotropical tree and liana species richness from soil and climatic data. *Biodiversity and Conservation*, **4**, 56–90.

Crook, K.A.W. (1981) The break-up of the Australian–Antarctic segment of Gondwanaland, in *Ecological Biogeography of Australia* (ed. A. Keast), Junk, the Hague, pp. 1–14.

Daily, G.C. and Ehrlich, P.R. (1995) Preservation of biodiversity in small rainforest patches: rapid evaluations using butterfly trapping. *Biodiversity and Conservation*, **4**, 35–55.

Darlington, P.J. (1965) *The Biogeography of the Southern End of the World*, Harvard University Press, Cambridge, Massachusetts, USA.

Delamare-Deboutteville, C. (1948) *Comptes rendus de l'Académie des Sciences*, **226**, 1544–6.

Erwin, T.L. (1982) Tropical forests: their richness in Coleoptera and other arthropod species. *The Coleopterists' Bulletin*, **36**, 74–5.

Erwin, T.L. (1991) How many species are there? Revisited. *Conservation Biology*, **5**, 1–4.

Erwin, T.L. and Scott, J.C. (1980) Seasonal and size patterns, trophic structure, and niches of Coleoptera in tropical arboreal ecosystems: the fauna of the tree *Luehea seemannii* Triana and Planch in the Canal Zone of Panama. *The Coleopterists' Bulletin*, **34**, 305–22.

Gaston, K.J. (1991) The magnitude of global insect species richness. *Conservation Biology*, **5**, 183–6.

Gentry A.H. and Dodson C. (1987) Contribution of non-trees to species richness of a tropical rain forest. *Biotropica*, **19**, 149–56.

Gillison, A.N. (1993) *Background notes on the TROFIS–Mt. Haig gradsect: Lamb Range, far north Queensland*, CSIRO, Atherton, Unpublished report.

Hammond, P.M., Kitching, R.L. and Stork, N.E. The composition and richness of the tree-crown Coleoptera assemblage in an Australian subtropical forest. *Ecotropica* (in press).

Hopkins, M.S. (1990) Disturbance – the forest transformer, in *Australian Tropical Rainforests: Science, Value, Meaning* (eds L.J. Webb and J. Kikkawa), CSIRO, East Melbourne, pp. 40–52.

Jones, D.L. (1988) *Native Orchids of Australia*, Reed, Wellington, Sydney.

Kemp, E.M. (1981) Tertiary palaeography and the evolution of Australian climate, in *Ecological Biogeography of Australia* (ed. A. Keast), Junk, The Hague, pp. 31–49.

Kitching, R.L. (1986) Predation and communities, in *Community Ecology – Pattern and Process* (eds J. Kikkawa and D. Anderson), Blackwell Scientific Publications, Oxford, pp. 147–76.

Kitching, R.L. (1994) Biodiversity and taxonomy: impediment or opportunity, in *Conservation Biology in Australasia and Oceania* (eds C. Moritz, J. Kikkawa and D. Doley), Surrey Beatty, Sydney, pp. 253–68.

Kitching, R.L. and Arthur, M. (1993) The biodiversity of arthropods from Australian rainforest canopies: summary of projects and the impact of drought. *Selbyana*, **14**, 29–35.

Kitching, R.L. and Theischinger, G. (1996) The biodiversity of arthropods from Australian rainforest canopies: Tipulidae with a description of the new species *Leptotarsus (Macromastix) alfie* Theischinger. *Entomologist* (in press).

Kitching, R.L. and Zalucki J. (1996) The biodiversity of arthropods from Australian rainforest canopies: some results on the role of the tree species, in *Tropical Rainforest Research: Current Issues* (eds D.S. Edwards, W.E. Booth and S.C. Choy), Kluwer, Amsterdam, pp. 21–8.

Kitching, R.L., Bergelson, J., Lowman, M.D., McIntyre, S. and Carruthers, G. (1992) The biodiversity of arthropods from Australian rainforest canopies: general introduction, methods, study sites and ordinal results. *Australian Journal of Ecology*, **18**, 181–91.

Kitching, R.L., Floater, G. and Mitchell, H. (1994) The biodiversity of arthropods from Australian rainforest canopies: ecological questions and management challenges, in *Biodiversity: Its Complexity and Role* (eds M. Yasuno and M.M. Watanabe), Global Environmental Forum, Tokyo, pp. 119–37.

Kress, W.J. (1986) The systematic distribution of vascular epiphytes; an update. *Selbyana*, **9**, 2–22.

LaSalle, J. and Gauld, I.D. (1993) Hymenoptera: their diversity, and their impact on the diversity of other organisms, in *Hymenoptera and Biodiversity* (eds J. LaSalle and I.D. Gauld), C.A.B. International, Wallingford, pp. 1–26.

Lawton, J.H. (1978) Host-plant influences on insect diversity: the effects of space and time, in *Diversity of Insect Faunas* (eds L.A. Mound and N. Waloff) Royal Entomological Society of London, London, pp. 105–25.

Lovejoy, T.E., Bierregaard, R.O., Rankin, J.M. and Schubart, H.O.R. (1983) Ecological dynamics of tropical forest fragments, in *Tropical Rain Forest: Ecology and Management* (eds S.L. Sutton, T.C. Whitmore and A.C. Chadwick), Blackwell Scientific Publications, Oxford, pp. 377–84.

Lowman, M.D. (1985) Temporal and spatial variability in insect grazing of the canopies of five Australian rain forest tree species. *Australian Journal of Ecology*, **10**, 7–24.

Lowman, M.D. (1986) Insect herbivory in Australian rain forests – is it higher than in the Neotropics? *Proceedings of the Ecological Society of Australia*, **14**, 109–19.

Lowman, M.D. (1992) Herbivory in Australian rain forests, with particular reference to the canopies of *Doryphora sassafras* (Monimiaceae). *Biotropica*, **24**, 263–72.

Lowman, M.D. and Box, J.R. (1983) Variation in leaf toughness and phenolic content among five species of Australian rain forest trees. *Australian Journal of Ecology*, **12**, 17–25.

Majer, J.D. (1993) Comparison of the arboreal ant mosaic in Ghana, Brazil, Papua New Guinea and Australia – its structure and influence on arthropod diversity, in *Hymenoptera and Biodiversity* (eds J. LaSalle and I.D. Gauld), C.A.B. International, Wallingford, pp. 115–41.

Majer, J.D. and Recher, H.F. (1988) Invertebrate communities on western Australian eucalypts: a comparison of branch clipping and chemical knock-down methods. *Australian Journal of Ecology*, **13**, 269–78.

Majer, J.D., Recher, H.F., Perriman, W.S. and Achutan, N. (1990) Spatial variation of invertebrate abundance within the canopies of two Australian eucalypt forests. *Studies in Avian Biology*, **13**, 65–72.

May, R.M. (1990) How many species? *Philosophical Transactions of the Royal Society, Series B*, **330**, 293–304.

McIntyre, S., Kitching, R.L. and Jessup, L.W. (1994) Structure and composition of tropical rain forest in plots at Cape Tribulation, Queensland. *Proceedings of the Royal Society of Queensland*, **104**, 25–41.

Nadkarni, N.M. and Longino, J.T. (1990) Invertebrates in canopy and ground organic matter in a neotropical montane forest, Costa Rica. *Biotropica*, **22**, 286–9.

Nadkarni, N.M. and Matelson, T.J. (1991) Biomass and nutrient dynamics of epiphytic litterfall in a neotropical montane forest. *Biotropica*, **24**, 24–30.

Olson, D.M. (1994) The distribution of leaf litter invertebrates along a Neotropical altitudinal gradient. *Journal of Tropical Ecology*, **10**, 129–50.

Paoletti, M.G., Taylor, R.A.J., Stinner, B.R., Stinner, D.H. and Benzing, D.H. (1991) Diversity of soil fauna in the canopy and forest floor of a Venezuelan cloud forest. *Journal of Tropical Ecology*, **7**, 373–83.

Souza, O.F.F. de and Brown, V.K. (1994) Effects of habitat fragmentation on Amazonian termite communities. *Journal of Tropical Ecology*, **10**, 197–206.

Stork, N.E. (1987a) Guild structure of arthropods from Bornean rain forest trees. *Ecological Entomology*, **12**, 69–80.

Stork, N.E. (1987b) Arthropod faunal similarity of Bornean rain forest trees. *Ecological Entomology*, **12**, 219–26.

Stork, N.E. (1993) How many species are there? *Biodiversity and Conservation*, **2**, 215–32.

Stork, N.E. and Brendell, M.J.D. (1990) Variation in the insect fauna of Sulawesi trees with season, altitude and forest type, in *Insects and the Rain Forest of South–east Asia (Wallacea)* (eds W.J. Knight and J.D. Holloway), Royal Entomological Society of London, pp. 173–90.

Stork, N.E. and Brendell, M.J.D. (1993) Arthropod diversity studies in lowland rain forest of Seram, Indonesia, in *Natural History of Seram, Maluku, Indonesia* (eds I.D. Edwards, A.A. Macdonald and J. Proctor), Intercept Press, Andover, pp. 115–30.

Truswell, E.M. (1990) Australian rainforests: the 100 million year record, in *Australian Tropical Rainforests: Science, Value, Meaning* (eds L.J. Webb and J. Kikkawa), CSIRO, East Melbourne, pp. 7–22.

Webb, L.J. (1968) The environmental relationships of the structural types of Australian rain forest vegetation. *Ecology*, **49**, 296–311.

Webb, L.J. (1969) Edaphic differentiation of some forest types of eastern Australia. II. Soil chemical factors. *Journal of Ecology*, **57**, 817–30.

Wayne, P.M. and Bazzaz, F.A. (1991) Assessing diversity in plant communities: the importance of within-species variation. *Trends in Ecology and Evolution*, **6**, 400–4.

8

Canopy arthropods of coastal Sitka spruce trees on Vancouver Island, British Columbia, Canada

N.N. Winchester

ABSTRACT

Arthropod biodiversity was investigated in the Carmanah Valley on Vancouver Island, British Columbia. One component of this study, the arboreal arthropods, were collected within the canopies of five old-growth Sitka spruce trees using branch-clipping and substrate coring. Arthropods associated with branches in the canopy were dominated by individuals in the phytophagous, predator and parasitoid guilds. Individual tree characteristics and seasonality both contributed significantly to the proportional structuring of the phytophagous and predator guilds. Vertical partitioning was not a significant factor in guild proportionality. Interaction effects were only significant for the phytophagous guild. Taxonomic distinctness was most pronounced in the canopy moss mats where oribatid mites are members of a distinct arboreal community. A total of 56 species were resident in the canopy, of which 15 undescribed species were canopy-specific. Comparisons between the high-canopy and three ground sites indicated that, overall, percentage species similarity was low. The importance of describing species assemblages, documenting their habitat preferences and including processes into the framework of arthropods in old-growth forests is a challenge that lies ahead. Recognizing these components should assist efforts in addressing the issues that surround the maintenance of biological diversity (form and function) in these old-growth forests.

INTRODUCTION

The ultimate goal in recording biological diversity is to build a factual foundation for answering basic questions about evolution and ecology

Canopy Arthropods. Edited by N.E. Stork, J. Adis and R.K. Didham. Published in 1997 by Chapman & Hall, London. ISBN 0 412 74900 9

(May, 1992). The setting for building this foundation is the natural landscape which represents a mosaic of geological, environmental, ecological and evolutionary processes. With increased human disturbance across virtually all natural landscapes, the focus to study and preserve biological diversity has been centred in the tropics. These areas, which are rapidly being lost, contain more than half of the world's species (Wilson, 1988, 1992; Ehrlich, 1988). Tropical biotopes most at risk are the species-rich forests. However, it is a global reality that forests throughout the world are being compromised by human-induced perturbations. In temperate zones some of the last remaining tracts of intact old-growth coniferous forests occur in the Pacific Northwest of North America (Franklin, 1988). The ongoing fragmentation of these landscapes has heightened the awareness for a need to understand the endemic fauna and flora and apply system-based conservation approaches across a wide range of forest types.

Historically, little research concerning the conservation of biodiversity has been done in the primeval old-growth forests of the Pacific Northwest (Winchester and Ring, 1996) and this research has generally failed to link results to ecosystem processes. In British Columbia these forests are thought to contain much of the biodiversity of the province (Fenger and Harcombe, 1989; Bunnell, 1990; Winchester and Ring, 1996). They often have diffuse boundaries with other ecosystems, and this temporal and spatial mosaic creates a dynamic and complex set of habitats that are utilized by a variety of species. The faunal elements associated with these old-growth forests form a heterogeneous group, and nowhere is this more evident than in the arthropods. Arthropods, primarily insects, are an integral part of most old-growth systems and may comprise 80–90% of the total species in these systems (Asquith *et al.*, 1990). They play a primary role in the functioning of natural ecosystems, may regulate nutrient cycling (Mattson and Addy, 1975; O'Neill, 1976) and are now frequently mentioned as important components of diversity that need to be identified (May, 1986; Wilson, 1988; di Castri *et al.*, 1992; Samways, 1994).

Within the forests of the Pacific Northwest one of the least explored habitats is the forest canopy. Only a handful of studies on old-growth forest canopy invertebrates of the Pacific Northwest have been completed to date (Voegtlin, 1982; Schowalter, 1986, 1989). These studies were carried out in the context of old-growth Douglas fir–hemlock surveys in Oregon. Given the importance of arthropods in these old-growth forests, coupled with the lack of taxonomic knowledge of the canopy, the objective of this paper is to present results from the canopy segment of a larger study (Winchester, 1993; Ring and Winchester, 1996) which documents the arthropod fauna from an old-growth Sitka spruce forest. Specifically, I use guild structure and habitat-specificity to address the following questions:

1. What is the proportional guild structure of arthropods in the Sitka spruce canopy?
2. What are the effects of tree, time and height on the proportions of individuals in guilds?
3. Are new species present in the canopy and is there evidence to support habitat-specificity?

STUDY AREA

The study area is located in the Upper Carmanah Valley drainage (48°44'N 124°37'W) on the south-west coast of Vancouver Island, British Columbia, Canada (Figure 8.1). This typical U-shaped coastal valley, approximately 6731 ha in extent, is situated between the villages of Port Renfrew and Bamfield. The entire valley lies within the Coastal Western Hemlock Biogeoclimatic Zone with the exception of two high-elevation areas (Meidinger and Pojar, 1991). The dominant conifers in the Carmanah drainage are Sitka spruce (*Picea sitchensis* (Bong) Carr.), western hemlock (*Tsuga heterophylla* (Rafn.) Sarg.), western red cedar (*Thuja plicata* D.Don) and Pacific silver fir (*Abies amabilis* (Dougl.) Forb.).

The sample area in the Upper Carmanah Valley drainage includes six study sites: old-growth canopy, old-growth forest floor (both old-growth sites are approximately 700 years old), transition zone (edge between old-growth and clear-cut), clear-cut, second-growth 1, approximately 10 years old and second-growth 2, approximately 60 years old. All study sites except the 60-year-old site are located adjacent to each other along an approximately 4 km transect. This watershed represents an intact ancient forest that has evolved since the Wisconsin glaciation. In 1985 the clear-cut site (approx. 4 ha) was harvested and is the only area in the entire Carmanah watershed to be logged. The 60-year-old site, located along the Rosander mainline, is not contained in the Carmanah Valley watershed but lies 30 km to the south-east of the Upper Carmanah Valley. Data from the second-growth sites are still being processed and are not included in this paper.

Canopy access

A 2000-m linear transect was placed along Carmanah creek, and all Sitka spruce trees higher than 30 m were recorded. From these trees, five were randomly chosen to be incorporated into an access system. Access to the Sitka spruce canopy was by means of a 2 : 1 mechanical advantage pulley system, by wearing a harness and being attached to a series of climbing lines. In this way five adjacent Sitka spruce trees could be sampled simultaneously. Four wooden platforms strapped onto the branches and trunk

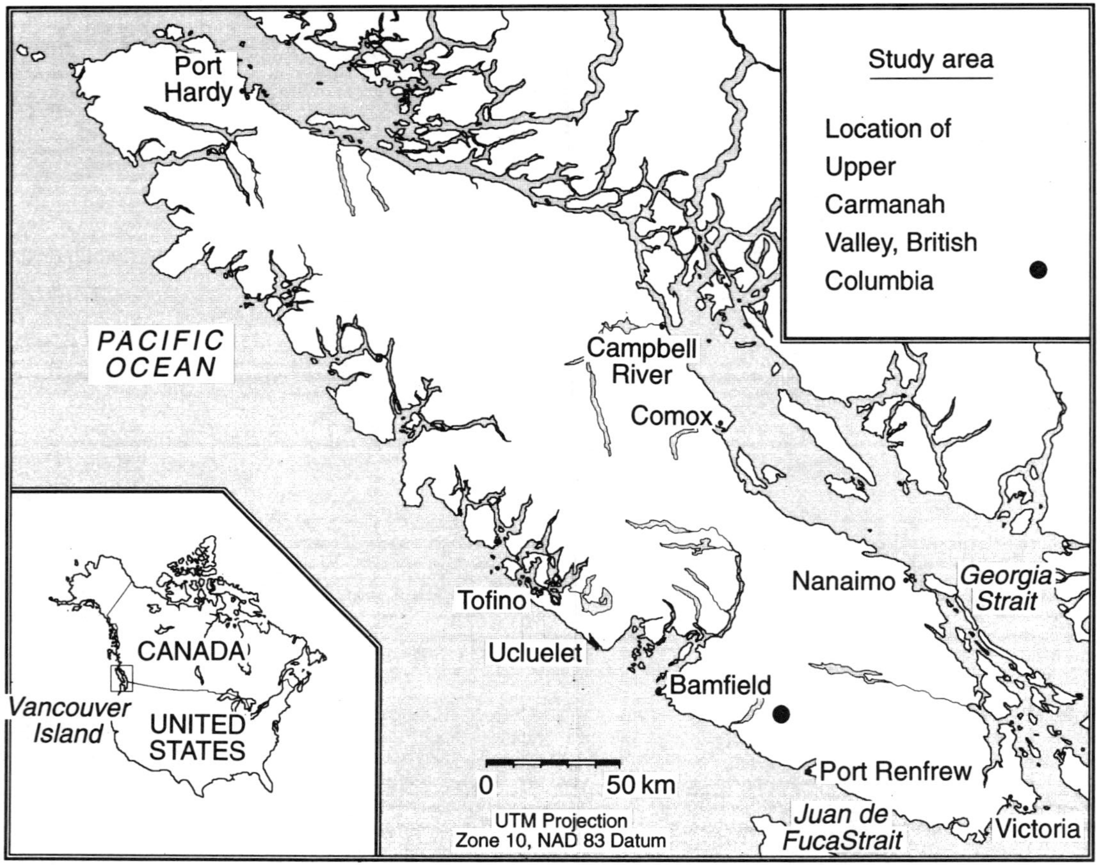

Figure 8.1 Map location of the Upper Carmanah Valley canopy research site, Vancouver Island, British Columbia, Canada.

of the main tree gave consistent heights (31 to 67 m) from which to sample. A series of burma bridges provided access to four other Sitka spruce trees, complete with platforms (Ring and Winchester, 1996). At the time of study, this station was the only permanent access system of this type available for long-term canopy work in northern temperate rainforests.

Survey design

Owing to the diversity of arthropods and their varied habits, no single survey method or sampling technique can be used for a complete study. The variety of techniques used in this study are summarized in Winchester and Scudder (1993) and the Biological Survey of Canada (1994). This paper will only deal with aspects of the 1991 sampling protocol which are listed below.

SAMPLING PROTOCOL

Branch clippings

The branch clipping programme was conducted in the five Sitka spruce trees contained in the fixed-access canopy system. The sampling procedure was modified after Schowalter (1989). In each tree, three samples were taken at each of three heights (33, 45 and 54 m). A total of 45 branches were collected for each of six sample periods. Samples were collected at 1-month intervals from May–October, 1991. All insects were removed from each sample and prepared for identification. Immature individuals of all species were reared to maturity. Single branches from each tree and level were run through Tullgren funnels to extract Collembola and Acari. After sorting, all branches were dried and total branch weight and needle weight were recorded.

Moss cores

A hand-held moss/soil corer (ca. 3 cm × 5 cm) was used to collect five moss/soil cores at random from each sample site once a month from May–October. A total of 120 cores were collected. Arthropods were extracted in the laboratory using Tullgren funnels for 48 hours. Samples were preserved in 75% ethanol. Volume displacement and dry weight were recorded for each core sample.

Table 8.1 Guild structure of the arthropod fauna collected from the branch-clipping programme. Data for tree, time and height were pooled and expressed as the mean number of individuals/kg dry plant material

Guild	*Percentage representation*
Phytophages	41.3
Predators	37.2
Parasitoids	11.8
Epiphyte fauna	8.3
Scavengers	1.0
Tourists	0.4

Sample sorting and data analysis

An informative view of canopy arthropods can be gained by placing them in guilds defined in terms of feeding habits. The guilds used in this study were structured after work by Root (1967, 1973) and further elaborated upon by Moran and Southwood (1982) and Stork (1987, 1988). The six guilds recognized in this study were: phytophages, epiphyte fauna, scavengers, predators, parasitoids and tourists. All arthropods except the Acarina and Collembola collected from the branch-clipping programme were identified to Family and arranged by guild. Guilds were expressed as a percentage of total individuals for the variables tree, time and height.

A mixed model three-way analysis of variance (SAS Institute Inc., 1982) was used to test for differences in the phytophage, predator, and parasitoid guilds. The low number of individuals in all other guilds (i.e. cells in the ANOVA) precluded statistical verification (see Simberloff, 1976, 1978). Data were expressed as mean number of individuals/kg dry plant material. Tree, time and height were the main effects (tree was random, time and height fixed) and a significance level of 0.05 was used. In factorial designs it is misleading to present significance tests that address the main effects if the interactions terms are significant (Krebs, 1989); therefore interaction terms are reported where significant.

Species level identifications were completed for most of the oribatid mites. Numerical relationships between oribatid species and the four study sites were calculated using all individuals from the moss cores, pooled over all collection times.

GUILD STRUCTURE OF THE COASTAL SITKA SPRUCE FOREST CANOPY

Canopy guild composition

Guild structure, expressed as mean biomass/kg dried plant material for all individuals from the branch-clipping sampling programme, is presented in Table 8.1. The canopy arthropod fauna in this study is dominated by the phytophagous (41.3%), predator (37.3%) and parasitoid (11.8%) guilds. Phytophagous guild representation is similar to the percentage contribution reported for temperate trees by Moran and Southwood (1982), but considerably higher than that reported by Stork (1987) for tropical trees. Predator and parasitoid guild proportions in our study were higher than that reported by Moran and Southwood (1982) and Stork (1987). Numerical dominance of functional groups in this study supported previous findings from coniferous forests by Schowalter and Crossley (1987) and Schowalter (1989). The phytophagous guild was composed of a small number of species (e.g. Lepidoptera, 13 species) which contain a large number of individuals (N. Winchester, unpublished data). This appears typical of plant-feeding species in this system and may relate to the reduced number of food options (mainly developing vegetative buds and female cones). The predator guild contains more species than the phytophagous guild and is composed primarily of 38 arachnid species which do not contain a large number of individuals (N. Winchester, unpublished data). Numerical dominance of spiders has been reported from other temperate studies (Nielsen, 1975; Ohmart and Voigt, 1981; Voegtlin, 1982; Bigot and Kabakibi, 1987; Basset, 1991a). The maintenance of high predator loading in a structurally and functionally diverse ecosystem such as the Carmanah Valley supports previous findings by Kareiva (1983), Risch (1981) and Schowalter (1986, 1989). The parasitoid guild is represented by a large number of species (e.g. Braconidae 118 species, N. Winchester, unpublished data) with low numbers of individuals. The main prey components of the parasitoids are species from the Lepidoptera and Aphididae. Species accumulation of parasitoids is not driven by taxonomic richness in the phytophagous guild, but is structured by the host-stage that is attacked. From the branch-clippings, several parasitoid species have been reared that attack the egg or larva or pupa of a variety of lepidopteran species. This variety of available host stages may contribute to an increase in parasitoid species that inhabit the canopy. Conspicuous by their absence in the canopy are Formicidae, with only four winged individuals collected during the entire study. Ants can exert considerable impact on other insects in arboreal habitats (Stork, 1987), although percentage contributions in temperate forests are generally low

(Southwood *et al.*, 1982; Basset, 1991a). The paucity of tourists is related to the transitory nature of these arthropods and branch-clipping does not adequately sample this faunal component. Basset (1991a) noted that interception traps collected many more vagile arthropods, whereas restricted canopy fogging yielded more sedentary, apterous and juvenile specimens. This point is supported by the use of canopy Malaise traps which collected 20 000 individuals, most of which were tourists (N. Winchester, unpublished data).

TREES/HEIGHT/TIME

Guild patterns have been shown to vary considerably, depending upon which variables or combination of variables are being considered (Southwood, 1960, 1961; Moran and Southwood, 1982; Kennedy and Southwood, 1984; Stork, 1987). Guild proportionality was explored for the phytophagous, predator and parasitoid guilds by using numbers of individuals and separating guild structure based on three factors: (i) tree individuality; (ii) vertical partitioning; and (iii) temporal sequencing.

Trees

Do trees act as individuals (*sensu* Moran and Southwood, 1982)? There is a significant effect of individual trees on the phytophagous guild ($F_{4,179} > 7.22$, $P < 0.0001$), and the predator guild ($F_{4,179} > 2.75$, $P < 0.05$), but no effect on the parasitoid guild ($F_{4,179} > 0.25$, n.s.). The number of insects on individual trees has been shown to vary with population size and proportional distribution among guilds (Southwood *et al.*, 1982). There appeared to be a remarkable consistency within guilds among the first three trees (Table 8.2). Where differences in proportional representation of insects in guilds between trees were evident, accumulations of single species in the phytophagous guild (usually aphids) occurred. The accumulation of single species, such as the aphid *Euceraphis punctipennis* (Zetterstedt), was recorded by Southwood *et al.* (1982). This difference was most pronounced in trees 4 and 5 where proportional representation of phytophages reached 46.3% and 60.6%, respectively. I conclude that although there is variation in guild proportionality between trees, this variation arises primarily from species which accumulate individuals in the phytophagous guild. The species in these trees are virtually identical (N. Winchester, unpublished data), confirming the observation by Moran and Southwood (1982) that the major guilds in the arboreal community are shaped by habitat characteristics of the tree which serve to impose a proportional consistency. Therefore, mature Sitka spruce present a habitat template that may dictate the guild composition of species, but not individuals. Individuals in the phytophagous and

Table 8.2 Percentage of arthropod individuals recorded from five Sitka spruce trees (time and height pooled) in the Carmanah Valley, arranged by guild

Guild	*Tree 1*	*Tree 2*	*Tree 3*	*Tree 4*	*Tree 5*
Phytophages	32.9	30.1	38.2	46.3	60.6
Predators	46.1	41.3	46.2	32.3	24.3
Parasitoids	9.7	18.8	7.6	8.2	10.1
Epiphyte fauna	9.7	8.5	6.3	12.3	4.6
Scavengers	1.4	1.3	0.4	0.2	0.0
Tourists	0.0	0.0	1.3	0.6	0.5

predator guilds exhibit non-uniformity between trees, which may be the result of a myriad of factors that are coupled with the physical characteristics of the tree. Factors may include plant chemistry (Southwood *et al.*, 1982), plant architecture (Lawton, 1983, 1986; Morse *et al.*, 1985), and plant health. Trees may act as individuals, in the sense that there are differences in guild proportionality and these differences are most evident in the phytophagous guild.

Height

Does tree height affect guild proportionality among individuals? Guild proportionality between heights, pooling trees and time, indicates that all guilds were similar (Table 8.3). The effect of height was not significant for the phytophagous, predator or parasitoid guilds ($P > 0.05$). This may reflect the ability of the phytophagous guild to track the availability of developing vegetative buds which occurred throughout the vertical profile of the canopy. Predators, comprising mainly web-constructing arachnids, also seemed able to utilize the entire vertical profile of the canopy. The guild proportionality of parasitoids is virtually identical between the high and mid-canopy zones, but is reduced, although not significantly, in the lower zone. This reduction may be a result of host-specificity, as the phytophage guild appeared to be composed of a higher proportion of Lepidoptera in the high and mid-canopy. The epiphytic guild has the highest guild proportionality in the lower canopy and may be associated with features of the habitat, including reduced moisture and a higher loading of the moss/lichen component. It is, however, known that certain groups studied here (e.g. oribatid mites) do segregate on a vertical gradient (Winchester, 1993), and one should approach with caution statements regarding vertical partitioning across a wide range of taxa.

Table 8.3 Percentage of arthropod individuals recorded from three heights (time and tree pooled) in the Carmanah Valley, arranged by guild

Guild	*High*	*Mid*	*Low*
Phytophages	42.7	42.7	36.2
Predators	39.1	35.2	41.0
Parasitoids	11.3	13.2	5.7
Epiphyte fauna	6.0	7.1	15.2
Scavengers	0.9	1.6	1.2
Tourists	0.0	0.0	0.7

Time

Does time affect guild proportionality among individuals? Time alludes to seasonality and has been shown to have an effect on species and individual composition (Erwin and Scott, 1980; Schowalter *et al.*, 1988). This is a factor that should be considered when addressing consistencies in guild proportionalities (Stork, 1988). Considerable differences are exhibited in guild proportionality through time, and these changes are similar in terms of direction but not magnitude (Table 8.4).

The significant effect of time on the phytophagous guild ($F_{5,20} > 6.91$, $P < 0.0001$) is related to the flush of vegetative buds and development of female cones, and supports the observation of seasonal structure for phytophagous insects noted by Lawton (1983). Early in the growing season (June to early July) the development of vegetative buds and female cones provides a food source that enables the accumulation of individuals in the phytophagous guild (54.6–57.7%) to occur. This pattern has also been noted in other canopy studies by Nielsen and Ejlersen (1977), Schowalter *et al.* (1988) and Basset (1991b). During late July, phytophagous numbers start to decline (32.8%) while the proportional representation of predators continues to increase until late September (53.25%). There is no significant effect of time on the predator guild ($P > 0.5$) which indicates that the recorded increased proportional representation from early July (23.88%) to late September (53.25%) is a reflection of changes in the phytophagous guild. A high proportion of the predator guild is composed of spiders which are present in relatively even numbers over the length of the growing season. Temperate spiders have been shown to be poorly synchronized with herbivore accumulations (Renault and Miller, 1972; Basset, 1991b) and may be able to wait or switch prey items based on availability. Input from the forest floor or adjacent riparian zones has been suggested by D. Voegtlin

Table 8.4 Percentage of arthropod individuals recorded from six time (date) intervals (tree and height pooled) in the Carmanah Valley, arranged by guild

Guild	*Time (date)**					
	1	*2*	*3*	*4*	*5*	*6*
Phytophages	54.6	57.8	32.8	24.9	24.9	35.7
Predators	28.5	23.9	41.4	42.3	53.3	52.9
Parasitoids	4.1	11.1	20.1	11.7	17.8	10.2
Epiphyte fauna	11.9	6.1	5.1	19.6	2.4	0.0
Scavengers	0.4	0.5	0.5	1.1	1.8	1.3
Tourists	0.7	0.6	0.0	0.4	0.0	0.0

* Times (dates) are: 1 = 4.6.91; 2 = 3.7.91; 3 = 30.7.91; 4 = 27.8.91; 5 = 21.9.91; 6 = 27.10.91

(unpublished data) as areas which provide a food source during times of low numbers of resident herbivores. The parasitoids are more difficult to follow and are closely associated with the number and stage of their hosts, principally Lepidoptera and Aphididae. Time has a significant effect on the parasitoid guild ($F_{5,20}$ >3.14, P <0.05). There appear to be two peaks of emergence for parasitoids, one that occurs in late July and the other in early September. Synchronization of emergence is closely associated with host biology (see Price, 1991; Hawkins, 1993). Reared material (N. Winchester, unpublished data) indicates that the first peak is composed of parasitoids that attack the larval stages of Lepidoptera, while the second peak appears to be related to parasitoids that attack the pupal stage.

Interaction effects

Interaction effects were only significant for the phytophagous guild (tree × height, $F_{8,179}$ >4.33, P <0.0001; tree × time, $F_{20,179}$ >2.63, P <0.0005; and tree × height × time, $F_{40,179}$ >1.87, P <0.005). Interaction terms are difficult to interpret and I present these results to indicate that these interactions need to be considered when addressing the significance attributed to the main effects. Further study is required to ascertain the biological meaning of these interactions. Currently I am detailing life histories of several species from the phytophagous guild in order to address the biological significance of the interaction terms. Interaction effects were not significant (P >0.05) for any of the other guilds.

ACARINA

Of the 7219 soil microarthropods collected by coring the thick, rich moss mats of the canopy, the numerically dominant group, both in terms of individuals (5937) and species (85+), is the Acarina. Although not specifically dealing with the canopy, mites were found to be one of the largest arthropod components in studies conducted in the rainforests of Peru (Beck, 1963), Nigeria (Madge, 1965) and Costa Rica (Nadkarni and Longino, 1990). Within the Acarina, the Oribatida ('beetle mites') is the dominant Suborder in our samples. A similar situation was found in the tropics by Beck (1963) and in mature northern temperate forests by Wallwork (1983) and Moldenke and Lattin (1990). Numerical relationships of the oribatid mite fauna (Figure 8.2) indicate that the canopy has the highest number of species (56), followed by the forest floor (48). The

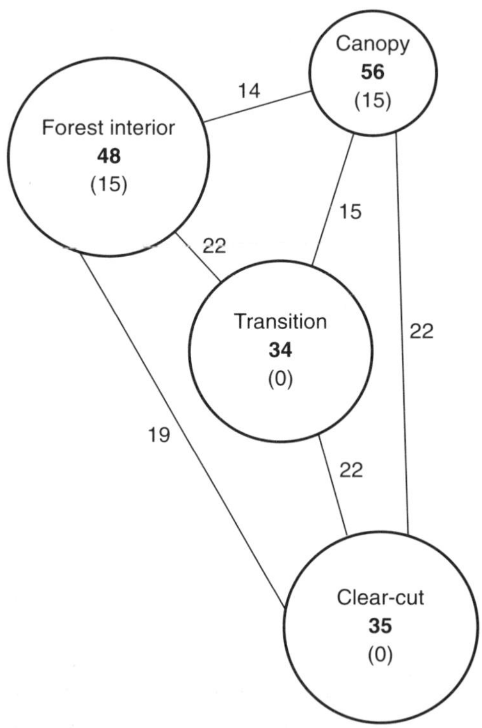

Figure 8.2 Numerical relationship between oribatid species from four study sites in the Upper Carmanah Valley. Data are pooled from all trap collections over all time intervals. Numbers along the lines represent those species in common between the sites. Numbers within the circles represent the number of species occupying a given site and numbers in brackets represent the presence of new species.

total number of species present in the transition zone (34) is similar to the clear-cut zone (35). Percentage similarity is lowest between the canopy and forest floor (18%) and highest between the canopy and clear-cut (41%). Overall, percentage similarity is generally high between any of the other pair-wise ground comparisons (range 42–88%). Of the 30 confirmed new species, 15 were specific to the forest floor and 15 were specific to the canopy.

CANOPY ORIBATIDA

Perhaps the most interesting and least explored habitat in the Sitka spruce canopy is the 4–28-cm deep moss mats which support a well-developed soil layer. These mats are primarily composed of three moss species, *Isothecium myosuroides* Brid., *Antitrichia curtipendula* (Hedw.) Brid, and *Dicranum fuscescens* Sm., which are also abundant on the forest floor. Soil microarthropods dominate this canopy soil/litter habitat, a fact which has not been well documented in these forests, but has been noted in other canopy studies (Nadkarni and Longino, 1990; Paoletti *et al.*, 1990). From the oribatid mites that have been processed to date, there is strong evidence that we are dealing with a distinct arboreal fauna. A high number of species with low percentage similarity to ground sites is consistent with the findings of Behan-Pelletier *et al.* (1993). The discovery of several new oribatid species is not surprising (see Behan-Pelletier, 1993) given the scope of this study. Fifteen undescribed species appear to be confined to habitats found only in the old-growth forest canopy. For example, *Dendrozetes* represents the first record for this genus in North America and this new species has modifications for an arboreal existence (V. Behan-Pelletier, personal communication). *Parapirnodus, Paraleius*, and *Anachipteria* are genera that are known to be arboreal (V. Behan-Pelletier, personal communication) and in this study each was represented by an undescribed, strictly arboreal species. Similarly, new species with unique habitat associations have been recorded in Oregon (Voegtlin, 1982), northern Venezuela (Behan-Pelletier *et al.*, 1993), Peru (Wunderle, 1992) and Australia (Walter *et al.*, 1994). The microhabitats associated with the canopy of the ancient Sitka spruce trees are not replicated in any second-growth forest canopies that we have surveyed to date, and it is unlikely that these habitat features will develop in second-growth forests that are in an 80–120-year rotation. I suggest that there are enough differences in canopy microhabitat conditions to promote the development of taxonomically discrete species assemblages that will be lost if these canopy habitats are not retained, or allowed to develop in second-growth forests. Canopy specificity indicates that the species richness of arboreal oribatids is not just a sub-set of the ground fauna. Studies from distinct geographic areas indicate that, in general, the overlap between arboreal oribatid species and their ground-dwelling counterparts is less than 40%.

SUMMARY

The resident canopy arthropod fauna in this study is dominated by the phytophagous and predator–parasitoid guilds, supporting previous studies in coniferous forests (Schowalter and Crossley, 1987; Schowalter, 1989). The phytophagous guild is composed mainly of species that are feeding on the developing vegetative buds and female cones. These species appear to have little effect on the host, with a negligible loss of developing plant tissue. I infer from this guild structure that herbivory in this mature, structurally complex forest is relatively insignificant and, through a series of checks and balances, large-scale herbivore damage (insect outbreaks) is negligible. This supports previous findings by Reichle *et al.* (1973), Nielsen (1978), Ohmart *et al.* (1983) and Schowalter (1989) who noted that herbivory was less than 10% of the standing crop in mature forests. The maintenance of a high predator loading in a structurally and functionally diverse ecosystem such as the Carmanah Valley supports previous findings by Kareiva (1983), Risch (1981) and Schowalter (1986, 1989).

I present evidence to suggest that several species – many new to science – exhibit habitat specificity that restricts their distribution to structural attributes contained only in the old-growth forest, both on the forest floor and in the canopy. Canopy specificity is most pronounced in the microarthropods that inhabit the thick, rich moss-mats. This arboreal community is dominated, in both numbers of individuals and species, by oribatid mites. Of all the arthropod groups that have been examined, the oribatids contain the greatest number of new species. Patterns of community structure on trees, examined at the guild level, indicate that in terms of number of individuals, the phytophage and predator–parasitoid guilds are numerically dominant. The high proportion of the predator–parasitoid complement indicates that herbivory in these mature, structurally complex forests is relatively insignificant. Members of the phytophage guild are primarily composed of species from the Lepidoptera and Aphididae which are associated with the developing vegetative buds and female cones. Guild proportionality exhibits a temporal sequencing over time. Vertical height, however, does not affect guild proportionality. Differences between trees, while not pronounced, are significant and relate to numerical accumulation in the phytophage guild. The summarizing of these key patterns and documentation of changes due to disturbance should identify ecological roles of arthropods that are at the heart of the biodiversity challenge. The arthropod specimens collected in this study will help provide an understanding of the diversity, habitat requirements and system processes that occur within these northern temperate old-growth rainforests.

Acknowledgements

I gratefully acknowledge the continued support, advice and cooperation given to me by Richard A. Ring. Thanks is extended to the Western Canada Wilderness Committee who made available the research facility and were instrumental in helping to establish the fixed canopy access system. This project was made possible with financial support by FRDA research grants from the B.C. Ministry of Forests, Research Branch. Acknowledgement is made to A. Mackinnon and B. Nyberg for their continued support. I am indebted to the following taxonomic experts for identifications, their contributions are invaluable and form the essence for understanding arthropod biodiversity in these old-growth forests: R.S. Anderson, G.E. Ball, V. Behan-Pelletier, R.G. Bennett, Y. Bousquet, D.E. Bright, F. Brodo, D. Buckle, J.F. Burger, J.M. Campbell, R.A. Cannings, D.S. Chandler, E.I. Coher, B.E. Cooper, A. Davies, R. Duncan, the late G. Eickwort, A.T. Finnamore, A. Fjellberg, B. Foottit, G.A.P. Gibson, H. Goulet, K.G.A. Hamilton, J. Huber, L.M. Humble, E.E. Lindquist, S.A. Marshall, L. Masner, E.L. Mockford, A.P. Nimmo, J.D. Oswald, S.J. Peck, D. Pollock, F. Rafi, J.H. Redner, M.J. Sharkey, R.M. Shelley, D. Shpeley, A. Smetana, J. Turgeon, V.R. Vickery, H.C.W. Walther, R. West, G.B. Wiggins, D.M. Wood. I also thank B. Lund, N. Prockiw, and a host of volunteers for assistance with the sorting and preparation of specimens, and K. Jordan and S. Hughes for providing technical expertise in the arboreal aspects of this project.

REFERENCES

Asquith, A., Lattin, J.D. and Moldenke, A.R. (1990) Arthropods, the invisible diversity. *Northwest Environmental Journal*, **6**, 404–5.

Basset, Y. (1991a) The taxonomic composition of the arthropod fauna associated with an Australian rainforest tree. *Australian Journal of Zoology*, **39**, 171–90.

Basset, Y. (1991b) The seasonality of arboreal arthropods foraging within an Australian rainforest tree. *Ecological Entomology*, **16**, 265–78.

Beck, L. (1963) Zur Okologie und Taxonomie der neotropischen Bodentiere. 1. Aur Oribatidenfauna Perus. *Zoologische Jahrbucher (Systematik)*, **90**, 299–392.

Behan-Pelletier, V.M. (1993) Diversity of soil arthropods in Canada: systematic and ecological problems, in Systematics and Entomology: Diversity, Distribution, Adaptation, and Application (eds G.E. Ball and H.V. Danks), *Memoirs of the Entomological Society of Canada*, **165**, 11–50.

Behan-Pelletier, V.M., Paoletti, M.G., Bissett, B. and Stinner, B.R. (1993) Oribatid mites of forest habitats in northern Venezuela. *Tropical Zoology, Special Issue*, **1**, 39–54.

Bigot, L. and Kabakibi, M. (1987) Evolution spatio-temporelle de la composition et de la structure du peuplement fromdicole sur chene-liege dans le Massif des Maures (Var.). *Bulletin d'Ecologie*, **18**, 157–68.

Biological Survey Of Canada (1994) Terrestrial arthropod biodiversity: planning a study and recommended sampling techniques. *Supplement to the Bulletin of the Entomological Society of Canada*, **26**, 1–33.

Bunnell, F.L. (1990) Forestry wildlife: w(h)ither the future? in *Wild and Managed: Differences and Consequences* (eds A.F. Pearson and D.A. Challenger), Proceedings of a Symposium on Forests, Students for Forestry Awareness, U.B.C., Vancouver, BC, pp. 163–76.

Di Castri, F., Vernhes, J.R. and Younes, T. (1992) Inventorying and monitoring of biodiversity. A proposal for an international network. *Biology International, Special Issue*, **27**, 1–27.

Ehrlich, P.R. (1988) The loss of diversity: causes and consequences, in *Biodiversity* (ed. E.O. Wilson), National Academy Press, Washington, DC, pp. 21–7.

Erwin, T.L. and Scott, J.C. (1980) Seasonal and size patterns, trophic structure, and richness of Coleoptera in the tropical arboreal ecosystem: the fauna of the tree *Luehea seemannii* in the Canal Zone of Panama. *The Coleopterists' Bulletin*, **34**, 305–55.

Fenger, M. and Harcombe. A. (1989) A discussion paper on old-growth forests, biodiversity, and wildlife in British Columbia. Prepared by the Wildlife Branch of the BC Ministry for the Environment, for the workshop 'Towards an Old-Growth Strategy', November 3–5, Parksville, BC. British Columbia Ministry for the Environment, Lands and Parks.

Franklin, J.F. (1988) Structural and functional diversity in temperate forests, in *Biodiversity* (ed. E.O. Wilson), National Academy Press, Washington DC, pp. 166–75.

Hawkins, B.A. (1993) Refuges, host population dynamics and the genesis of parasitoid diversity, in *Hymenoptera and Biodiversity* (eds J. LaSalle and I.D. Gauld), The Natural History Museum and CAB International, Wallingford, UK, pp. 235–56.

Kareiva, P. (1983) Influence of vegetation texture on herbivore populations: resource concentration and herbivore movement, in *Variable Plants and Herbivores in Natural and Managed Systems* (eds R.F. Denno and M.S. McClure), Academic Press, New York, pp. 259–89.

Kennedy, C.E.J. and Southwood, T.R.E. (1984) The number of species associated with British trees: a re-analysis. *Journal of Animal Ecology*, **53**, 455–78.

Krebs, C.J. (1989) *Ecological Methodology*, Harper Collins, London.

Lawton, J.H. (1983) Plant architecture and diversity of phytophagous insects. *Annual Review of Entomology*, **28**, 23–39.

Lawton, J.H. (1986) Surface availability and insect community structure: the effects of architecture and fractal dimension of plants, in *Insects and the Plant Surface* (eds B.E. Juniper and T.R.E. Southwood), Edward Arnold, London, pp. 317–31.

Madge, D.S. (1965) Leaf fall and litter disappearance in a tropical forest. *Pedobiologia*, **5**, 273–88.

Mattson, W.J., and Addy, N.D. (1975) Phytophagous insects as regulators of forest primary production. *Science*, **190**, 515–22.

May, R.M. (1986) The search for patterns in the balance of nature: advances and retreats. *Ecology*, **67**, 1115–26.

May, R.M. (1992) How many species inhabit the earth? *Scientific American*, **267**, 42–8.

Meidinger, D. and Pojar, J. (1991) *Ecosystems of British Columbia*, BC Ministry of Forests, Victoria, British Columbia.

Moldenke, A.R. and Lattin, J.D. (1990) Density and diversity of soil arthropods as 'biological probes' of complex soil phenomena. *The Northwest Environmental Journal*, **6**, 2.

Moran, V.C. and Southwood, T.R.E. (1982) The guild composition of arthropod communities in trees. *Journal of Animal Ecology*, **51**, 289–306.

Morse, D.R., Lawton, J.H., Dodson, M.M. and Williamson, M.H. (1985) Fractal dimension of vegetation and the distribution of arthropod body lengths. *Nature*, **314**, 731–3.

Nadkarni, N.M. and Longino, J.T. (1990) Invertebrates in canopy and ground organic matter in a neotropical montane forest, Costa Rica. *Biotropica*, **22**, 286–9.

Nielsen, B.O. (1975) The species composition and community structure of the beech canopy fauna in Denmark. *Videnskabelige Meddelelser fra Dansk Naturhistorisk Forening*, **138**, 137–70.

Nielsen, B.O. (1978) Food resource partition in the beech leaf-feeding guild. *Ecological Entomology*, **3**, 193–201.

Nielsen, B.O. and Ejlersen, A. (1977) The distribution pattern of herbivory in a beech canopy. *Ecological Entomology*, **2**, 293–9.

Ohmart, C.P. and Voigt, W.G. (1981) Arthropod communities in the crown of the natural and planted stands of *Pinus radiata* (Monterey pine) in Canada. *Canadian Entomologist*, **113**, 673–84.

Ohmart, C.V., Stewart, L.G. and Thomas, J.R. (1983) Phytophagous insect communities in the canopies of three *Eucalyptus* forest types in south-eastern Australia. *Australian Journal of Ecology*, **8**, 395–403.

O'Neill, R.V. (1976) Ecosystem perturbation and heterotrophic regulation. *Ecology*, **57**, 1244–53.

Paoletti, M.G., Stinner, B.R., Stinner, D., Benzing, D. and Taylor, R. (1990) Diversity of soil fauna in the canopy of Neotropical rainforest. *Journal of Tropical Ecology*, **7**, 135–45.

Price, P.W. (1991) Evolutionary theory of host and parasitoid interactions. *Biological Control Theory and Application in Pest Management*, **1**, 83–93.

Reichle, D.E., Golstein, R.A., Van Hook, R.I. Jr and Dodson, G.J. (1973) Analysis of insect consumption in a forest canopy. *Ecology*, **54**, 1076–84.

Renault, T.R. and Miller, C.A. (1972) Spiders in a fir–spruce biotype: abundance, diversity, and influence on spruce budworm densities. *Canadian Journal of Zoology*, **50**, 1039–46.

Ring, R.A. and Winchester, N.N. (1996) Coastal temperate rainforest canopy access systems in British Columbia, Canada. *Selbyana*, **17**, 22–6.

Risch, S.J. (1981) Insect herbivore abundance in tropical monocultures and polycultures: an experimental test of two hypotheses. *Ecology*, **62**, 1325–40.

Root, R.B. (1967) The niche exploitation pattern of the Blue-gray Gnatcatcher. *Ecological Monographs*, **37**, 317–50.

Root, R.B. (1973) Organization of a plant–arthropod association in simple and diverse habitats: the fauna of collards (*Brassica oleracea*). *Ecological Monographs*, **43**, 95–124.

Samways, M.J. (1994) *Insect Conservation Biology*, Chapman & Hall, London, UK.

SAS Institute, Inc. (1982) *SAS User's Guide: Statistics*, 1982 edition, SAS Institute, Inc., Cary, North Carolina.

Schowalter, T.D. (1986) Ecological strategies of forest insects: the need for a community-level approach to reforestation. *New Forests*, **1**, 57–66.

Schowalter, T.D. (1989) Canopy arthropod community structure and herbivory in old-growth and regenerating forests in western Oregon. *Canadian Journal of Forestry Research*, **19**, 318–22.

Schowalter, T.D. and Crossley, D.A. Jr (1987) Canopy arthropods and their response to forest disturbance, in *Forest Hydrology and Ecology at Coweeta* (eds W.T. Swank and D.A. Crossley Jr), Springer-Verlag, New York, pp. 207–18.

Schowalter, T.D., Stafford, S.G. and Slagle, R.L. (1988) Arboreal arthropod community structure in an early successional coniferous forest ecosystem in western Oregon. *Great Basin Naturalist*, **47**, 327–33.

Simberloff, D.S. (1976) Experimental zoogeography of islands: effects of island size. *Ecology*, **50**, 629–48.

Simberloff, D.S. (1978) Colonization of islands by insects: immigration, extinction and diversity, in *Diversity of Insect Faunas* (eds L.A. Mound and N. Waloff), Symposium of the Royal Entomological Society of London, No. 9, Blackwell Scientific Publications, London, pp. 139–53,

Southwood, T.R.E. (1960) The abundance of the Hawaiian trees and the number of their associated insect species. *Proceedings of the Hawaiian Entomological Society*, **17**(2), 299–303.

Southwood, T.R.E. (1961) The number of species of insect associated with various trees. *Journal of Animal Ecology*, **30**, 1–8.

Southwood, T.R.E., Moran, V.C. and Kennedy, C.E.J. (1982) The assessment of arboreal insect fauna: comparisons of knockdown sampling and faunal lists. *Ecological Entomology*, **7**, 331–40.

Stork, N.E. (1987) Guild structure of arthropods from Bornean rain forest trees. *Ecological Entomology*, **12**, 61–80.

Stork, N.E. (1988) Insect diversity: facts, fiction and speculation. *Biological Journal of the Linnean Society*, **35**, 321–37.

Voegtlin, D.J. (1982) *Invertebrates of the H.J. Andrews Experimental Forest, Western Cascade Mountains, Oregon: A Survey of Arthropods Associated with the Canopy of Old-growth Pseudotsuga menziesii*. Forest Research Laboratory, Oregon State University, Corvallis, pp. 1–29.

Wallwork J.A. (1983) Oribatids in forest ecosystems. *Annual Review of Entomology*, **28**, 109–30.

Walter, D.E., O'Dowd, D.J. and Barnes, V. (1994) The forgotten arthropods: foliar mites in the forest canopy. *Memoirs of the Queensland Museum*, **36**(1), 221–6.

Wilson E.O. (1988) The current state of biological diversity, in *Biodiversity* (eds E.O. Wilson and F.M. Peter), National Academy Press, Washington DC, pp. 3–17.

Wilson, E.O. (1992) *The Diversity of Life*, Belknap Press, Cambridge, Massachusetts.

Winchester, N.N. (1993) Coastal Sitka spruce canopies: conservation of biodiversity. *Bioline*, **11**(2), 9–14.

Winchester, N.N. and Ring, R.A. (1996) Centinelan extinctions: extirpation of northern temperate old-growth rainforest arthropod communities. *Selbyana*, **17**, 50–7.

Winchester, N.N., and Scudder, G.G.E. (1993) *Methodology for Sampling Terrestrial Arthropods in British Columbia*, Resource Inventory Committee, B.C. Ministry of Environment, Lands and Parks.

Wunderle, I. (1992) Die baum – und bodenbewohnenden Oribatiden (Acari) im Tieflandregenwald von Panguana, Peru. *Amazoniana*, **12**(1), 119–42.

9

The beetle fauna of different tree species in forests of Rwanda and East Zaire

T. Wagner

ABSTRACT

The arthropod fauna of trees was sampled by canopy fogging in several different forest types at the end of the local dry season in October–November 1993 in Central Africa. On four trees of *Lannea fulva* (Anacardiaceae) in a dry forest in East Rwanda, a total of 84 beetle species were found. Two species of Anthicidae were dominant, representing 75.2% of total beetle abundance. There were 224 beetle species on eight trees of *Teclea nobilis* (Rutaceae) in a nearby gallery forest. Phytophagous beetles dominated with 53% of all individuals. In particular, five species of Alticinae (Chrysomelidae) represented nearly 30% of total phytophage abundance. The fauna on the different trees was relatively similar, the species accumulation curve approaching an asymptote indicating that the fauna was well sampled. In a montane rainforest in West Rwanda 393 beetle species were collected from nine trees of *Carapa grandiflora* (Meliaceae). Fungivorous species dominated. Species accumulation and the corresponding rarefaction curve did not plateau, i.e. the beetle fauna on nine trees of this species was not completely inventoried. In an upper lowland rainforest in the Kivu Province, Zaire, 233 beetles of 137 species were collected from five trees of *Carapa grandiflora*. This extremely sparse distribution may be caused by a high abundance of ants in the Kivu forest.

INTRODUCTION

Studying arthropod communities on trees has been made much easier through the introduction of insecticide fogging. Using this method it has been possible to obtain quantitative samples of arthropods for

Canopy Arthropods. Edited by N.E. Stork, J. Adis and R.K. Didham. Published in 1997 by Chapman & Hall, London. ISBN 0 412 74900 9

investigations of species composition, distribution patterns and the diversity of arthropod communities living in the tree canopy. As a result, theoretical aspects of the formation and development of animal communities have been studied more intensively. Canopy fogging has led to remarkable findings concerning global diversity (Erwin, 1982; Stork, 1991). However, most of this research has been carried out in South America (Erwin and Scott, 1980; Farrell and Erwin, 1988) and south-east Asia (Stork, 1987, 1991; Morse *et al.*, 1988), while there is a lack of comparable studies in the African tropics.

This paper presents the results of a study of the arthropod fauna of trees in several types of forest in Central Africa (Appendix 9A), and essentially deals with the beetle fauna of named species of trees. Furthermore, the effect of ant abundance on the abundance and distribution of beetles is discussed.

MATERIALS AND METHODS

Research was carried out in October and November 1993 in Rwanda and in the Kivu Province of Zaire at the end of the local dry season. For different types of forests a widespread and abundant tree species was chosen. In East Rwanda (Rusumo, Ibanda Makera, 1450 m above sea level) four trees of *Lannea fulva* (Anacardiaceae) in a dry forest, and eight trees of *Teclea nobilis* (Rutaceae) in a gallery forest near the Akagera river were fogged with knockdown insecticides. Nine trees of *Carapa grandiflora* (Meliaceae) were investigated in the montane rainforests of Forêt de Nyungwe, and in those of the nearby isolated area of Cyamudongo, both located in the Zaire/Nile watershed (1750–2200 m a.s.l.). Five trees of *C. grandiflora* were also fogged at the Irangi research station (Kivu, Zaire, 950 m a.s.l.).

Trees 6–9 m in height were each fogged from the ground for about 4 minutes with natural pyrethrum (1% active ingredient) in the early morning using a Swingfog SN-50. Falling arthropods were collected on eight 1-m^2 sheets hung near the trunk under the canopy. The majority of arthropods fell within a few minutes after fogging. All individuals which fell during a drop time of 1½ hour were collected and preserved in alcohol. The samples were sorted to higher taxonomic units, the individuals counted and the beetles sorted to morphospecies or determined to known species. Diversity and distribution patterns described by Shannon–Wiener diversity (H_S) and evenness (e) indices are discussed only briefly here. The more detailed description obtained by means of species-accumulation curves and the corresponding 'Shinozaki' curves are used. The latter is a practical rarefaction method for the description, analysis and comparison of diversity and community features in biotopes with different habitat heterogeneity (Magurran, 1988; Achtziger *et al.*, 1992).

RESULTS

Approximately 30 000 arthropods, including 8500 beetles, were collected from the 26 trees investigated. There was an average of 30.25 species and 221.5 individual beetles, and a total of 84 beetle species, in the canopy of *Lannea fulva* ($H_S = 2.02$, $e = 0.57$). *Formicomus schimperi* Pic and *F. spatulus* van Hille (Anthicidae) dominated (67.6–82.5%).

In samples from *Teclea nobilis*, an average of 85.25 species and 575 individuals were collected from individual tree-crowns, with a total of 224 beetle species on all trees combined (Figure 9.1, $H_S = 3.34$, $e = 0.77$). The phytophagous Alticinae, Galerucinae, Bruchidae, Apionidae and Curculionidae formed the bulk of individuals in each tree, with an average of 53% of all beetles. Most of the species in these families and subfamilies were found on all of the trees. A few species of *Longitarsus*, *Kenialtica* and *Phyllotreta* (Alticinae) reached an average of 29.3% of total beetle abundance (further details in Wagner, 1994). On four trees *Pseudoleptaleus unifasciatus* (Desbrochers) (Anthicidae) and on five trees an undescribed species of *Sphinginopalpus* (Malachiidae) showed higher percentage representation.

The average number of species in *Carapa grandiflora* tree-crowns in the montane rainforest was equal to that of *T. nobilis* ($S = 89.0$), but the number of individuals was much lower ($n = 311$, Figure 9.2, $H_S = 3.75$, $e = 0.89$). Consequently, the total number of beetle species was higher ($S = 393$). The dominant species on the single trees belonged to various groups: *Palaminus* sp. (Paederinae, Staphylinidae, 13.0%), *Epuraea*

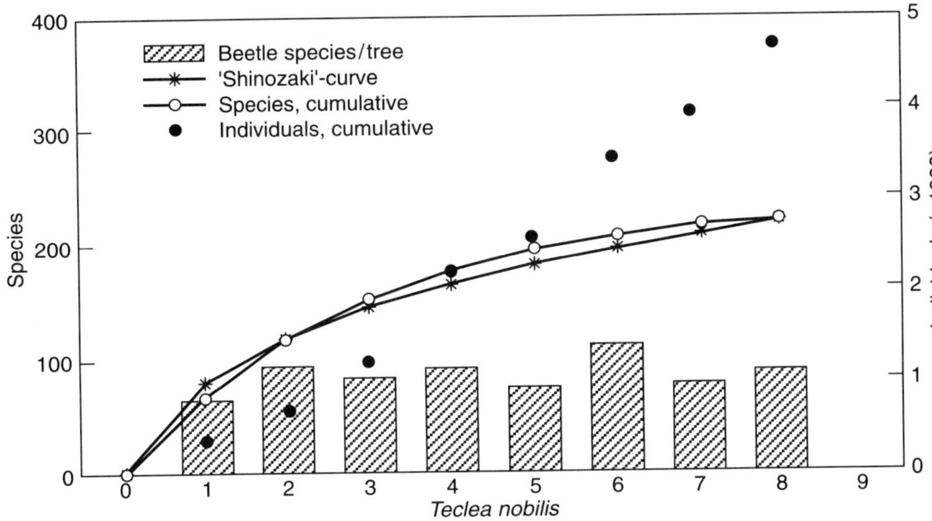

Figure 9.1 Cumulative species curve and abundance patterns in *Teclea nobilis* (gallery forest, East Rwanda).

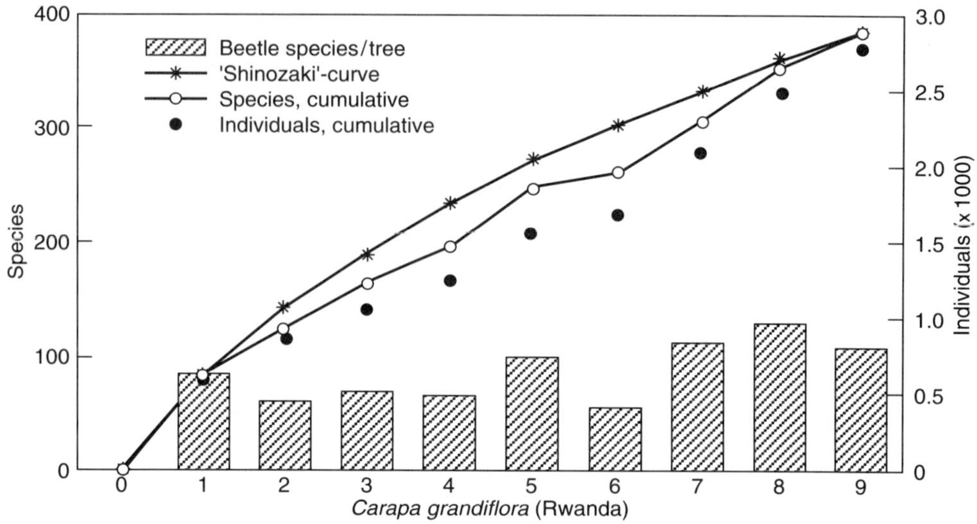

Figure 9.2 Cumulative species curve and abundance patterns in *Carapa grandiflora* (montane rainforest, West Rwanda).

sp. (Nitidulidae, 25.5%), *Melanophthalma* sp. (Lathridiidae, 7.2%) and a species of Melandryidae (9.8%). On two trees, two unidentified species of Corylophidae dominated, with 5.8% and 24.1% of total beetle abundance, and on three trees one undescribed *Prosthaptus* sp. (Cantharidae) dominated with 10.7%, 14.3%, and 16.3% of total abundance, respectively. In *C. grandiflora,* the groups with the greatest constancy were the fungivorous Corylophidae and Lathridiidae.

Samples from *C. grandiflora* in the upper lowland rainforest in Zaire showed a totally different compositional pattern for beetles (Figure 9.3, $H_S = 3.47$, $e = 0.96$). An average of only 37.4 species and 46.2 individuals were found on individual trees. Most beetle species were only found once on a single tree. Of the 137 species collected, only two species were found on all five of the trees sampled. These species were *Fustigerinus* sp. (Clavigerinae, Pselaphidae) and an unidentified Scolytidae.

DISCUSSION

Distribution patterns and diversity

C. grandiflora showed the highest species richness in the montane rainforest (393 species). Species accumulation and 'Shinozaki' curves do not reach a plateau after nine trees, indicating that more samples would

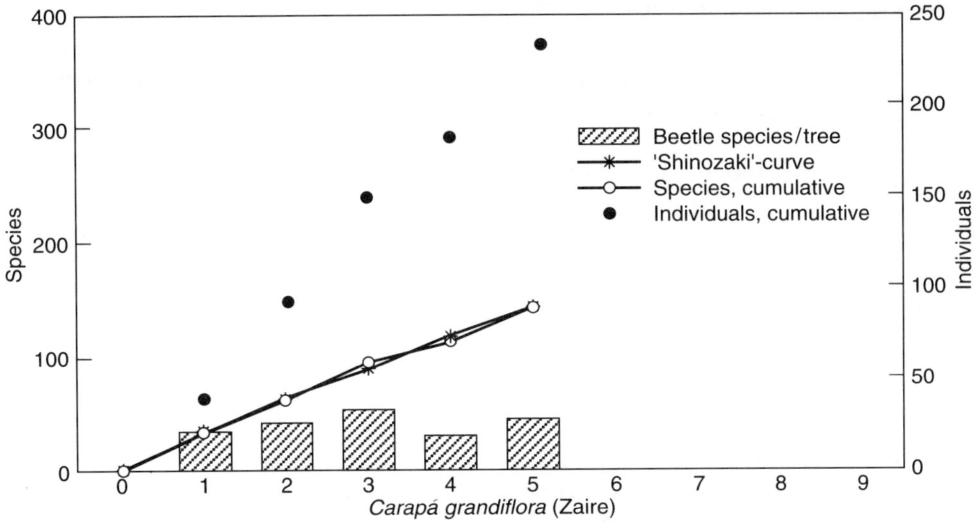

Figure 9.3 Cumulative species curve and abundance patterns in *Carapa grandiflora* (upper lowland rainforest, Kivu, Zaire).

have been necessary in order to collect all beetle species in the area. This pattern of a high number and random distribution of species, together with low population levels, is typical of non-interactive communities with a high beta-diversity (Schoener, 1986; Cornell and Lawton, 1992). The biotope is not saturated with species, i.e. there are numerous vacant niches which may be available for occupation by randomly immigrating species, or these niches may be repeatedly vacated as a result of the extinction of small local populations.

The distribution pattern for *T. nobilis* was quite the opposite. Samples from eight trees showed saturation at a level of 224 beetle species. This indicates the presence of an interactive community of beetles on this species of tree. The number of coexistent species is limited, nearly all niches are occupied and some successful species have high population levels. There are strong interactions between species on the same trophic level which leads to narrow niches (Cornell and Lawton, 1992). There are indications of interspecific competition, especially between phytophagous beetle species on *T. nobilis*. This is supported by Schoener (1983) who found evidence for interspecific competition within this feeding guild, but stands in contradiction to the hypothesis of Jermy (1985) who maintains that there is normally a lack of competition between phytophagous species. In *T. nobilis* there was a relatively constant number of species of Chrysomelidae, Bruchidae, Apionidae and

Curculionidae, with 37 to 50 species in each tree, while the abundance of other species varied greatly. *Longitarsus* sp. (Alticinae) was the dominant phytophagous species on three trees. It was rare in two other trees where *Kenialtica* sp. (Alticinae) may fill the equivalent niche (see Wagner, 1996).

T. nobilis has a much denser canopy than *C. grandiflora* and many more, smaller leaves. The available 'leaf resource' of *T. nobilis* is several times greater than that of *C. grandiflora,* and one should expect a greater abundance of beetles. However, the random distribution and low number of phytophagous beetles in *C. grandiflora* may indicate that the 'leaf resource' is incompletely exploited.

The patterns of arthropod communities on *C. grandiflora* in the upper lowland rainforest in Zaire are very different to those in the montane rainforest. The distribution of beetles is random and near to the theoretical maximum of a non-interactive community, where there are many rare, singleton species. This distribution may be caused by ants, in particular a species of Myrmicinae which is ubiquitous and very abundant throughout the entire forest. Strong dominance of ants was also found by Floren and Linsenmair (1994) in a lowland rainforest in Malaysia. The high dominance of ants in lowland rainforests and their decrease in abundance with altitude seems to be a general phenomenon. Montane forests are cold, making foraging inefficient for ants (Brown, 1973; Stork and Brendell, 1990). It is possible that the high abundance of ants at lower altitudes prevents permanent settlement of beetles and other arthropods (see Table 9.1). Among the beetles, only species with special adaptations or biology, like the myrmecophagous *Fustigerinus* sp. (Clavigerinae, Pselaphidae), and *Zyras* sp. (Alaeocharinae, Staphylinidae) or wood boring Scolytidae and Platypodidae which were found on nearly all trees in Zaire, appear to be able to coexist with these ants. The high dominance of one ant species led to a decrease in diversity in the lowland rainforest, which is in contradiction to the studies of, for example, Hammond (1990) and Stork and Brendell (1990).

Next to ants, Diptera, especially Sciaridae, Chironomidae, Tipulidae and Ceratopogonidae, were very abundant on all trees investigated in the upper lowland rainforest. These groups are also very abundant in the lowland rainforests of Sulawesi (Stork and Brendell, 1990), where Diptera represented 43.0–82.9% of all arthropods in single fogs. In Brunei, Diptera represented 21.5% of all arthropods (Stork, 1991). In agreement with the results presented here from Central Africa, the proportion of Diptera in higher altitude (1760 m) Sulawesi fog samples drops to 23%.

Despite the different distribution of arthropod taxa on *C. grandiflora* in the two types of forests, the total number of individuals per tree was nearly the same (Table 9.2). Trees of *T. nobilis* with the same height as

C. grandiflora had higher numbers of individuals because of the denser foliage. In *L. fulva*, species richness and diversity were lowest in situations where a few insect species were dominant in abundance. This would suggest an interactive community with a very restricted number of niches. Unfortunately, the samples obtained from this tree species are too small and too heterogeneous to allow further extrapolation.

Possible explanations for different distribution patterns

T. nobilis is a typical tree of gallery forests in East Africa and is adapted to changes in water level and soil moisture. At the time of investigation the soil was very dry (and the herb-layer withered) and hence the number of species may have been reduced (cf. Erwin and Scott, 1980). The low moisture levels may explain the lack of many fungivorous beetles in this tree. The influence of drought is even stronger for the arthropod community on *L. fulva*. Tree growth is limited to the short rainy season, and the leaves are tough and densely haired. This can be interpreted as an adaptation against drought and herbivorous insects, and indeed there were few such insects.

While the habitats of *T. nobilis* and *L. fulva* in East Rwanda are strongly influenced by the dry season, *C. grandiflora* is found in more humid areas. Higher humidity and/or higher temperature may, in principle, cause a higher diversity in the rainforest types investigated. The annual precipitation of about 1600 mm in the montane rainforest of West Rwanda is more than twice that in East Rwanda, but the average annual temperature is comparatively low at 14°C (Fischer and Hinkel, 1992). In spite of the short dry season in montane rainforest, the low temperature causes a permanent humid climate. The upper lowland rainforest has a typical hot, wet tropical climate at 2500–3000 mm precipitation per year and an average annual temperature of 25°C.

The area at Irangi is a primary forest, while some sites of montane rainforests (Cyamudongo) are influenced by humans. The area investigated was never intensively logged, but the composition of forest trees has changed through the selective felling of timber and grazing of cattle in the past (Fischer and Hinkel, 1993). It seems that selective felling and grazing have no negative influence on the diversity of arthropods in the tree canopies investigated. Compared to the other sites of *C. grandiflora* in primary montane rainforest of Nyungwe there is no significant difference in diversity and distribution patterns. The values from both primary and secondary types of montane rainforests are higher than that from primary upper-lowland rainforest where the system is strongly dominated by one ant species.

Another difference between the montane rainforest and the upper-lowland rainforest is their historical development. The region around

Table 9.1 Distribution of number of individuals and proportional representation of different arthropod taxa on *Carapa grandiflora* in (a) montane rainforest (Cyamudongo and Nyungwe, West Rwanda, 1750–2200 m), and (b) upper lowland rainforest (Irangi in East Zaire 950 m a.s.l.). Other arthropods not shown include (in descending order of individuals): Blattodea, Isopoda, Planipennia, Diplopoda, Mantodea, Dermaptera, Ephemeroptera, Trichoptera, Caelifera, Phasmoptera, Zygentoma and Strepsiptera.

(a) **Montane rainforest**

	Tree number								
Group	*1*	*2*	*3*	*4*	*5*	*6*	*7*	*8*	*9*
Coleoptera	207	204	295	393	304	596	255	168	398
(%)	24.4	23.7	19.2	31.7	25.7	44.7	19.2	43.6	28.7
Formicidae	34	250	128	27	67	31	33	3	39
(%)	4.01	29.0	8.34	2.18	5.66	2.32	2.48	0.78	2.81
Other Hymenoptera	157	123	280	157	223	187	247	37	218
(%)	18.5	14.3	18.2	16.6	18.8	14.0	18.6	9.61	15.7
Diptera	43	44	109	110	47	109	223	19	67
(%)	5.07	5.12	7.10	8.87	3.97	8.18	16.8	4.93	4.83
Lepidoptera	31	18	55	33	31	27	35	18	66
(%)	3.65	2.09	3.58	2.66	2.62	2.03	2.63	4.67	4.83
Homoptera	61	37	110	118	34	175	208	16	103
(%)	7.19	4.30	7.17	9.52	2.87	13.1	15.7	4.15	7.42
Heteroptera	47	29	69	41	19	53	30	15	59
(%)	5.54	3.37	4.49	3.31	1.60	3.97	2.26	3.90	4.25
Thysanoptera	62	49	98	64	290	47	73	21	76
(%)	7.31	5.70	6.38	5.16	24.5	3.52	5.49	5.45	5.48
Psocoptera	5	15	44	39	12	65	105	6	33
(%)	0.59	1.74	2.87	3.15	1.01	4.88	7.91	1.56	2.38
Ensifera	25	14	12	12	17	6	2	9	27
(%)	2.95	1.63	0.78	0.97	1.43	0.45	0.15	2.33	1.95
Collembola	68	21	94	55	12	7	115	29	60
(%)	8.02	2.44	6.12	4.43	1.01	0.52	8.66	7.53	4.32
Arachnida	39	38	206	112	104	27	64	44	219
(%)	4.60	4.42	13.4	9.03	8.79	2.03	4.82	11.4	15.8
Other arthropods	69	18	35	39	25	3	2	2	22
(%)	8.14	2.09	2.28	3.14	2.11	0.22	0.15	0.52	1.59
Total (*n*)	848	860	1535	1240	1183	1333	1328	385	1387

(b) **Upper lowland rainforest**

	Tree number				
Group	*1*	*2*	*3*	*4*	*5*
Coleoptera	39	52	58	32	52
(%)	3.92	3.18	3.06	4.33	5.11

Table 9.1 continued

Group	Tree number 1	2	3	4	5
Formicidae	399	974	1305	394	373
(%)	40.1	59.6	68.9	53.3	36.6
Other Hymenoptera	76	80	61	29	51
(%)	7.64	4.90	3.22	3.92	5.01
Diptera	306	391	226	158	221
(%)	30.8	23.9	11.9	21.4	21.7
Lepidoptera	5	9	11	5	5
(%)	0.50	0.55	0.58	0.67	0.49
Homoptera	48	28	31	61	220
(%)	4.83	1.71	1.63	8.25	21.6
Heteroptera	9	13	4	0	11
(%)	0.90	0.80	0.21	-	1.08
Thysanoptera	20	17	50	9	8
(%)	2.01	1.04	2.64	1.22	0.78
Psocoptera	14	8	42	14	10
(%)	1.41	0.49	2.22	1.89	0.98
Ensifera	6	6	5	2	3
(%)	0.60	0.37	0.26	0.27	0.29
Collembola	12	2	45	7	11
(%)	2.21	0.12	2.38	0.94	1.08
Arachnida	39	41	51	28	29
(%)	3.92	2.51	2.69	3.79	2.85
Other arthropods	11	12	4	0	23
(%)	1.11	0.73	0.21	–	2.26
Total (*n*)	994	1633	1893	739	1017

Irangi belongs to the Central African refuge forest area. Conditions conducive to forests have existed there for at least 80 000 years (Hamilton, 1982). There was only a change in forest type during the last ice age. The montane rainforest at the Zaire–Nile watershed is much younger. During the last glacial maximum, between 25 000 and 15 000 years before present, there was an average temperature decrease of 4°C, and a decrease of 500 to 700 mm precipitation per year (Bonnefille *et al.*, 1990). These values imply a lowering of the tree-line by 1000 to 1500 m. Hence, the present montane rainforest vegetation is not older than about 10 000 years. During the last glaciation there was *Hagenia*-forest and afro-alpine vegetation in this area. Evidently, there is high arthropod diversity of tree canopies in geologically young forest systems which are also

Table 9.2 Distribution of individuals of different arthropod taxa for the tree species sampled. n, no. of individuals in all trees investigated; n/t, mean of individuals per tree; %, mean of percentage per tree

	Tree species							
Group	*Lannea fulva**		*Teclea nobilis⁺*		*Carapa grandiflora‡*		*Carapa grandiflora§*	
Coleoptera								
n	886		4594		2820		233	
n/t	221	(127)	575	(245)	311	(133)	46.2	(10.7)
%	38.5	(17.8)	41.6	(10.1)	29.0	(9.48)	3.92	(0.85)
Formicidae								
n	1035		1252		612		3445	
n/t	258	(238)	156	(94.4)	68.0	(76.8)	689	(428)
%	36.9	(22.5)	11.8	(5.91)	6.40	(8.76)	51.7	(14.4)
Other Hymenoptera								
n	86		1176		1629		297	
n/t	21.5	(9.88)	147	(56.7)	181	(72.9)	59.4	(20.6)
%	3.34	(0.98)	10.4	(4.80)	16.0	(3.04)	4.94	(1.68)
Diptera								
n	47		1432		771		1302	
n/t	11.7	(6.89)	179	(97.6)	85.7	(61.6)	260	(89.9)
%	2.37	(2.40)	12.3	(5.20)	7.21	(3.97)	21.9	(6.77)
Lepidoptera								
n	50		114		314		35	
n/t	12.5	(13.4)	14.3	(8.55)	34.9	(15.9)	7.00	(2.83)
%	2.00	(2.35)	0.96	(0.45)	3.19	(1.04)	0.56	(0.07)
Homoptera								
n	202		343		862		388	
n/t	50.5	(47.0)	42.9	(21.6)	95.8	(65.6)	77.6	(80.7)
%	8.15	(7.02)	3.32	(1.85)	7.93	(4.24)	7.60	(8.28)
Heteroptera								
n	51		200		362		37	
n/t	20.2	(14.2)	25.0	(17.4)	40.2	(18.3)	7.40	(5.32)
%	3.40	(1.95)	1.97	(1.74)	3.63	(1.18)	0.60	(0.46)
Thysanoptera								
n	29		298		780		104	
n/t	7.25	(8.06)	37.2	(56.7)	86.7	(79.2)	20.8	(17.1)
%	1.29	(1.56)	2.13	(2.58)	7.66	(6.39)	1.54	(0.77)
Psocoptera								
n	20		686		324		88	
n/t	5.00	(4.24)	85.7	(69.0)	36.0	(32.7)	17.6	(13.9)
%	1.15	(1.51)	6.18	(4.72)	2.90	(2.27)	1.40	(0.69)
Ensifera								
n	8		137		124		22	
n/t	2.00	(0.81)	17.1	(13.5)	13.8	(8.21)	4.40	(1.82)
%	0.35	(0.19)	1.43	(1.26)	1.40	(0.91)	0.36	(0.14)

Table 9.2 continued

Group	Tree species							
	*Lannea fulva**		*Teclea nobilis*+		*Carapa grandiflora*‡		*Carapa grandiflora*§	
Collembola								
n	5		93		461		87	
n/t	1.25	(1.50)	11.6	(11.6)	51.2	(37.4)	15.4	(17.0)
%	0.18	(0.21)	0.81	(0.64)	4.78	(3.04)	1.35	(0.94)
Arachnida								
n	35		606		853		188	
n/t	8.75	(2.23)	75.7	(37.0)	94.8	(73.0)	37.6	(9.48)
%	1.75	(1.21)	5.26	(1.12)	8.25	(4.65)	3.15	(0.65)
Other arthropods								
n	9		219		215		50	
n/t	2.25	(2.63)	26.6	(14.6)	23.8	(21.7)	10.0	(8.80)
%	0.52	(0.73)	2.36	(2.32)	2.25	(2.43)	0.86	(0.89)
Total (*n*)	2492		11092		10127		6279	
Arthropods/ tree	623		1386		1125		1256	

* Dry forest, East Rwanda, $n = 4$
+ Gallery forest, East Rwanda, $n = 8$
‡ Montane rainforest, West Rwanda, $n = 9$
§ Upper lowland rainforest, East Zaire, $n = 5$
Values in parentheses are standard deviations.

partly influenced by humans, while on the same tree species in a much older primary forest the arthropod diversity is lower and the community strongly dominated by one species.

Acknowledgements

Many thanks to helpful friends in Rwanda and Zaire, especially to E. Fischer, H. Hinkel, J. Munyabasungu and M. Schmidt, and to A. Melzer, K. Rüsch, B. Schartmann, G.E. Schmitz and A. Volkwein for their assistance sorting and counting the samples and R. Achtziger for his beneficial discussion of the rarefaction method. Thanks also to M. Onofrietto and N.E. Stork for their kind linguistic help and M. Krämer and M. Schmitt for critical comments. This study was partly supported by 'Deutscher Akademischer Austauschdienst', 'Ministerium des Inneren und für Sport (Rheinland-Pfalz)' and 'Stifterverband für die Deutsche Wissenschaft'.

REFERENCES

Achtziger, R., Nigman, U. and Zwölfer, H. (1992) Rarefaction-Methoden und ihre Einsatzmöglichkeiten bei der zooökologischen Zustandsanalyse und Bewertung von Biotopen. *Z. Ökologie u. Naturschutz*, **1**, 89–105.

Bonefille, R., Roeland, J.C. and Guiot, J. (1990) Temperature and rainfall estimates for the past 40,000 years in equatorial Africa. *Nature*, **346**, 347–9.

Brown, M.V. (1973) A comparison of the Hylean and Congo-West African rainforest ant faunas, in *Tropical Forest Ecosystems in Africa and South America: A Comparative Review* (eds B.J. Meggers, E.S. Ayensu and W.D. Duckworth), Smithsonian Institute Press, Washington DC, pp. 161–85.

Cornell, H.V. and Lawton, J.H. (1992) Species interactions, local and regional processes, and limits to the richness of ecological communities: a theoretical perspective. *Journal of Animal Ecology*, **61**, 1–12.

Erwin, T.L. (1982) Tropical forests: their richness in Coleoptera and other arthropod species. *The Coleopterists' Bulletin*, **36**, 74–5.

Erwin, T.L. and Scott, J.C. (1980) Seasonal and size patterns, trophic structure, and richness of Coleoptera in the tropical arboreal ecosystem: the fauna of the tree *Luehea seemannii* Triana and Planch in the Canal Zone of Panama. *The Coleopterist's Bulletin*, **34**, 305–22.

Farrell, B.D. and Erwin, T.L. (1988) Leaf-beetles community structure in an amazonian rainforest canopy, in *Biology of Chrysomelidae* (eds P. Jolivet, E. Petitpierre and T.H. Hsiao), Kluwer, Amsterdam, pp. 73–89.

Fischer, E. and Hinkel, H. (1992) Natur Ruandas. Materialien zur Partnerschaft Rheinland-Pfalz/Ruanda 1992/1, Ministerium des Innern und für Sport, Rheinland-Pfalz, Mainz.

Fischer, E. and Hinkel, H. (1993) Zur Flora und Fauna der Wälder Ruandas/Zentralafrika. I. Der Cyamudongo-Wald. *Mainzer naturwiss. Archiv.*, **31**, 239–75.

Floren, A. and Linsenmair, K.E. (1994) Zur Diversität und Wiederbesiedlungsdynamik von Arthropoden auf drei Baumarten in einem Regenwald in Sabah, Malaysia. *Andrias*, **13**, 23–8.

Hamilton, A.C. (1982) *Environmental History of East Africa: A Study of the Quaternary*, Academic Press, London.

Hammond, P. (1990) Insect abundance in the Dumoga-Bone National Park, North Sulawesi, with special reference to the beetle fauna of lowland rainforest in the Toraut region, in *Insects and the Rainforest of South East Asia (Wallacea)* (eds W.J. Knight and J.D. Holloway), Royal Entomological Society of London, London, pp. 197–254.

Jermy, T. (1985) Is there competition between phytophagous insects? *Zeitschrift für Zoologische Systematik und Evolutionsforschung*, **23**, 275–85.

Magurran, A.E. (1988) *Ecological Diversity and its Measurements*, Croom Helm, London.

Morse, D.R., Stork, N.E. and Lawton, H.J. (1988) Species number, species abundance and body length relationship of arboreal beetles in Bornean lowland rainforest trees. *Ecological Entomology*, **13**, 25–37.

Schoener, T.W. (1983) Field experiments on interspecific competition. *American Naturalist*, **122**, 240–85.

Schoener, T.W. (1986) Overview: kinds of ecological communities – ecology becomes pluralistic, in *Community Ecology* (eds J. Diamond and T.J. Case), Harper and Row, New York, pp. 467–79.

Stork, N.E. (1987) Guild structure of arthropods from Bornean rainforest trees. *Ecological Entomology*, **12**, 69–80.

Stork, N.E. (1991) The composition of the arthropod fauna of Bornean lowland rainforest trees. *Journal of Tropical Ecology*, **7**, 161–80.

Stork, N.E. and Brendell, M.J.D. (1990) Variation in the fauna of Sulawesi trees with season, altitude and forest type, in *Insects and the Rainforest of South East Asia (Wallacea)* (eds W.J. Knight and J.D. Holloway), Royal Entomological Society of London, London, pp. 173–90.

Wagner, T. (1996) Verteilungsmuster und Artenvielfalt von kronenbewohnenden Käfern auf verschiedenen Baumarten in Zentralafrika. *Verhandlungen Westdeutscher Entomologentag*, **7**, 79–87.

Appendix 9A Individuals and (morpho)-species of beetles collected from different tree species. All beetles morphotyped by the author. Taxonomic examination by various specialists has been completed for the following groups: Buprestidae (H. Mühle), Cleridae (R. Gerstmeier), Coccinellidae (H. Fürsch), Anthicidae (G. Uhmann), Cerambycidae (K. Adlbauer), Bruchinae (J. E. Decelle). Many other taxonomists are currently working on various other groups. a, individuals; b, (morpho)-species. For details of tree species, see Table 9.2

	Lannea fulva		*Teclea nobilis*		*Carapa grandiflora (Rwanda)*		*Carapa grandiflora (Zaire)*	
	a	*b*	*a*	*b*	*a*	*b*	*a*	*b*
Larvae	–	–	6	–	41	–	4	–
Carabidae	2	1	88	3	11	8	2	2
Staphyliniformia								
Hydrophilidae	–	–	–	–	–	–	1	1
Histeridae	–	–	–	–	2	1	2	2
Hydraenidae	–	–	2	1	–	–	–	–
Ptiliidae	–	–	–	–	3	3	4	2
Scydmaenidae	–	–	–	–	8	5	9	6
Staphylinidae								
Aleocharinae	2	2	3	3	68	13	14	6
Osoriinae	–	–	–	–	–	–	1	1
Oxytelinae	–	–	4	1	49	6	1	1
Paederinae	9	1	102	1	113	5	8	4
Scaphidiinae	–	–	–	–	2	2	6	3
Staphylininae	–	–	–	–	7	2	–	–
Tachyporinae	–	–	–	–	6	3	4	3
Trichophyinae	–	–	1	1	–	–	–	–
Xantholininae	–	–	–	–	3	2	–	–
Pselaphidae	–	–	–	–	3	2	9	4
Eucinetiformia								
Scirtidae	2	1	10	2	6	2	1	1
Clambidae	–	–	6	2	19	5	1	1

Appendix 9A continued

	Lannea fulva		*Teclea nobilis*		*Carapa grandiflora (Rwanda)*		*Carapa grandiflora (Zaire)*	
	a	*b*	*a*	*b*	*a*	*b*	*a*	*b*
Scarabaeiformia								
Ceratocanthidae	–	–	–	–	–	–	1	1
Scarabaeidae								
Aphodinae	–	–	–	–	2	1	–	–
Melolonthinae	–	–	–	–	1	1	1	1
Elateriformia								
Buprestidae	5	2	11	4	17	6	1	1
Ptilodactylidae	–	–	–	–	24	1	2	2
Eucnemidae	–	–	–	–	1	1	–	–
Throscidae	–	–	–	–	1	1	–	–
Elateridae	2	2	15	3	56	13	11	5
Lycidae	–	–	2	2	1	1	2	2
Cantharidae	–	–	–	–	133	4	8	3
Bostrichiformia								
Dermestidae	1	1	2	2	–	–	–	–
Anobiidae								
Anobiinae	60	6	123	6	18	10	–	–
Ptininae	–	–	7	2	–	–	–	–
Cucujiformia/Cleroidea								
Trogossitidae	–	–	–	–	3	3	–	–
Cleridae	3	2	5	3	24	7	1	1
Malachiidae	9	2	314	4	77	8	1	1
Cucujiformia/Cucujoidea								
Sphindidae	–	–	–	–	–	–	1	1
Nitidulidae	1	1	9	4	124	5	4	2
Rhizophagidae	–	–	–	–	1	1	–	–
Silvanidae	–	–	–	–	2	1	–	–
Laemophloeidae	–	–	10	1	6	5	–	–
Phalacridae	9	3	167	10	132	8	4	2
Cryptophagidae	–	–	–	–	3	2	1	1
Languriidae	–	–	9	2	8	1	1	1
Erotylidae	–	–	–	–	2	2	1	1
Discolomidae	–	–	–	–	8	2	5	3
Endomychidae	–	–	–	–	1	1	1	1
Coccinellidae								
Coccinellinae	18	14	64	12	215	18	–	–
Epilachninae	–	–	–	–	28	15	–	–
Corylophidae	6	3	72	6	385	8	9	4
Lathridiidae	–	–	53	4	194	8	1	1

Appendix 9A continued

	Lannea fulva		*Teclea nobilis*		*Carapa grandiflora (Rwanda)*		*Carapa grandiflora (Zaire)*	
	a	*b*	*a*	*b*	*a*	*b*	*a*	*b*
Cucujiformia/Tenebrionoidea								
Mycetophagidae	1	1	–	–	1	1	1	1
Ciidae	2	1	-	-	10	5	–	–
Melandryidae	1	1	2	1	9	2	4	2
Mordellidae	2	1	20	6	72	8	4	3
Tenebrionidae								
Alleculinae	1	1	40	2	60	4	4	3
Lagriinae	5	2	5	3	25	5	1	1
Tenebrioninae	–	–	18	4	6	3	6	3
Oedemeridae	–	–	1	1	2	2	–	–
Meloidae	–	–	–	–	–	–	1	1
Salpingidae	–	–	–	–	11	2	–	–
Anthicidae	682	2	535	6	–	–	1	1
Aderidae	9	3	32	9	42	6	3	3
Scraptiidae	4	3	84	2	74	6	3	2
Cucujiformia/Chrysomeloidea								
Cerambycidae	4	2	14	7	21	9	3	3
Chrysomelidae								
Alticinae	11	4	1525	23	139	19	10	5
Bruchinae	4	3	521	8	67	6	–	–
Cassidinae	–	–	8	4	5	2	–	–
Chrysomelinae	–	–	1	1	9	2	–	–
Clytrinae	1	1	1	1	1	1	–	–
Criocerinae	–	–	1	1	–	–	–	–
Cryptocephalinae	–	–	6	4	16	10	–	–
Eumolpinae	5	3	6	4	9	4	12	7
Galerucinae	12	6	274	13	40	14	10	6
Hispinae	–	–	–	–	2	1	1	1
Zeugophorinae	–	–	1	1	–	–	–	–
Cucujiformia/Curculionoidea								
Nemonychidae	–	–	–	–	1	1	–	–
Anthribidae	–	–	56	8	55	10	1	1
Attelabidae	–	–	–	–	2	2	–	–
Brentidae	1	1	–	–	3	3	–	–
Apionidae	3	2	234	12	80	9	4	3
Scolytidae	2	1	4	4	18	5	12	6
Platypodidae	–	–	–	–	1	1	3	3
Curculionidae	7	5	130	20	231	56	26	11
Total	886	84	4594	224	2820	393	233	137

10

Tree-crown beetles in context: a comparison of canopy and other ecotone assemblages in a lowland tropical forest in Sulawesi

P.M. Hammond, N.E. Stork and M.J.D. Brendell

ABSTRACT

The surface-dwelling beetle assemblage occupying tree-crowns in a relatively uniform 500-ha tract of lowland forest in North Sulawesi, Indonesia, was sampled by knockdown insecticide fogging. Sample data for beetles from all strata of the study area, including tree-crown samples collected by methods other than fogging, are employed to evaluate the tree-crown beetle assemblage. In particular, these data are used to determine which of the species represented in tree-crown samples are likely, at least as adults, to be regular denizens of the tree-tops. Further, these 'resident' species are assigned to two categories: *specialists* (only or mostly found in tree-crowns) and *generalists* (of regular occurrence both in tree-crowns and at lower levels). The remaining species found in tree-crown samples are regarded as *tourists* or, in cases of particular uncertainty, are not assigned to any of the three categories.

The principal findings are that:

1. Samples so far evaluated from the 500-ha study area contain 4026 beetle species, and the area is estimated to contain at least 6425 beetle species.
2. Fogging samples from the study area contain 1355 beetle species, belonging to 91 different 'family-groups'.
3. Most beetle species resident in tree-crowns in the study area were collected by fogging.
4. Some 39.3% of species present in the fogging samples were not present in any other samples from the study area.
5. Some 20.1% of the species present in the fogging samples can be categorized as 'tourists' and 49.9% as tree-crown residents, with 30.1% of species uncategorized.

Canopy Arthropods. Edited by N.E. Stork, J. Adis and R.K. Didham. Published in 1997 by Chapman & Hall, London. ISBN 0 412 74900 9

Less than half of the tree-crown residents (21.9% of the species overall) are categorized as tree-crown specialists.

6. The best represented beetle family-groups in the fogging samples, with more than 50 species each, are (in order of species richness) Curculionidae, Aleocharinae (Staphylinidae), Anthribidae, Corylophidae, Aderidae, Scolytidae, Lamiinae (Cerambycidae) and Mordellidae.
7. The best represented feeding group in the fogging samples is the fungivore group with 27.5% of the species, followed by herbivores (25.2%), predators (17.4%), xylophages (16.1%) and saprophages (13.8%).
8. Compared with similar data-sets for other tropical sites, the tree-crown assemblage studied here is especially rich in Anthribidae, Mordellidae, Aderidae and Eucnemidae and, in feeding group terms, in fungivores and xylophages.
9. Sample and inventory data from all ecotones suggest that between 20% and 30% of the beetle species present in the study area may be tree-crown residents as adults, but most of these are unlikely to be exclusively so.
10. The same data suggest that between 8% and 13% of the study area's beetle species are more or less exclusively tree-crown dwelling as adults.

INTRODUCTION

New methods of gaining access to the crowns of tall forest trees (Basset *et al.*, 1997, Chapter 2, this volume; Stork and Hammond, 1997, Chapter 1, this volume) and, in particular, the use of one 'remote method' (insecticide fogging), have led to a clear demonstration that tree-crowns of tropical moist forests are extremely rich in arthropod species (Roberts, 1973; Wolda, 1979; Southwood *et al.*, 1982a,b; Stork, 1987a,b, 1991; Basset, 1990, 1991; Stork and Brendell, 1990; Basset and Kitching, 1991; Basset and Arthington, 1992; Allison *et al.*, 1993; Floren and Linsenmair, 1994; Guilbert *et al.*, 1994; Hofer *et al.*, 1994; Russell-Smith and Stork, 1994; Guilbert and Chazeau, in press). Various studies, many of them focusing especially on Coleoptera (Erwin and Scott, 1980; Erwin, 1982, 1983a; Stork, 1987b, 1991; Mawdsley, 1994), but some also dealing with groups such as ants (Formicoidea) or bugs (Hemiptera), have shown that single tree-crown fog samples or small quadrat fogs (e.g. 12×12 m) from tropical sites may harbour upwards of 2000 arthropod species.

How these numbers relate to numbers at larger spatial scales, however, remains obscure, despite some attempts at extrapolation (Erwin, 1982; Stork, 1988). Unfortunately, the data necessary to establish the empirical relationship between tropical tree-crown arthropod species numbers at the smallest scales (e.g. a single tree) and much larger (e.g. regional) spatial scales are currently lacking. The equally interesting question of how the numbers of species found in tree-crowns relate to numbers found elsewhere in a forest has been little investigated, although assumptions about this relationship have been made frequently, and sometimes stated explicitly. For example, Erwin (1982) adopted a 2 : 1 canopy species to ground level species ratio as a working principle for tropical forests, while Stork (1993) suggested that a 1 : 1 ratio, or one somewhat favouring ground level species richness, might be more realistic. However, Hammond (1992) noted that data available for temperate forests suggest relatively weak stratification, a very small canopy specialist component and a 'typical' canopy to ground arthropod species ratio of 1 : 10 (or more). The same author suggested that substantial variation in this ratio is to be expected in tropical forests, but observed that a ca. 1 : 5 canopy to ground ratio was indicated for one tropical forest area in Sulawesi, that dealt with in the present contribution.

The suggestion that the greater part of tropical forest diversity is to be found in the tree-tops is also implicit in the way that canopies have been described, for example as 'the heart of biotic diversity' (Erwin, 1983b, 1988). Against this background it should be clear that, like other studies of species assemblages, those concerning tropical tree-crowns will derive a number of benefits from being put in context. First, this is generally necessary in order to make any serious evaluation of tree-crown assemblage composition: how many of the species recorded are tree-crown specialists; how many are generalists that are found at all levels in a forest; and how many species are casual visitors (vagrants or tourists) to the higher levels of the forest? Second, the significance of the tree-crown arthropod assemblage in terms of the forest as a whole can only be fully appreciated once its relative richness in comparison with other forest arthropod assemblages has been established.

The aim of the present paper is to place the extensively investigated tree-crown beetle assemblage of a relatively uniform 500-ha area of lowland rainforest in Sulawesi (Hammond, 1990; Stork and Brendell, 1990) in its context. This represents the first attempt to evaluate a tropical tree-crown assemblage in this way. There is an urgent need for similar studies to be carried out at other moist tropical forest sites, notably in the Neotropics, and with other major arthropod groups. For the moment, however, this paper includes the only well-documented account of a tropical tree-crown assemblage whose relationship, in terms of species richness, with the species occupying other forest strata has been established. The questions we address and provide at least partial or provisional answers to, are:

1. What part of the total forest beetle assemblage is to be found in tree-crowns?
2. How do these species use the tree-crowns?
3. How 'stratified' are this forest's beetle species?
4. To what extent does fogging give a reasonable picture of the canopy beetle assemblage?

METHODS

Study site

The investigations reported on in this paper were carried out during Project Wallace, the Royal Entomological Society of London/Indonesian Institute of Sciences joint expedition to N. Sulawesi, Indonesia, in 1985. Insect samples were taken at a number of sites in and near the Dumoga-Bone National Park, Sulawesi Utara. However, the principal focus of investigations was a roughly 4 × 1 km tract of relatively uniform lowland rainforest situated in the Dumoga-Bone National Park at the western end of the Dumoga Valley, at approximately 0°34'N and 123°54'E. A full description of the lowland forest study area, including comment on its geology, vegetation and weather, is provided by Hammond (1990), Stork and Brendell (1990), and others (see Knight and Holloway, 1990).

Field methods

The programme of sampling, the most extensive to have been reported for any major insect group in a restricted area of tropical forest, is briefly summarized below. Detailed descriptions are provided by Hammond (1990) and Stork and Brendell (1990). The principal objectives of the fieldwork were to: (i) obtain as complete a representation as possible of the beetle species present in the lowland forest study area, especially those of a number of specially targeted groups; and (ii) obtain a range of quantitative samples so that the proportional representation of the study area's beetle assemblages could be estimated for various patterns of sampling and levels of sampling 'effort'. In pursuit of the first objective, special efforts were made to obtain species of selected groups, such as Carabidae, certain staphylinid subfamilies (e.g. Oxytelinae, Steninae), Scarabaeoidea, Cleroidea, Coccinellidae and Tenebrionidae (see Hammond, 1990 for a full list).

Sampling regimes

Field work was carried out between mid-January and early December 1985. Insect samples were taken at a number of locations although most, some 2693 samples, were obtained from the lowland forest study area, at elevations of ca. 200–400 m. A wide range of collecting and sampling techniques were employed with the aim of compiling as complete an inventory as possible of the beetle species resident in the study area. Quantitative sampling was concentrated on three 1-ha plots. Details of the regime adopted for sampling at ground level are provided by Hammond (1990). The regime adopted for sampling at tree-crown level is described (for fogging) by Stork and Brendell (1990) and (for other methods) by Hammond (1990).

Sampling methods

For ground level sampling extensive use was made of Malaise traps, large-area flight interception traps, pitfall traps, yellow-pan traps, baited traps and light traps. Many samples were also obtained by hand collecting, e.g. by sweeping, from vegetation and from aquatic habitats, and many samples were taken from substrates such as soil, leaf litter and decaying wood. Special collections were made from many other microhabitats known to harbour specialist beetle species, such as fungal fruiting-bodies, and the nests of ants and birds. Details of sampling methods and how they were applied are provided by Hammond (1990).

Sampling at tree-crown level involved an extensive programme of insecticide fogging using a synthetic pyrethroid, Reslin E. In the lowland forest study area fogging was carried out on nine dates between 5 February and 2 December 1985, at a total of four sites. On each occasion the trees above five separate 12×12 m quadrats were fogged. Samples were collected into 1-m^2 trays, with a minimum of 20 trays used for each quadrat. Details of the fogging protocol are provided by Stork and Brendell (1990).

Other sampling at tree-crown level in the study area included the use of Malaise traps suspended within tree-crowns at three sites in close proximity to the fogging quadrats. These traps, identical in design to ones employed at the same sites at ground level, operated continuously from February to December 1985, except for a few intervals when the traps were grounded. Baited traps, identical to those used at ground level (see Hanski and Niemelä 1990), and employing a variety of baits, were operated on lines suspended from branches in the canopy on several occasions in February and March 1985. A few samples were also obtained from actinic light traps suspended within tall trees in the study area in October 1985. Once again, details of all of these methods and protocols for their use are provided by Hammond (1990).

Sample sorting and analysis

Samples from the lowland forest study area contained an estimated 831 000 individual adult Coleoptera, belonging to 127 'family-groups' (Appendices 10A and 10B). For 88 of these family-groups an attempt was made to locate, prepare and sort representatives of all species present in samples from the study area. For 38 of these 88 well-studied groups the attempt to locate all species was deemed successful, while in the other 50 groups at least 90% of the species present in the samples are assumed to have been located and sorted (see Appendix 10A). The raw samples of the remaining 39 family-groups were less comprehensively scrutinised and *less* than 90% of species in these groups are likely to have been extracted and sorted. For 11 family-groups (Ptiliidae, Histeridae, Leiodidae, Scydmaenidae, Euaesthetinae, Aleocharinae, Pselaphidae, Scaphidiidae, Corylophidae, Aderidae and Curculionidae) the proportion of species extracted from the samples is likely to be especially low (see Appendices 10A and 10B).

Species sorting

For the type of study that forms the basis of the present paper, due rigour in species sorting is of paramount importance (Hammond, 1994). In order to achieve an appropriate level of accuracy the best skills available were employed for species sorting, and help was received from many sources (see Acknowledgements). As sorting proceeded, a voucher collection of carefully prepared and labelled specimens bearing family and species codes was gradually assembled. Care was taken to arrange and progressively annotate this collection so that it best fulfilled its function as an identification tool. Particular emphasis was placed on resolving problems of species sorting in well represented, but 'difficult', groups. For example, to ensure accurate species assignments in the arboreal staphylinid genus *Paluminus*, dissections of male and female genitalia and terminalia were made for all of the more than 1500 specimens collected.

Despite these efforts, uncertainties remain for a few small groups of species or individual species in various family-groups. These doubtful cases include some Ptiliidae, notably *Acrotrichis*, Scydmaenidae, notably *Euconnus*, Scaphidiidae, notably *Scaphisoma*, some Aleocharinae, e.g. *Gyrophaena*, some Ptilodactylidae, and a few members of other families. For three family-groups the species sorting problems that remain outstanding are more extensive. First, for Phalacridae, satisfactory sorting has not been achieved for what are here regarded as some six species, although specimens of these encompass the major part of the phalacrid material collected by fogging. Second, for some groups of species in the family Corylophidae, including some of those most abundant in the fogging samples, species limits have not been satisfactorily resolved. Finally, the sorting of the numerous species of Aderidae present in fogging samples on the one hand, and those present in other quantitative samples on the other, was done as two separate exercises, and the work of matching the two resulting sets of voucher specimens remains incomplete.

Descriptive taxonomic work on a number of groups of Coleoptera considered in this paper has now been completed, and published accounts are available for Rhysodidae (Bell and Bell, 1988), Cicindelinae (Cassola, 1991), Sphaeridiinae Omicrini (Bameul, 1993), Leiodidae Agathidiini (Angelini and Cooter, 1993), a few groups of Staphylinidae (G. de Rougemont, unpublished data; Kistner and Zimmerman, 1986; Kistner and McNairn, 1991), Scarabaeidae Aphodiinae in part (Krikken and Huijbregts, 1987; K. Edelman and J. Krikken, unpublished data), Clambidae (Endrödy-Younga, 1995), Cantharidae (Wittmer, 1989, 1992), Melyridae Malachiinae (Wittmer, 1990), and Pterogeniidae (Burckhardt and Löbl, 1992).

Allocation of species to stratum categories

The use of canopy samples for analysing the composition of tree-crown assemblages is generally compromised by the presence in these samples of many species that are not in any sense residents of the upper strata of a forest. However, in those temperate regions of the world, such as the UK and nearby North European countries, whose insect faunas are relatively well known, establishing which of the insect species sampled from tree-crowns are residents and which are merely 'tourists' or vagrants is often feasible (Gaston *et al.*, 1993; see also Hammond and Owen, in press; Stork and Hammond, 1997, Chapter 1, this volume). In the Tropics the position is very different, as knowledge of the biology of many species is limited, and their usual pattern of occurrence in different ecotones often unknown. As a consequence, there have been few previous attempts to evaluate the species of any major insect group found in tropical tree-crowns in this way. Only groups represented by relatively modest numbers of species (in comparison with the Coleoptera) at any one tropical site have been successfully dealt with, e.g. Acrididae (grasshoppers) (Amedegnato, 1997, Chapter 14, this volume). Uniquely, the results of the comprehensive programme of sampling at ground and tree-crown levels in the Sulawesi lowland forest study area during Project Wallace furnish sufficient background information on the distribution of beetle species within the forest for an attempt to evaluate the status of most of the tree-crown occurring beetle species to be made. Most of the evidence used to allocate these species to stratum categories is drawn from the pattern of occurrence of the species themselves in tree-crown level (fogging, Malaise traps, baited traps) and other samples. Occurrences of species in actinic light-traps operated in the crowns of trees were given little weight, as such traps (used by Sutton, 1983, 1989) are known to attract large numbers of aquatic, riparian and other ground-level species that, apart from dispersal or mating flights, are very unlikely to enter or pass through tree-crowns.

Particular emphasis was placed on evidence obtained from directly comparable samples, e.g. from identical Malaise or baited traps set at different forest levels. However, care was taken to discount or place little weight on the occurrence of species at ground level where this was in close proximity to freshly fallen trees or in newly formed forest gaps. Apart from the pattern of occurrence in samples, knowledge of the general biology (especially the feeding habits and habitat associations usual for members of particular groups) was also of service in determining whether some species belonged to the tree-crown assemblage proper or were merely tourists from lower levels. Finally, the varying patterns and intensities of sampling at different levels in the study area were also taken into account.

The categorization of tree-crown occurring species adopted here is a simple three-fold one: (i) tree-crown specialists; (ii) stratum generalists; and (iii) lower level or 'ground' specialists. The two 'specialist' categories encompass species which are effectively confined, at least as adults, to one level in the forest, and do not occur elsewhere except as tourists. The generalist category, on the other hand, includes those species that regularly occur, as adults, at both tree-crown and lower levels. A disadvantage of this simple framework is that it ignores the fact that some stratum specialists are more clearly or completely confined to one particular stratum than are other 'specialists', and that 'generalist' species may exhibit varying degrees of bias towards one stratum or another. In addition, although not directly relevant to the principal purposes to which the categorizations are put in this paper, it also ignores the position of larval and pupal stages. For a number of the species categorized here as tree-crown specialists on the basis of the fidelity of adults to this stratum, the larvae are almost certainly ground specialists. The same applies to a number of the species which as adults are categorized as generalists. Nevertheless, using this simple categorization, the majority of beetle species could be allocated, with varying degrees of confidence, to one of the three categories. A number of species were referred to a fourth, 'uncertain' category, especially in cases where ground-level samples were less fully investigated (e.g. Aleocharinae and Curculionidae, see Appendix 10B). Most of these 'uncertain' species are ones represented in tree-crown samples by only one or two individuals and are also of rare occurrence in, or absent from, the ground-level samples that have so far been fully scrutinised. Experience with samples from better-known temperate sites suggests that many of these species are, in fact, likely to be ground-level specialists or generalists. An unknown proportion, on the other hand, are likely to be 'rare' canopy specialists.

Allocation of species to feeding groups

Using a five-fold system of herbivores, xylophages, fungivores, saprophages and predators, all beetle species represented in tree-crown samples were allocated to feeding groups on an individual basis. In assigning species, the principles and methods described by Hammond (1990) were adopted, except that the 'xylomycetophagous guild' is incorporated here in the fungivore category. Most species could be assigned confidently to one of the five groups. Some instances of particular uncertainty are noted in the Discussion below.

RESULTS

The beetle fauna of the lowland forest study area as a whole

The number of beetle species so far identified from the lowland forest samples stands at 4026, an increase of 385 species over the total recognized at an earlier stage of this study (Hammond, 1990). The proportional representation of these species among superfamilies is indicated in Figure 10.1 and that of family-groups is detailed in Appendix 10A. For 44 family-groups the number of species recognized remains the same as in 1990. However, for 21 of the family-groups the number of species recognized has been reduced slightly as a result of further taxonomic work. The number has been substantially reduced (from 132 to 84) for one family-group, Tenebrionidae in major part, the original figure given in 1990 having been the result of a typographical error. For 66 family-groups the figures given here show increases over those recorded in 1990, largely as a result of the preparation and study of additional material. For some family-groups the increases are large, e.g. from 98 to 177 species of Scolytidae, 24 to 39 species of Platypodidae, 241 to 305 species of Curculionidae (in major part), and 187 to 253 species of Aleocharinae (see Appendix 10A for further details).

The number of additional species estimated to be present in unsorted samples overall is 836. Most of these belong to groups such as the Aleocharinae, Pselaphidae and Curculionidae for which many samples have not been systematically searched. To arrive at an estimate of the number of beetle species actually present in the study area, a further allowance was made for species likely to have been missed by the programme of sampling. Although this allowance cannot be expected to be at all accurate, individual assessments were made for each family-group (see Appendix 10A), and care was taken to use all available data on the patterns of occurrence of members of these groups in samples of various types. As a result, the conclusion that some 75.5% of the species present were, in fact, obtained by the extensive and varied programme of sampling is unlikely to represent a gross over- or underestimate.

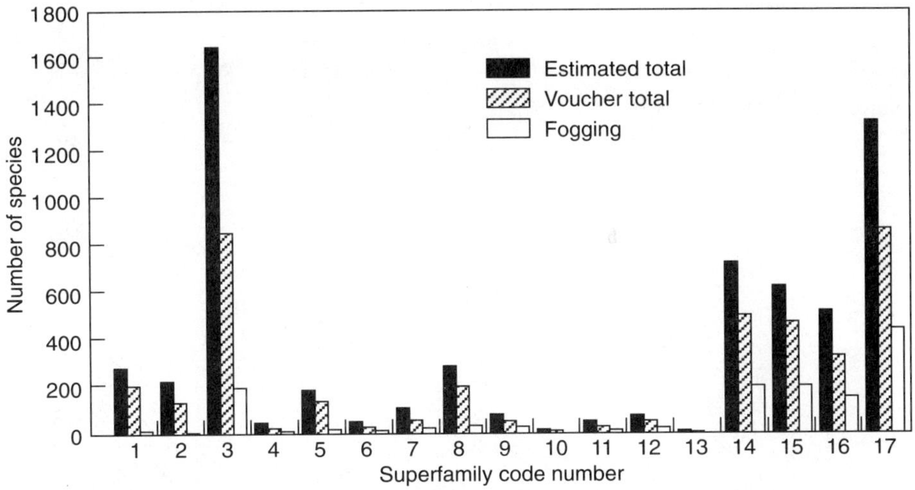

Figure 10.1 Proportional representation of Coleoptera species by superfamily in a 500-ha area of lowland rainforest in Sulawesi: estimated total number of species, total number of species in voucher collection, and number of species present in canopy fogging samples. The superfamily codes are: 1, Caraboidea (Adephaga); 2, Hydrophiloidea; 3, Staphylinoidea; 4, Eucinetoidea; 5, Scarabaeoidea; 6, Dryopoidea; 7, Buprestoidea; 8, Elateroidea; 9, Cantharoidea; 10, Dermestoidea; 11, Bostrichoidea; 12, Cleroidea; 13, Lymexyloidea; 14, Cucujoidea; 15, Tenebrionoidea; 16, Chrysomeloidea; 17, Curculionoidea.

Fogging samples

The 1106 m^2 of fogging samples obtained from the lowland forest study area between February and December 1985 contained ca. 18 000 individuals of 1355 beetle species. The distribution of these species among superfamilies is indicated in Figure 10.1. The fogging samples contained representatives of all 17 superfamilies, but relatively few species of some, notably Caraboidea, Hydrophiloidea and Scarabaeoidea (Figure 10.2), compared with species totals for the study area as a whole. By far the best represented superfamily was the Curculionoidea, followed (in order) by Cucujoidea, Tenebrionoidea, Staphylinoidea and Chrysomeloidea (Figures 10.1 and 10.3).

The overall family-group representation of species, as well as the representation of species found only in fogging samples, and species falling into each of the stratum categories is detailed in Appendix 10B. The best-represented family-groups were Curculionidae (excluding Otiorhynchinae and Rhyncophorinae) with 140 species, Aleocharinae with 91 species, Anthribidae with 87 species, Aderidae with ca. 80

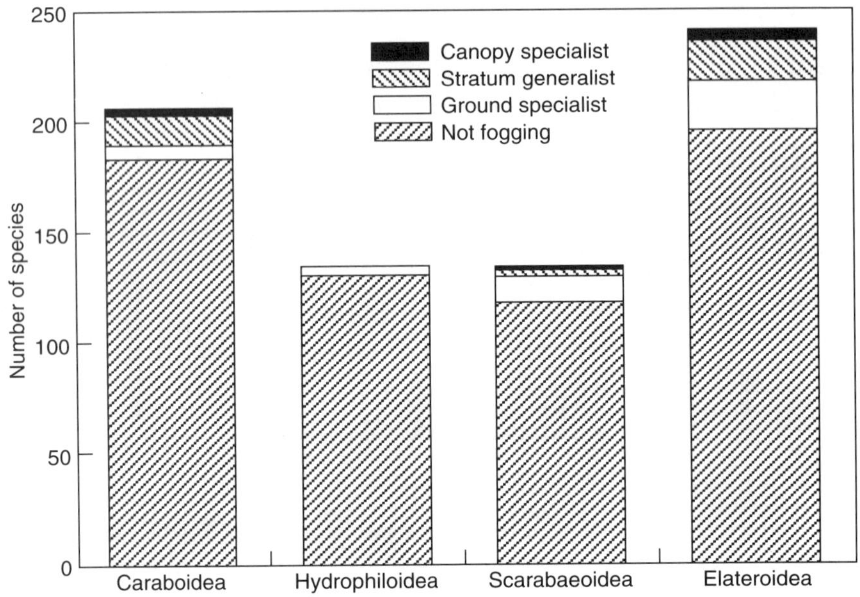

Figure 10.2 Numbers of species in four well-represented superfamilies (Caraboidea, Hydrophiloidea, Scarabaeoidea and Elateroidea) in the lowland forest study area. The upper three portions of each column (where present) show the numbers of species in each stratum category that are present in fogging samples.

species, Corylophidae with ca. 79 species, Scolytidae with 57 species, Lamiinae with 56 species, and Mordellidae with 51 species. Compared with the lowland forest beetle data-set as a whole, notably well-represented family-groups in the fogging samples were Ptinidae, Ptilodactylidae, Propalticidae, Phalacridae, Corylophidae, Scraptiidae, Attelabidae and Otiorhynchinae (see Tables 10.1 and 10.2 for further details). Notably poorly represented family-groups were Dytiscidae, Clambidae, Lucanidae, Cerylonidae, Biphyllidae and Hydrophilinae (all with no species), and Histeridae, Staphylininae and Leiodidae (see Tables 10.1 and 10.3 for further details).

More than one-third (39.3%) of the beetle species represented in these fogging samples were not found to be present in any other samples from the study area. A high proportion (53%) of the species recorded only from fogging samples were present in only one 1-m^2 sample, making conclusions as to which stratum category they should be assigned difficult to reach (see above). Nevertheless, 70% of the 1355 species present in the fogging samples were allocated to stratum categories: 21.9% as canopy

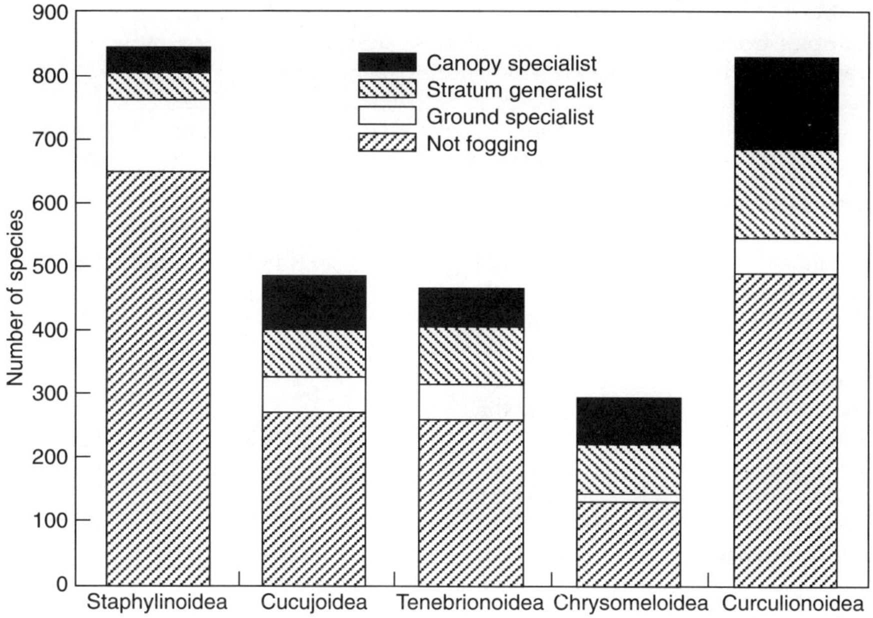

Figure 10.3 Numbers of species in the five best represented superfamilies (Staphylinoidea, Cucujoidea, Tenebrionoidea, Chrysomeloidea and Curculionoidea) in the Sulawesi lowland forest study area. The upper three portions of each column show the numbers of species in each stratum category that are present in fogging samples.

specialists, 28.0% as stratum generalists, and 20.1% as ground-level specialists, or tourists in terms of the tree-crowns where they were found; 30.0% could not confidently be assigned to any of these three categories (see Discussion below). High proportions of fungivores and xylophages and low proportions of saprophages and predators proved to be tree-crown specialists. Unlike other feeding groups, most species of herbivores were categorized as stratum generalists (Figure 10.4).

Overall, 623 (46.0%) of the species represented in the fogging samples were present in only one 1-m^2 sample. This proportion of 'singletons' is not exceptionally high, and does not necessarily indicate that the fogging programme missed a large proportion of the 'resident' tree-crown species present in the study area. This point is reinforced when the allocations to stratum category of the 623 species found in only one fogging sample are considered. Very few (7.9%) were categorized as tree-crown specialists; 14.6% were categorized as stratum generalists, and a more substantial 27.8% as ground specialists. Although almost half of the species (49.9%) could not be assigned with any confidence to one of these

Table 10.1 Representation of Coleoptera species in tree-crown fogging samples from the Sulawesi lowland forest study area: family-groups represented by an estimated >100 species in the study area

	No. of spp. fogging/ estimated total	*Voucher collection total (%)*	*Estimated total (%)*
Corylophidae*	79/155	78.2	51.0
Aderidae*	76/170	73.8	44.7
Lamiinae	56/175	43.4	32.0
Mordellidae	51/185	40.5	27.6
Curculionidae+	140/520	45.9	26.9
Anthribidae	87/340	38.7	25.6
Scolytidae	57/230	32.2	24.8
Elateridae	24/105	32.4	22.9
Buprestidae	25/120	36.2	20.8
Aleocharinae	91/480	36.0	19.0
Paederinae	23/135	24.7	17.0
Nitidulidae	18/110	24.7	16.4
Tenebrionidae+	16/110	19.0	14.6
Carabidae+	20/200	13.4	10.0
Pselaphidae	19/205	20.9	9.3
Scaphidiidae	8/100	15.7	8.0
Ptiliidae	9/130	19.1	6.9
Scydmaenidae	8/130	17.4	6.2
Histeridae	2/160	2.2	1.2

* Figures approximate as sorting of voucher collection species for these groups remains incomplete.
+ For the scope of these family-groups, see Appendix 10A.

three categories, most of these unassigned species belong to groups in which ground-level samples were not at all thoroughly scrutinised. For example, in the Aleocharinae, 20 out of 41 species fogged as singletons were not assigned, in the Corylophidae 18 out of 28, in the Aderidae 11 out of 22, in the Anthribidae 25 out of 45, and in the Curculionidae 46 out of 72 species. Except for the latter two families, relatively few of the unassigned species in such groups are likely to be canopy specialists. In the Anthribidae and Curculionidae, as well as several more fully evaluated groups (e.g. Buprestidae, Elateridae, Scolytidae and various chrysomelid subfamilies), a greater proportion of unassigned species are likely to be true tree-crown specialists that are simply 'rare' in the samples taken.

Table 10.2 Representation of Coleoptera species in tree-crown fogging samples from the Sulawesi lowland forest study area: proportionally well-represented beetle family-groups (excluding those represented by less than 10 species in the study area samples)

	No. of spp. fogging/ estimated total	*Voucher collection total (%)*	*Estimated total (%)*
Ptinidae	7/10	100.0	70.0
Ptilodactylidae	9/14	90.0	64.3
Propalticidae	12/22	80.0	54.6
Phalacridae	18/22	81.8	51.4
Corylophidae*	79/15	78.2	51.0
Scraptiidae	5/10	100.0	50.0
Attelabidae	15/30	75.0	50.0
Otiorhynchinae	16/26	61.5	48.5
Aderidae*	80/170	76.7	47.1
Melyridae	14/30	60.9	46.7
Coccinellidae	39/66	59.1	45.9
Anobiidae	19/25	76.0	45.2
Alleculinae	10/23	75.0	43.5
Galerucinae	41/10	70.7	41.0
Cantharidae	8/20	61.6	40.0

* Figures are approximate as sorting of voucher collection species for these groups remains incomplete.

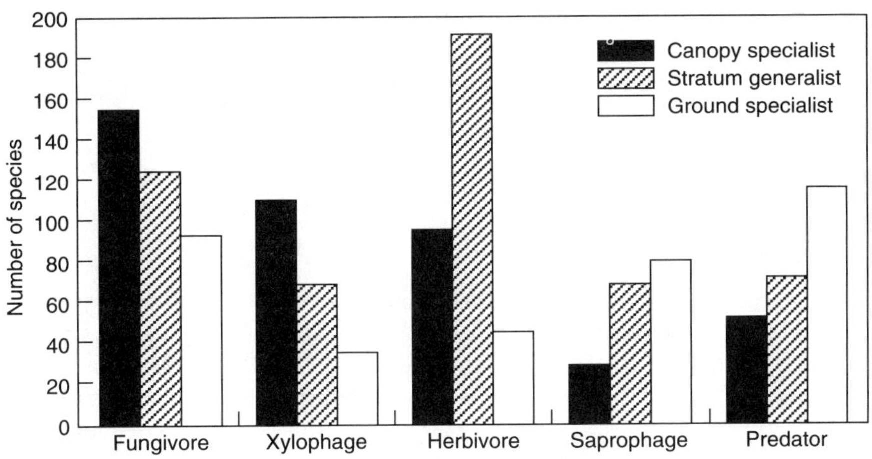

Figure 10.4 Proportional representation of species present in canopy fogging samples, categorized as canopy specialists, stratum generalists, and 'ground' specialists (see text for explanation), among five broad feeding groups: fungivores, xylophages, herbivores, saprophages and predators (for definitions, see Hammond, 1990). Species not readily assigned to any of the three stratum categories were allocated on a (largely) *pro rata*, group by group, basis (see text).

Table 10.3 Representation of Coleoptera species in tree-crown fogging samples from the Sulawesi lowland forest study area: proportionally least well represented family-groups (excluding those represented by less than 10 species in the study area samples)

	No. of spp. fogging/ estimated total	*Voucher collection total (%)*	*Estimated total (%)*
Dytiscidae	0/28	0.0	0.0
Clambidae	0/11	0.0	0.0
Lucanidae	0/11	0.0	0.0
Cerylonidae	0/19	0.0	0.0
Biphyllidae	0/11	0.0	0.0
Hydrophilinae	0/23	0.0	0.0
Histeridae	2/89	2.2	1.2
Staphylininae	2/62	3.2	2.1
Leiodidae	2/32	6.2	2.9
Tachyporinae	2/41	4.9	3.1
Rutelinae	1/22	4.5	3.6
Scarabaeinae	2/38	5.3	4.4
Sphindidae	1/14	7.1	5.0
Scydmaenidae	8/46	17.4	6.2
Colydiidae	3/35	8.6	6.7
Ptiliidae	9/47	19.1	6.9
Scaphidiidae	8/51	15.7	8.0
Pselaphidae	19/91	20.9	9.3
Erotylidae	4/32	12.5	10.0
Carabidae*	20/149	13.4	10.0

* Excludes Cicindelinae, Paussinae and Rhysodidae.

Other tree-crown samples

Malaise traps

Three small Malaise traps were operated in the crowns of tall trees at Plots A, B and C for a combined total of 81 trap–weeks. The results of this trapping for all insect orders, in terms of abundance and variations with season and plot, have been discussed by Hammond (1990). In comparison with ground-level trapping, catch rates and numbers of species obtained were uniformly low for all major insect groups. The three traps obtained 1116 individual beetles of 327 species; 76 of these (23.2%) were recorded from no other samples, the majority (48 species) belonging to just three families (Cerambycidae, Anthribidae and Curculionidae). Half (50.2%) of the 327 species were not present in any fogging samples from the nearby quadrats in the lowland forest study area. Of the 49.8% of the species also present in fogging samples, more than one-third (16.8% of the total) were

Table 10.4 Representation of beetle 'family-groups' (following the system of Hammond, 1990) in three Malaise traps operated in tree-crowns for a total of 81 trap-weeks in the lowland forest study area, Sulawesi (see Hammond, 1990 for details of trapping). 1 = total species, 2 = number of species also present in lowland forest fogging samples (number of additional species present in other fogging samples, i.e. not from the lowland forest study area, in parentheses), 3 = number of species otherwise found only in fogging samples, 4 = number of species found only in tree-crown Malaise traps

	1	*2*	*3*	*4*
Carabidae	3	–	–	1
Histeridae	1	–	–	1
Ptiliidae	2	–	–	1
Omaliinae	1	(1)	–	–
Oxytelinae	4	4	–	–
Paederinae	7	6+(1)	–	–
Staphylininae	1	(1)	–	–
Aleocharinae	6	2	–	2
Pselaphidae	4	3	–	–
Scirtidae	1	1	–	–
Scarabaeinae	1	–	–	–
Cetoniinae	1	–	–	–
Ptilodactylidae	3	3	–	–
Limnichidae	2	2	–	–
Buprestidae	2	–	–	–
Elateridae	8	7	–	–
Throscidae	4	1	–	1
Eucnemidae	12	3+(1)	1	4
Lycidae	1	1	–	–
Cantharidae	2	2	–	–
Dermestidae	1	–	–	–
Anobiidae	4	2	–	2
Ptinidae	1	1	–	–
Cleridae	18	8	1	4
Melyridae	2	2	–	–
Lymexylidae	1	1	–	–
Nitidulidae	4	–	–	1
Rhizophagidae	1	–	–	–
Phalacridae*	4	3	–	–
Laemophloeidae	2	1	1	–
Propalticidae	1	–	–	–
Biphyllidae	1	–	–	–
Erotylidae	1	–	–	1
Languriidae	1	–	–	–
Corylophidae*	15	10	7	2
Coccinellidae	13	9	4	–
Bothrideridae	2	–	–	2

Table 10.4 continued

	1	2	3	4
Mycetophagidae	1	1	–	–
Mordellidae	12	8	1	–
Anthicidae	3	2	–	–
Pedilidae	1	1	–	–
Aderidae*	5	5	3	2
Othniidae	2	1	–	–
Lagriidae	2	1	1	–
Alleculinae	1	1	–	–
Tenebrionidae	3	2	–	1
Cerambycinae	12	–	–	7
Lamiinae	41	14+(3)	3	14
Eumolpinae	11	11	1	–
Galerucinae	5	4	–	1
Alticinae	3	2	–	–
Anthribidae	53	18+(2)	6	17
Brentidae	1	–	–	1
Otiorhynchinae	1	1	–	–
Rhyncophorinae	1	–	–	–
Curculionidae	21	8	1	10
Scolytidae	8	3+(1)	1	1
Total	327	154+(10)	33	76

* Values given for Phalacridae, Corylophidae and Aderidae are provisional as sorting of voucher collection species for these groups remains incomplete.

present only in these two types of sample (fogging and tree-crown Malaise traps). Overall, 37.6% of the species present in the tree-crown Malaise trap samples are categorized here as canopy specialists, 33.0% as stratum generalists, and 19.0% as ground-level specialists; 10.4% were not assigned to any of the three categories. The proportion of species considered to be tree-crown 'residents' as adults is higher than in the fogging samples. The Malaise traps also obtained proportionally more herbivore species (see below). Further details of the species composition of the tree-crown Malaise trap samples are provided in Table 10.4.

Baited traps

At all heights of more than 0.5 m above ground level, the abundance and species richness of Coleoptera in baited traps were very low (Hammond, 1990). Only 78 beetle species were trapped at heights of 9.5–25 m, where catch-rates of beetles were uniformly less than 1% of those at ground level. Twenty of the 78 species (25.6%) were not found in any other tree-crown or ground level samples, and 7.7% otherwise only in tree-crown samples of other types. Most (66.7%) were also found in ground-level samples. Overall, 13 of the 78 species (16.7%) are categorized here as canopy specialists, 35.9% as stratum generalists, and 25.6% as ground specialists; 21.8% were not assigned to a stratum category. Only six of the 13 canopy specialists and only one of the unassigned species were present in fogging samples.

Actinic light-traps

All of the beetle species obtained were found also at ground level. Most individuals belonged to ground specialist species associated with waterside habitats of the type that frequently dominate catches in light-traps suspended well above ground level (Sutton, 1983, 1989).

DISCUSSION

Characteristics of the tree-crown beetle assemblage

The picture of the study area's tree-crown beetle assemblage presented here is based primarily on the results obtained from a comprehensive programme of insecticide fogging. The results of sampling at tree-crown level by other means furnish additional insights and help to provide a more complete picture. Two features, taxonomic and feeding-group representation, are emphasized, although data on other characteristics, notably microhabitat affiliation and body size (N. Stork and P. Hammond, unpublished data), have also been assembled. As already noted, for any discussion of tree-crown arthropod samples to be truly informative about the assemblage of species present in the tree-tops, it is important to tease out the various faunal elements represented in the samples. In this way the 'noise' introduced by species that are merely tourists at tree-crown level may be minimized.

Taxonomic composition

The study area as a whole – like many forested regions of Sulawesi – is especially rich in species of Eucnemidae, Mordellidae, Aderidae and Anthribidae compared with most other regions of the moist tropics (Hammond, 1990; also unpublished data). Not surprisingly, this is reflected in the composition of the tree-crown beetle assemblage. In addition, compared with comparable forests in other parts of south-east Asia, the Sulawesi lowland forest tree-crown assemblage is relatively poor in species of several families, such as Elateridae, Tenebrionidae, Chrysomelidae, Pselaphidae and Dryopidae (Stork, 1991; Mawdsley,

1994). Compared with the Neotropics, Sulawesi tree-crown assemblages are poor in species of several other families, such as Apionidae, Bruchidae, Anobiidae, Ptilodactylidae and Coccinellidae (Erwin and Scott, 1980; Davies *et al.*, 1997, Chapter 5, this volume). Based on fogging sample data, the tree-crown beetle assemblages of higher elevation sites (1100–1760 m) in N. Sulawesi are poorer in species of a number of family-groups, notably Buprestidae, Eucnemidae, Anobiidae, Lagriidae, Cerambycinae, Platypodidae and Scolytidae, and are richer in Carabidae, Staphylininae, Elateridae, Cisidae, Melandryidae and Alticinae (N. Stork and P. Hammond, unpublished data). Such inter-regional and elevational differences in canopy beetle assemblages generally correlate with overall faunistic differences (Hammond *et al.*, in press), although many individual factors undoubtedly have some influence. For example, most forest-dwelling species of Bruchidae and Apionidae are associated with leguminous trees, and are well represented in forests in which many suitable hosts occur, such as in the Neotropics.

Taxonomic 'clumping' in tree-crown beetle assemblages

Removing from consideration the obvious ground specialists (tourists in terms of tree-crowns), it is apparent that the great majority of species taken by fogging or in other tree-crown samples belong to a relatively small number of groups that are taxonomically, as well as trophically (see below), clumped. This pattern is even clearer when the tree-crown specialists and/or the more abundant species in fogging samples are considered alone.

Although 95 of the 127 family-groups recorded from the study area are represented in fogging or other tree-crown samples, the majority of species in a number of these family-groups are generalist or ground specialist species. For example, all Sphaeridiinae, Histeridae, Hydraenidae, Leiodidae, Oxytelinae, Euaesthetinae, Staphylininae, Aphodiinae, Scarabaeinae, Rutelinae, Chelonariidae, Lampyridae, Lymexylidae and Sphindidae present in fogging samples are probably ground specialist species (see Appendix 10B). Only 34 of the 91 family-groups represented in fogging samples contain more than one species allocated to the canopy specialist category, and only 15 contain more than five species regarded as canopy specialists. More than half (55.6%) of the canopy specialist species from fogging samples belong to just seven family-groups, Curculionidae (with 38 species), Lamiinae (32 species), Anthribidae (29 species), Corylophidae (28 species), Coccinellidae (13 species), Aderidae (13 species) and Paederinae (12 species). Taxonomic clumping of canopy specialist species is also evident at lower taxonomic levels, as most species belong to a relatively small number of genera, or groups of related genera. For example, most of the canopy specialist Anthribidae belong to the tribe Choragini, the Propalticidae all belong to the genus *Propalticus*, almost all of the Paederinae to *Palaminus*, most of the Otiorhynchinae to the tribe Celeuthetini, and most of the Coccinellidae to *Pseudoscymnus* and allied genera.

A somewhat different picture is obtained when the abundance or frequency of individual species in fogging samples is considered. The most abundant 113 species (those represented in the fogging samples by more than 16 individuals) are distributed through 30 family-groups. Most, however, belong to just nine groups: Aderidae (10 species), Paederinae (9 species), Eumolpinae (8 species), Galerucinae (7 species), Otiorhynchinae (7 species), Corylophidae (6 species), Alticinae (6 species), Anthribidae (6 species) and Curculionidae (6 species). Not surprisingly, very few (only one) of these 113 species is categorized as a ground specialist, but most (ca. 60%) are stratum generalists rather than tree-crown specialists. When the 32 most abundant species (those represented by more than 64 individuals) are considered, the proportional representation of stratum generalists is even greater (ca 80% of the species). Thus, few of the canopy specialist species are exceptionally abundant in fogging samples, although many of the stratum generalist species that predominate belong to the same narrow taxonomic groupings as many of the tree-crown specialist species, e.g. *Palaminus* (Paederinae), Celeuthetini (Otiorhynchinae), *Rhyparida* and related genera (Eumolpinae), and *Glypostenoda* (Mordellidae).

Feeding group representation

The best-represented feeding group in the fogging samples was the fungivore guild (27.5% of species), including at least one slime-mould associated species and some 60 wood-inhabiting xylomycophagous species. Herbivores were also well represented (25.2%), while predators (17.4%), xylophages (16.1%) and saprophages (13.8%) were less so. However, compared with samples from trees in Borneo (Stork, 1991; Mawdsley, 1994) and from elsewhere in the Tropics, the proportion of herbivores was low, and that of fungivores particularly high (Table 10.5; P. Hammond *et al.*, in preparation). The significance of these results depends on the reliability with which feeding group assignments are made. Accurate allocation of the members of any tropical beetle assemblage to feeding groups is difficult. Even where the allocation is done on a species by species basis, as in the present study, imprecision and uncertainties remain. Particular problems are presented by family-groups in which more than one type of feeding is found, such as Elateridae, Nitidulidae, Phalacridae, Mordellidae, Tenebrionidae and Anthribidae. However, reasonably confident feeding-group assignments

Table 10.5 Proportional representation of Coleoptera species among feeding 'guilds' (as defined by Hammond, 1990) in fogging samples from the Sulawesi lowland forest study area, compared with fogging samples from other sites (data for other sites from Hammond *et al.*, in press)

	Herbivores	*Xylophages*	*Fungivores*	*Saprophages*	*Predators*	*No. of spp. total*
Sulawesi	25.1	16.1	27.5	13.8	17.4	1355
Brunei	34.6	7.7	18.5	15.5	23.7	875
Australia	19.8	21.1	23.1	8.4	27.3	454
UK	23.5	10.0	26.5	11.0	29.0	200

were possible for many species once a more precise assessment of taxonomic affinities had been made. For example, 34 of the 87 species of Anthribidae in the fogging samples are small members of the tribe Choragini, most of which are likely to feed on ascomycete fungi found on twigs and branches. Many of the other Anthribidae in the tree-crown samples, belonging to the subfamily Anthribinae, are also of genera in which species are likely to be fungivores. The 36 species of Cryptorhynchinae could be fairly confidently categorized as wood-feeders, dorcatomine Anobiidae as fungivores, most Buprestidae (*Agrilus* species) as twig-girdlers, and Aleocharinae of the genus *Gyrophaena* as fungivores. Most Lathridiidae belong to genera of Corticariinae in which species are mostly pollen-feeders, most Phalacridae to herbivorous genera, most Mordellidae to genera in which larvae are presumed to be stem-borers, and all Coccinellidae to genera in which predacious habits are the rule.

Stratification

The results (Appendix 10B) indicate that a significant proportion of the beetle species occurring in the study area are effectively restricted, as adults, to one stratum (or series of strata) of the forest. Most – probably between 70–80% – are effectively specialized, as adults, to the lower (below tree-crown) forest strata. Between 8% and 13% are probably effectively specialized to tree-crowns, and between 10% and 16% of species are to be found regularly as adults both at the higher and lower levels of the forest (see Tables 10.6 and 10.7; Figures 10.2, 10.3, 10.5 and 10.6). The proportion of high-stratum specialists and the degree of stratification, as indicated by the relatively low proportion of stratum generalists, are both likely to be distinctly greater than in temperate forests where tree-crown assemblages have been examined in the context

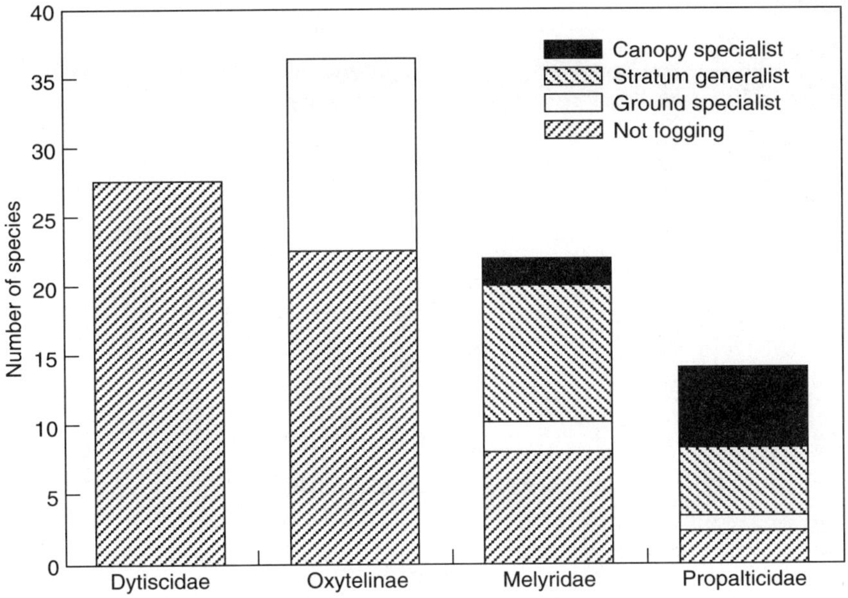

Figure 10.5 Numbers of species present in each stratum category (upper three portions of each column, where present) for canopy fogging samples from the Sulawesi lowland forest study area. A selection of taxonomic groupings (Dytiscidae, Oxytelinae, Melyridae, Propalticidae) is presented to illustrate differences in patterns.

Table 10.6 Representation of Coleoptera species in tree-crown fogging and other samples from the Sulawesi lowland forest study area, based on (A) the family-groups for which the study area is considered to have been fairly completely (>70%) inventoried, (B) a larger dataset including also family groups considered to have been moderately well (>55%) inventoried, and (C) the full study area data-set for 128 family-groups, including 12 for which inventories are well short of completeness (see Appendix 10A and text for explanation)

Occurrence	*A*	*B*	*C*	*Mean*
No. of spp.	2641	3228	4862	
Percentage in:				
Fogging samples	27.1	28.3	27.9	27.8
Other samples	90.2	90.0	89.2	89.8
Fogging samples only	9.8	10.0	10.8	10.2
Other samples only	72.9	71.7	72.5	72.4
Fogging and other samples	17.3	18.3	16.9	17.5

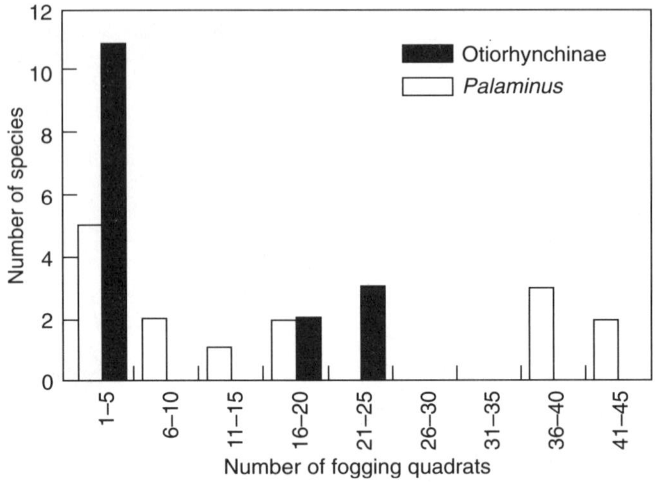

Figure 10.6 Frequency of otiorhynchine Curculionidae and *Palaminus* (Paederinae) species in fogging samples from the Sulawesi lowland forest study area. In fogging samples, both taxonomic groupings are represented almost entirely by canopy residents (i.e. canopy specialists and stratum generalists), but exhibit different frequency distributions. Nine frequency classes of 12 × 12 m fogging quadrats (out of 45 quadrats sampled) illustrate the restricted distributions of some species and the widespread distributions of others.

Table 10.7 Proportional representation of Coleoptera species in forest stratum categories from lowland forest study area fogging samples. Figures are provided for three data-sets coded A, B and C, as in Table 10.6.

Data-set		*Category**			
	No. of spp.	*Canopy*	*General*	*Ground*	*Uncertain*
A	762	23.0	28.0	22.2	26.8
B	915	22.5	30.8	20.4	26.3
C	1355	21.9	28.0	20.0	30.1

* Canopy, tree-crown specialists; General, stratum generalists; Ground, ground specialists.

of the species present in the forest as a whole (Hammond and Owen, in press; Stork and Hammond, 1997, Chapter 1, this volume, P. Hammond and N. Stork, unpublished data). The beetle assemblages occurring at the higher elevation sites in Sulawesi investigated during Project Wallace are also likely to be less stratified (Hammond, 1990; Stork and Brendell, 1990). However, as suggested by Hammond (1992, 1995), more marked stratification and a proportionally larger tree-crown specialist component may well be characteristic of some other forests, especially in the Tropics and Subtropics.

Frequency and distribution

The frequency of species in the fogging and other tree-crown samples has already been mentioned in passing. This, and both the spatial and seasonal distribution of species are clearly of considerable relevance to any assessment of the completeness of sampling, and thus to any conclusions concerning the richness of the tree-crown resident beetle assemblage. A brief comment on these topics is included here. The position with respect to seasonal distribution may be summarized succinctly: except for a very small number of species, mostly herbivores, evidence for any restriction to particular seasons is lacking.

The spatial distribution and frequency of species in the fogging samples may be gauged at four levels, namely those of: (i) the four forest sites where fogging was carried out; (ii) the nine 'fogs' carried out at these four sites; (iii) the 45 quadrats fogged; and (iv) the 1106 1-m^2 samples taken from these quadrats. The extent to which the distributions and frequencies of tree-crown species versus tourists differ from each other depends greatly on the scale at which they are examined. In general, the greatest differences are apparent at the smallest spatial scales. Differences between different groups of canopy specialists may also be more or less apparent depending on scale. An example is provided by the guild of specialist, plant-climbing, predacious Staphylinidae of the genus *Palaminus*, composed of 15 species, and that comprising broad-nosed weevils, mostly belonging to the Celeuthetini, composed of 17 species. Adults of the latter group are leaf-chewers while the larvae are (generally polyphagous) root-feeders. The *Palaminus* guild includes seven species regarded here as tree-crown specialists, and eight species regarded as stratum generalists, four of which show a bias towards tree-crowns. Most of the species are remarkably uniformly distributed through the samples, with five species present in 38 or more of the 45 quadrats, and nine species present at all four fogging sites. Although frequent in fogging samples, members of the phyllophage weevil guild are evidently more spatially restricted than *Palaminus* at larger scales. Only two species were found at all four fogging sites and

Table 10.8 Representation by frequency classes (number of 1-m^2 tray samples) of beetle species in fogging samples from the lowland forest study site, by superfamilies, stratum categories and feeding groups

	No. of trays									
	1	*2*	*3–4*	*5–8*	*9–16*	*17–32*	*33–64*	*65–128*	*129–256*	*257–514*
Superfamilies										
Caraboidea	10	–	7	1	3	1	–	–	–	–
Hydrophiloidea	3	1	–	–	–	–	–	–	–	–
Staphylinoidea	102	24	30	17	9	7	2	2	4	–
Eucinetoidea	2	3	1	1	–	–	1	–	–	–
Scarabaeoidea	8	3	1	1	–	–	1	–	–	–
Dryopoidea	7	–	5	1	1	2	1	–	–	1
Buprestoidea	12	8	3	2	–	–	–	–	–	–
Elateroidea	25	6	5	6	1	–	1	–	–	–
Cantharoidea	16	1	9	–	2	1	–	1	–	–
Dermestoidea	2	1	–	–	–	–	–	–	–	–
Bostrichoidea	11	5	8	2	–	1	–	–	–	–
Cleroidea	13	4	7	3	2	–	–	1	–	–
Lymexyloidea	1	–	–	–	–	–	–	–	–	–
Cucujoidea	96	16	22	23	17	7	6	2	3	1
Tenebrionoidea	67	34	26	33	24	12	3	3	1	–
Chrysomeloidea	63	27	23	16	11	8	7	3	4	1
Curculionoidea	182	52	49	28	16	16	1	4	–	–
Stratum categories										
Canopy specialists	45	58	57	56	34	33	7	5	2	–
Stratum generalists	91	45	83	58	43	23	16	12	10	3
Ground specialists	175	37	33	16	9	1	–	–	–	–
Uncertain category	326	46	24	6	1	1	–	–	–	–
Feeding groups										
Fungivores	185	49	46	35	22	12	6	2	2	1
Xylophages	104	35	29	13	6	2	1	–	1	–
Herbivores	134	50	50	46	25	25	8	6	5	1
Saprophages	78	28	26	27	17	12	3	2	–	1
Predators	134	25	46	15	17	7	5	7	4	–
Total	635	186	197	136	88	58	23	17	12	3

four species at three of the four sites. This difference, however, is much more marked at the level of the fogging quadrat (Figure 10.6). Each guild includes five 'common' species, but while the common *Palaminus* were present in almost all quadrats, each of the common phyllophage weevil species was found in only about half of all quadrats.

As the pattern of occurrence in fogging samples was one source of evidence in making strata categorizations, it is not surprising that species from different strata exhibit, on average, differences in abundance and frequency in fogging samples (i.e. number of plots, fogging quadrats or trays in which they are present). As expected, tree-crown specialists are generally more frequent and abundant in fogging samples than are lower-level (ground) specialists (Table 10.8). However, as already indicated, the 'very frequent' category is dominated by stratum generalists, with some 80% of the 32 most abundant species in the fogging samples. The frequency of tree-crown specialist species in fogging samples varies considerably between taxonomic groups. Family-groups such as Paederinae, Lathridiidae, Pedilidae, Alleculinae, Otiorhynchinae are represented by many individuals in fogging samples, but not a great number of species, while a number of other groups are at the opposite extreme. Buprestidae, Lamiinae, Anthribidae and Scolytidae, in particular, are rich in canopy specialist species, although none of them is represented in fogging samples by more than a few individuals. The canopy specialist species of several other family groups, such as Anobiidae, Cisidae, Brentidae, Omaliinae and Cerambycidae, also occur at very low frequency, many as singletons, in fogging samples. The distribution of species among frequency classes (based on the number of 1-m^2 trays in which species were found) is summarized for taxonomic groups (superfamilies), stratum categories and feeding groups in Table 10.8.

CONCLUDING REMARKS

Confidence that the results presented and discussed above provide an accurate picture of the tree-crown beetle assemblage examined is likely to depend very much on whether answers to the following questions can be provided and, if so, what these answers are:

1. How complete is the sampling of the lowland forest tree-crowns by fogging?
2. How representative is the sampling of lowland forest tree-crowns by fogging?
3. Can the resident and non-resident elements of the tree-crown assemblage be distinguished reliably?
4. What part of the tree-crown beetle assemblage is missed by fogging and by other methods of sampling?

Any attempt to answer these questions fully is beyond the scope of this paper, but the issues involved are discussed briefly below.

Completeness

Several lines of evidence may be employed. First, using data from the fogging samples themselves the pattern of species accumulation with sampling effort may be examined, most usefully at (i) the level of individual plots and (ii) for the lowland forest dataset as a whole. Extrapolative methods, such as the jack-knife, detailed by Colwell and Coddington (1994), may be used to estimate how many additional beetle species would be obtained *by fogging* with increased sampling effort. However, as with many other types of sample the proviso that fogging does not sample a single assemblage (vagrants from other assemblages are included) obtains. Second, some use may be made of inter-plot differences in beetle species obtained by fogging. Third, the composition of other tree-crown level samples in relation to those obtained by fogging may be used to estimate the size of the tree-crown assemblage component not sampled by fogging. Finally, the patchiness of occurrence of tree-crown resident species, as revealed by fogging, gives some indication of the proportion of species that the fogging programme is likely to have missed.

In reality, two separate questions are to be addressed here. First, what portion of the tree-crown beetle assemblage does fogging (as a technique) not sample? Second, given the spatial and temporal pattern of the fogging programme conducted and the amount of fogging done, what portion of the study area tree-crown fauna obtainable by fogging is likely to have been missed?

Representativeness

Assuming that the fogging programme did not obtain all beetle species present in tree-crowns in the study area, were those obtained a representative cross-section of the full tree-crown beetle assemblage in terms of taxonomic groupings, trophic groups, microhabitat associations, body size and other attributes? Most importantly, are they representative of the tree-crown assemblage in terms of the stratum categories (tree-crown specialist, generalist, ground specialist) adopted for discussion here? Would a complete list of tree-crown beetle species contain a greater or smaller proportion of canopy specialists?

In general terms, these questions may be answered by reference to fogging data from previous studies (especially those in areas where insect faunas are well known, e.g. Stork and Hammond, 1997, Chapter 1, this volume), and from knowledge of how species of various strata generally accumulate with sampling effort (Hammond, 1994). The patterns of accumulation of species in different strata in the lowland forest dataset itself will give a more direct indication of the bearing of sampling effort on their proportional representation. Again, using the jack-knife or other methods, separate projections may be made for each stratum. A general problem here, as already noted above, is how to deal with unclassified species (those not referred to one of the three stratum categories). Where taxonomic groups have been well sampled in the study area as a whole, and also well studied, it may reasonably be assumed that a high proportion of species not classified because they were present as singletons in the fogging samples, but were not present in other samples, are in fact canopy specialists. They are simply rare, or at least rare in fogging samples. The same, however, is unlikely to be true of taxonomic groups that were not also well sampled at lower forest levels. In such instances, many species present as singletons in fogging samples, but not as yet detected in lower level samples, are likely to be generalists or ground specialists. The safest procedure to adopt with the unclassified species may be to initially allocate them to the three stratum categories in proportion to species already classified.

The size of the tree-crown beetle assemblage

The number of species estimated to be present in the ca. 500-ha lowland forest study area is 6425. Reasons for believing that a figure of this order is reasonably accurate are discussed above and by Hammond (1990). Some 1550 beetle species, including 1355 taken by fogging, were present in samples taken at tree-crown level. Only half (49.9%) of the species collected by fogging were confidently categorized as true members of the tree-crown assemblage (tree-crown specialists and stratum generalists). These 677 species probably include most beetles that occur commonly, as adults, at tree-crown level. However, as discussed above, the tree-crown beetle assemblage as a whole is likely to contain a substantially larger number of species. Not all species are collected by fogging, many of the species in tree-crown samples that were not assigned to a stratum category are sure to be tree-crown residents, and the sampling programme will have missed an (uncertain) number of tree-crown residents. Making due allowance for all of these factors, a minimum of 1290 tree-crown resident species is reached, i.e. ca. 20% of the number of species estimated to be present in the study area overall. If a much greater proportion of the species present in tree-crown samples are, in fact, tree-crown residents, and a much greater allowance for species as yet uncollected is made, this could justify an increase on this figure of up to 50%.

Significance of the Sulawesi results

In referring species from tree-crown level samples to a particular stratum category (tree-crown specialist, stratum generalist or ground specialist), all available data on the occurrence of species in samples from all levels were brought into play. These data are not inconsiderable as they concern, in all, some 831 000 individual beetles distributed through 3000 samples. However, the usefulness and significance of these forest stratification data also depend on how representative the area studied is of tropical lowland forest in general, and how representative beetles are of arthropods as a whole in such forests. Any attempt to deal with the latter question is bound to be speculative, as data on tropical assemblages of the other exceptionally species-rich groups, Hymenoptera, Diptera and Acari, necessary to gauge the extent to which the Coleoptera are representative, are lacking. Tropical forests that are richer overall in beetle species, that are more stratified, and that have a higher proportion of their beetle species in the tree-tops than the Sulawesi study area are sure to exist. Nevertheless, gross departures from the pattern described here for a lowland rainforest in Sulawesi are unlikely. Ecological patterns of this scale tend, in large measure, to be relatively consistent. Even in the most stratified of tropical moist forests, the number of arthropod species resident in the crowns of trees is likely to be substantially lower than the number that reside at lower levels.

Acknowledgements

We are indebted to a number of fellow coleopterists for their valuable help in sorting beetle samples to species: R.T. Thompson (Aderidae, Curculionoidea, *Palaminus*), R.J.W. Aldridge (Cerambycidae), W. Wittmer (Cantharidae, Melyridae), E. Lewis (Phalacridae), S.L. Shute (Chrysomeloidea), R.T. Bell (Rhysodidae), M. Brancucci (Chauliognathinae), F. Cassola (Cicindelinae), J. Liebherr (Agonini), C. Johnson (Ptiliidae), M.L. Cox (Alticinae), R.G. Booth (Coccinellidae), J. Cooter and F. Angelini (Leiodinae), A. Kirejtschuk (Nitidulidae), J. Muona (Eucnemidae), R.A. Beaver (Scolytidae, Platypodidae), J. Krikken (Scarabaeinae, Aphodiinae), A. Slipinski (Cerylonidae, Colydiidae), D.G. Halstead (Silvanidae), L. Bomans (Lucanidae), M. Franciscolo (Mordellidae), I. Löbl (Pterogeniidae, Passandridae, etc.), F. Bameul (Omicrini), M. Hansen (Hydrophilidae), J. Huijbregts (some Sphaeridiinae and Aphodiinae), B. Valentine (Anthribidae), P. Eggleton (Brentidae), B. Levey (Buprestidae), E. Ricchiardi (Trichiinae, Valginae), S. Endrödy-Younga (Clambidae), R.B. Madge (Silphidae), M. Uhlig (*Erichsonius*), H. Schillhammer (*Gabrius*), G. de Rougemont (Paederinae), M.E. Bacchus (Cetoniinae), and C. Scholtz (Trogidae).

We gratefully acknowledge the assistance provided to us in our fieldwork in Sulawesi, and subsequently the advice and help given during the long process of analysing the results of our work on Sulawesi beetles by many others. These are duly acknowledged in previous papers (Hammond, 1990; Stork and Brendell, 1990).

REFERENCES

Allison, A., Samuelson, G.A. and Miller, S.E. (1993) Patterns of beetle species diversity in New Guinea rain forest as revealed by canopy fogging: preliminary findings. *Selbyana*, **14**, 16–20.

Amedegnato, C. (1997) Diversity of an Amazonian canopy grasshopper community in relation to resource partitioning and phylogeny, in *Canopy Arthropods* (eds N.E. Stork, J. Adis and R.K. Didham), Chapman & Hall, London, pp. 281–319 (this volume).

Angelini, F. and Cooter, J. (1993) New species of *Agathidium* Panzer and *Liodopria* Reitter (Col., Leiodidae) from Sulawesi. *Entomologist's Monthly Magazine*, **129**, 67–79.

Bameul, F. (1993) Omicrini from Sulawesi Utara, Indonesia (Coleoptera: Hydrophilidae: Sphaeridiinae). *Invertebrate Taxonomy*, **7**, 751–78.

Basset, Y. (1990) The arboreal fauna of the rainforest tree *Argyrodendron actinophyllum* as sampled with restricted canopy fogging: composition of the fauna. *The Entomologist*, **109**, 173–83.

Basset, Y. (1991) The taxonomic composition of the arthropod fauna associated with an Australian rainforest tree. *Australian Journal of Zoology*, **39**, 171–90.

Basset, Y. and Arthington, A.H. (1992) The arthropod community of an Australian rainforest tree: abundance of component taxa, species richness and guild structure. *Australian Journal of Ecology*, **17**, 89–98.

Basset, Y. and Kitching, R.L. (1991) Species number, species abundance and body length of arboreal arthropods associated with an Australian rainforest tree. *Ecological Entomology*, **16**, 391–402.

Basset, Y., Springate, N.D., Aberlenc, H.P. and Delvare, G. (1997) A review of methods for sampling arthropods in tree canopies, in *Canopy Arthropods* (eds N.E. Stork, J. Adis and R.K. Didham), Chapman & Hall, London, pp. 27–52 (this volume).

Bell, R.T. and Bell, J.R. (1988) Rhysodini of Sulawesi and nearby islands (Coleoptera: Carabidae or Rhysodidae). *Journal of the New York Entomological Society*, **96**, 7–15.

Burckhardt, D. and Löbl, I. (1992) A review of the Pterogeniidae (Coleoptera). *Zoological Journal of the Linnean Society*, **104**, 243–91.

Cassola, F. (1991) Studi sui Cicindelidi. LXIII. I Cicindelidae (Coleoptera) dell'isola di Sulawesi, Indonesia. *Annali del Museo Civico di Storia Naturale 'G. Doria'*, **88**, 481–664.

Colwell, R.K. and Coddington, J.A. (1994) Estimating terrestrial biodiversity through extrapolation. *Philosophical Transactions of the Royal Society of London, Series B*, **345**, 101–18.

Davies, J.G., Stork, N.E., Brendell, M. and Hine, S.J. (1997) Beetle species diversity and faunal similarity in Venezuelan rainforest tree canopies, in *Canopy Arthropods* (eds N.E. Stork, J. Adis and R.K. Didham), Chapman & Hall, London, pp. 85–103 (this volume).

Endrödy-Younga, S. (1995) Clambidae of Sulawesi with reference to the biogeography of the Sunda Islands (Coleoptera, Eucinetoidea). *Systematic Entomology*, **20**, 1–16.

Erwin, T.L. (1982) Tropical forests: their richness in Coleoptera and other species. *The Coleopterist's Bulletin*, **36**, 74–5.

Erwin, T.L. (1983a) Beetles and other insects of tropical forest canopies at Manaus, Brazil, sampled by insecticidal fogging, in *Tropical Rainforest Ecology and Management* (eds S.L. Sutton, T.C. Whitmore and A.C. Chadwick), Blackwell Scientific Press, Oxford, pp. 59–79.

Erwin, T.L. (1983b) Tropical forest canopies: the last biotic frontier. *Bulletin of the Entomological Society of America*, **29**(1), 14–19.

Erwin, T.L. (1988) The tropical forest canopy. The heart of biotic diversity, in *Biodiversity* (eds E.O. Wilson and F.M. Peter), National Academic Press, Washington, DC, pp. 123–9.

Erwin, T.L. and Scott, J.C. (1980) Seasonal and size patterns, trophic structure and richness of Coleoptera in the tropical arboreal ecosystem: the fauna of the tree *Luehea seemannii* Triana and Planch in the Canal Zone of Panama. *The Coleopterist's Bulletin*, **34**, 305–21.

Floren, A. and Linsenmair, K.E. (1994) Zur Diversität und Wiederbesiedelungs-dynamik von Arthropoden auf drei Baumarten in einem Regenwald in Sabah, Malaysia. *Andrias*, **13**, 23–8.

Gaston, K.J., Blackburn, T.M., Hammond, P.M. and Stork, N.E. (1993) Relationships between abundance and body size: where do tourists fit? *Ecological Entomology*, **18**, 310–14.

Guilbert, E. and Chazeau, J. Arthropod communities in two primary forest types of New Caledonia sampled by fogging. *Bonner Zoologische Beitrage* (in press).

Guilbert, E., Chazeau, J. and Bonnet de Larbogne, L. (1994) Canopy arthropod diversity of New Caledonian forests sampled by fogging: preliminary results. *Memoirs of the Queensland Museum*, **36**, 77–85.

Hammond, P.M. (1990) Insect abundance and diversity in the Dumoga-Bone National Park, N. Sulawesi, with special reference to the beetle fauna of lowland rain forest in the Toraut region, in *Insects and the Rain Forests of South East Asia (Wallacea)* (eds W.J. Knight and J.D. Holloway), Royal Entomological Society of London, London, pp. 197–254.

Hammond, P.M. (1992) Species inventory, in *Global Biodiversity, Status of the Earth's Living Resources* (ed. B. Groombridge), Chapman & Hall, London, pp. 17–39.

Hammond, P.M. (1994) Practical approaches to the estimation of the extent of biodiversity in speciose groups. *Philosophical Transactions of the Royal Society of London, Series B*, **345**, 119–36.

Hammond, P.M. (1995) Described and estimated species numbers: an objective assessment of current knowledge, in *Microbial Diversity and Ecosystem Function* (eds D. Allsopp, R.R. Colwell and D.H.L. Hawksworth), C.A.B. International, Wallingford, pp. 29–71.

Hammond, P.M. and Owen, J.A. The beetles of Richmond Park SSSI – a case study. *English Nature Science* (in press).

Hammond, P.M., Kitching, R.L. and Stork, N.E. Biodiversity of arthropods from Australian rain forest canopies: Coleoptera from subtropical tree-crowns. *Ecotropica* (in press)

Hanski, I. and Niemelä, J. (1990) Elevational distributions of dung and carrion beetles in northern Sulawesi, in *Insects and the Rain Forests of South East Asia (Wallacea)* (eds W.J. Knight and J.D. Holloway), Royal Entomological Society of London, London, pp. 145–52.

Hofer, H., Brescovit, A.D., Adis, J. and Paarmann, W. (1994) The spider fauna of Neotropical tree canopies in Central Amazonia: first results. *Studies on Neotropical Fauna and Environment*, **29**, 23–32.

Kistner, D.K. and McNairn, M.J. (1991) A new genus and two new species of predacious Staphylinidae associated with *Pheidolegeton* in Sulawesi (Coleoptera: Aleocharinae) with notes on their behaviour. *Sociobiology*, **18**, 305–21.

Kistner, D.K. and Zimmerman, R. (1986) A new genus of ant mimic from Sulawesi and its relationship to *Pheigotoxenus* (Coleoptera: Staphylinidae). *Sociobiology*, **11**, 325–38.

Knight, W.J. and Holloway, J.D. (eds) (1990) *Insects and the Rain Forests of South East Asia (Wallacea)*, Royal Entomological Society of London, London.

Krikken, J. and Huijbregts, J. (1987) Southeast Asian *Termitobius*: a taxonomic review: with descriptions of four new species (Coleoptera: Aphodiinae). *Zoologische Mededelingen, Leiden*, **61**, 97–111.

Mawdsley, N. (1994) *Community Structure of the Coleoptera Assemblage of a Bornean Tropical Forest*. Unpublished PhD Thesis, University of London.

Roberts, H.R. (1973) Arboreal Orthoptera in the rain forest of Costa Rica collected with insecticide: a report on the grasshoppers (Acrididae) including new species. *Proceedings of the Academy of Natural Sciences of Philadelphia*, **125**, 49–66.

Russell-Smith, A. and Stork, N.E. (1994) Abundance and diversity of spiders from the canopy of tropical rainforests with particular reference to Sulawesi, Indonesia. *Journal of Tropical Ecology*, **10**, 545–58.

Southwood, T.R.E., Moran, V.C. and Kennedy, C.E.J. (1982a) The richness, abundance and biomass of the arthropod communities on trees. *Journal of Animal Ecology*, **51**, 635–49.

Southwood, T.R.E., Moran, V.C. and Kennedy, C.E.J. (1982b) The assessment of arboreal insect fauna: comparison of knockdown sampling and faunal lists. *Ecological Entomology*, **7**, 331–40.

Stork, N.E. (1987a) Guild structure of arthropods from Bornean rain forest trees. *Ecological Entomology*, **12**, 69–80.

Stork, N.E. (1987b) Arthropod faunal similarity of Bornean rain forest trees. *Ecological Entomology*, **12**, 219–26.

Stork, N.E. (1988) Insect diversity: facts, fiction and speculation. *Biological Journal of the Linnean Society*, **35**, 321–37.

Stork, N.E. (1991) The composition of the arthropod fauna of Bornean lowland rain forest trees. *Journal of Tropical Ecology*, **7**, 161–80.

Stork, N.E. (1993) How many species are there? *Biodiversity and Conservation*, **2**, 215–32.

Stork, N.E. and Brendell, M.J.D. (1990) Variation in the insect fauna of Sulawesi trees with season, altitude and forest type, in *Insects and the Rain Forests of South East Asia (Wallacea)* (eds W.J. Knight and J.D. Holloway), Royal Entomological Society of London, London, pp. 173–90.

Stork, N.E. and Hammond, P.M. (1997) Sampling arthropods from tree-crowns by fogging with knockdown insecticides: lessons from studies of oak tree beetle assemblages in Richmond Park (UK), in *Canopy Arthropods* (eds N.E. Stork, J. Adis and R.K. Didham), Chapman & Hall, London, pp. 3–26 (this volume).

Sutton, S.L. (1983) The vertical distribution of flying insects in lowland rain forests of Panama, Papua New Guinea and Brunei. *Zoological Journal of the Linnean Society*, **78**, 287–97.

Sutton, S.L. (1989) The spatial distribution of flying insects, in *Tropical Rain Forest Ecosystems* (eds H. Leith and M.J.A. Werner), Elsevier Science Publisher, Amsterdam, pp. 427–36.

Wittmer, W. (1989) Die Gattung *Polemiosilis* Pic auf Sulawesi (Coleoptera, Cantharidae). *Tijdschrift voor Entomologie*, **132**, 129–34.
Wittmer, W. (1990) Die Familie Malachiidae (Coleoptera) auf Sulawesi. *Tijdschrift voor Entomologie*, **133**, 107–20.
Wittmer, W. (1992) Beitrag zur Kenntnis der indo-malaiischen Cantharidae (Col.). *Mitteilungen aus der Entomologischen Gesellschaft, Basel*, **42**, 98–111.
Wolda, H. (1979) Abundance and diversity of Homoptera in the canopy of a tropical forest. *Ecological Entomology*, **4**, 181–90.

Appendix 10A Representation of beetle 'family-groups' (following the system of Hammond, 1990) in a ca. 500-ha tract of fairly uniform lowland rainforest in North Sulawesi, Indonesia. The figures given in bold type are for the number of species from the study area: (1) represented in the voucher collection; (2) estimated to be present in the collection (from the study area) taken during Project Wallace; and (3) estimated to occur in the area (see Discussion). At a previous stage of the study (Hammond, 1990) counts and estimates were made that differ from the current ones (see Discussion); these earlier figures are given in plain type for comparison

	1		*2*		*3*	
Curculionidae[1]	**305**	241	**390**	375	**520**	460
Aleocharinae	**253**	187	**370**	320	**480**	460
Anthribidae	**225**	198	**275**	250	**340**	320
Scolytidae	**177**	98	**190**	125	**230**	160
Pselaphidae	**91**	83	**150**	130	**205**	180
Carabidae[2]	**149**	151	**154**	160	**200**	215
Mordellidae	**126**	122	**135**	150	**185**	190
Lamiinae	**129**	132	**135**	135	**175**	160
Aderidae	**103***	73	**125***	140	**170***	180
Histeridae	**89**	78	**123**	110	**160**	165
Corylophidae	**101***	74	**120***	90	**155***	125
Eucnemidae	**107**	108	**115**	115	**150**	150
Paederinae	**93**	97	**105**	105	**135**	135
Ptiliidae	**47**	47	**93**	95	**130**	150
Scydmaenidae	**46**	42	**90**	90	**130**	160
Buprestidae	**69**	71	**75**	75	**120**	120
Tenebrionidae[3]	**84**	132[8]	**88**	135[8]	**110**	160[8]
Nitidulidae	**73**	71	**80**	80	**110**	110
Elateridae	**74**	71	**80**	80	**105**	105
Osoriinae	**53**	53	**70**	65	**100**	100
Scaphidiidae	**51**	47	**75**	75	**100**	100
Galerucinae	**58**	51	**75**	70	**100**	95
Staphylininae	**62**	59	**75**	65	**95**	90
Alticinae	**52**	53	**70**	65	**95**	90
Coccinellidae	**66**	62	**70**	68	**85**	85
Leiodidae	**32**	27	**50**	50	**70**	90

Appendix 10A continued

	1		2		3	
Brentidae[4]	**45**	41	**50**	50	**70**	90
Lycidae	**44**	36	**50**	50	**70**	65
Tachyporinae	**41**	41	**48**	45	**65**	60
Eumolpinae	**30**	40	**45**	50	**70**	75
Endomychidae	**46**	38	**50**	40	**60**	50
Platypodidae	**39**	24	**45**	30	**55**	40
Oxytelinae	**37**	38	**40**	38	**48**	50
Scarabaeinae	**38**	33	**40**	35	**45**	43
Cleridae	**36**	36	**38**	36	**45**	45
Colydiidae	**35**	34	**38**	37	**45**	45
Cerambycinae	**31**	29	**33**	30	**45**	35
Anobiidae	**25**	26	**30**	28	**42**	35
Sphaeridiinae	**24**	21	**30**	40	**40**	55
Laemophloeidae	**28**	21	**30**	28	**40**	38
Erotylidae	**32**	29	**34**	30	**40**	38
Anthicidae	**24**	24	**30**	28	**40**	40
Dytiscidae	**28**	27	**30**	27	**38**	40
Hydrophilinae	**23**	22	**26**	25	**35**	35
Phalacridae	**22***	14	**27***	18	**35***	25
Silvanidae	**24**	20	**28**	25	**35**	35
Otiorhynchinae	**26**	22	**27**	24	**33**	30
Scirtidae	**19**	20	**22**	25	**30**	33
Melyridae	**23**	20	**24**	20	**30**	30
Cerylonidae	**19**	18	**23**	20	**30**	28
Lagriidae	**22**	20	**23**	24	**30**	30
Attelabidae	**20**	17	**23**	20	**30**	30
Rhynchophorinae	**23**	22	**24**	20	**30**	30
Euaesthetinae	**12**	12	**20**	15	**28**	28
Aphodiinae	**19**	18	**20**	20	**28**	25
Rutelinae	**22**	16	**23**	16	**28**	23
Melolonthinae	**19**	18	**20**	18	**26**	28
Throscidae	**17**	18	**20**	20	**25**	25
Biphyllidae	**11**	11	**18**	18	**25**	25
Languriidae	**18**	15	**20**	18	**25**	25
Cisidae	**13**	13	**16**	20	**25**	30
Alleculinae	**15**	25	**18**	18	**23**	25
Propalticidae	**15**	16	**17**	16	**22**	20
Chysomelidae[5]	**14**	8	**18**	10	**22**	15
Limnichidae	**13**	13	**15**	16	**20**	20
Cantharidae	**13**	9	**14**	10	**20**	18
Sphindidae	**14**	16	**17**	18	**20**	25
Cicindelinae	**15**	16	**15**	16	**18**	20
Omaliinae	**11**	12	**12**	12	**18**	20
Clambidae	**11**	7	**12**	10	**16**	13

Appendix 10A continued

	1		*2*		*3*	
Lucanidae	**11**	11	**11**	11	**16**	15
Bothrideridae	**11**	9	**12**	9	**16**	15
Lathridiidae	**9**	11	**12**	16	**16**	25
Cetoniinae	**11**	9	**11**	9	**15**	13
Dermestidae	**8**	10	**10**	10	**15**	14
Rhizophagidae	**11**	11	**12**	11	**15**	14
Mycetophagidae	**8**	9	**10**	12	**15**	16
Apionidae	**8**	9	**9**	10	**15**	15
Ptilodactylidae	**10**	7	**12**	10	**14**	12
Hispinae	**10**	10	**11**	10	**14**	12
Megalopininae	**8**	8	**8**	8	**12**	12
Steninae	**8**	8	**8**	8	**12**	12
Ptinidae	**7**	7	**8**	7	**10**	10
Melandryidae	**6**	7	**7**	8	**10**	10
Meloidae	**2**	8	**3**	8	**10**	10
Scraptiidae	**5**	7	**7**	7	**10**	10
Passalidae	**5**	5	**5**	5	**8**	8
Elmidae	**5**	4	**5**	4	**8**	7
Lymexylidae	**6**	6	**6**	6	**8**	8
Pedilidae[6]	**5**	5	**6**	5	**8**	8
Othniidae[7]	**5**	2	**6**	2	**8**	3
Oedemeridae	**5**	4	**6**	4	**8**	6
Paussinae	**5**	4	**5**	4	**7**	6
Hydraenidae	**4**	4	**4**	4	**7**	7
Lampyridae	**4**	4	**5**	5	**7**	7
Gyrinidae	**4**	4	**4**	4	**6**	6
Dryopidae	**3**	4	**4**	4	**6**	7
Acanthoceridae	**3**	3	**3**	3	**6**	6
Rhipiphoridae	**4**	4	**4**	4	**6**	6
Salpingidae	**3**	3	**4**	3	**6**	6
Leiochrini	**5**	4	**5**	4	**6**	6
Noteridae	**4**	3	**4**	3	**5**	5
Valginae	**3**	3	**3**	3	**5**	5
Dynastinae	**4**	3	**4**	3	**5**	5
Chelonariidae	**3**	3	**3**	3	**5**	5
Lophocateridae	**4**	4	**4**	3	**5**	4
Passandridae	**2**	2	**2**	2	**5**	5
Rhysodidae	**3**	2	**3**	2	**4**	3
Silphidae	**3**	3	**3**	3	**4**	4
Bostrichidae	**3**	1	**3**	2	**4**	4
Mycteridae	**3**	2	**3**	2	**4**	3
Geotrupidae	**2**	2	**2**	2	**3**	3
Psephenidae	**2**	2	**2**	2	**3**	3
Callirhipidae	**1**	0	**1**	0	**3**	3

Appendix 10A continued

	1		*2*		*3*	
Nosodendridae	**3**	2	**3**	2	**3**	3
Cryptophagidae	**2**	2	**3**	2	**3**	3
Discolomidae	**2**	2	**3**	2	**3**	3
Pterogeniidae	**2**	2	**2**	2	**3**	3
Inopeplidae	**2**	1	**2**	1	**3**	3
Prioninae	**0**	0	**0**	0	**3**	3
Georyssidae	**1**	1	**1**	1	**2**	2
Proteininae	**1**	1	**1**	1	**2**	2
Eucinetidae	**1**	1	**1**	1	**2**	2
Trogidae	**1**	0	**1**	0	**2**	1
Hybosoridae	**1**	1	**1**	1	**2**	2
Trichiinae	**1**	1	**1**	1	**2**	2
Cassidinae	**1**	1	**1**	1	**2**	2
Cerophytidae	**1**	0	**1**	0	**1**	0
Total	**4026**	3641	**4862**	4532	**6425**	6236

* Figures for these 'family-group' are approximate as sorting of voucher collection species remains incomplete.

[1] Curculionidae *sensu lato*, but excluding Otiorhynchinae and Rhyncophorinae for which totals are given separately.

[2] Carabidae *sensu lato*, but excluding Rhysodidae, Cicindelinae and Paussinae for which totals are given separately.

[3] Tenebrionidae *sensu lato*, but excluding Lagriidae, Alleculinae and Leiochrini for which totals are given separately

[4] Excludes Apionidae.

[5] Chrysomelid subfamilies excluding Hispinae, Cassidinae, Eumolpinae, Galerucinae and Alticinae, for which totals are given separately.

[6] Pedilidae are often included in Anthicidae.

[7] Excludes Salpingidae for which a total is given separately.

[8] The figures given for Tenebrionidae by Hammond (1990) are the result of a typographical lapse and should have been substantially lower.

Appendix 10B Representation of beetle 'family-group' (following the system of Hammond, 1990) in tree-crown fogging samples from a ca. 500-ha tract of fairly uniform lowland rainforest in North Sulawesi, Indonesia. (1) Total number of species represented in fogging samples. 2. Number of these species found only in fogging samples. 3. Number of tree-crown specialists (see text for explanation of stratum categories). 4. Number of stratum generalists. 5. Number of ground specialists. 6. Number of species of uncertain stratum category. 7. Total number of species represented in the voucher collection from all strata. 8. Estimated completeness in the voucher collection of species collected by all sampling methods: A, >90% B, 70–90%; C, <70%

	1	*2*	*3*	*4*	*5*	*6*	*7*	*8*
Rhysodidae	–	–	–	–	–	–	3	A
Paussinae	–	–	–	–	–	–	5	A
Cicindelinae	2	–	–	2	–	–	15	A
Carabidae[1]	20	4	2	10	6	2	149	A
Noteridae	–	–	–	–	–	–	4	A
Dytiscidae	–	–	–	–	–	–	28	A
Gyrinidae	–	–	–	–	–	–	4	A
Georyssidae	–	–	–	–	–	–	1	A
Sphaeridiinae	4	–	–	–	4	–	24	B
Hydrophilinae	–	–	–	–	–	–	23	B
Histeridae	2	–	–	–	2	–	89	C
Ptiliidae	9	3	–	1	6	2	47	C
Hydraenidae	2	1	–	–	2	–	4	A
Leiodidae	2	–	–	–	2	–	32	C
Scydmaenidae	8	4	–	2	3	3	46	C
Silphidae	–	–	–	–	–	–	3	A
Proteininae	–	–	–	–	–	–	1	A
Omaliinae	4	3	4	–	–	–	11	A
Osoriinae	11	3	–	5	4	2	53	B
Oxytelinae	14	1	–	–	13	1	37	A
Megalopininae	–	–	–	–	–	–	8	A
Steninae	1	–	–	1	–	–	8	A
Euaesthetinae	3	–	–	–	3	–	12	C
Paederinae	23	8	12	6	2	3	93	A
Staphylininae	2	–	–	–	2	–	62	A
Tachyporinae	2	1	–	–	1	1	41	A
Aleocharinae	91	45	7	9	29	46	253	C
Pselaphidae	19	11	4	1	7	7	91	C
Scaphidiidae	8	3	1	2	2	3	51	C
Clambidae	–	–	–	–	–	–	11	A
Eucinetidae	–	–	–	–	–	–	1	A
Scirtidae	8	–	–	6	2	–	19	B
Lucanidae	–	–	–	–	–	–	11	A
Trogidae	–	–	–	–	–	–	1	A
Acanthoceridae	1	–	1	–	–	–	3	A

Appendix 10B continued

	1	2	3	4	5	6	7	8
Passalidae	–	–	–	–	–	–	5	A
Geotrupidae	–	–	–	–	–	–	2	A
Hybosoridae	–	–	–	–	–	–	1	A
Aphodiinae	4	–	–	–	4	–	19	A
Scarabaeinae	2	–	–	–	2	–	38	A
Melolonthinae	8	–	–	3	5	–	19	A
Trichiinae	–	–	–	–	–	–	1	A
Valginae	1	–	–	1	–	–	3	A
Rutelinae	1	–	–	–	1	–	19	A
Dynastinae	–	–	–	–	–	–	4	A
Cetoniinae	–	–	–	–	–	–	11	A
Ptilodactylidae	9	1	1	–	7	1	10	B
Chelonariidae	1	–	–	–	1	–	3	A
Psephenidae	–	–	–	–	–	–	2	A
Elmidae	–	–	–	–	–	–	5	A
Dryopidae	–	–	–	–	–	–	3	B
Limnichidae	8	1	1	4	3	–	13	B
Buprestidae	25	14	6	10	2	7	69	A
Callirhipidae	–	–	–	–	–	–	1	A
Cerophytidae	–	–	–	–	–	–	1	A
Elateridae	24	4	1	12	6	5	74	A
Throscidae	4	–	–	1	3	–	17	B
Eucnemidae	17	3	2	1	10	4	107	B
Lycidae	18	1	–	10	4	4	44	B
Lampyridae	3	–	–	–	3	–	4	A
Cantharidae	8	1	–	2	4	2	13	A
Dermestidae	3	1	–	1	–	2	8	A
Nosodendridae	–	–	–	–	–	–	3	A
Bostrichidae	1	–	–	1	–	–	3	A
Anobiidae	19	5	7	7	1	4	25	B
Ptinidae	7	5	3	2	–	2	7	A
Lophocateridae	1	1	–	–	–	1	4	A
Cleridae	15	2	3	6	5	1	36	A
Melyridae	14	5	1	7	1	5	23	A
Lymexylidae	1	–	–	–	1	–	6	A
Nitidulidae	18	4	–	6	7	5	73	B
Rhizophagidae	2	–	1	–	1	–	11	A
Phalacridae*	18	8	6	4	2	6	22	B
Sphindidae	1	–	–	–	1	–	14	B
Laemophloeidae	9	5	1	1	1	6	28	B
Passandridae	–	–	–	–	–	–	2	A
Silvanidae	5	–	–	1	3	1	24	A
Cryptophagidae	–	–	–	–	–	–	2	A
Propalticidae	12	7	3	3	1	5	15	A

Appendix 10B continued

	1	*2*	*3*	*4*	*5*	*6*	*7*	*8*
Biphyllidae	–	–	–	–	–	–	11	B
Languriidae	5	2	–	2	1	2	18	B
Erotylidae	4	2	1	–	2	1	32	A
Cerylonidae	–	–	–	–	–	–	19	B
Corylophidae*	79	39	28	18	7	26	101	C
Endomychidae	9	7	2	2	–	5	46	A
Coccinellidae	39	11	13	9	10	7	66	A
Discolomidae	1	–	–	1	–	–	2	A
Lathridiidae	7	2	2	2	1	2	9	B
Colydiidae	3	2	–	–	–	3	35	A
Bothrideridae	–	–	–	–	–	–	9	A
Mycetophagidae	3	–	1	2	–	–	8	B
Cisidae	6	5	1	–	1	4	13	B
Pterogeniidae	–	–	–	–	–	–	1	A
Melandryidae	1	–	1	–	–	–	6	A
Mordellidae	51	11	8	21	11	11	126	B
Rhipiphoridae	3	–	–	2	–	1	4	A
Oedemeridae	2	1	–	1	–	1	5	A
Mycteridae	3	1	1	2	–	–	3	A
Anthicidae	6	2	–	3	1	2	24	B
Pedilidae[2]	5	2	2	2	1	–	6	B
Aderidae*	80	29	13	22	16	29	103	C
Meloidae	–	–	–	–	–	–	2	A
Scraptiidae	5	4	4	1	–	–	5	A
Othniidae[3]	2	1	–	1	–	1	5	A
Salpingidae	1	–	–	–	–	1	3	A
Inopeplidae	–	–	–	–	–	–	2	A
Lagriidae	8	2	1	4	2	1	22	A
Alleculinae	10	–	4	5	1	–	15	A
Leiochrini	–	–	–	–	–	–	5	A
Tenebrionidae[4]	16	4	4	6	3	3	84	A
Prioninae	–	–	–	–	–	–	–	A
Cerambycinae	5	5	3	–	–	2	31	A
Lamiinae	56	28	32	7	4	13	129	A
Hispinae	4	1	–	1	1	2	10	A
Cassidinae	–	–	–	–	–	–	1	A
Eumolpinae	26	8	9	15	–	2	30	B
Galerucinae	41	7	2	27	1	11	58	B
Alticinae	25	14	7	7	3	8	52	B
Chrysomelidae[5]	8	7	3	–	–	5	14	A
Anthribidae	87	45	29	16	7	35	225	B
Attelabidae	15	3	3	10	1	1	20	A
Apionidae	4	2	1	2	–	1	8	A
Brentidae[6]	14	11	5	1	2	6	45	A

Appendix 10B continued

	1	*2*	*3*	*4*	*5*	*6*	*7*	*8*
Otiorhynchinae	16	5	7	9	–	–	26	A
Rhyncophorinae	3	1	–	–	1	2	23	A
Curculionidae[7]	140	77	38	42	9	51	305	C
Scolytidae	57	33	4	6	14	33	177	A
Platypodidae	11	3	1	3	3	4	39	A
Totals	1355	532	297	380	272	406	4026	

*Figures for these family-groups are approximate, as sorting of voucher collection species remains incomplete.
[1]Carabidae *sensu lato*, but excluding Rhysodidae, Cicindelinae and Paussinae for which totals are given separately.
[2]Pedilidae are often included in Anthicidae.
[3]Excludes Salpingidae for which a total is given separately.
[4]Tenebrionidae *sensu lato*, but excluding Lagriidae, Alleculinae and Leiochrini, for which totals are given separately.
[5]Chrysomelid subfamilies excluding Hispinae, Cassidinae, Eumolpinae, Galerucinae and Alticinae for which totals are given separately.
[6]Excludes Apionidae.
[7]Curculionidae *sensu lato*, but excluding Otiorhynchinae and Rhyncophorinae for which totals are given separately.

11

Patterns of beetle species diversity in *Castanopsis acuminatissima* (Fagaceae) trees studied with canopy fogging in mid-montane New Guinea rainforest

A. Allison, G.A. Samuelson and S.E. Miller

ABSTRACT

Canopy fogging with a pyrethrum-based insecticide was used to study the structure and diversity of beetle communities in eight *Castanopsis acuminatissima* (Bl.) A.DC. (Fagaceae) trees at elevations of 1200–1400 m in the Wau Valley, Papua New Guinea. Arthropods were collected in 1-m^2 trays suspended beneath the trees, sorted to Order and all beetles were identified to morphospecies. The eight trees yielded 3977 individual beetles representing 418 morphospecies in 53 families. The mean number of beetle morphospecies per tree was 114 (range 82–155). Of these morphospecies, 199 (47.6%) were represented by single individuals, and only 62 (14.8%) were represented by 10 or more individuals. Approximately 25% of the species on each tree were tree-specific, although those species represented only ca. 5–10% of the individuals. The similarity of the beetle fauna between trees, as measured by the Jaccard coefficient, ranged from 12 to 31%. This study is part of a much larger canopy fogging project in which 51 trees at study sites of 500 m, 1200–1400 m and 2100–2250 m have been fogged. Preliminary analysis of these samples indicates that diversity is similar at sites at 500 m and 1200–1400 m, but higher at 2100–2250 m.

Canopy Arthropods. Edited by N.E. Stork, J. Adis and R.K. Didham. Published in 1997 by Chapman & Hall, London. ISBN 0 412 74900 9

INTRODUCTION

Canopy fogging techniques were originally developed to control pests. Gagné was one of the first to recognize the applicability of these techniques to study the diversity of insects in native forest trees (Gagné, 1979). This led to further similar studies and in 1980 Erwin reported on his canopy studies in Panama (Erwin and Scott, 1980). The debate that this work sparked on the magnitude of global species richness (Erwin, 1982; Stork, 1988, 1993; May, 1990; Gaston, 1991, Hodkinson and Casson, 1991; Hammond, 1992; 1994; Hodkinson, 1992) prompted us in 1987 to commence a canopy fogging study of rainforest trees along an altitudinal transect from 500–2200 m in New Guinea.

In order to put patterns of diversity into clear perspective it is necessary to measure both within-habitat diversity (alpha-diversity) and between-habitat diversity, also known as turnover (beta-diversity) (Whittaker, 1975). To achieve this our experimental design emphasized a single species of tree, *Castanopsis acuminatissima* (Fagaceae). We fogged this species, together with related (*Lithocarpus* and *Nothofagus*) and unrelated genera (of Burseraceae, Clusiaceae, Dipterocarpaceae, Elaeocarpaceae, Grossulariaceae, Juglandaceae, Lauraceae, Phyllocladaceae, Rosaceae, Sapindaceae) at study sites located at 500 m (Oomsis), 1200–1400 m (Wau Valley) and 2100–2200 m (Biaru Road). We have now fogged a total of 51 trees and obtained more than 45 000 beetle specimens. The preliminary analysis of beetle samples from the first eight trees has been completed (two trees of *C. acuminatissima* from each of the three study sites and two trees of *Lithocarpus celebicus* at 500 m) (Allison *et al.*, 1993b). In this paper we report on the preliminary analysis of insect samples from eight *C. acuminatissima* trees (including two trees from the earlier analysis) at the same elevation in the Wau Valley.

STUDY SITES

Field work was based at the Wau Ecology Institute in Papua New Guinea. Part of the field work was conducted in remnant *Castanopsis* forest approximately 1 km east of Wau Ecology Institute on the road to the former New Guinea Goldfields mine at Nami (1250 m). The remaining work was conducted on the lower slopes of Mt Missim on a ridge between Poverty and Sandy Creeks (1250–1350 m).

The Wau Valley generally receives ca. 1900 mm of rainfall annually. There is a slight seasonal pattern with ca. 60% of the annual total occurring from November to April under the influence of the north-west monsoons. The period from May to October, when the south-west tradewinds form the dominant weather pattern, is sometimes very dry (<25 mm of rainfall per month). The mean annual temperature at

Table 11.1 Details of total number of species and individuals and mean body length (± S.D.) of beetles collected from a total of 126 1-m^2 trays suspended beneath trees

Tree	*Date*	*Total species*	*Total individuals*	*Mean (S.D.) body length (mm)*	*Total no. of trays used*
3*	15–10–87	82	335	2.66 (1.62)	13
4	04–11–87	86	300	2.51 (1.74)	13
9*	07–08–88	99	404	2.84 (1.74)	13
10	18–08–88	155	800	2.89 (2.46)	17
11	18–08–88	154	547	2.55 (1.63)	16
12	19–08–88	129	696	2.30 (1.35)	15
13	19–08–88	132	493	2.40 (1.65)	16
14	20–08–88	106	402	2.39 (1.28)	23

* Trees 3 and 9 are the same tree sampled 10 months apart.

1230 m (Wau Ecology Institute) in the Wau Valley is 22°C with little annual variation (Gressitt and Nadkarni, 1978). The vegetation in the area is generally classified as mid-montane rainforest (van Valkenburg and Ketner, 1994).

Study trees

We fogged eight trees of *Castanopsis acuminatissima* over a 2-year period (Table 11.1). *C. acuminatissima* occurs throughout New Guinea from nearly sea level to at least 2200 m and, like many tropical fagaceous trees, is found mainly on ridge crests (Soepadmo, 1972; Streimann, 1983; van Valkenburg and Ketner, 1994). It is extremely abundant in the Wau Valley and occurs in both primary and secondary forest. In some areas it appears as a monospecific dominant. It is the only species of *Castanopsis* in New Guinea, although there are a large number of related species to the west in Indonesia and south-east Asia, including a number of species that occupy subtropical and temperate environments (Soepadmo, 1972).

Our original sampling design called for 10 trays per tree, but to guard against the possibility of samples being lost, we used at least 13 trays for each tree and obtained a total of 126 tray samples (Table 11.1).

METHODS

Our methods are described in full in Allison *et al.* (1993b) and are summarized here. We selected trees for fogging that seemed healthy, had no epiphytes or understorey, did not have flowers or fruit, and had a canopy that did not intermingle with adjacent trees. Each of the trees was 20–25 m tall and had a canopy volume of ca. 3000–4000 m^3.

Trees were fogged at 06:00 h (when the air was almost still) for 15 minutes using Pyranone® mixed with kerosene to form a solution of 5% active ingredient that was delivered by a Curtis Dynafog model 2610 fogger with the fog density dial set on '5'. The fogger was operated manually at canopy level from an adjacent tree. No living insects were found on branches subsequently removed from fogged trees.

Insect samples were collected in shallow plastic trays (126 in total), each 1 m^2 in area, suspended approximately 1 m above the ground beneath each tree. Each tray was numbered individually and had an 8-cm diameter aluminium funnel inserted in the centre, flush with the tray surface. A 125-ml Nalgene® bottle approximately half full of 70% ethanol was attached to the base of each funnel. Paint brushes (ca. 5 cm wide) were used to sweep the insect samples into the bottles for preservation. Trays were left in place for 2 hours after fogging although our impression was that most of the insects dropped from the tree within 20–30 minutes of fogging. Each sample bottle was labelled with tray and tree number. To ensure good preservation, the alcohol (70% ethanol) was changed at least four times in the month following collection.

In the laboratory all beetle specimens were removed, mounted and identified to morphospecies. Where possible our morphospecies concepts have been verified by specialists in each family. All information was entered into a computer database and reconciled with labelled specimens to ensure accuracy. All beetles and tree voucher specimens are deposited in the Bishop Museum, Hawaii. Beetles were measured from the anterior-most part of the head to the apex of the elytra, except for species with hidden heads for which the anterior edge of the pronotum was used and weevils for which the anterior edge of the eye was used. For species with more than a few individuals, several individuals, determined by inspection to be of average size, were measured. Mass was calculated from the equation derived by Schoener (1980) for Costa Rican rainforest beetles.

RESULTS

The eight trees yielded 3977 individual beetles representing 418 morphospecies in 53 families (Table 11.1). The total number of species on each of the trees ranged from 82 to 155. Four families had more than 25 species each: Chrysomelidae (39), Staphylinidae (39), Curculionidae (37) and Coccinellidae (29), together comprising 34% of total species richness. They each had more than 200 individuals and together were represented by 1885 individuals (47% of the total). The only other family with more than 200 individuals was Attelabidae which had 431 individuals (but only three species), more than Curculionidae (206), Coccinellidae (229), but less than Chrysomelidae (601) and Staphylinidae (849). These five families comprised 58% of total individuals. Eleven families were represented by single species (number of individuals in parentheses): Lampyridae (1), Merophysiidae (1), Rhizophagidae (1), Sphindidae (1), Hydrophilidae (2), Trogositidae (2), Alleculidae (3), Bostrichidae (4), Endomychidae (6), Eucnemidae (6) and Mycetophagidae (114).

There was no correlation between the number of collecting trays used beneath a tree and the number of individual beetles recorded from that tree (simple linear regression, $R^2 < 0.04$, $n = 8$). In other words, the variance in numbers of individuals per tree masked any effect from differing numbers of trays. Therefore, to simplify analysis we simply report on the total insect sample recorded from each tree.

Slightly fewer than half the species (199, 47.6%) were represented by single individuals ('singletons') and 19.9% (83) were represented by two individuals, with declining numbers represented by three or more individuals. Only 62 species (14.8%) were represented by 10 or more individuals (Figure 11.1). More than half the beetle species (55.3%) were found on only one tree and fewer than one-third were found on three or more trees (Table 11.2).

A separate analysis was performed on chrysomelids (herbivores) and staphylinids (predominantly predators; four species of omaliines, thought to be pollen-feeders, were excluded from this analysis). Patterns were similar to the overall trend described above, except that singletons comprised only 33.3% of chrysomelids (13 of 39 species) and 51.4% of staphylinids (18 of 35 species) (Figure 11.1).

Table 11.2 Distribution of beetle species on eight *Castanopsis acuminatissima* trees from ca. 1200–1400 m elevation in the Wau Valley, Papua New Guinea

Total no. of trees on which beetle species were recorded	*No. (%) of beetle species*
1	232 (55.3)
2	70 (16.8)
3	41 (9.8)
4	30 (7.2)
5	14 (3.4)
6	7 (1.7)
7	13 (3.1)
8	12 (2.9)

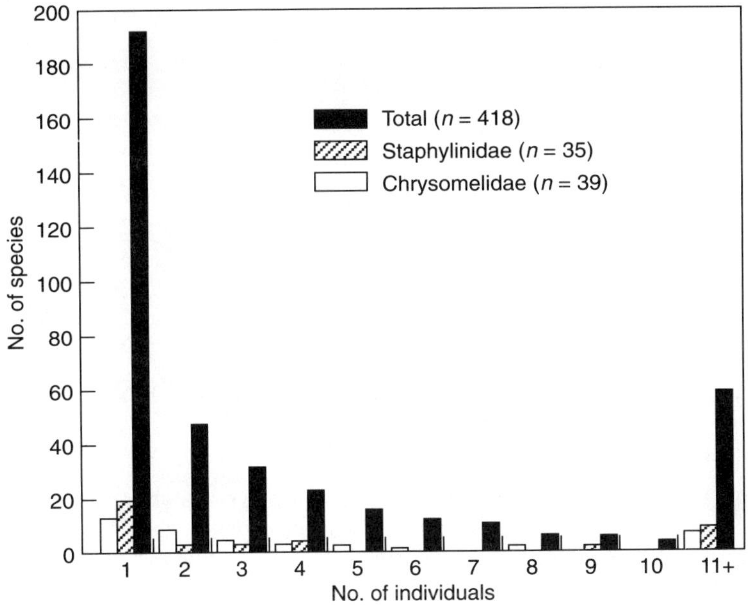

Figure 11.1 Comparison of numbers of individuals per species for all eight trees at 1200–1400 m. Categories are all beetles, Staphylinidae (except Omaliinae) and Chrysomelidae.

Although there was a large number of singleton species, they comprised only 5.0% of total individuals. The percentage of singletons on individual trees ranged from 10.1 to 17.2% (Table 11.3). These higher figures include species occurring as singletons on individual trees. However, about half these species occur on more than one tree and therefore are not treated as singletons in the overall sample (Table 11.2).

The species diversity of the tree samples as measured by the Q-statistic ranged from 32.31 to 68.72 (Table 11.3). (Q is used instead of α because the underlying species-abundance distribution did not follow the log-series; the Q-statistic is a non-parametric distribution based on the inter-quartile slope of the cumulative species abundance curve; Magurran, 1988.) Higher measures of diversity reflect lower total numbers of individuals in the sample.

The similarity in the beetle faunas of the various trees, as measured by the Jaccard coefficient, C_J (Magurran, 1988), ranged from 0.12 to 0.37 (Table 11.4). Trees 3 and 9, the same individual tree fogged about 10 months apart, had a Jaccard coefficient of 0.30, similar to the 0.31 for trees 3 and 4 (fogged 20 days apart and separated by a distance of ca. 50 m). The greatest similarity between samples, and the only samples

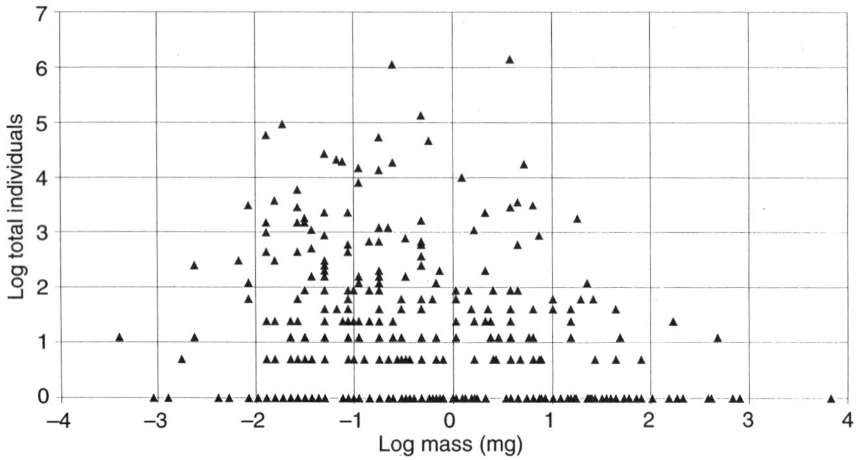

Figure 11.2 Double logarithmic plot of body mass versus abundance for beetles from eight trees at 1200–1400 m. Body mass was calculated from length using the formula for Costa Rican rainforest beetles derived by Schoener (1980).

with a Jaccard coefficient above 0.31, were from trees 10 and 11. These trees were ca. 15 m apart and were fogged at the same time.

Mean length of beetle species in the overall sample was 2.86 mm and ranged from 0.50 to 22.0 mm (SD = 2.23). Mean length of beetles in each of the tree samples ranged from 2.30 to 2.89 mm (Table 11.1). In a double logarithmic plot of abundance–body mass, the data points fall within an approximately triangular-shaped zone, as observed for many other assemblages (Gaston *et al.*, 1993a,b) (Figure 11.2). Blackburn *et al.* (1990) examined similar samples from Brunei (Morse *et al.*, 1988) and concluded that this pattern was caused by a preponderance of middle-sized species. However, it is not clear whether the observed patterns of size and abundance result from biological processes, or from statistical or sampling properties common to distributions of any collection of objects subject to optimal design constraints (see also Gaston *et al.*, 1993b).

DISCUSSION

Our results are similar to those obtained from other canopy fogging studies (Erwin and Scott, 1980; Erwin, 1983; Stork and Brendell, 1990; Stork, 1991; Allison *et al.*, 1993b). A prominent feature common to these studies is that there is a strong relationship between the number of individuals and the number of species, primarily because a high percentage (47.6% in our study) of the species are singletons. Our figure is lower than

Table 11.3 Details of number and percentage of beetle species represented by single individuals ('singletons')

Tree No.	*Total singletons*	*Total individuals*	*Singletons (%)*	*Q-statistic*
3	51	336	15.2	32.31
4	48	300	16.0	38.23
9	44	404	10.9	38.23
10	85	800	10.6	68.72
11	58	547	10.6	46.89
12	70	696	10.1	45.81
13	78	493	15.8	58.26
14	69	402	17.2	42.78

Table 11.4 Similarity of the beetle faunas of eight *Castanopsis acuminatissima* trees in the Wau Valley, Papua New Guinea using the Jaccard coefficient

	Tree 4	*Tree 9*	*Tree 10*	*Tree 11*	*Tree 12*	*Tree 13*	*Tree 14*
Tree 3	0.31	0.30	0.19	0.22	0.19	0.19	0.16
Tree 4		0.23	0.19	0.17	0.20	0.14	0.12
Tree 9			0.22	0.23	0.19	0.18	0.18
Tree 10				0.37	0.29	0.30	0.26
Tree 11					0.29	0.32	0.32
Tree 12						0.32	0.32
Tree 13							0.29

that reported by Morse *et al.* (1988) from Brunei (57.8%), but higher than that reported by Basset and Kitching (1991) from Australia (35.7%). The reasons for this variation are undoubtedly complex and probably reflect differences in the insect communities as well as differences is sampling methodology and effort. Our results and those of Basset and Kitching (1991) are based on single tree species. Morse *et al.* (1988) sampled single individuals of 10 species of lowland trees and their samples may include greater beta-diversity. In any case, it is the generally high percentage of singleton species in canopy fogging samples (assumed to include the total insect fauna of the tree canopy) that led to the extremely high estimates of the diversity of the rainforest canopy insect fauna advanced by Erwin and others (Erwin, 1982, 1988; Stork, 1988). However, in our study, of the species recorded from individual trees as singletons, approximately half occurred on other trees, sometimes in reasonably large numbers.

The ecology of singleton species can be better understood by documenting their presence and absence on other related and unrelated species of trees. Our experimental design allows for this comparison. At 1200–1400 m we fogged four *Lithocarpus* trees and one tree each of *Engelhardia mersingensis, Callophyllum* sp., *Prunus* sp., and *Carpodetus arboreus*. Beetle species that have no particular affinity to *C. acuminatissima* or fagaceous trees in general should show up on some of these unrelated species of trees. Other species may occur only on fagaceous trees. Analysis of these data is necessary to better assess the true number of species that are restricted to *C. acuminatissima*. Our preliminary results indicate that herbivores (the leaf-chewing family Chrysomelidae) are far less likely to occur as singletons than are predators (non-omaliine Staphylinidae) (Figure 11.1). General insect predators might be expected to occur widely throughout the forest on any available leaf substrate while leaf-chewers are more likely to be restricted to a given species of tree because of specialized adaptation to leaf chemistry and other factors. Farrell and Erwin (1988), in a study of Peruvian canopy beetles, found that Chrysomelidae were more closely associated with individual 'forest types' than were Staphylinidae (singletons were excluded from their analysis). On balance, we would therefore predict that predators would be more likely to occur as singletons. This is consistent with our results but is something that requires further analysis.

Montane forests in Papua New Guinea are not stable ecosystems but are constantly subjected to natural disturbances, including landslides, drought, fire and storms (Johns, 1986; van Valkenburg and Ketner, 1994). *Castanopsis acuminatissima* appears to be well adapted to this successional regime (van Valkenburg and Ketner, 1994). Insects associated with successional forests that include *Castanopsis* would therefore be expected to have a high dispersal capacity. A high percentage of the singleton species recorded in this study may represent dispersing individuals that are not otherwise associated with this tree species.

Another important component of diversity is turnover (beta-diversity). Our experimental design allows us to compare the canopy insect fauna of *C. acuminatissima* at different elevations. Preliminary results suggest that turnover is very high between the three elevational study sites (Allison *et al.*, 1993b). Pair-wise Jaccard coefficients between *C. acuminatissima* trees at different elevations (500 m, 1200–1400 m and 2100 m) averaged ca. 0.06 compared with 0.24 for the eight trees sampled at 1200–1400 m, a four-fold difference.

Comparison of beetle diversity (Q-statistic) in our study trees at three elevations confirms the conclusion of our previous altitudinal transect (Allison *et al.*, 1993b). Table 11.5 shows that diversity is similar at the 500 m and 1200–1400 m sites and higher at the 2100 m site. This relationship between diversity and altitude needs more examination (see

Table 11.5 Comparison of the beetle faunas of *Castanopsis acuminatissima* at different elevations using the Q-statistic. (See Allison *et al.*, 1993a for details of trees 1, 2, 7 and 8.)

Elevation (m)	*Tree no. (s)*	*Q-statistic*
2100	1	103.31
2100	2	67.36
1200–1400	3, 4, 9, 10–14	32–68 (see Table 11.2)
500	7	43.24
500	8	53.25

Allison *et al.*, 1993b). It is possible that it represents phenological differences between trees at the various sites, but could also be because of greater topographical and habitat heterogeneity at higher altitudes (N. Stork, personal communication).

As mentioned earlier, one tree was fogged twice and is included in this analysis as trees 3 and 9. The total number of individuals and species increased slightly in the second fogging, presumably indicating that beetles had rapidly recolonized the tree. The fauna of the recolonized tree was as similar to the original fauna (Jaccard coefficient = 0.30) as the original tree (no. 3) was to tree 4 located close by (Jaccard coefficient = 0.31) and fogged 20 days later. These findings are consistent with predictions of island biogeography theory (MacArthur and Wilson, 1967) that species number remains fairly constant but that species composition changes during turnover events.

With the exception of Erwin and Scott (1980), Gagné (1979) and several temperate studies (Barnard *et al.*, 1986; Blackburn *et al.*, 1993; Gaston *et al.*, 1993a), published canopy fogging studies are not based on multiple foggings of the same tree species at the same site, making statistical evaluation difficult. Also, early studies did not segregate the overall samples from individual trees (the insects were collected on large sheets rather than multiple replicate small trays). Thus, our study and others in this volume (Floren and Linsenmair, 1997, Chapter 16; Mawdsley and Stork, 1997, Chapter 6; Stork and Hammond, 1997, Chapter 1; Wagner, 1997, Chapter 9, this volume) are among the first to report data from multiple individual conspecific trees from the same site with subsampling by replicate trays.

To our knowledge the only other species of *Castanopsis* that has been fogged was that reported by Stork (1991) from lowland Borneo. One tree yielded 396 individuals representing 103 species. This compares with an average of 497 individuals (range 300–800) and 117 species (range 82–155) from our eight trees. This is slightly higher than the mean of

112 species that we earlier reported from our 500-m study site (two trees) and lower than the mean of 192 species for the 2100–2200 m study site (two trees) (Allison *et al.*, 1993b).

The distributional patterns of individual species are extremely complex and will require detailed analysis to understand and summarize. That is beyond the scope of this paper and will be forthcoming as we complete analysis of additional samples. Elsewhere, we have discussed the implications of this issue to overall estimates of species numbers (Basset *et al.*, 1996a).

To further refine our understanding of the role of singletons it will be necessary to study the feeding ecology of individual species. Such work has been undertaken by our colleague, Yves Basset, as part of a comprehensive study of the feeding ecology of the insect fauna on 10 species of trees in the Wau Valley, including *C. acuminatissima* (Basset, 1994, 1997, Chapter 12, this volume; Basset *et al.*, 1996b and unpublished data).

Acknowledgements

This study is an expanded version of a paper presented to the New Guinea Biological Society but published only as an abstract (Allison *et al.*, 1993a). It was partially funded by the New England Biolabs Foundation and undertaken in collaboration with Wau Ecology Institute. Harry Sakulas facilitated logistics at Wau Ecology Institute. We are greatly indebted to many systematists who have verified morphospecies assignments; they will be acknowledged fully in a subsequent paper. Candida Cardenas, Karin Kami, June Ibara, David Preston and Valerie Hedlund prepared the beetles for study. Gordon Nishida prepared the illustrations. Our colleagues Y. Basset, J. Coddington, T.L. Erwin, and N.E. Stork provided useful discussion of some of the ideas presented here.

REFERENCES

Allison, A., Miller, S.E. and Samuelson, G.A. (1993a) Patterns of beetle species diversity in *Castanopsis acuminatissima* (Fagaceae) trees studied with canopy fogging techniques in mid-montane New Guinea rainforest, in *Proceedings of the Biological Society of New Guinea.* (ed. R. Höft), Wau Ecology Institute Wau, Papua New Guinea, p. 89.

Allison, A., Samuelson, G.A. and Miller, S.E. (1993b) Patterns of beetle species diversity in New Guinea rainforest as revealed by canopy fogging: preliminary findings. *Selbyana*, **14**, 16–20.

Barnard, P.C., Brooks, S.J. and Stork, N.E. (1986) The seasonality and distribution of Neuroptera, Raphidioptera and Mecoptera on oaks in Richmond Park, Surrey, as revealed by insecticide knock-down sampling. *Journal of Natural History*, **20**, 1321–31.

Basset, Y. (1994) Palatability of tree foliage to chewing insects: a comparison between a temperate and a tropical site. *Acta Oecologia*, **15**, 181–91.

Basset, Y. (1997) Species abundance and body size relationships in insect herbivores associated with New Guinea forest trees, with particular reference to insect host-specificity, in *Canopy Arthropods* (eds N.E. Stork, J. Adis and R.K. Didham), Chapman & Hall, London, pp. 237–64 (this volume).

Basset, Y. and Kitching, R.L. (1991) Species number, species abundance and body length of arboreal arthropods associated with an Australian rainforest tree. *Ecological Entomology*, **16**, 391–402.

Basset, Y., Samuelson, G.A., Allison, A. and Miller, S.E. (1996a) How many species of host-specific insects on a species of tropical tree? *Biological Journal of the Linnean Society* (in press).

Basset, Y., Samuelson, G.A. and Miller, S.E. (1996b) Similarities and contrasts in the local insect faunas associated with ten forest tree species of New Guinea. *Pacific Science*, **50**, 157–83.

Blackburn, T.M., Harvey, P.H. and Pagel, M.D. (1990) Species number, population density and body size in natural communities. *Journal of Animal Ecology*, **59**, 335–46.

Blackburn, T. M., Gaston, K.J. and Stork, N.E. (1993) Temporal dynamics of body size of beetles on oaks: a cautionary tale. *Ecological Entomology*, **18**, 399–401.

Erwin, T.L. (1982) Tropical forests: Their richness in Coleoptera and other arthropod species. *The Coleopterist's Bulletin*, **36**, 74–5.

Erwin, T.L. and Scott, J.C. (1980) Seasonal and size patterns, trophic structure, and richness of Coleoptera in the tropical arboreal ecosystem: the fauna of the tree *Luehea seemannii* Triana and Planch in the Canal Zone of Panama. *The Coleopterist's Bulletin*, **34**, 305–22.

Erwin, T.L. (1988) The tropical forest canopy: The heart of biotic diversity, in *Biodiversity* (eds E.O. Wilson and F.M. Peter), National Academy Press, Washington, DC, pp. 123–9.

Erwin, T.L. (1983) Beetles and other insects of tropical forest canopies at Manaus, Brazil, sampled by insecticidal fogging, in *Tropical Rain Forest: Ecology and Management* (eds S.L. Sutton, T.C. Whitmore and A.C. Chadwick), Blackwell Scientific Publications, Oxford. pp. 59–75.

Farrell, B.D. and Erwin, T.L. (1988) Leaf-beetle community structure in an Amazonian rain forest canopy, in *Biology of the Chrysomelidae* (eds P. Jolivet, E. Petitpierre and T. Hsiao), Kluwer Academic Publishers, Dordrecht. pp. 73–90.

Floren, A. and Linsenmair, K.E. 1997. Diversity and recolonization dynamics of selected arthropod groups on different tree species in a lowland rainforest in Sabah, Malaysia, in *Canopy Arthropods* (eds N.E. Stork, J. Adis and R.K. Didham), Chapman & Hall, London, pp. 344–81 (this volume).

Gagné, W.C. (1979) Canopy-associated arthropods in *Acacia koa* and *Metrosideros* tree communities along an altitudinal transect on Hawaii Island. *Pacific Insects*, **21**, 56–82.

Gaston, K.J. (1991) The magnitude of global insect species richness. *Conservation Biology*, **5**, 283–96.

Gaston, K.J., Blackburn, T.M., Hammond, P.M. and Stork, N.E. (1993a) Relationships between abundance and body size: where do tourists fit? *Ecological Entomology*, **18**, 310–14.

Gaston, K.J., Blackburn, T.M. and Lawton, J.H. (1993b) Comparing animals and automobiles: a vehicle for understanding body size and abundance relationships in species assemblages? *Oikos*, **66**, 172–9.

Gressitt, J.L. and Nadkarni, N. (1978) *Guide to Mt Kaindi: Background to New Guinea Montane Ecology*, Wau Ecology Institute Handbook, Wau, Papua New Guinea.

Hammond, P.M. (1992) Species inventory, in *Global Biodiversity: Status of the Earth's Living Resources* (ed. B. Groombridge), Chapman & Hall, London, pp. 17–39.

Hammond, P.M. (1994) Practical approaches to the estimation of the extent of biodiversity in speciose groups. *Philosophical Transactions of the Royal Society of London, Series B*, **345**, 119–36.

Hodkinson, I.D. (1992) Global insect diversity revisited. *Journal of Tropical Ecology*, **8**, 505–8.

Hodkinson, I.D. and Casson, D. (1991) A lesser predilection for bugs: Hemiptera (Insecta) diversity in tropical rain forests. *Biological Journal of the Linnean Society*, **43**, 101–9.

Johns, R.J. (1986) The instability of the tropical ecosystem in New Guinea. *Blumea*, **31**, 341–71.

MacArthur, R.H. and Wilson, E.O. (1967) *The Theory of Island Biogeography*, Monographs in Population Biology, Princeton University Press, Princeton, New Jersey.

Magurran, A.E. (1988) *Ecological Diversity and its Measurement*, Princeton University Press, Princeton.

Mawdsley, N.A. and Stork, N.E. (1997) Host-specificity and the effective specialization of tropical canopy beetles, in *Canopy Arthropods* (eds N.E. Stork, J. Adis and R.K. Didham), Chapman & Hall, London, pp. 104–30 (this volume).

May, R.M. (1990) How many species? *Philosophical Transactions of the Royal Society of London, Series B*, **330**, 293–304.

Morse, D.R., Stork, N.E. and Lawton, J.H. (1988) Species number, species abundance and body length relationships of arboreal beetles in Bornean lowland rain forest trees. *Ecological Entomology*, **13**, 25–37.

Schoener, T.W. (1980) Length-weight regressions in tropical and temperate forest-understorey insects. *Annals of the Entomological Society of America*, **73**, 106–9.

Soepadmo, E. (1972) Fagaceae. *Flora Malesiana*, **7**(2), 265–403.

Stork, N.E. (1988) Insect diversity: facts, fiction and speculation. *Biological Journal of the Linnean Society*, **35**, 321–37.

Stork, N.E. (1991) The composition of the arthropod fauna of Bornean lowland rain forest trees. *Journal of Tropical Ecology*, **7**, 161–80.

Stork, N.E. (1993) How many species are there? *Biodiversity and Conservation*, **2**, 215–32.

Stork, N.E. and Brendell, M.J.D. (1990) Variation in the insect fauna of Sulawesi trees with season, altitude and forest type, in *Insects and the Rain Forests of South East Asia (Wallacea)* (eds W.J. Knight and J.D. Holloway), Royal Entomological Society of London, London. pp. 173–90.

Stork, N.E. and Hammond, P.M. (1997) Sampling arthropods from tree-crowns by fogging with knockdown insecticides: lessons from studies of oak tree beetle assemblages in Richmond Park, UK, in *Canopy Arthropods* (eds N.E. Stork, J. Adis and R.K. Didham), Chapman & Hall, London, pp. 3–26 (this volume).

Streimann, H. (1983) *The Plants of the Upper Watut Watershed of Papua New Guinea*, National Botanic Garden, Canberra.

van Valkenburg, J.L.C.H. and Ketner, P. (1994) Vegetation changes following human disturbance of mid-montane forest in the Wau area, Papua New Guinea. *Journal of Tropical Ecology*, **10**, 41–54.

Wagner, T. (1997) The beetle fauna of different tree species in forests of Rwanda and East Zaire, in *Canopy Arthropods* (eds N.E. Stork, J. Adis and R.K. Didham), Chapman & Hall, London, pp. 169–83 (this volume).

Whittaker, R.H. (1975) *Communities and Ecosystems*, 2nd edn, MacMillan, New York.

12

Species abundance and body size relationships in insect herbivores associated with New Guinea forest trees, with particular reference to insect host-specificity

Y. Basset

ABSTRACT

Chewing insects feeding externally on 10 species of forest trees were collected by hand-collecting, beating, branch-clipping, flight intercept traps and pyrethrum knockdown during a 1-year period at Wau, Papua New Guinea. Insect host-specificity was assessed by feeding trials in captivity. The analyses considered 6130 individuals representing 704 species of adult Coleoptera and larvae of Lepidoptera. The proportion of rare species (defined as being represented by only a single individual in the collections) was high but significantly lower for specialists than for generalists. As a result, the proportion of rare species was significantly lower in the compound community (i.e. the insect community supported by the 10 tree species considered together) than in the component communities (i.e. the insect communities supported by each tree species). In other terms, increasing sampling effort to different habitats (tree species) resulted in better estimates of overall insect abundance, because it increased the probability of discovering the 'true' abundance of some species, these often being generalists. Species abundance distributions within the component communities were analysed by fitting regressions through plots of species biomass (i.e. number of individuals multiplied by average body weight) against species rank. The number of insect species within the community was related to an estimate of the number of young leaves available year-long on each tree species. The maximum insect biomass which could be achieved on each tree species was also related to an estimate of leaf expansion rates. However, the slopes of the regressions, as well as average insect body size within the community,

Canopy Arthropods. Edited by N.E. Stork, J. Adis and R.K. Didham. Published in 1997 by Chapman & Hall, London. ISBN 0 412 74900 9

could not be predicted by either these or other host-tree attributes. Insect biomass, and, presumably, food resources, tended to be more evenly distributed in species-rich than in species-poor communities. This suggests that host attributes influence to some extent the structure of these chewing insect communities, but that the way that food resources are shared within these communities may have more to do with intrinsic community characteristics (e.g. vagility of constituent species, or resistance to invasion, competition, etc.) or historical factors.

INTRODUCTION

In recent years, there has been considerable interest in the relationship between species number, species abundance and body size as reflecting the structure of animal communities (recent review in Tokeshi, 1993). Since arthropod communities are often species-rich and their constituent species exhibit a wide range of body sizes, they are particularly amenable for such analyses. For example, some studies have focused on the communities of arthropods associated with temperate trees (Blackburn *et al.*, 1993a) or with tropical/subtropical trees (Morse *et al.*, 1988; Basset and Kitching, 1991; Stork and Blackburn, 1993). In the latter case, the trophic relation between the tree species sampled and the arthropods collected from it, particularly insect herbivores, has rarely been ascertained.

Many authors studying species abundance and body size relationships have tested the hypothesis that energy availability limits the abundance of organisms of different size differently (Griffiths, 1992; Blackburn *et al.*, 1993a). To date, the results are ambiguous (Blackburn *et al.*, 1993a,b; Gaston *et al.*, 1993a). The biological interpretation of the patterns reported so far is not straightforward because of two major uncertainties. First, these patterns may result more from general statistical properties of large collections of objects than from real biological processes (Blackburn *et al.*, 1993b; Gaston *et al.*, 1993b). Second, and perhaps representing an allied problem, the ways that the data have been gathered for these analyses are not exempt from flaws and limitations. Data obtained from the compilation of literature may yield different results from these obtained in field studies (Blackburn *et al.*, 1993a). For the latter, the influence of sampling methods and protocol have a significant effect on the perception of community structure (Basset and Kitching, 1991). Data obtained from the field represent a considerable investment in time and effort and, as pointed out by Kitching and Arthur (1993), many workers prefer to re-analyse the few data-sets available. A possible consequence of this situation is the risk of developing theories based on a few data-sets whose sampling limitations and biases have not been fully assessed and recognized. On the other hand, compilation of the literature may underestimate the number of rare and small species (Blackburn *et al.*, 1993a).

With particular reference to studying the relationships between the species abundance and body size of insect herbivores associated with tree species growing in diverse habitats, such as tropical rainforests, two difficulties may complicate the analyses of field data. First, the boundaries of arboreal communities are not obvious (Stork, 1987, 1991; Basset, 1993) because the diversity of surrounding habitats increases the probability of collecting transient species (Basset and Kitching, 1991). This, and other sampling artefacts, may explain in part why the number of apparently rare insect species collected on tropical foliage is so high (Morse *et al.*, 1988; Basset and Kitching, 1991; Allison *et al.*,1993). The role of transient, or 'tourist' species has been investigated recently for adult beetles associated with oak trees in Britain by Gaston *et al.* (1993a). These authors showed that despite tourists disproportionately inflating the number of rare species, they do not alter fundamentally the relationship between abundance and body size in the system studied. Currently, a similar data-set, in which the relationships between insect variables can be examined for insects of different affinities with the habitat (tree species) sampled, is not available for any tropical system. Both the poor knowledge of tropical faunas and the difficulty of inferring the host range of many tropical herbivorous insects account for this situation.

Second, many studies have analysed their data by pooling all insects collected from different tree species, despite, at least in one instance (Morse *et al.*, 1988), significant differences in species abundance among tree species. This procedure may be unsound because communities are usually ruled by different resource bases, with often fairly different consequences for community structure and species abundance patterns (Price, 1992; Tokeshi, 1993). For example, Schoener and Janzen (1968) reasoned that if a dispersed energy source is available for only a short period of time, a good strategy would be to divide it among many small individuals, each with a relatively short maturation time. Comparing tropical dry forests, tropical rainforests and temperate forests, they showed that insect size increases with the length of growing season in each forest type. Consistent with this, Opler (1978), showed that leaf-mining moth species associated with deciduous *Quercus* in North America are smaller, but occur at much higher densities, than their counterparts associated with evergreen *Quercus*.

The effect of successional status on the structure of insect communities has often been emphasized (Southwood *et al.*, 1979). In early stages of succession, the density and biomass of insect herbivores are often high, but species richness is low (Opler, 1978; Southwood *et al.*, 1979). This may represent another example in which primary production may directly affect community structure. In short, it is possible, but not certain, that by pooling data obtained from tree species with different patterns of leaf production and successional status, interesting patterns in species number, abundance and body size are confounded or obscured, thus resulting in overall pictures showing no clear biological processes.

The aims of the present study were to describe and contrast the local relationships between species number, species abundance and body size for chewing insects associated with 10 tree species in Papua New Guinea. These relationships included: (i) the species-abundance distribution based on the number of individuals; (ii) the species-abundance distribution based on biomass; (iii) the species-body length distribution; and (iv) the relationship between population abundance and body weight. Analyses were restricted to a subset of the insect community, the chewing insects, because the affinity of these insects for the habitat (tree species) in which they were collected could be ascertained in most cases (see Methods). The present contribution is more empirical than theoretical and sought to test: (i) whether the above relationships determined for the compound community studied (i.e. the overall insect community supported by the ten tree species studied, *sensu* Root, 1973) were different from those determined for the different component communities (i.e. the insect communities supported by each of the tree species studied); and (ii) whether relationships were different among the different component communities and were related to host attributes, particularly patterns of leaf production and successional status. Other insect variables of particular relevance to the system studied, such as the predictors of local species richness on each tree species and the representativeness of sampling (Basset, 1996), the intraspecific variability in insect body-size in relation to feeding ecology (Basset *et al.*, 1994) and the identity of the insects collected and the similarity in the faunal composition of the different tree species (Basset *et al.*, 1996), are presented elsewhere.

METHODS

Study site and insect collecting

The study was performed on the slopes of Mt Kaindi, near and within the grounds of the Wau Ecology Institute, Wau, Papua New Guinea (7°24'S, 146°44'E). Altitude of collecting ranged from 1100–2362 m (summit), but was mostly confined around 1200–1400 m. Mt Kaindi has been locally cleared leaving a mosaic of grasslands and forest patches dominated by secondary forest (van Valkenburg and Ketner, 1994). The main forest formations encountered on the slopes include lower and mid-montane rainforest (Johns, 1982). The climate is 'humid to perhumid mesothermal with little or no water deficit' (McAlpine *et al.*, 1983). The study area is detailed further by Gressitt and Nadkarni (1978) and van Valkenburg and Ketner (1994).

Insects were collected from 10 plant species representing native forest trees and shrubs with different successional status and patterns of leaf production (Table 12.1). For the analyses tree species were often considered in two categories of successional status (pioneer and persistent) and of patterns of leaf production (non-flushing and flushing) (Table 12.1).

Chewing insects feeding externally were collected from the foliage of the study trees by hand-collecting, foliage-beating, branch-clipping, flight intercept traps and pyrethrum knockdown. The first four of these methods were used both during day- and night-time, whereas pyrethrum knockdown was only performed early in the morning. Further, living specimens from the first three methods were used in feeding trials (see next section). Hand-collecting and foliage-beating represented, for each tree species, about 50 hours of hand-collecting activity and 300 beating samples distributed among different individual trees. Branch-clipping (as in Basset *et al.*, 1992) represented, for each tree species, 55 samples of about 33 m^2 of leaf surface, obtained from different individuals. One flight intercept trap (as described in Springate and Basset, 1996) was set up in the middle of the crown of one individual of each tree species and run continuously through one year. Insects were removed from these traps every 11 days. One individual of each tree species was sampled using pyrethrum knockdown (solution of 5% Pyranone and kerosene), using 12–20 trays of 1 m^2 surface area, depending on tree size (total 159 trays used for all tree species). The protocol of Allison *et al.* (1993) was used, except that the trees were fogged by climbing directly into them, rather than being fogged from adjacent trees.

Active sampling was performed from February to July 1992 and from November 1992 to April 1993 and the traps were run from April 1992 to April 1993. Field data have thus been gathered over more than a year and account for the seasonal variation in insect diversity and abundance at the Wau site. When the foliage could not be sampled from the ground, access was gained by single rope techniques (Perry, 1978). Sampling effort was the same for each tree species and all material thus derived has been considered for subsequent analyses. For simplicity, the combined data obtained by hand-collecting, foliage-beating and branch-clipping are referred to as 'foliage samples'.

Morphospecies assignment and assessment of ecology

Chewing insects were assigned to morphospecies, on the basis of external characters, in Papua New Guinea. The taxa considered include adults of Lagriidae, certain Cerambycidae (which sometimes chew leaves for

Table 12.1 Study trees, their successional status (Pi = pioneer, Pe = persistent), phenology of leaf production, number of days required for young leaves to mature (s.e.) and the number of individuals collected with all sampling methods for proven feeders only and for all chewing insects

Hosts	*Plant family*	*Status*	*Leaf production*	*Leaf expansion*	*Feeders*	*All*
Elmerrillia tsiampacca (L.) Dandy	Magnoliaceae	Pe	Continuous	29.6 (1.8)	74	147
Cinnamomum cf *culilaban* (L.) Presl	Lauraceae	Pe	Intermittent leaf flushes*	35.5 (0.3)	211	510
Piper plagiophyllum K. Sch. and Laut.	Piperceae	Pi	Continuous	20.7 (1.5)	335	514
Ficus nodosa Teys. and Binn.	Moraceae	Pi	Deciduous, leaf flushes*	24.2 (0.4)	508	741
Pipturus argenteus Wedd.	Urticaceae	Pi	Continuous	32.5 (1.0)	625	1241
Castanopsis acuminatissma A. DC	Fagaceae	Pe	Intermittent leaf flushes*	26.5 (0.6)	403	633
Caldcluvia brassii Hoogl.	Cunioniaceae	Pi	Continuous	70.0 (3.3)	445	812
Aleurites moluccana Willd.	Euphorbiaceae	Pe	Continuous	34.9 (4.0)	114	208
Melicope denhamii (Seem.) T. Hartley	Rutaceae	Pe	Intermittent leaf flushes*	26.2 (0.9)	314	944
Cordia dichotoma Forst.	Boraginaceae	Pi	Deciduous, leaf flushes*	19.8 (0.4)	261	389

* Apparently 2–3 times a year, as a synchronous event.

maturation feeding), Chrysomelidae, Curculionoidea, Scarabaeidae Melolonthinae, larvae of Lepidoptera, adults and larvae of Orthoptera and Phasmatoptera. All Curculionoidea collected on the foliage were tested for feeding on leaves but some morphospecies were later assigned to the wood-feeding guild (see below). Only mature larvae of Lepidoptera were assigned to morphospecies. In this case additional characters were used for this purpose, such as feeding behaviour (i.e. feeding tracks), locomotive and general behaviour, and coloration. As far as possible, caterpillars were reared to obtain the adults and to refine the morphospecies assignment.

Since knowledge of the ecology of most Papuan insects is fragmentary, insect specialization was assessed from laboratory feeding trials. Live insects were stored in plastic vials at room temperature and in conditions of near-saturated relative humidity. They were provided with fresh foliage of the tree species from which they were collected until they either died or accepted food. In the latter case, they were then tested in random order for 24-hour periods on the foliage of the nine other study species. Feeding damage was scored visually, relative to insect body size, on a logarithmic scale, as follows: 0, no feeding; 1, attempting to feed; 10, moderate feeding; 100, extensive feeding. This procedure emphasized regular feeding as compared with food-probing. Insects were assigned to four leaf-feeding categories according to the results of these tests: (a) 'leaf-feeder specialists', i.e. insects tested on three or more plant species but which only fed on the plant they were collected from (sum of feeding scores <100); (b) 'leaf-feeder generalists', i.e. insects tested on three or more plant species and which fed on two or more plants belonging to different plant families (sum of feeding scores ≥100); (c) 'leaf-feeder, unknown specialization', i.e. insects which, because of death, could not be tested on more than two plant species; and (d) 'incidentals', i.e. insects which did not feed in the trials. Only beetle morphospecies were considered in this last assignment, in order to avoid non-feeding parasitized or moulting caterpillars. Categories (a), (b) and (c) were further referred to as 'proven feeders'. Further, weevil morphospecies that were collected dead (i.e. by flight intercept traps or pyrethrum knockdown) were assigned to the categories 'additional, leaf feeder' or 'additional, wood borer' using the information provided by experienced weevil taxonomists (E.C. Zimmerman and R.S. Anderson, personal communication). In general, there was a good correspondence between the food preferences as assessed in feeding trials and insect presence in particular tree species (see further discussion of the results obtained with feeding trials in Basset, 1994).

As far as possible mean body size of each morphospecies was determined from a series of 10 measurements, taken, in most cases, in alcohol, and to the nearest 0.1 mm. For caterpillars, the highest measurement of the series was considered, in order to approach the size of the last instar. The relation between body size (mm) and body weight (mg dry weight) were obtained using the equations provided by Schoener (1980) for tropical insects.

Once all measurements had been taken, all adult insects were later dry mounted at the Bishop Museum, Honolulu. Morphospecies assignment, hereafter 'species' for the sake of simplicity, was checked and updated by several taxonomists (see Acknowledgements). Beetle species, but not lepidopteran species unless adults were reared and could be compared, were cross-checked among tree species. The material has been deposited in the collections of the Bishop Museum.

Statistical methods

The number of rare species collected in the field is likely to depend on sample size and sampling effort. It may therefore be difficult to compare the results obtained with different sampling methods, and so the ratio number of rare species : total number of species collected was preferred. A rare species was defined as a singleton (i.e. species represented by a single individual in the collections).

Species dominance within the insect communities studied was assessed in several ways. First, the Berger–Parker index (Magurran, 1988) was calculated for communities supported by each tree species, using the 'proven-feeders' data. Second, rank-abundance plots were drawn for different subsets of the data. Since biomass probably reflects resource requirements more accurately than do numbers (Tokeshi, 1993), species abundance was defined not only as the number of individuals in each species but also as the biomass represented by each species (i.e. no. of individuals × mean body weight). Differences in the various abundance distributions obtained were tested by comparing cumulative abundance plots using ranked species (Kolmogorov–Smirnov two-sample test, as advocated by Tokeshi, 1993). To compare dominance/evenness among the insect communities supported by the different tree species, ordinary least squares regressions (OLS) of log(species rank) against log(abundance) of proven chewers were calculated and the slopes compared. Fit of the species-abundance data to a particular species-abundance model was only attempted for the biomass data (see Results section). Since the impact on community structure of a high number of rare species is arguable (see Results section, proportion of rare species), biomass analyses were further restricted to species which, when ranked in order of decreasing abundance, contributed to 95% of the biomass (see Tokeshi, 1990). These species are hereafter termed 'dominant species'. In addition, the index of evenness E, recently proposed by Bulla (1994), was calculated for these same data-sets. However, since this index has not yet gained wide acceptance in the scientific community, the analyses and the discussion emphasize the OLS regressions rather than this measurement.

Species body length (instead of body weight) distributions were analysed using body length classes as in Morse *et al.* (1988), for ease of comparison with other studies. However, the low number of insect species collected on each tree species, as well as their inconsistent body length distributions, prevented fitting a regression line through the upper tail of the distribution (i.e. the part to the right of the mode) in most cases. Instead, the skewness of the body size distributions was calculated for each tree species.

To analyse the relationships between body size and abundance, the methodology developed by Blackburn *et al.* (1992) was followed. This included plotting the log(no. of individuals) against the log(body weight) and fitting an OLS regression through the highest points in each of eight equal size classes. This procedure is known as the 'negative upper bound slope' (NUBS, Blackburn *et al.*, 1992). For this analysis, body weight was used instead of body length because the former is more likely to be related to food resources than the latter. Since some have argued (Griffiths, 1992) that regression slopes for these analyses should be calculated using the reduced major axis method (RMA) instead of OLS regression, both types of regression were calculated. However, the emphasis in the results is on OLS regressions as the error variance in body weight is likely to be much smaller than the error variance in abundance.

RESULTS

The proportion of rare species

The analyses concerned 6130 individuals distributed in 704 species. Coleoptera dominated the samples (4696 individuals and 391 species), followed by Lepidoptera (1361 individuals and 286 species). However, since morphospecies assignment for Lepidoptera only involved mature larvae, it is possible that the relative number of species of Lepidoptera has been underestimated compared with that for Coleoptera. The other herbivorous insects were largely Orthoptera and Phasmatoptera.

Table 12.2 details, by categories, the number of chewing insect species collected on each tree species, along with the number of rare (singleton) species. When the data were pooled for all tree species, rare species represented 39% of the total number of species collected. However, the proportion of rare species depended significantly on insect category (G-test with specialist, generalist, incidental and additional categories, $G = 30.7$, $P < 0.001$): the lowest proportion was in the specialist category

Table 12.2 Total number of species (T) and number of rare species (R) collected from each tree species, for the following categories: specialists, generalists, all proven feeders, incidentals, additional and beetles only (see text). The last entry refers to pooled data for all tree species, with the percentage of rare species in each category

Tree	*Specialists*		*Generalists*		*Proven chewers*		*Incidentals*		*Additionals*		*Beetles*	
	T	*R*	*T*	*R*	*T*	*R*	*T*	*R*	*T*	*R*	*T*	*R*
Elmerrillia	4	1	7	4	20	10	9	5	24	18	41	27
Cinnamomum	16	6	17	8	37	16	24	10	55	31	83	43
Piper	6	2	10	3	18	6	16	5	45	29	69	37
Ficus	36	14	21	7	61	23	9	1	52	31	83	38
Pipturus	24	4	16	5	52	16	15	7	37	18	74	30
Castanopsis	53	13	20	8	94	34	21	10	38	21	64	30
Caldcluvia	11	3	25	10	39	14	34	12	63	42	112	55
Aleurites	3	1	16	7	25	9	7	4	32	22	46	30
Melicope	13	9	20	8	36	20	24	9	43	27	76	41
Cordia	19	3	23	10	45	16	8	5	33	22	45	27
All trees	185	48	155	61	399	148	103	30	202	106	391	154
(% rare sp.)	–	(26)	–	(39)	–	(37)	–	(29)	–	(52)	–	(39)

(26%), whereas the highest proportion was in the additional category (52%, Table 12.2). Of the 202 species assigned to the latter category, 112 were probably wood-eaters. The proportion of rare wood-eaters was 51% (57 rare species). Further, the proportion of rare species was significantly higher in the generalist than in the specialist category (G = 5.78, $P < 0.05$).

When each tree species was considered separately (i.e. arthropod communities considered as component communities), the proportion of rare species within proven feeders and for all chewing insect categories ranged from 33% (*Piper, Pipturus*) to 56% (*Melicope*), and from 39% (*Pipturus*) to 62% (*Elmerrillia*), respectively. G-tests indicated that the proportion of rare species was uniformly distributed among tree-species for both the proven feeder category (G = 7.85, $P = 0.55$) and for all insects (G = 15.0, $P = 0.09$). There was no obvious effect of successional status or patterns of leaf production on the proportion of rare species recorded on each tree species for all chewing insects (G = 2.4, $P = 0.12$ and G = 0.34, $P = 0.56$, respectively). Further, no obvious effect of wide range of host attribute (e.g. leaf nitrogen, leaf water content, leaf palatability, etc., see Basset, 1994 and 1996 for a full list) on the proportion of rare species collected was detected (analyses with all chewing insects and proven feeders).

When all tree species were considered together, the proportion of rare species was significantly lower than when the tree species were considered separately (only the beetle data were used for this analysis, since all species of adult beetles could be cross-checked with confidence among tree species: separate data, 358 rare species out of 693 (52%), pooled data, 154 rare species out of 391 (39%), G = 15.2, $P < 0.001$).

Lastly, the possible influence of sampling method on the proportion of rare species collected was examined (Table 12.3). The proportion of rare species apparently did not depend on the sampling methods used (G = 0.06, $P = 0.80$). Further, when the results from the three sampling methods were pooled, the proportion of rare species was not significantly different from when the results from each method were considered separately (G = 0.25, $P = 0.88$). The results of this section suggest that the sampling methods and tree species did not greatly influence the proportion of rare species recorded, but that the ecology of the insects and the number of habitats (tree species) sampled did.

Species-abundance distribution: numbers

There was a weak tendency for species-rich communities of proven feeders to include more individuals than species-poor communities (Table 12.1, $r_s = 0.65$, $P = 0.05$). The Berger–Parker dominance index varied from 0.107 (*Castanopsis*, Table 12.4) to 0.650 (*Melicope*), without

Table 12.3 Number of total species and rare species collected with the different sampling methods (each tree species has been considered separately for this analysis)

Method	*Total no. of species*	*No. of rare species*	*Percentage*
Foliage samples	610	300	49
Flight intercept traps	450	233	52
Pyrethrum knockdown	209	103	49
All three methods pooled	994	489	49

Table 12.4 Characteristics of the communities of chewing insects on their hosts (data for proven feeders only): Berger–Parker dominance index; average (s.e.) body length (mm) and average (s.e.) body weight (mg dry weight) of chewing species; skewness of the species body length distribution (with body length classes as in Figure 12.3); negative upper bound slope (s.e.) and coefficient of correlation for this regression (see text)

Host	*Dominance*	*Length*	*Weight*	*Skewness*	*NUBS*	*r*
Elmerrillia	0.351	13.8	4.31	1.955	n.s.	0.67
		(2.4)	(0.68)			
Cinnamomum	0.336	18.8	12.62	0.765	n.s.	0.54
		(2.0)	(5.96)			
Piper	0.561	11.0	4.07	1.389	–1.595	0.89
		(1.8)	(1.01)		(0.365)	
Ficus	0.179	12.8	9.87	0.254	–0.723	0.85
		(1.3)	(5.31)		(0.202)	
Pipturus	0.180	17.0	20.05	0.546	–0.568	0.85
		(1.8)	(5.85)		(0.143)	
Castanopsis	0.107	15.0	5.34	0.565	–0.696	0.97
		(1.00)	(1.99)		(0.073)	
Caldcluvia	0.173	14.0	21.61	0.670	n.s.	0.58
		(1.7)	(5.63)			
Aleurites	0.254	19.0	31.05	1.273	–0.497	0.81
		(3.7)	(21.11)		(0.179)	
Melicope	0.650	19.5	12.53	1.073	n.s.	0.65
		(1.9)	(3.58)			
Cordia	0.134	16.0	3.81	0.391	n.s.	0.05
		(1.5)	(0.40)			
All trees	0.062	15.6	11.76	0.107	–0.561	0.80
		(0.6)	(1.86)		(0.172)	

NUBS, negative upper bound slope; n.s., not significant

any obvious trends related to host attributes. In particular, the influence of successional status and leaf production patterns was not significant (t = 0.79, P = 0.45 and t = 0.18, P = 0.86, respectively). There was, however, a significant trend for species-poor communities to be dominated by particular species (Spearman rank correlation coefficient between the number of species of proven feeders and the Berger–Parker index, $r_s = -0.806$, $P < 0.01$). Not unexpectedly, dominance appeared lower in the compound community than in the separate component communities (Table 12.4).

Rank-abundance plots for various insect categories revealed striking differences (Figure 12.1), all of them significant (cumulative abundance plots are not shown here). In particular, the overall distribution (all chewing insects) was significantly different from that of the specialists (Kolmogorov–Smirnov two-sample test, D = 0.599, $P < 0.001$), of the 'additional' category (D = 0.806, $P < 0.001$), and of the generalists (D = 0.773,

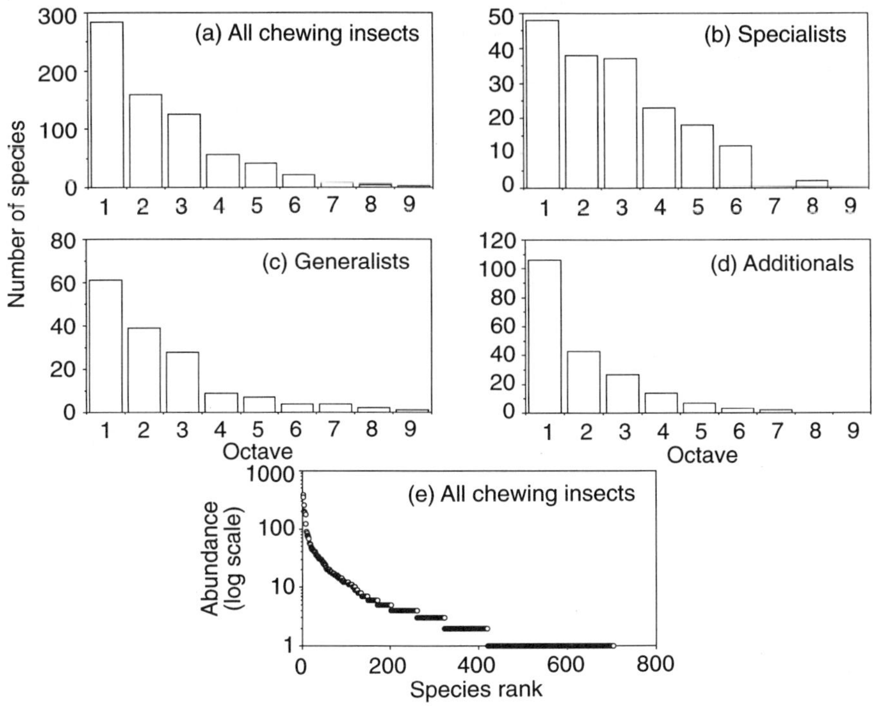

Figure 12.1 Species-abundance distribution plots (no. of individuals) in the compound community. Species abundance distributed in octaves for (a) all chewing insects, (b) chewing specialists, (c) chewing generalists, and (d) additionals. Octaves represent abundance classes in base 2 (i.e. octave 1 = 1 individual, octave 2 = 2–3 individuals, octave 3 = 4–7 individuals, octave 4 = 8–15 individuals, etc.). Abundance ranked by species for all chewing insects in (e).

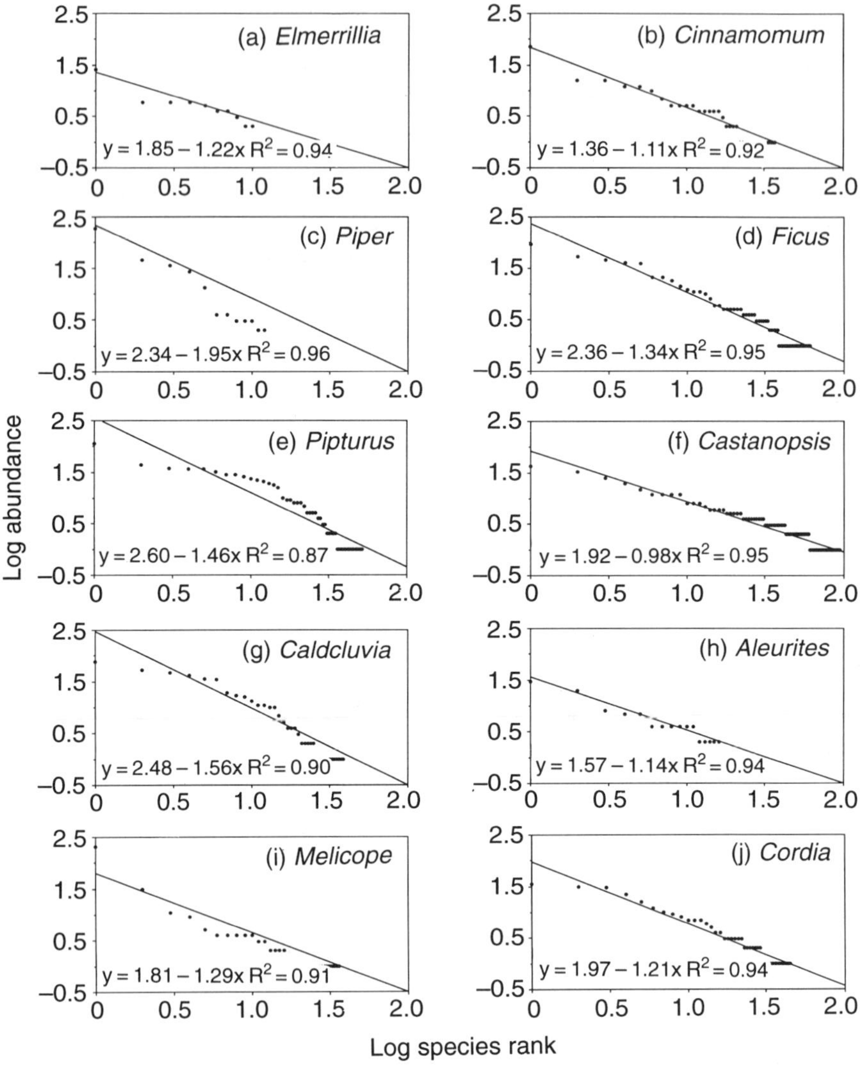

Figure 12.2 Abundance (no. of individuals) ranked by species of proven feeders in the different component communities studied. The curve fitted to the data is indicated in each case.

P <0.001). The distribution of the specialists was significantly different from that of the 'additionals' (D = 0.220, P <0.001) and of the generalists (D = 0.266, P <0.001). On average, the population levels of specialists appeared significantly higher than those of generalists (Mann–Whitney U test = 16 506.5, P <0.01). As previously found (Morse *et al.*, 1988; Basset and Kitching, 1991), the species-abundance distribution resembled a log-normal distribution, with a high number of rare species and with a hidden mode (see Figure 12.1e).

To check whether the species-abundance distribution differed between tree species, curves were fitted to log-log plots for proven feeders (Figure 12.2). Slopes varied from –1.95 (*Piper*) to –0.98 (*Castanopsis*). These steepest and shallowest slopes were significantly different from each other (t = 14.22, P <0.001, see Figure 12.2). The corresponding regression fitted to the overall compound community was y = 3.21 – 1.25 x (± 0.013), R^2 = 0.96, F = 8537.6, P <0.001, n = 398. In this case, the slope was within the highest range (shallower slopes) of the slopes for the separate component communities. Interestingly, pioneer trees showed on average significantly steeper slopes than did persistent trees (t = 2.63, P <0.05). This suggests that the communities supported by pioneer trees were more dominated by a few species than the communities supported by persistent trees. There was no obvious influence of leaf production patterns in this regard (t = 1.45, P = 0.20) and none of the host attributes was obviously correlated with the slopes of the regressions, or with the maximum specific number of individuals achieved on each tree species. In short, species-abundance distributions, as expressed by the number of individuals, could differ significantly between particular tree species but it was difficult to explain these differences from host attributes.

Species-abundance distribution: biomass

Dominant species (i.e. representing together 95% of biomass on each tree species when ranked in order of decreasing abundance) included on average 53.8% (s.e. = 2.75, range = 38–65%) of the species of proven feeders (Figure 12.3). The number of dominant species supported by each tree species was positively correlated with an estimate of food resources available year-long (number of newly emerged leaves/leaflets recorded in branch-clipping samples; Basset, in press, r_s = 0.770, P <0.05). The maximum and minimum biomass achieved by individual species on each tree species were both positively correlated with an estimate of leaf expansion rate (average number of days required to expand from bud burst to full mature size as measured by tagging leaves in the field, data presented in Table 12.1, r_s = 0.685, P <0.05 and r_s = 0.818, P <0.01, respectively).

Slopes fitted through species rank and biomass of dominant species varied from –1.52 (*Piper*) to –1.04 (*Caldcluvia*). These steepest and shallowest slopes were significantly different from each other (t = 2.84, P <0.01). Values of the index of evenness E varied from 0.41 (*Aleurites*) to 0.62 (*Caldcluvia*), with significant differences (see 95% C.L., Figure 12.3). The corresponding regression fitted to the overall compound

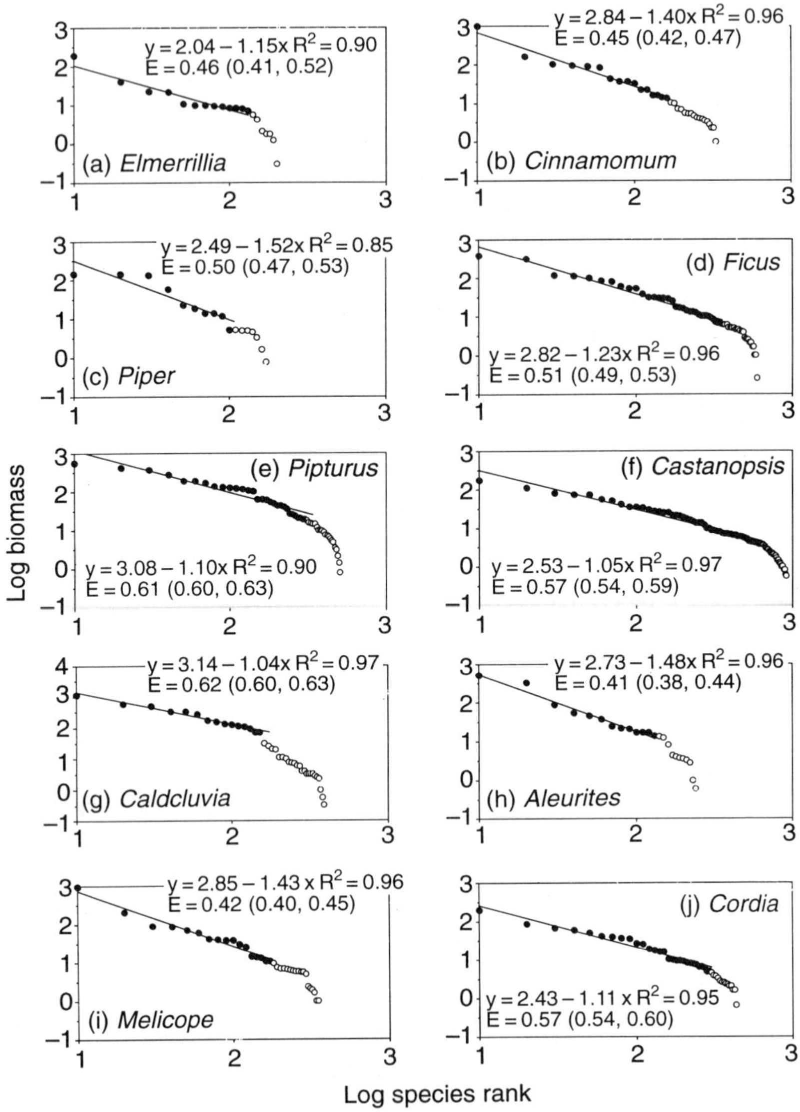

Figure 12.3 Abundance (biomass) ranked by species of proven feeders in the different component communities studied. Dominant species are represented with filled circles. The curve fitted to the data for dominant species is indicated in each case (see text), as well as the value of the index of evenness (E) calculated for the dominant species (95% C.L. in brackets).

community was y = 4.16 – 1.38 x (± 0.019), $R^2 = 0.95$, $F = 5216.3$, $P < 0.001$, $n = 298$. As for the species-abundance distribution (above), the slope was higher (shallower) than the steepest slopes of the component communities. The slopes calculated for each tree species were correlated with the Berger–Parker dominance index calculated for each tree species ($r_s = -0.76$, $P < 0.05$), but not with the slopes fitted through the number of individuals ranked by species (previous section, $r_s = 0.13$, $P > 0.50$). There was also a weak positive relationship between the biomass slopes and the number of species of proven feeders within the community ($r_s = 0.685$, $P < 0.05$), but not with average body mass of community members (see Table 12.4, $r_s = 0.07$, $P > 0.50$). No host attribute was obviously correlated with these slopes and there was no detectable influence of successional status ($t = 0.84$, $P = 0.42$) or leaf production patterns ($t = 0.12$, $P = 0.91$). When the data were re-analysed using a different criteria to define dominant species, in this case representing on average 90% of biomass on each tree species, patterns were slightly different. Slopes calculated with these dominant species ranged from –1.50 (*Aleurites*) to –0.69 (*Pipturus*) and were correlated with the proportion of specialists within the community (i.e. the ratio no. of specialist species: no. of specialist + generalist species, $r_s = 0.67$, $P < 0.05$).

Testing for goodness-of-fit to statistically-oriented models (e.g. log-series, log-normal) was not considered here because the biological meaning of such models is not straightforward (see reviews in Magurran, 1988; Tokeshi, 1993). Most niche-oriented models which rely on stochastic events could not be considered either because species richness varied tremendously among replicates (see Tokeshi, 1993). Instead, the data were compared with the geometric series, a deterministic niche-oriented model (Magurran, 1988; Tokeshi, 1993). None of the distributions of biomass in dominant species associated with the different tree species fitted this model (examples of the steepest and shallowest curves as determined above: *Piper*: k = 0.3311, χ^2 for goodness-of-fit = 50.3, $P < 0.001$; *Caldcluvia*: k = 0.1493, $\chi^2 = 434.8$, $P < 0.001$). In summary, species-abundance distribution as expressed by biomass could to some extent be predicted (the maximum biomass achieved on each tree species) from host attributes (leaf expansion rate) but the way that resources were shared within the community (the slopes of the regressions fitted) could not.

Species-body length distribution

The distribution of body length for the compound community is plotted in Figure 12.4. Overall, chewing insects were larger than members of other arboreal guilds (compare with e.g. Morse *et al.*, 1988; Basset and Kitching, 1991). The distribution of body length in specialists was skewed to the left and, consequently, on average, specialist species were

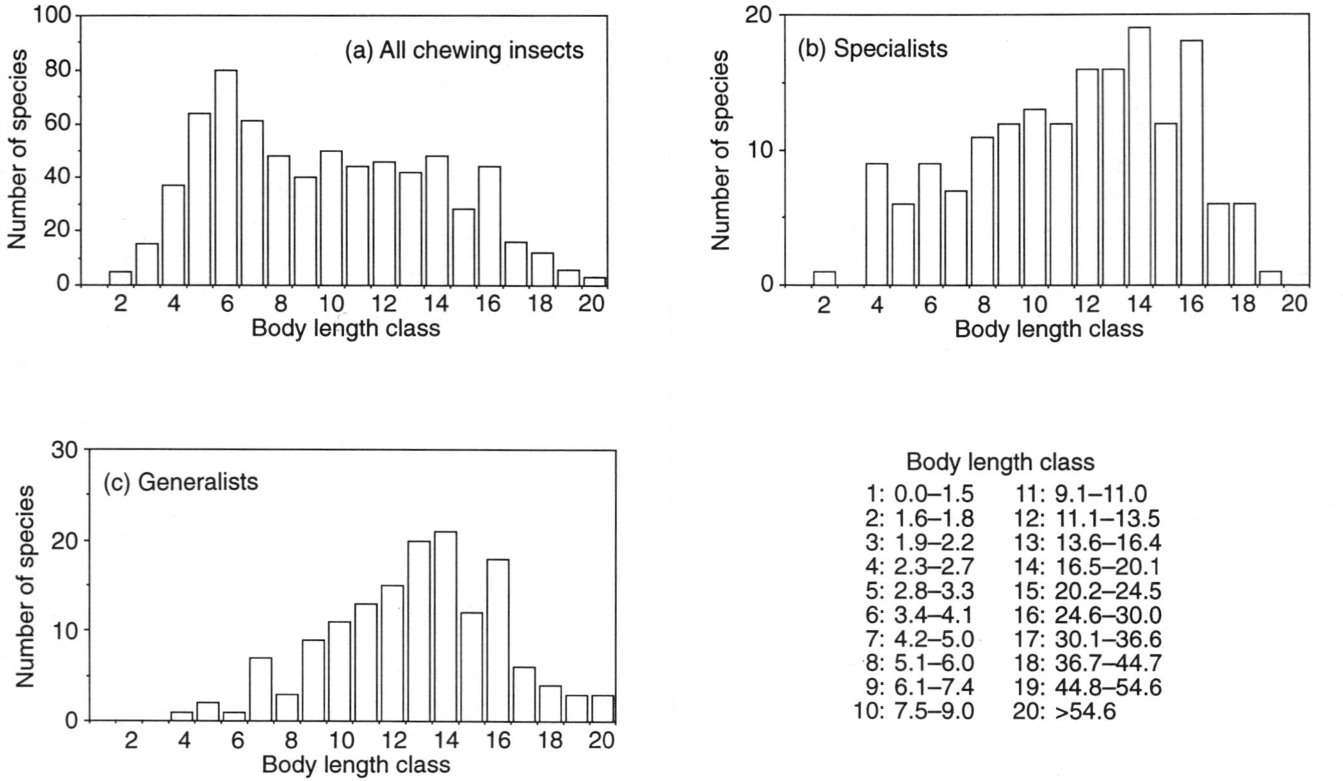

Figure 12.4 Distribution of body length within the compound community for (a) all chewing insects, (b) specialists, and (c) generalists. Body length classes are indicated in the lower right (units are in mm).

significantly smaller than generalist species (Mann–Whitney *U*-test = 15 588.5, $P < 0.01$). This result was similar when the data-set was restricted to Coleoptera ($U = 1059.0$, $P < 0.001$) but not when it was restricted to Lepidoptera ($U = 6438.0$, $P = 0.71$).

Despite significant differences in the average body length (Kruskal–Wallis, $W = 19.01$, $P < 0.05$) or body weight ($W = 30.82$, $P < 0.001$) of chewing species supported by the tree species studied (Table 12.4), these appeared unrelated to host attributes. In particular, there was no significant effect of successional status and leaf production patterns on average body length, body weight or on the skewness of distribution for species body length (Mann–Whitney *U*-tests, all tests $P > 0.11$). Similarly, no significant correlation was noted between host attributes, notably leaf nitrogen, leaf size or specific leaf weight (a measurement related to leaf toughness), and the three above variables. However, the two deciduous hosts (*Ficus* and *Cordia*) presented much less skewed distributions (Table 12.4) than the other tree species, suggesting that insect species on these trees were relatively well represented in lower body length classes. Restricting the analyses to specialists did not result in different conclusions. Most of these results suggest that the species: body length distribution of chewing insects on particular tree species is only weakly controlled by plant-related attributes.

Relationship between population abundance and body weight

As found in other studies sampling whole insect communities (Morse *et al.*, 1988), plots of body weight against abundance on logarithmic scales formed an apparent triangular area for the compound community (Figure 12.5). There was no significant relationship between body weight and abundance when all data points were considered (e.g. for all chewing insects $n = 689$, $R^2 = 0.02$, $F = 1.26$, $P = 0.26$). However, considering only proven feeders, it was possible to fit an OLS regression through the data ($n = 384$, $y = 0.60 - 0.14x$ ($\pm$ 0.053), $R^2 = 0.017$, $F = 6.55$, $P < 0.05$; corresponding RMA slope = –1.04). Still, the variance in abundance explained by body weight was very low, as found in other studies (Morse *et al.*, 1988).

Restricting attention to the upper bound of the triangular area and fitting regressions using the NUBS method, resulted in significant regressions for all chewing insects, specialists and generalists (Figure 12.5). Corresponding RMA slopes were –0.708, –0.987 and –0.848, respectively. NUBS were not significantly different between specialists and generalists ($t = 0.172$, $P > 0.50$). The NUBS calculated for proven feeders (Table 12.4, entry 'all trees') was less steep, but not significantly so, than that calculated for all chewing insects ($t = 0.24$, $P > 0.50$; $n = 384$ and $n = 689$, respectively; the corresponding RMA slope for proven feeders was

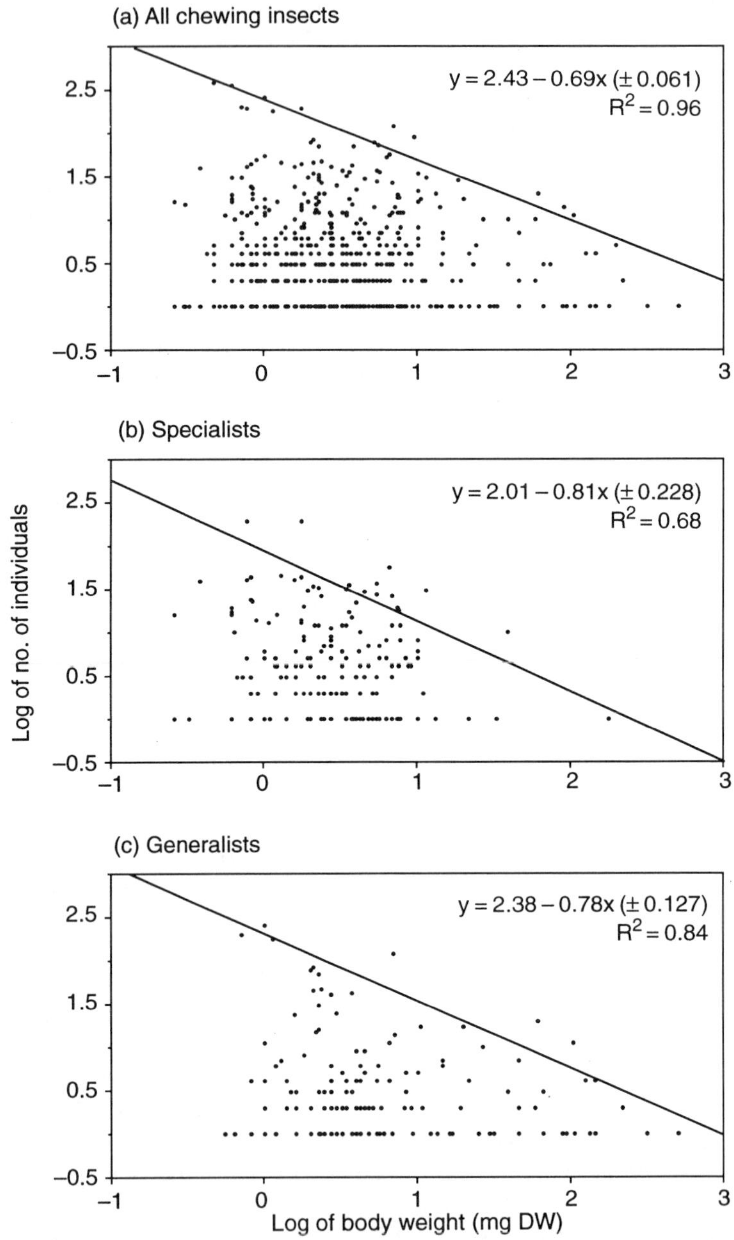

Figure 12.5 Log of body weight (mg dry weight, DW) plotted against log of number of individuals in the compound community for (a) all chewing insects, (b) specialists, and (c) generalists. The curves and the equations fitted represent the negative upper bound slope in each case, with, in brackets, the standard error of the slope.

–0.70). The high standard errors of most slopes hinder a reliable comparison of the different slopes calculated.

NUBS were also compared between the proven feeders of Coleoptera and Lepidoptera. It was possible to fit an OLS regression for the former (slope = –0.821 (± 0.119), R^2 = 0.89, F = 47.7, P <0.001, n = 86) but not for the latter (R^2 = 0.003, F = 0.015, P = 0.96, n = 274). Lepidoptera were rarer on the foliage of study trees than Coleoptera (on average about five times rarer; Mann–Whitney U = 17 806.5, P <0.001) and this, along with underestimation of the body weight of the final stage of several species, may have prevented fitting a NUBS to the lepidopteran data-set.

NUBS were also calculated for each tree species, using the data-sets of proven feeders (Table 12.4). Regressions could only be fitted to *Piper, Ficus, Pipturus, Castanopsis* and *Aleurites*. For these trees, NUBS ranged from –1.60 (*Piper*) to –0.50 (*Aleurites*) and corresponding RMA slopes ranged from –1.79 to –0.61. The steepest and shallowest slopes were significantly different from each other (t = 3.24, P <0.01). Increasing the number of body weight classes (Blackburn *et al.*, 1992) did not result in better curve fitting. Since only half of the insect communities associated with the tree species studied could be described with a NUBS, it is difficult to comment on the causative factor of these differences and why curves could be fitted on some tree species and not on others. However, a trend was noted for a positive correlation between leaf expansion rates and the NUBS (r_s = 1.00, P = 0.05, n = 5). The analyses performed in this section suggest that calculations of NUBS may be influenced by how we define the boundaries of the community under study.

DISCUSSION

The compound community

It is of interest to contrast the results obtained for the compound community and those for the different component communities. The slope fitted to the species-abundance distribution for the compound community, for numbers as well as for biomass, was often shallower than the corresponding slopes for the component communities. In particular, the proportion of rare species was significantly lower in the compound community than when component communities were considered alone. The causal effect of this observation appears to have more to do with the ecology of the insects and with the number of habitats (tree species) sampled than with the sampling methods used or the particular habitats sampled. In rainforest habitats, increasing sampling effort with a particular method may increase the proportion of rare species collected, but increasing the number of habitats sampled may decrease this proportion because this procedure increases the probability of finding the 'true' habitat of some insect species. A. Allison, G.A. Samuelson and S.E. Miller (personal communication) found a similar pattern for beetle species when fogging several individuals of *Castanopsis acuminatissima* in the Wau Valley, near the present study site (Allison *et al.*, 1997, Chapter 11, this volume).

In the present system, a high proportion of 'rare' species appear to be rare because they were presumably sampled in 'marginal' habitats in comparison with their 'true' habitats (the 'mass effect' of Shmida and Wilson, 1985). This applies to many beetles in the 'additional' category which are in fact wood-borers, to many generalist leaf-feeders, and probably also to some species in the 'incidental' category (note, in this case, that some species may be feeding on non-leafy parts of the study trees). Considering other members of the arboreal guilds (see Morse *et al.*, 1988; Basset and Kitching, 1991) instead of chewing insects, may further increase this trend because many arboreal predators, fungal-feeders and scavengers are less likely to be restricted to a particular tree species (see review in Basset, 1992), but may still exhibit some preferences (Didham, 1997, Chapter 15, this volume). In short, the proportion of rare insect species collected on the vegetation of rainforest trees may be partly explained by high habitat diversity, the fact that a certain proportion of insect species are able to exploit different habitats and by the limitation of sampling procedures; it does not automatically imply that insect populations are extremely low in rainforests.

This has important implications for our perception of the structure of insect communities associated with rainforest trees. For example, in the present system, population abundances of generalists appear to be significantly lower than those of specialists (as judged by their species-abundance distributions). However, this may only be due to the sampling artefact discussed above and this matter cannot be resolved unless the true population size of most generalist species present in the system can be estimated. With regard to studies analysing energy availability and the abundance of organisms of different size, it should be emphasized that some species which are large and appear to be rare (i.e. many generalists, as found in this study) may be found in habitats not sampled by the investigator. This will result in biased slopes (i.e. steeper slopes than the actual ones) describing species abundance patterns or 'negative upper bound slopes', because some species will fit in the 'wrong place' in the distribution. This effect is likely to be more marked for component than compound communities, providing the latter include a sufficient number of habitats. Further, this could prevent unravelling some interesting patterns regarding the way that food resources are shared within particular component communities. Clearly, we need to pay more attention to how we can define the boundaries of a particular community of arboreal insects in the tropics. For component communities this could translate to defining 'dominant' species and restricting analyses to these members of the community. Although it is tempting to discuss the NUBS determined for the compound community of the present study (and particularly the increase in steepness from the series: all chewing insects – generalists – specialists, see Figure 12.5) in relation to the energetic equivalence rule (see discussions in, for example, Damuth, 1981; Griffiths, 1992; Blackburn *et al.*, 1993a), the above considerations should serve as a cautionary note.

The component communities

Many of the relationships between species number, abundance and body size differed significantly among the different component communities studied. One must keep in mind that the interpretation of these differences may be obscured by the problems exposed in the above section. It was noted that on the deciduous tree species (*Cordia* and *Ficus*) the distribution of body size was skewed to lower body size classes. This appears consistent with the prediction of Opler (1978) that the reduced availability in time of young foliage on deciduous tree species should select for low body size of associated insects. However, rigorous tests of this observation are difficult and should be restricted to intrageneric comparisons of lepidopteran species (see the argument for root-feeding beetle larvae below). This was impossible with the present state of knowledge of the material collected. Thus, the overall distribution of species body length for chewing insects on particular tree species appeared to be only weakly influenced by host-related attributes. Two lines of explanation are possible. First, among beetles, many species (e.g. some Eumolpinae, Alticinae, Leptopiinae, Otiorhynchiinae, etc.) are root feeders as larvae and their body size as adults may not be much influenced by the availability or other characteristics of young foliage of the host on which the adults feed. Second, body size may be better explained by phylogenetic constraints and radiation of particular insect lineages on particular host-plants than by intrinsic characteristics of host-trees.

The few negative upper bound slopes which could be fitted to the relationship between population abundance and body weight are quite different from each other. Interestingly, they appear to be related to the leaf expansion rates of host-trees. Although both the low sample size and the weak correlation between the variables increase the probability that this observation may be due to chance only, it is consistent with the interaction noted between leaf expansion rates and the maximum insect biomass achieved on each tree species (see further discussion below).

Species-abundance distributions, particularly as expressed by biomass, appear more interesting in the sense that they could be predicted to some extent. First, the number of species ('dominant' or not) within the component community appears to be related to a year-long estimate of food resources; the number of young leaves produced by each tree species (Basset, 1996). Second, the maximum biomass achieved by any insect species on each tree species appears to be influenced by the leaf expansion rates of young leaves: insects were more likely to attain high biomass on young foliage which grows over prolonged periods.

Shallow slopes in log-log plots of biomass against species rank correspond to a more even distribution of biomass among members of the insect community, and, presumably, of food resources and energy. These slopes cannot be predicted from host attributes and appear more influenced by some as yet unidentified intrinsic community characteristics (probably related, to some extent, to species richness and the proportion of specialists within the community). For example, these characteristics might include the vagility of constituent species, resistance to invasion, taxonomic isolation of community members, competition, predation and historical factors. In addition, year-to-year variation in insect populations and other non-measured host attributes such as the genetic variability of the host, the fractal nature of foliage, the predictability of food resources among years and other factors influencing more directly root-feeding larvae could also influence how the resources are shared within herbivore communities.

Interestingly, slopes were, on average, shallower when the community was species-rich. Pimm (1991) emphasized that increasing species richness and connectance (i.e. the extent to which species are interconnected in food webs) both make animal communities harder to invade (see also Cotgreave and Harvey, 1994). Clearly, these relationships are complex, and simplistic models such as the geometric series do not help in this regard. Once problems inherent to the definition of the boundaries of an arboreal community of rainforest insects can be solved, stochastic niche-oriented models (Tokeshi, 1993) could be worthwhile investigating. In summary, for the present system, the abundance of food resources available to the insect community influences the insect species-richness and the maximum biomass attainable, but does not directly influence how the resources themselves are shared between community members.

To close on this matter, the influence of the successional status of the host-tree on the various relationships describing the component communities studied is not obvious. Although the effects of succession on insect communities have been well studied on a large scale, such as from fallow field to woodland (Southwood *et al.*, 1979), these effects may be more difficult to elucidate for insect communities associated with individual tree-species. For example, in the present case, some tree species were difficult to assign to a particular successional stage (e.g. *Aleurites* can be assigned either to the pioneer or to the persistent category).

CONCLUSIONS

We need to define strict rules to delimit the boundaries of particular communities of arboreal insects in the tropics. Without this, we risk analysing patterns derived from loosely defined 'assemblages' of species, with little interesting biological meaning to be gained. Perhaps this can be achieved by setting a minimum level of affinity between the presumed members of the community and the host-tree. Computer simulations could help in choosing an 'ideal' criterion and its critical value, but this goes much beyond the scope of the present paper.

When community boundaries in the present system were restricted to species which were proven to feed and which together contributed to at least 95% of insect biomass, food resources appeared much more evenly shared between the insect members of the community on certain tree species (e.g. *Caldcluvia, Castanopsis, Pipturus*), than on others (e.g. *Piper, Aleurites*). Cotgreave and Harvey (1994) examined the evenness of abundance in 90 bird communities world-wide. They showed that more species-rich communities are often more even and suggested that when niche partitioning has occurred in the evolutionary past, it has led to reduced current competition and a more even distribution of abundance. This brings to mind a possible correspondence with the classification of communities of parasites in a host into interactive and isolationist types (Holmes and Price, 1986). Perhaps a different terminology should be used here to avoid confusion. For example, 'interactive' refers originally to a type of parasite community with a high frequency of interspecific interactions (Holmes and Price, 1986). In contrast, the term has also been used to indicate that communities of arboreal arthropods could share some species and thus interact with each other (Basset, 1993; see also Gilbert, 1980). Perhaps the terms 'equitable' and 'non-equitable' could be used to characterize communities of tropical insect herbivores with particular reference to how food resources are shared. These terms are rather neutral and do not make assumptions regarding the causative factor (e.g. competition, predation, etc.) behind such a dichotomy. In analogy with Holmes and Price (1986), equitable communities of chewing insects on their tropical host-trees could be characterized as being typically species-rich, with a relatively even share of biomass among insect species, attaining high biomass per unit foliage and having reached an 'evolutionary equilibrium' (i.e. most species are well adapted to conditions and food resources). Non-equitable communities would exhibit the converse characteristics. More data on different component communities of insect herbivores associated with tropical trees could greatly help to determine the extent of this dichotomy and to refine its characteristics.

Acknowledgements

I thank Patrick Basset, Nathan Daniel, Robert Höft, Martin Hutten, Martin Kasbal, Neil Springate and George Weiblen for their help in the field, as well as Harry Sakulas and the staff of Wau Ecology Institute for everyday support. J.D. Holloway (CAB International, London), M. Horak, E.C. Zimmerman (CSIRO Division of Entomology, Canberra), S.E. Miller, G.A. Samuelson (Bishop Museum, Honolulu), R.T. Thompson, K. Tuck, K. Sattler (Natural History Museum, London) and E.G. Munroe checked my morphospecies assignments. The manuscript benefited from comments by Robert Cowie, Scott Miller, Nigel Stork and M. Tokeshi. The study was funded by the Swiss National Science Foundation.

REFERENCES

Allison, A., Samuelson, G.A. and Miller, S.E. (1993) Patterns of beetle species diversity in New Guinea rain forest as revealed by canopy fogging: preliminary findings. *Selbyana*, **14**, 16–20.

Allison, A., Samuelson, G.A. and Miller S.E. (1997) Patterns of beetle species diversity in *Castanopsis acuminatissima* (Fagaceae) trees studied with canopy fogging in mid-Montane New Guinea rainforest, in *Canopy Arthropods* (eds N.E. Stork, J. Adis and R.K. Didham), Chapman & Hall, London, pp. 224–236 (this volume).

Basset, Y. (1992) Host specificity of arboreal and free-living insect herbivores in rain forests. *Biological Journal of the Linnean Society*, **47**, 115–33.

Basset, Y. (1993) Arthropod species-diversity and component communities in the rainforest canopy: lessons from the study of an Australian tree. *Tropical Zoology, Special Issue*, **1**, 19–30.

Basset, Y. (1994) Palatability of tree foliage to chewing insects: a comparison between a temperate and a tropical site. *Acta Oecologica*, **15**, 181–91.

Basset, Y. (1996) Local communities of arboreal herbivores in Papua New Guinea: predictors of insect variables. *Ecology* (in press).

Basset, Y. and Kitching, R.L. (1991) Species number, species abundance and body length of arboreal arthropods associated with an Australian rainforest tree. *Ecological Entomology*, **16**, 391–402.

Basset, Y. , Aberlenc, H.-P. and Delvare, G. (1992) Abundance and stratification of foliage arthropods in a lowland rain forest of Cameroon. *Ecological Entomology*, **17**, 310–18.

Basset, Y., Springate, N.D. and Samuelson, G.A. (1994) Feeding habit and range of body size: a case study in Papua New Guinea using arboreal leaf-beetles and weevils (Coleoptera: Chrysomelidae, Curculionidae). *Mitteilungen der Schweizerischen Entomologischen Gesellschaft*, **67**, 347–61.

Basset, Y., Samuelson, G.A. and Miller, S.E. (1996) Similarities and contrasts in the local insect faunas associated with ten forest tree species of New Guinea. *Pacific Science*, **50**, 157–83.

Blackburn, T.M., Lawton, J.H. and Perry, J.N. (1992) A method of estimating the slope of upper bounds of plots of body size and abundance in natural animal assemblages. *Oikos*, **65**, 107–12.

Blackburn, T.M., Brown, V.K., Doube, B.M., Greenwood, J.J.D., Lawton, J.H. and Stork, N.E. (1993a) The relationship between abundance and body size in natural animal assemblages. *Journal of Animal Ecology*, **62**, 519–28.

Blackburn, T.M., Lawton, J.H. and Pimm, S.L. (1993b) Non-metabolic explanations for the relationship between body size and animal abundance. *Journal of Animal Ecology*, **62**, 694–702.

Bulla, L. (1994) An index of evenness and its associated diversity measure. *Oikos*, **70**, 167–71.

Cotgreave, P. and Harvey, P.H. (1994) Evenness of abundance in bird communities. *Journal of Animal Ecology*, **63**, 365–74.

Damuth, J. (1981) Population density and body size in mammals. *Nature*, **290**, 699–700.

Didham, R.K. (1997) Dipteran tree-crown assemblages in a diverse southern temperate rainforest, in *Canopy Arthropods* (eds N.E. Stork, J. Adis and R.K. Didham), Chapman & Hall, London, pp. 320–43 (this volume).

Gaston, K.J., Blackburn, T.M., Hammond, P.M. and Stork, N.E. (1993a) Relationships between abundance and body size: where do tourist fit? *Ecological Entomology*, **18**, 310–14.

Gaston, K.J., Blackburn, T.M. and Lawton, J.H. (1993b) Comparing animals and automobiles: a vehicle for understanding body size and abundance relationships in species assemblages? *Oikos*, **66**, 172–9.

Gilbert, L.E. (1980) Food web organization and the conservation of neotropical diversity, in *Conservation Biology* (eds M.E. Soulé and B.A. Wilcox), Sinauer, Sunderland, pp. 11–33.

Gressitt, J.L. and Nadkarni, N. (1978) *Guide to Mt. Kaindi. Background to Montane New Guinea Ecology*, Wau Ecology Institute, Handbook no. 5, Wau, Papua New Guinea.

Griffiths, D. (1992) Size, abundance, and energy use in communities. *Journal of Animal Ecology*, **61**, 307–15.

Holmes, J.C. and Price, P.W. (1986) Communities of parasites, in *Community Ecology. Pattern and Process* (eds J. Kikkawa and D.J. Anderson), Blackwell, Oxford, pp. 187–213.

Johns, R.B. (1982) Plant zonation, in *Biogeography and Ecology of New Guinea* (ed. J.L. Gressitt), Junk, The Hague, pp. 309–30.

Kitching, R.L. and Arthur, M. (1993) The biodiversity of arthropods from Australian rain forest canopies: summary of projects and the impact of drought. *Selbyana*, **14**, 29–35.

Magurran, A.E. (1988) *Ecological Diversity and its Measurement*, Croom Helm, London.

McAlpine, J.R., Keig, G. and Falls, R. (1983) *Climate of Papua New Guinea*, CSIRO, Canberra.

Morse, D.R., Stork, N.E. and Lawton, J.H. (1988) Species number, species abundance and body length relationships of arboreal beetles in Bornean lowland rain forest trees. *Ecological Entomology*, **13**, 25–37.

Opler, P.A. (1978) Interaction of plant life history components as related to arboreal herbivory, in *The Ecology of Arboreal Folivores* (ed. G.G. Montgomery), Smithsonian Institution Press, Washington, pp. 23–31.

Perry, D.R. (1978) A method of access into the crowns of emergent and canopy trees. *Biotropica*, **10**, 155–7.

Pimm, S.L. (1991) *The Balance of Nature? Ecological Issues in the Conservation of*

Species and Communities, The University of Chicago Press, Chicago.

Price, P.W. (1992) The resource-based organization of communities. *Biotropica*, **24**, 273–82.

Root, R.B. (1973) Organization of a plant–arthropod association in simple and diverse habitats: the fauna of collards (*Brassica oleracea*). *Ecological Monographs*, **43**, 95–124.

Schoener, T.W. (1980) Length–weight regressions in tropical and temperate forest-understorey insects. *Annals of the Entomological Society of America*, **73**, 106–9.

Schoener, T.W. and Janzen, D.H. (1968) Notes on environmental determinants of tropical versus temperate insect size patterns. *The American Naturalist*, **102**, 207–24.

Shmida, A. and Wilson, M.V. (1985) Biological determinants of species diversity. *Journal of Biogeography*, **12**, 1–20.

Southwood, T.R.E., Brown, V.K. and Reader, P.M. (1979) The relationships of plant and insect diversities in succession. *Biological Journal of the Linnean Society*, **12**, 327–48.

Springate, N.D. and Basset, Y. (1996) Diel activity of arboreal arthropods associated with Papua New Guinean trees. *Journal of Natural History*, **30**, 101–12.

Stork, N.E. (1987) Guild structure of arthropods from Bornean rain forest trees. *Ecological Entomology*, **12**, 69–80.

Stork, N.E. (1991) The composition of the arthropod fauna of Bornean lowland rain forest trees. *Journal of Tropical Ecology*, **7**, 161–80.

Stork, N.E. and Blackburn, T. M. (1993) Abundance, body size and biomass of arthropods in tropical forest. *Oikos*, **67**, 483–9.

Tokeshi, M. (1990) Niche apportionment or random assortment: species abundance patterns revisited. *Journal of Animal Ecology*, **59**, 1129–46.

Tokeshi, M. (1993) Species abundance patterns and community structure. *Advances in Ecological Research*, **24**, 111–86.

Valkenburg, J.L.C.H. van and Ketner, P. (1994) Vegetation changes following human disturbance of mid-montane forest in the Wau area, Papua New Guinea. *Journal of Tropical Ecology*, **10**, 41–54.

13

Arthropod biodiversity in the canopy of New Caledonian forests

E. Guilbert

ABSTRACT

Canopy arthropod communities in two primary forests in New Caledonia were sampled by insecticide fogging. One site is located in dense sclerophyllous forest (Pindaï) and the other in dense evergreen forest in Rivière Bleue Provincial Park. This study describes the community composition at each site from season to season. The two sites are characterized here by five families: Clubionidae, Theridiidae (Araneae) and Formicidae at Pindaï, and Cecidomyiidae and Chironomidae (Diptera) at Rivière Bleue. The sites show further functional characteristics when the families are rearranged into trophic guilds; tourists characterize Rivière Bleue forest and predators, including ants, characterize Pindaï forest.

INTRODUCTION

Canopy arthropod communities were little understood until the middle of this century because of problems with access. They have attracted increasing interest in the last 30 years due to the canopy fogging technique first used by Martin (1966). Such communities are now sampled world-wide using this method (Martin, 1966; Gagné, 1979; Wolda, 1979; Erwin, 1982, 1983, 1989; Southwood *et al.*, 1982; Hijii, 1983, 1986; Stork, 1987b; Basset, 1990, 1991). Samples of arthropods from canopy fogging have revealed an unexpected richness of canopy arthropods and have provided the basis for new estimates of the number of all living species on Earth (Erwin, 1982; Stork, 1988).

Canopy Arthropods. Edited by N.E. Stork, J. Adis and R.K. Didham. Published in 1997 by Chapman & Hall, London. ISBN 0 412 74900 9

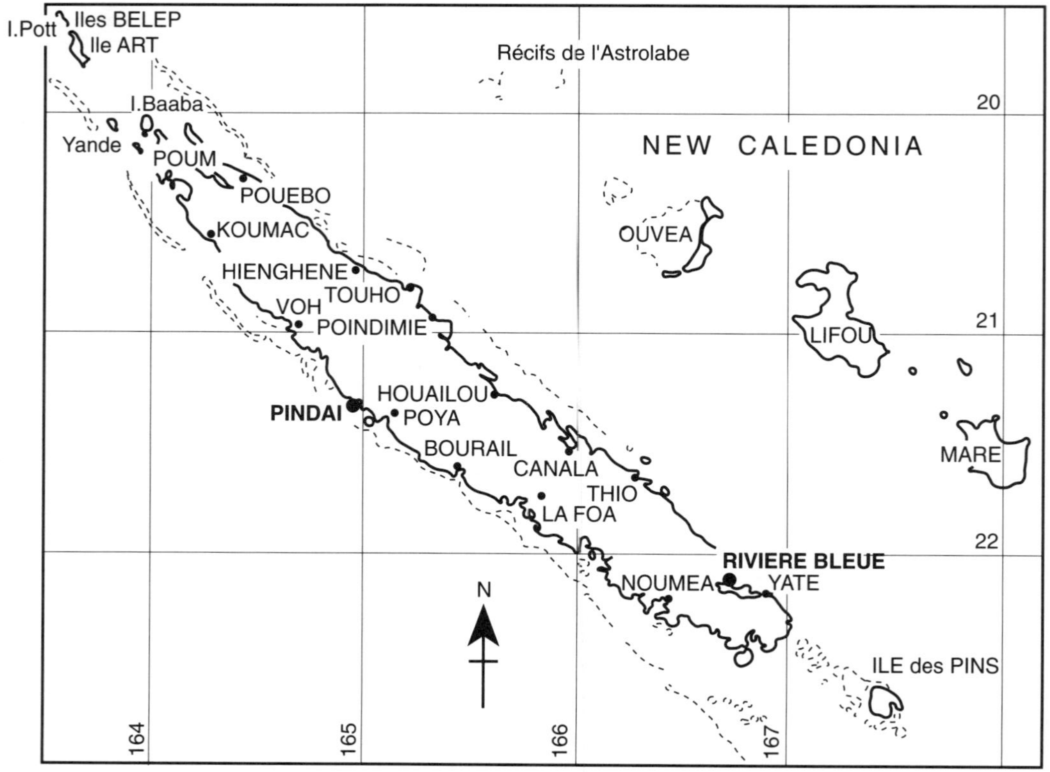

Figure 13.1 Map of New Caledonia, showing the sites where canopy arthropods were sampled: Rivière Bleue dense evergreen forest and Pindaï sclerophyll forest.

New Caledonia is one of the 10 biodiversity 'hot spots' defined by Myers (1988). Its insect fauna is partly known from several collections made since Montrouzier in 1860 (Cohic, 1950). Several different research programmes have contributed to the knowledge of its flora and fauna and New Caledonia is now a priority of the French national programme 'Diversitas' 1993–1995. However, the insect fauna of the canopy is still largely unknown.

In this study, canopy arthropods were sampled by knockdown insecticide fogging at two sites of different forest type. The aim of this paper is to identify the principal characteristics of community composition at the different sites, both at the family level and by trophic guild analysis.

SAMPLING SITES

Canopy arthropods were sampled at two sites in New Caledonia (Figure 13.1). One is located on the West coast, close to the sea, in the relict sclerophyll forest of Pindaï (North Province, 30 m above sea level) on limestone and conglomerates. The other is located in dense evergreen forest to the north of the Grand Lac de Yaté, in Rivière Bleue Provincial Park (South Province, 160 m a.s.l.) on ultramafic alluvium. These two sites have very different vegetation (Jaffré and Veillon, 1990; Jaffré *et al.*, 1993) and prevailing climate, especially rainfall.

A portable fogging machine (Dynafog Golden Eagle Backpack 2980) was used to release a fast-killing pyrethrin-based fog (Cyfluthrin, water and polyhydric alcohols). The fogger was manipulated from the ground with the insecticide fog rising into the trees. The maximum height of the canopy is around 25 m in Rivière Bleue and 10–15 m in Pindaï. Both sites were fogged early in the morning.

The samples were collected on 40 randomly placed collecting trays organized in four neighbouring plots of 10 trays each. The trays were 1 m^2 white plastic sheets raised 0.5–0.8 m above the ground to guarantee a horizontal surface. The total sampling site was 350–400 m^2 of forest and each plot covered 30–40 m^2 in area. At each site all trees were individually fogged for 4 minutes and, hence, each site was fogged for about 20–30 minutes in total. A 2-hour drop time was allowed subsequent to fogging. After this time arthropods which had fallen onto the sheets were collected by washing the sheets with water and a wetting agent (tipol) and filtering the liquid through a double screen mesh (0.6 and 0.3 mm). Samples were stored in 95% alcohol and later sorted to taxonomic order. The most speciose groups were sorted to family: Araneae, Coleoptera, Diptera, Hemiptera, Hymenoptera and Orthoptera. All arthropod taxa were assigned to trophic guilds as defined by Moran and Southwood (1982) and Stork (1987a).

Table 13.1 Weather conditions during sampling at Rivière Bleue and Pindaï: date, temperature, humidity and monthly rainfall. Temperature and humidity were taken on the day of sampling. Monthly rainfall is the total rainfall in the month of sampling.

*Samples**	*Date*	*Temperature (°C)*	*Humidity (%)*	*Rainfall (mm)*
R. Bleue 1	16.07.92	11	94	342
R. Bleue 2	21.10.92	13	96	411
R. Bleue 3	21.01.93	22	94	192.7
R. Bleue 4	08.05.93	16	96	100.8
Pindaï 1	30.06.92	15	92	5.4
Pindaï 2	06.10.92	14	92	12.8
Pindaï 3	05.01.93	20	94	31.2
Pindaï 4	14.04.93	17	90	6

* 1, 2, 3 and 4 represent seasons.

Each site was sampled four times a year to cover seasonal variation (see Table 13.1 for dates and climatic data).

RESULTS

A total of 174 103 individuals belonging to 30 orders or other higher taxa were collected from the eight fogs (Table 13.2). Some 37% were collected in the rainforest of Rivière Bleue and 63% in Pindaï sclerophyll forest. This corresponds to an average density of 399.56 and 688.58 individuals/m^2 for the two sites, respectively. Total numbers of individuals varied from season to season. More than half of the specimens collected were accumulated between April and May. Seasonal variations were found to be higher at Pindaï, as 60% of the collection from this site was amassed between April and May. A total of 194 families were collected, of which 130 were found to be common to both sites. There were 161 and 159 families from Rivière Bleue and Pindaï, respectively, 25% of which were Coleoptera. Diptera, Hemiptera and Hymenoptera made up around 20% of the families each. Araneae numbered 14% and Orthoptera only 2% of the total number of arthropod families found. Results of abundance among the different orders are shown in Appendix 13A.

Collembola were prevalent at Rivière Bleue, comprising 37% of all individuals at this site, whereas they represented only 13.2% at Pindaï. In contrast, Psocoptera were prevalent at Pindaï (28.4%), but represented only 4.8% at Rivière Bleue. Diptera and Acarina were also well represented at Rivière Bleue (14.1% and 11.4% of total abundance,

Table 13.2 Abundance of arthropod taxa at Rivière Bleue and Pindaï, all seasons combined

Taxa	*R. Bleue*	*Pindaï*
Acarina	7291	9130
Amphipoda	11	0
Araneae	1671	6730
Chilopoda	6	48
Coleoptera	4633	4443
Collembola	23652	14557
Dermaptera	6	0
Dictyoptera	254	867
Diptera	8994	4959
Embioptera	1	0
Ephemeroptera	14	0
Hemiptera	2807	10290
Hymenoptera	4463	17152
Isopoda	7	22
Lepidoptera	78	483
Megaloptera	33	62
Mollusca	11	11
Neuroptera	33	215
Opilionida	1	0
Orthoptera	208	310
Phasmida	9	27
Pseudoscorpionida	61	557
Psocoptera	3076	31323
Raphidioptera	1	0
Strepsiptera	1	0
Symphipleona	1	5
Thysanura	0	75
Thysanoptera	5653	7674
Trichoptera	167	0
Insect larvae	787	1233
Total	63930	110173

respectively), and Hymenoptera were well represented in Pindaï (15.6%). Thysanoptera were well represented at both sites (8.8% at Rivière Bleue and 7.0% at Pindaï). Other orders such as Coleoptera, Hemiptera and Araneae showed lower percentages.

All of these groups displayed seasonal variation at both sites as shown in Figure 13.2. Collembola were prevalent at Rivière Bleue, but only between January and May (28.2% and 67.0%). Diptera was the prevalent

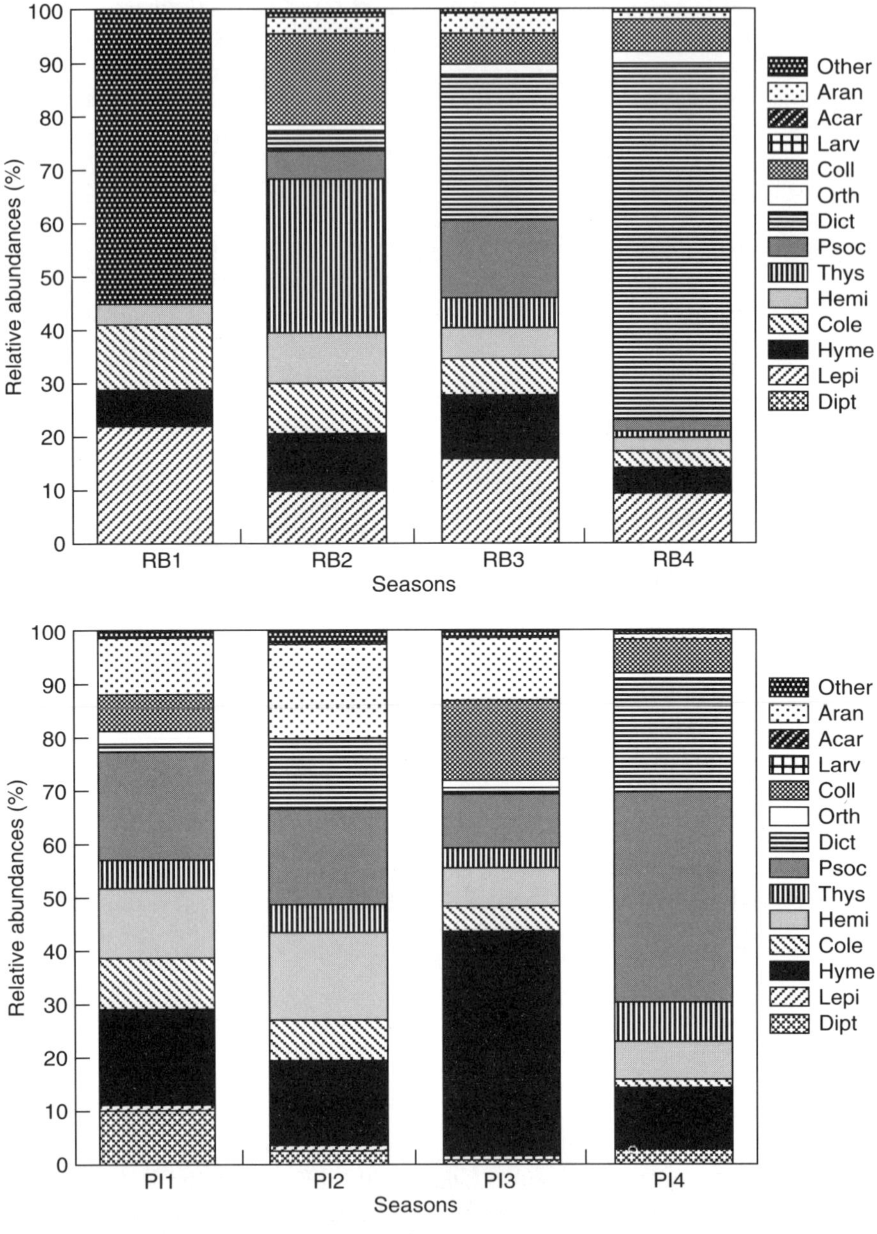

Figure 13.2 Relative abundance of different orders collected at Rivière Bleue (RB) and Pindaï (PI) in each of the four seasons (see Table 13.1 for details). Dipt, Diptera; Lepi, Lepidoptera; Hyme, Hymenoptera; Cole, Coleoptera; Hemi, Hemiptera; Thys, Thysanoptera; Psoc, Psocoptera; Dict, Dictyoptera; Orth, Orthoptera; Coll, Collembola; Larv, Larvae; Acar, Acarina; Aran, Araneae.

Order in July (22.6%), while Thysanoptera prevailed in October (29.3%). At Pindaï, Psocoptera were the most prevalent group in May, but Hymenoptera were prevalent in January. These two orders were both prevalent at Pindaï in October (around 18% each), while Araneae dominated in July.

Araneae

Clubionidae (23.3% and 35.8% of Araneae at Rivière Bleue and Pindaï, respectively) and Theridiidae (23.3% and 34.4%, respectively) were the prevalent families of Araneae at both sites. The Salticidae were well represented at Rivière Bleue (11.7%) whereas they represented only 3.5% at Pindaï. Although the Philodromidae represented 9.5% at Pindaï, there was only a single specimen collected at Rivière Bleue. Oonopidae were quite abundant at both sites with 4.9% and 7.5% at Rivière Bleue and Pindaï, respectively. Other families represented less than 5% each of total Araneae abundance. Most of these families show variation in abundance between seasons.

Coleoptera

Curculionidae were prevalent at both sites (19.3% and 15.8% at Rivière Bleue and Pindaï, respectively). Chrysomelidae and Corylophidae were also well represented with percentages between 6.9% and 11.7%. Staphylinidae and Pselaphidae were well represented at Rivière Bleue (15.2% and 8.8%, respectively), but represented no more than 2.0% at Pindaï. In contrast, Phalacridae (11.0%) and Coccinellidae (8.2%) were better represented at Pindaï than at Rivière Bleue, as well as Ciidae, Scolytidae and Cerambycidae, but with lower percentages. The other numerous families did not exceed 4.0% in total. All these Families showed higher seasonal variation in abundance than the spider Families.

Diptera

There were more families of Brachycera than Nematocera (30 as opposed to 12), but the most abundant dipteran families were Nematocera, in particular Chironomidae, Ceratopogonidae, Cecidomyiidae and Sciaridae. At Rivière Bleue, their percentages varied from 22.3% to 28.8%, except Sciaridae with 6.7%. At Pindaï, Chironomidae (33.5%) and Ceratopogonidae (20.3%) were prevalent, while Cecidomyiidae and Sciaridae were less important (9.4% and 7.8%, respectively). Chloropidae were the only important brachyceran family at both sites (8.9% and 3.6% at Pindaï and Rivière Bleue, respectively), while Stratiomyidae was prevalent at Pindaï (5.0%) and Dolichopodidae at Rivière Bleue (2.8%). Most other families showed percentages lower than 1.0% of total abundance (mostly Brachycera). Seasonal variation was not great at Rivière Bleue compared with that at Pindaï.

Hemiptera

Prevalent families at the two sites were Cicadellidae and Psyllidae. However, Cicadellidae represented 11.2% at Rivière Bleue and 35.3% at Pindaï, while Psyllidae represented 22.3% and 19.5% at these sites. Heteroptera were better represented at Rivière Bleue than at Pindaï; for example, Miridae (6.4% at Rivière Bleue, 3.1% at Pindaï). Nevertheless, there were a lot of juveniles of Heteroptera of which the family was not determined, hence heteropteran families were perhaps more important than these figures suggest. All dominant families showed high seasonal variation, except Cicadellidae.

Hymenoptera

Formicidae were prevalent at both sites, representing 71.2% of Hymenoptera at Pindaï and 20.7% at Rivière Bleue. Consequently, in percentage terms, other hymenopteran families were poorly represented at Pindaï. Aphelinidae, which represented 16.5% of hymenopteran abundance at Rivière Bleue, represented just 5.7% at Pindaï, but was actually more numerous. Platygasteridae, Encyrtidae, Trichogrammatidae, Eulophidae and Sphecidae were also well represented at Rivière Bleue, but with low percentages (6.0–8.0%). Some of these latter families were more abundant at Pindaï, but showed lower percentages, except Sphecidae, which was more abundant at Rivière Bleue. If one excludes Formicidae, the dominant families were essentially chalcidoids. The Aphelinidae, Eulophidae and Trichogrammatidae showed quite similar proportions at both sites (20.3%, 14.9% and 7.6%, respectively).

Trophic guilds

Higher taxa may be arranged by trophic guild in order to give a functional view of community composition.

The most abundant trophic guild at both sites was the epiphyte grazers (38.7% and 38.3% at Pindaï and Rivière Bleue, respectively), comprising Psocoptera and Collembola. Predators, sap suckers and parasites represented 14.7%, 13.8% and 13.0%, respectively, at Pindaï and 13.0%, 11.5% and 13.0%, respectively, at Rivière Bleue. The most abundant predators were Acarina and Araneae (Clubionidae and Theridiidae). Sap suckers were represented by Thysanoptera and some Hemiptera (Cicadellidae and Psyllidae), especially at Pindaï. Parasites were mostly Hymenoptera

(Scelionidae and Chalcidoidea, Trichogrammatidae) at both sites, Aphelinidae at Rivière Bleue and Encyrtidae at Pindaï. Ants represented 10.2% at Pindaï, but less than 1.0% at Rivière Bleue. In contrast, tourists represented 11.6% at Rivière Bleue, but only 3.7% at Pindaï. Most of the tourists were Diptera, Nematocera such as Ceratopogonidae, Chironomidae and Cecidomyiidae. The other guilds did not exceed 4.0% at each site. Chewers were mostly Chrysomelidae and Curculionidae at both sites. There were also Scolytidae, Cerambycidae, Acrididae and Lepidoptera at Pindaï. Scavengers were mostly represented by Corylophidae, Sciaridae and Dictyoptera. Insect predators were represented by Staphylinidae, Dolichopodidae and Sphecidae at Rivière Bleue and by Coccinellidae, Colydiidae and Sphecidae at Pindaï. Some taxa were not assigned to any guilds, they were mostly juveniles of Hemiptera and larvae.

As with the different families, trophic guilds showed seasonal variation in proportional representation. For example, epiphyte grazers, which were the dominant guild at both sites, numbered only 6.0% of the individuals at Rivière Bleue in October, while at Pindaï they represented 10.2% and 8.7% in October and January, respectively. In contrast, they showed high abundances in May at both sites (69.0% and 55.3% at Rivière Bleue and Pindaï, respectively).

DISCUSSION AND CONCLUSIONS

There were differences in the proportions of families represented at the two sites and these differences defined the principal characteristics of community composition in the two forest types. For example, Formicidae, Clubionidae and Theridiidae were better represented in the sclerophyll forest of Pindaï, and Cecidomyiidae, Ceratopogonidae and Chironomidae were better represented in the rainforest of Rivière Bleue. Similarly, some families showed greater seasonal variation than others. For example, Agaonidae, Cecidomyiidae and Psyllidae showed greatest variation between July and January and Formicidae, Clubionidae and Theridiidae between October and May, at Pindaï. Other families showed similar but less dramatic variation between sites and seasons, suggesting that they may have less impact on group composition at each site and in each season. Thus, the dominant families characterize the sites: Ceratopogonidae, Cecidomyiidae and Chironomidae characterize Rivière Bleue rainforest, which is subjected to frequent and heavy rains, and Formicidae, Clubionidae and Theridiidae characterize Pindaï sclerophyll forest, which is dry and subjected to greater climatic stress.

Similarly, some trophic guilds characterized the different sites and seasons. Over the total sample period tourists were better represented at Rivière Bleue than at Pindaï, while predators, including ants, were better represented at Pindaï. However, epiphyte grazers were more abundant in May and characterized this season at both sites. The epiphyte grazer guild was dominated by Psocoptera and Collembola which were more abundant at Rivière Bleue and Pindaï at this time.

Acknowledgements

I am indebted to J. Chazeau for field work assistance and to J. Najt and S. Tillier for their support. This study is part of the Muséum National d'Histoire Naturelle programme: 'Biodiversité terrestre en Nouvelle-Calédonie' and the ORSTOM programme: 'Caractérisation des peuplements des forêts et maquis non anthropisés'.

REFERENCES

Basset, Y. (1990) The arboreal fauna of the rainforest tree *Argyrodendron actinophyllum* as sampled with restricted canopy fogging: composition of the fauna. *The Entomologist*, **109**(3), 173–83.

Basset, Y. (1991) The taxonomic composition of the arthropod fauna associated with an Australian rainforest tree. *Australian Journal of Zoology*, **39**, 171–90.

Cohic, F. (1950) Activité saisonnière de la faune économique néocalédonienne. *Revue Française d'Entomologie*, **17**, 83–7.

Erwin, T.L. (1982) Tropical forests: their richness in Coleoptera and other species. *The Coleopterist's Bulletin*, **36**, 74–5.

Erwin, T.L. (1983) Tropical forest canopies: the last biotic frontier. *Bulletin of the Entomological Society of America*, **29**, 14–19.

Erwin, T.L. (1989) Canopy arthropod biodiversity: a chronology of sampling techniques and results. *Revista Peruana de Entomologia*, **32**, 71–7.

Gagné, W.C. (1979) Canopy-associated arthropods in *Acacia koa* and *Metrosideros* tree communities along an altitudinal transect on Hawaii island. *Pacific Insects*, **21**(1), 56–82.

Hijii, N. (1983) Arboreal arthropod fauna in a forest: seasonal fluctuations in density, biomass, and faunal composition in a *Chamaecyparis obtusa* plantation. *Japanese Journal of Ecology*, **33**, 435–44.

Hijii, N. (1986) Density, biomass, and guild structure of arboreal arthropod as related to their inhabited tree size in a *Cryptomeria japonica* plantation. *Ecological Research*, **1**, 97–118.

Jaffré, T. and Veillon, J.-M. (1990) Etude floristique et structurale de deux forêts denses humides sur roches ultrabasiques en Nouvelle Calédonie. *Adensonia*, **12**(3–4), 243–73.

Jaffré, T., Morat, P. and Veillon, J.-M. (1993) Etude floristique et phytogéographique de la forêt sclérophylle de Nouvelle Calédonie. *Adensonia*, **15**(1–4), 107–46.

Martin, J.L. (1966) The insect ecology of red pine plantations in central Ontario. IV. The crown fauna. *Canadian Entomologist*, **98**, 10–27.

Moran, V.C. and Southwood, T.R.E. (1982) The guild composition of arthropod communities in trees. *Journal of Animal Ecology*, **51**, 289–306.

Myers, N., (1988) Threatened biotas: 'hot spots' in tropical forests. *The Environmentalist*, **8**, 187–208.

Southwood, T.R.E., Moran, V.C. and Kennedy, C.E.J. (1982) The richness, abun-

dance and biomass of the arthropod communities on trees. *Journal of Animal Ecology*, **51**, 635–49.
Stork, N.E. (1987a) Guild structure of arthropods from Bornean rain forest trees. *Ecological Entomology*, **12**, 69–80.
Stork, N.E. (1987b) Arthropod faunal similarity of Bornean rain forest trees. *Ecological Entomology*, **12**, 219–26.
Stork, N.E. (1988) Insect diversity: facts, fiction and speculation. *Biological Journal of the Linnean Society*, **35**, 321–37.
Wolda, H. (1979) Abundance and diversity of Homoptera in the canopy of a tropical forest. *Ecological Entomology*, **4**, 181–90.

Appendix 13A Abundance by family of the six most diverse arthropod orders at Rivière Bleue and Pindaï, all seasons combined

Araneae

Families	*R. Bleue*	*Pindaï*
Araneae	56	178
Clubionidae	643	2421
Ctenidae	13	11
Filistatidae	0	3
Gnaphosiidae	59	37
Hahniidae	2	3
Linyphiidae	57	7
Micropholcommatidae	2	3
Mimetidae	5	2
Oonopidae	78	512
Philodromidae	1	645
Salticidae	187	302
Segetriidae	3	16
Symphytognatidae	14	0
Tetragnathidae	18	29
Theridiidae	371	2327
Thomisidae	37	217
Uloboridae	36	19
Undetermined	12	36
Total	1594	6768

Coleoptera

Families	*R. Bleue*	*Pindaï*
Aderidae	144	32
Anobiidae	2	1
Anthicidae	20	0
Anthribidae	111	220
Apionidae	11	0
Attelabidae	0	29
Bostrychidae	0	1
Bruchidae	0	149
Byrrhidae	101	0
Cantharidae	14	5
Carabidae	2	0
Cerambycidae	63	323
Cerylonidae	0	9
Chrysomelidae	374	309
Ciidae	72	228
Cleridae	10	6
Coccinellidae	150	368
Colydiidae	102	118
Corylophidae	428	528
Cucujidae	111	36
Curculionidae	895	713
Dermestidae	29	38
Elateridae	6	3
Endomychidae	50	0
Erotylidae	10	5
Eucnemidae	1	0
Languriidae	150	99
Lathridiidae	108	147
Melandryidae	73	1
Melyridae	1	1
Merophisiidae	156	2
Mordellidae	48	10
Mycetophagidae	0	1
Nitidulidae	3	124
Oedemeridae	0	1

Appendix 13A continued

Families	*R. Bleue*	*Pindaï*
Phalacridae	11	494
Phloeophilidae	2	0
Phycosecidae	0	4
Propalticidae	8	4
Pselaphidae	407	21
Ptiliidae	9	2
Salpingidae	24	9
Scarabaeidae	0	5
Scolytidae	125	232
Scraptiidae	50	6
Scydmaenidae	14	41
Staphylinidae	703	77
Sylvanidae	5	0
Tenebrionidae	25	94
Trogrossitidae	0	6
Total	4628	4502

Diptera: Nematocera

Families	*R. Bleue*	*Pindaï*
Bibionidae	24	0
Cecidomyiidae	2004	463
Ceratopogonidae	2171	1652
Chaeboridae	1	15
Chironomidae	2599	1003
Culicidae	16	5
Mycetophilidae	25	77
Psychodidae	59	9
Scatopsidae	41	0
Sciaridae	609	384
Simulidae	10	4
Tipulidae	79	102
Total	7638	3714

Diptera: Brachycera

Families	*R. Bleue*	*Pindaï*
Agromyzidae	20	12
Anthomyzidae	43	89
Asilidae	8	2
Asteiidae	15	2
Bombyliidae	0	1
Calliphoridae	1	11
Chloropidae	321	440
Chyromyidae	1	0
Conopidae	0	1
Cryptochetidae	3	14
Dolichopodidae	252	18
Drosophilidae	100	11
Empididae	123	16
Helcomyzidae	1	1
Heleomyzidae	0	4
Lauxaniidae	173	174
Lonchopteridae	0	7
Milichiidae	33	30
Muscidae	33	6
Phoridae	44	28
Pipunculidae	114	21
Sarcophagidae	0	18
Sciomyzidae	14	20
Sphaeroceridae	1	0
Stratiomiidae	26	246
Syrphidae	4	2
Tachinidae	12	24
Tephritiidae	8	20
Therevidae	1	3
Undetermined	12	0
Total	2363	1221

Hemiptera: Homoptera

Families	*R. Bleue*	*Pindaï*
Achilidae	44	8
Aleyrodiidae	122	275
Aphididae	4	14
Formes aptères	297	918
Cicadellidae	313	3623
Cicadidae	0	2
Cixiidae	79	98
Delphacidae	41	226
Derbiidae	18	0
Diaspidiidae	37	104
Eurymelidae	1	0

Appendix 13A continued

Families	*R. Bleue*	*Pindaï*
Flattidae	17	38
Fulgoroidea	196	367
Machaerotidae	57	0
Margarodidae	1	87
Meenoplidae	15	0
Nogodinidae	19	20
Psyllidae	621	2007
Ricanidae	0	7
Total	1882	7794

Hemiptera: Heteroptera

Families	*R. Bleue*	*Pindaï*
Antochoridae	6	270
Aradidae	59	194
Coreidae	20	2
Disparidae	0	1
Lygaeidae	95	94
Miridae	179	323
Nymphs	382	1050
Pentatomidae	11	38
Plataspidae	0	2
Reduviidae	19	53
Schizopteridae	5	1
Tessaratomidae	3	0
Tingidae	58	455
Undetermined	6	0
Total	906	2483

Hymenoptera

Families	*R. Bleue*	*Pindaï*
Agaonidae	213	39
Aphelinidae	740	977
Austroniidae	0	1
Bethylidae	0	23
Braconidae	156	263
Ceraphronidae	106	31
Chalcididae	2	0
Chrysididae	16	44
Diaspididae	62	0
Dryinidae	3	2
Elasmidae	3	17
Embolemidae	1	0
Encyrtidae	296	807
Eucharitidae	14	11
Eulophidae	358	464
Eupelmidae	96	48
Eurymelidae	6	9
Figitidae	1	0
Formicidae	927	12240
Gasteruptiidae	1	0
Heloridae	0	1
Ichneumonidae	44	19
Myrmaridae	196	516
Perilampidae	7	18
Platygasturidae	266	226
Pompilidae	0	2
Proctotrupidae	4	0
Pteromalidae	101	132
Scelionidae	202	781
Siricidae	1	0
Sphecidae	350	104
Stephanidae	2	0
Tetracamidae	2	1
Torymidae	33	31
Trichogrammatidae	271	377
Undetermined	0	2
Total	4480	17186

Part Three

Community Structure of Non-Coleopteran Assemblages

14

Diversity of an Amazonian canopy grasshopper community in relation to resource partitioning and phylogeny

C. Amedegnato

ABSTRACT

Advances in the inventory of Amazonian canopy grasshoppers and knowledge of their systematics allow an understanding of some of the determinants of their diversity in relation to space and food partitioning. A sample of the canopy grasshopper fauna from 60 trees of an upland rainforest has been analysed. This analysis has been considered in the comparative framework of samples taken from the same region, as well as from other regions, in the Amazon basin, in both forest and secondary-growth formations. The results obtained show the following: (i) the community is organized, in density and diversity, by the composition and structure of the vegetation, giving rise to stable faunal nuclei (one-half to two-thirds of the species, dominating the canopy) which are controlled by the local plant association and are surrounded by more variable canopy species; (ii) plant families are strongly divergent in their ability to sustain grasshopper populations, with the most abundant plant families more strongly determining the overall organization of the grasshopper community; (iii) the diversification of taxonomic groups relative to space and food exploitation gives rise to functional guilds, a first definition of which is given; and (iv) taxonomic groups are not equivalent in terms of the modalities of their canopy occupation, nor in terms of their participation in guilds, this being linked to behavioural preadaptations and to the historical development of ecosystems during recent climatic variations, as well as to the duration of their establishment in the South American biota (historic and phyletic components of diversity).

Canopy Arthropods. Edited by N.E. Stork, J. Adis and R.K. Didham. Published in 1997 by Chapman & Hall, London. ISBN 0 412 74900 9

INTRODUCTION

Tropical forests have long been attractive to naturalists. However, it is only recently that the canopy has been intensively prospected, using a variety of methods (Roberts, 1973; Adis *et al.*, 1984; Hammond, 1990). Even so, the ignorance of canopy biodiversity, and especially community organization, remains great. Even though the diversity and abundance of insects are still very poorly known, their determinants are nevertheless one of the primary topics in tropical ecology.

Grasshoppers are a well-known, widely ranging and abundant group in all grassy or bushy tropical communities. In the Neotropics, they display a high level of diversity in the canopy, their main life-zone in the forest. Typical arboreal adaptations are known from all continents, particularly Madagascar (Dirsh, 1962; Descamps and Wintrebert, 1966; Dirsh and Descamps, 1968) and south-east Asia (Riede, 1993), including posterior tarsal elongation, concavity of the sternum, tibial and femoral modifications and epi- or endophytic egg pods. However, they are especially important in tropical America.

Despite its great richness, the neotropical canopy acridofauna has been intensively studied only during the past two decades. The number of known acridid species is currently approximately 2200, distributed in 520 genera, as compared with the 900 previously described (Descamps, 1972). Approximately 850 of these are canopy or subcanopy species, hence, at current estimate, almost 40% of the neotropical acridid species are arboreal. What is intriguing in the Amazon forest canopy is the high number of very closely related sympatric species. Sometimes they are morphologically indistinguishable, yet live on the same tree. For example, nine species of Ophthalmolampae are found at one site in the north-west Amazon, with the number of species cohabiting on the same tree varying between two and four. Consequently there is direct confrontation with the problem of the diversity of the canopy acridofauna, of ascertaining its determinants and of understanding its possible organization.

Generally speaking, all facets of insect diversity (systematic, functional, biogeographical and ecological) are not always well understood. However, various studies carried out at different sites in the Amazon basin now allow insights into some ecological and geographical aspects of acridid diversity and provide a general framework within which to record specific results. It is hoped that these new data on one of the most important phyllophagous groups in the forest canopy will help to define more precisely studies that should be undertaken in order to progress in the study of canopy arthropod faunas.

METHODS

In the study of canopy grasshoppers, the greatest difficulties consisted of: (i) obtaining the material to study; and (ii) understanding the forest mosaic and the spatial stratification of insects. After considerable preliminary experimentation, a set of three complementary practices was determined.

Practice 1

It was opted for a complete separation of what were already considered relatively discrete biotic compartments of the forest canopy, such as palm trees (C. Amedegnato and S. Poulain, unpublished data), and of which only the general results will be mentioned.

Practice 2

For the other components of the canopy, sampling was undertaken by the felling of trees, at random, as well as by forest clearance, according to the specific forest type.

In a first comparative study, analysing samples obtained by the felling of forest hectares (surfaces felled by Indians for their 'chacras') and samples obtained from isolated trees, the latter method was found to be preferable. It provided better knowledge of the fauna of individual tree species and the results of individual tree sampling were entirely comparable from one place to another, notably between totally different biogeographical regions (Amedegnato and Descamps, 1980a,b). A better understanding of the forest mosaic was thus revealed by the second method.

The present study is mainly based on the analysis of a central sample of 60 trees between 30–35 m in height (15–20 m to the first branch), and varying in diameter from 60–120 cm. Specimens of leaves, fruit and bark were retained and a description taken (especially for the presence of latex, taste and so on). These tall trees allowed study of the grasshopper community in the upper stratum. An analysis of the second layer community was undertaken by means of a simultaneous study of two or three closely overlapping closed trees or from surface felling.

For secondary formation canopies, young or advanced (studies were undertaken from initial burning to 40-year-old restoration at localities in Peru), surface felling is the only option (surfaces of 500–1000 m^2). However, the method has the following limitations:

- Following a felling, there is the problem of separating arboreal and lower level faunas, although the latter is well known, of low abundance in forest and morphologically distinct. The only acridids that could have been mistaken for new canopy forms were unknown gap species bearing arboreal morphological adaptations, notably some Caloscirtae. In fact, few were found in samples and they were of low abundance. On the other hand, the palm fauna (Copiocerae) whose refuges when not feeding are the surrounding trees, has not been considered.

- The supracanopy fauna (large, highly mobile acridids) largely escapes this type of collection. This is important to note because, given their size and the importance of their gregarious nymphal bands, their role in the canopy is probably relatively important and does not correspond to their poor representation in the samples.

Practice 3

This involves the study of gut contents, comparing observations of the ingested epidermis with those of host-tree leaves, allowed a precise definition of the utilization of the host-tree as a food source.

The object of this work was the study of the acridofauna from a primary upland forest. However, the characteristics of the canopy, because of its heterogeneous aspects, could not be completely separated from those of restoration canopies, especially late restorations. Such knowledge is essential to understand the presence, abundance or rarity of some species in undisturbed forest.

The main study area was located in the Ampiyacu basin (Peru), on the hilly upland regions of Colonia, Estiron and Brillo Nuevo. The Rio Ampiyacu is a small river that joins the Amazon at Pebas. Its basin is, therefore, located between the Amazon, Napo and Putumayo rivers, a little north-west of Pebas. The samples analysed were collected at Estiron. The study range for trees or secondary formations has been estimated at about 2.5–3.0 km, in a homogeneous topographic formation. Local comparisons will focus on similar samples taken during 3 years of study in the same locality at the same period (November–December), as well as on earlier studies undertaken in the same region (M. Descamps), in July–August in the Yubineto area, and in secondary formations of the Ampiyacu region. For the latter, individual trees' populations were not isolated, but general results are known (Amedegnato and Descamps, 1980a,b). Regional or inter-regional comparisons were made using collections in the National Museum of Natural History in Paris. Localities mentioned for comparative purposes are in various biogeographical regions of the Amazon basin and are specified where necessary.

The egg-laying mode was tentatively established through the observation of ovipositor valves (especially when not regressed, as they bore soil traces), the location of very young nymphs and, when possible, the location of egg pods (laying females). Traps set on trunks produced no results in forest (Proscopids and *Poecilocloeus* nymphs in secondary formations).

All the insects, bearing labels with indication of the tree number (referring to the botanical samples) and of the dissection number (referring to the alcohol collection of guts and ovaries), as well as epidermis preparations, are kept in the Paris Museum.

RESULTS

Analysis of a forest canopy acridofauna

The case of palms

Although an integral part of the forest canopy, it is more by their particular architecture and by their close relation to monocotyledonous Graminaceae and Cyperaceae (Williams *et al.*, 1973), that palms appear to form a distinct canopy habitat. Palms were found to be occupied by two adaptive groups of different phyletic origins, Copiocerae for Acrididae and Leguini for Romaleidae. Leguini, which are large, rapidly flying insects with brilliantly coloured wings, seem to be supracanopy species. Copiocerae, on the contrary, have been found to be more intimately linked to the structure of the forest canopy. Copiocerae were collected regularly in tree fellings. They pass the major part of their life in trees surrounding the palms on which they feed. Their egg-pod is epiphytic (compared with the Leguini, which oviposit in the soil) and their nymphs develop on palms. In the region considered, the spatial occupation of palms by the Acrididae has been found to involve two genera and five species. Two species were typical of upland forests and were numerically dominant: *Copiocera nigricans* and *Eumecacris colombiana*.

The canopy excluding palms

Density of acridids in the forest

Of the 60 trees sampled only two were nearly empty (two insects captured per tree). The less populated trees carried a little less than 10 individuals, while for the most populated trees more than 100 individuals were taken. The average abundance was 21.7 per tree (σ = 19.8). Two other samples from the same locality (20 and 25 trees) had higher abundances, 36 (σ = 26) and 40 (σ = 28.8) individuals per tree, respectively. The average number of species per tree in the present study varied between about five and 20, with a mean of 10 species per tree.

The lower and medium canopy strata appeared to be a great deal less populated. In composite fellings, secondary trees were only well-populated in a few cases where trees belonged to the main tree families of the upper stratum. Similarly, some secondary trees belonging to Families common in late restoration canopies (20–40 years: Flacourtiaceae, Annonaceae) were also found to be well populated. Data from forest hectares (four samples) seemed to indicate the same thing. Thus, by simultaneously taking into account the number of specimens and species captured, there appeared to be no doubt that the population concerned was mainly that of the tallest trees, and that the second layer was almost unpopulated.

For trees with an average surface area estimated at approximately 80 m^2 and a mean population of 21.7 or more grasshoppers per tree, forest canopy grasshopper density would be 0.2–0.5/m^2 with a maximum of 0.7–0.8/m^2 , and occasionally 1.0/m^2 (Lauraceae, for example). These data are comparable with successional densities (cf. below).

Such generalizations belie the fact that there are strong inequalities in tree occupation. Modes of tree occupation appeared to be the result of two factors: (i) the trees themselves and their epiphytes; and (ii) the adaptations of different grasshopper groups.

Occupation of the vegetation, with special attention to grasshoppers

General composition of the canopy acridofauna

The basic fauna constituted approximately 160 species (the number of known local species, including low and open biotope species, being around 250–300), making this region the richest in Amazonia. These species are largely restricted to undisturbed forest (70%), while other species are more characteristic of late regeneration secondary growth (stages from 15/20 to 40 years), small gaps, or are wide-ranging supracanopy species. The composition of this community, as well as the main species characteristics, are given in Table 14.1 (a systematic list is given in Appendix 14A).

Despite the probable under-representation of large supracanopy species, the family Romaleidae largely dominated the fauna with 51.5% of individuals (26 genera, 38 species). The other important family of the Acridoidea was the Acrididae, with 35.6% of the specimens distributed in two main subfamilies, Proctolabinae (16.5%) and Ommatolampinae (14.0%) , the second being characterized by a higher species richness. The other great superfamily, Eumastacoidea (Proscopiidae and Eumastacidae), represented only 9.5% of the individuals and a small number of species which were clearly less adapted to the canopy. These values are similar to other Amazon forest samples: in general, Romaleidae 45–50%, Proctolabinae 15–18%, Ommatolampinae 20–30% (the weaker value apparently being linked to the more closed character of the forest sampled), Eumastacoidea 6–10% (Amedegnato and Descamps, 1980a,b and unpublished data).

From a functional and quantitative viewpoint, specialists proved to represent 71% of Romaleidae (12 species) on various hosts, 45% of Proctolabinae (two to three species), mainly on Lauraceae (and Vochysiaceae), and 43% of Ommatolampinae (special kinds of behaviour). For Acrididae, notably, must be added the specialists of palms which form a small proportion of the insects present in the crowns of trees and virtually the whole palm acridofauna.

A comparison with successional canopies, notably from 15–20 years, has shown a greater proportion of Romaleidae in undisturbed forest while, on the contrary, Proctolabinae were clearly more characteristic of regenerations (especially on Flacourtiaceae). With regard to Ommatolampinae, after a representation of 20–50% at the beginning of regeneration, they decline in abundance to forest levels after about 30–40 years. Like Proctolabinae, Proscopiidae are important during the first decades of regeneration (15–17%), but decrease in abundance at approximately 40 years. In general, the Eumastacoidea seem more or less constant at around 10% of total abundance in successional and forest habitats.

Vegetation utilization by the grasshopper groups

Romaleidae are the main forest crown inhabitants. There are a number of close, coexisting species, notably among Ophthalmolampae. Of 11 known species in the area, only two, *Ophthalmolampis condita* and *O. praeferox,* were not represented in the present samples. The nine species present were totally sympatric. To this set should also be added *Trybliophorus,* which is closely related. Examination of individual tree samples has shown that two or three species were commonly found in individual tree-crowns, although it was more frequent that a single species was present, most often in large colonies. Among Nautiae (six local species), four species frequently coexisted (often two *Pseudonautia* and two *Euprepacris*).

The considerable number of coexisting sibling-species raises questions about the origin of forest acridofauna diversity and concerns trophic resource partitioning among the most abundant species (see Tables 14.1 and 14.2). Ophthalmolampae (except *Nothonautia,* a genus mimetic of Nautiae) are distributed in four groups, corresponding to the genera *Ophthalmolampis, Adrolampis, Apophylacris* and *Peruviacris.*

Table 14.1 The canopy acridid community of an upland forest sample (rio Ampiyacu, Peru): list, phyletic organization, abundance and main behavioural traits

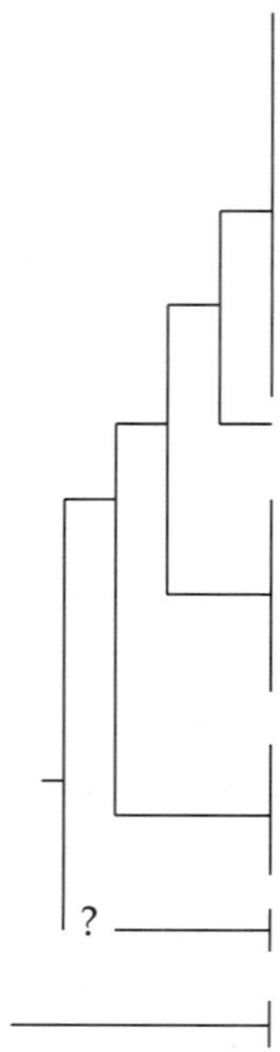

Identity	*CT*	*N*	*GA*	*LT*	*EgLT*	*RT*	
ROMALEIDAE							
Ophthalmolampae (Ophthalmolampini)							
1- *Apophylacris incondita*	F	0.69	M/R	IC	Ep	PyF	4 families
2- *Adrolampis contumax*	F	**10**	Ab	IC	Ep	SF	Lauraceae (2/3 gen)
3- *Adrolampis limbatipes*	F	0.23	Ab/M	IC	Ep	SF	Lauraceae
4- *Ophthalmolampis brunneiceps*	FS	1.07	M	IC	Ep	PyF/O	(S: Flacourtiaceae)
5- *Ophthalmolampis sigillata*	F	0.38	Ab/M	IC	Ep	PyF	
6- *Ophthalmolampis elaborata*	FS	0.46	M/R	IC	Ep	SF	Annonaceae: *Guatteria*
7- *Ophthalmolampis truculenta*	F	**4.07**	Ab	IC	Ep	SF	Myristicaceae: *Osteophloeum*
8- *Peruviacris cerciata*	F	0.54	Rd	IC	Ep	S	?
9- *Nothonautia splendens*	F	0.77	R	IC	Ep	PyFEl	
Nautiae							
10- *Euprepacris multilipennis*	F	1.38	M	IC	Ep	SEl	Bignoniaceae
11- *Euprepacris piperacipes*	F	**4.5**	M	IC	Ep	SEl	Bignoniaceae
12- *Pseudonautia seducta*	F	0.84	M	IC	Ep	SEl	Bignoniaceae
13- *Pseudonautia maculipes*	F	**4.3**	Ab	IC	Ep	SEl	Bignoniaceae
Lagarolampae							
14- *Habrolampis bicolor*	F	**3.1**	Ab	IC	Ep	PyEl	
15- *Helolampis coloniana*	FS	1.9	M	IC	Ep	PyFEl	
16- *Lagarolampis amazonica*	S	0.07	R	IC	S	PyF	
17- *Helicopacris viridans*	FG	0.15	Ab/R	?	Ep	?	
Other tribes							
18- *Trybliophorus octomaculatus*	F	**10.5**	Ab	IC	S	O	Combretaceae: *Terminalia*

Table 14.1 continued

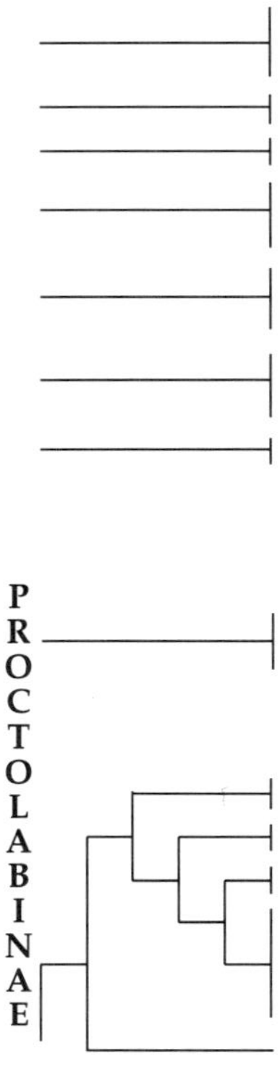
PROCTOLABINAE

Identity	CT	N	GA	LT	EgLT	RT	
19- *Eurostacris puncticrus*	F	1.6	M	IC	S?	SEl	?
20- *Pseudeurostacris valida*	F	1.76	M	IC	S?	SEl	id
21- Bactrophorini (4 g, 4 sp.)	F	0.6	M/R	IC	S	PyF	
22- *Taeniophora caqueta*	S	0.23	R	?	S	?	
23- *Hisychius nigrispinus*	F	1.46	Ab	CS	S	PyFEl	
24- *Pseudhisychius nigroornatus*	S	0.15	R	CS	S	PyFEl	
25- *Prionacris cantrix*	F	0.3	Rt	CS/SC	S	?	
26- *Aprionacris coerulescens*	F	0.07	Rt	CS/SC	S	?	
27- *Titanacris* (2 sp.)	FG	0.07	Mt	CS/SC	S	S?	*Inga* ?
28- *Tropidacris cristata*	FG	0.07	Mt	CS/SC	S	PyF	
29- *Xomacris nuptialis*	?	0.38	?	?	S	?	obs. on *Mendoncia*
ACRIDIDAE							
Coscineutini							
30- *Coscineuta pulchripes*	FS	0.46	R/Ab	CS	S	PyFEl	
31- *Coscineuta cicatricosa*	F	1.54	M	CS	S	PyFEl	
Proctolabini							
Proctolabae							
32- *Halticacris spinifer*	F	0.84	R	IC	S?	?	
33- *Witotacris concinna*	F	0.23	Rs	IC	S	SF	Vochysiaceae: *Qualea*
34- *Dendrophilacris lorenzi*	F	**6.3**	Ab	IC	S?	SF	Lauraceae: *Beilschmieda*
35- *Poecilocleus ferus*	F	1.3	M	IC	S	PyFEl	
36- *Poecilocleus estironana*	S	0.38	R	IC	S	O	Flacourtiaceae (2 sp.)
37- *Poecilocleus uncinatus*	S	2.9	M	IC	S	Py/O	Flacourtiaceae
38- *Loretacris fascipes*	F	0.14	M	IC	Ep	O	?

Table 14.1 continued

OMMATOLAMPINAE

Identity	CT	N	GA	LT	EgLT	RT	
Saltonacrae							
39- *Saltonacris avellinoi*	FS	**5.07**	Ab	IC	Ep	PyEl/O	?
40- other Saltonacrae	F	0.14	R	IC	Ep	?	
Eucephalacrae							
41- *Eucephalacris miguelangeli*	F	0.3	R/M	IC	S		
42- *Pareucephalacris luridicrus*	F	0.23	R	IC	S	BkF	
Rhytidochrotinae, Leptysminae							
43- *Galidacris eckardtae*	SG	0.92	M/R	CS	S	PyF	
44- *Galidacris purmae*	S	0.07	R	CS	S	S/O	?
45- *Stenopola* (2 sp.)	FG	0.23	R	?	En	?	
Caloscirtae							
46- *Calohippus sp*	F	**8.2**	Ab	IC	Ep	PyEl	(+ Caryocaraceae)
47- *Stigacris rubropicta*	S	0.07	R	IC	S	O	Melastomataceae
48- other Caloscirtae (4 g, 4 sp.)	FG	0.53	R	IC	S/Ep		
49- **Syntomacrae** (2 g, 2 sp.)	FG	1	R	IC	S/En		
Oulenotacrae							
50- *Eurybiacris luteoguttata*	F	0.46	M	IC	S	PyFEl	
51- Other Oulenotacrae (3 g, 5 sp.)	F	**2.15**	M/R	IC	S	Bk	
Vilernae							
52- *Hypsipages dives*	F	0.76	M	TB	S	Bk+M	
53- *Bryophilacris muscicolor*	F	0.23	M/R	T	S	SM	
54- *Rhabdophilacris sylvatica*	F	0.07	M/R	T	S	SM	
55- *Agenacris subbrevis*	F	0.07	R	?	En	SEp?	
56- *Sciponacris amazonica*	F	0.3	R	IC	En	SEp	

Table 14.1 continued

Identity	*CT*	*N*	*GA*	*LT*	*EgLT*	*RT*	
EUMASTACOIDEA							
Proscopiidae							
57- *Apioscelis* sp.	FS	**4.22**	Ab	CS	S	PyFEl	
58- *Proscopia* (2 sp.)	F	0.38	R	CS	S	PyFEl	
Eumastacidae							
59- *Pseudomastax personata*	FS	1.6	M	CS	?	PyFEl	
60- *Eumastacops caligo*	S	2.2	M	CS	?	O	?
61- *Pseudeumastacops militaris*	F	1.07	M	CS	?	PyFEl	
62- *Eumastax vittata*	SF	2.7	Ab	CS	?	PyFEl	
63- Other Eumastacids (2 g. 5 sp.)	SF	0.7	RM	CS	?	PyFEl	

CT, main canopy type inhabited by the species; F, forest; S, succession; G, gap
N, frequency in the sample studied
GA, general abundance in the forest: Ab, high; M, medium; R, rare; Rd, resource dependent on non-dominating family (?), Rs, rare because of seasonality; Rt/Mt, large territorial species of medium or high abundance but found with rare or medium frequency
LT life-type: IC, intracanopy; SC, supracanopy; CS, canopy surface; TB; trunk, branches
EgLT egg-laying type: Ep, epiphyte; En, endophyte; S, in the soil
RT, resource type: S, specialist: Py, polyphagous; F, on tree foliage; El, on epiphytic lianas; Ep, on epiphytes of another type; O, oligophagous: Bk, on bark; M, on moss.

Table 14.2 Resource partitioning among main canopy Romaleids*

	Plants*												
Insects	*Co*	*La*	*Ly*	*My*	*Sp*	*Cy*	*An*	*UF*	*Ot*	*Bg*	*uUL*	*dUL*	*Observations*†
Ophthalmolampae													
Apophilacris incondita	7.2		35.7	35.7		21.4							14 + x
Adrolampis contumax		99.4	0.6										147
Adrolampis limbatipes		98.1										1.9	52
Ophthalmolampis sigillata	22.2	5.5	55.5						16.8				18
Ophthalmolampis brunneiceps			3.7						Fl			81.5	27 + 16
Ophthalmolampis elaborata							100						4 + 82
Ophthalmolampis truculenta				100									100
Peruviacris cerciata								100					7
Nothonautia splendens		57.1	0.8							42.1			7 + x
Nautiae													
G.= 2; sp.= 4										100			122
Lagarolampae													
Habrolampis bicolor												100	48
Helolampis coloniana	12	4	16	4	4	24			36				25
Other													
Trybliophorus octomaculatus	92.9		2.6						4.5				113
Eurostacrini (G = 2; sp.= 2)											100		69
Hisychius nigrispinus			12.5	6.2					56.3	25			16

* Co, Combretaceae; La, Lauraceae; Ly, Lecythidaceae; My, Myristicaceae; Sp, Sapotaceae; Cy, Caryocaraceae; An, Annonaceae; Ot, other families; UF, unknown family; Bg, Bignoniaceae liana; uUL, unique unknown liana; dUL, diverse unknown lianas

† Number of observations and supplementary ones from samples of the same locality (determined numbers, included in the results), or of other localities (x, not included in results).

Ophthalmolampis
This is a polyvalent genus with four important species: one polyphagous tree foliage feeder, one polyphagous epiphyte feeder (feeding on Flacourtiaceae in successional forest) and two other species which are specialists that feed on particular families, one on Myristicaceae in forest and one on Annonaceae in forest and in old secondary formations.

Adrolampis
This comprises two species specializing on Lauraceae (*Beilshmieda* sp. and *Aniba* sp.)

Apophylacris
This is a monospecific genus, polyphagous on tree foliage.

Peruviacris
This is a southern genus, specializing on an unknown plant.

Trybliophorus is not an Ophthalmolampae but belongs to a phyletically close group of Romaleidae. *Trybliophorus* is apparently specialized on Combretaceae (*Terminalia* sp.) although it is sometimes found on Lecythidaceae without developing colonies. It was thus found that of this set of nine (eight Ophthalmolampae species excluding *Nothonautia*, but including *Trybliophorus*), six were exclusively dependent on particular tree families (except Lecythidaceae and Sapotaceae), all of which rank among the most important in the ecosystem. Consequently, the majority of co-occurrences on a single tree may result from the partial overlap of different trees.

The complementary host-tree use of different genera and species seems to result from an adaptive radiation of the group Ophthalmolampae on the main tree families. In contrast, the four species of Nautiae were all found to be specialists of the same family, Bignoniaceae, and to occur simultaneously on the same individual host-plant. Possible modes of coexistence in this group are unknown.

Like Romaleidae, canopy Acrididae have been found to be represented mainly by intracanopy forms. Within this family has evolved a kind of behaviour that is exclusive to the group: trunk and branch exploitation.

Among Proctolabinae, *Coscineuta*, with two sibling species, was found to be polyphagous. Species of Proctolabae, the largest group within the subfamily (140 species distributed in 13 genera), seemed to occupy most spatial or trophic levels. In the forest, *Dendrophilacris lorenzi* was found to coexist on Lauraceae with the Ophthalmolampae specialist on *Beilschmieda*, while *Witotacris*, a seasonal genus, specialized on Vochysiaceae (here *Qualea rosea*). The genus *Poecilocloeus*, an important component of the Proctolabae with 64 known species in Amazonia is represented here by three species (among 10 local species and 16 species known for the north-western dispersal centre). *P. ferus* appeared to be polyphagous in undisturbed forest, while the two other species, *P. estironana* and *P. uncinatus*, were oligophagous on Flacourtiaceae in secondary successional canopies.

Very little precise data are available for the little-known, derived groups Saltonacrae and Eucephalacrae. The former is one of the most characteristic, comprising species with endophytic and epiphytic egg-laying types. *Saltonacris avellinoi*, one of the dominant species of the canopy, is a polyphagous epiphyte feeder in the forest, but has also been found in large colonies on an unknown plant in late secondary formations. Eucephalacrae, a group more typical of the Amazon periphery, seem to be corticolous.

In contrast, Ommatolampinae and the closely related subfamilies Rhytidochrotinae and Leptysminae (Table 14.1) have a far higher generic diversity and a variety of behaviours unknown in other groups. The most abundant, *Calohippus* (Syntomacrini), has been observed exploiting a large variety of different epiphytes, as well as the foliage of Caryocaraceae (*Caryocar* sp.). Most other Caloscirtae and Syntomacrae, although present in canopy communities, appeared only rarely in undisturbed forest. Syntomacrae and *Galidacris* sp. (Rhythidochrotinae) are well represented in 5–10 m trees at the forest edge. Similarly, another species, *Stigacris rubropicta* (Caloscirtae), is found characteristically in young (10–15-year) successional canopies, on the foliage of Melastomataceae.

Ommatolampini (Vilernae and Oulenotacrae) almost exclusively exploit tree trunks, branches and their annexes. They only rarely live on canopy foliage (e.g. *Eurybiacris luteoguttata*). Most of the Vilernae found on trunks and branches appeared to be dependent on mosses. In *Hypsipages dives* a good proportion of bark seemed to enter the diet as well. Oulenotacrae (except *Eurybiacris*) also appeared to feed on tree bark, although along with bark remains in the gut were found tiny epidermal leaf fragments. This group may to some extent be detritophagous on branches.

Agenacris and *Sciponacris* are phyletically distinct, displaying an endophytic egg-laying type, on epiphytes on which they probably specialize (for *Sciponacris* an unknown succulent plant with very large cells and cavities). There are no data available for *Agenacris*.

Eumastacids and Proscopids, although proportionally less abundant, nevertheless appeared to be an important component of the canopy acridofauna. No fundamental differences in species composition have been noted between successional and forest canopies, although there is a higher relative abundance of Proscopids in successions. Eumastacids have been found to be very clearly favoured by the presence of Leguminosae, notably Mimosaceae, which they nevertheless do not always and exclusively consume, being strongly polyphagous.

These groups appeared, in general, to be relatively mobile at the surface of tree-crowns and exhibited a behaviour similar to that shown by other species at the surface of lower canopies (successions, gaps), or even at the surface of shrub communities. They are, however, true forest species and in no way thamnophilous species (which would make them wide-ranging on the South American continent). Most of these Eumastacids and Proscopids species have never been found at the lower level, under normal conditions.

Occupation of the vegetation with special attention to tree families

The 60 trees sampled yielded 1302 grasshoppers. The main features of the fauna have been summarized in Table 14.3, based on insect distribution and dietary analysis. In order to provide this information, grasshoppers have been classified into the following categories, which partly correspond to guilds (cf. below):

- tree foliage feeders
- trunk and branch insects, that feed on bark and/or vegetal remains and small epiphytes
- epiphyte feeders (a complex category)

Epiphyte populations have proven to be complex and difficult to understand; however, a large part of this community appeared to be linked to a dominant family of lianas, the Bignoniaceae (Schnell, 1987; Foster, 1990; Prance, 1990). Moss-eating arboreal acridids have been classed here as trunk and branch insects. They are characterized by very low abundance, but high diversity (six genera, eight species from a known local fauna of seven genera and 17 species; Descamps, 1979a,b; Amedegnato, 1985). Tree families were very different in their acridid populations.

Lauraceae, Combretaceae and Myristicaceae supported large numbers of individuals and species. These tree families had different grasshopper specialists. In other Amazonian forest ecosystems where Vochysiaceae is a dominant tree family, it too supports a diverse grasshopper fauna. Indeed, species of Vochysiaceae in south-western Amazonia (Jenaro-Herrera, Peru, on sandy soil) carried important populations of a specialist, *Witotacris annulicrus*, mainly on *Vochysia* but also on *Qualea* and *Ruizterania*. This specialized genus (*W. concinna* in the north-west) was represented in the present samples by a few isolated individuals, due to seasonality. In other parts of north-western Amazonia *W. concinna* was well represented during July–August (Rio Yubineto, Peru)

Lauraceae (303 individuals, 70% of which feed on foliage) appeared to be characterized by three specialists, two species of *Adrolampis*

Table 14.3 Distribution of the acridid arboreal community in relation to tree families and their epiphytes and lianas. Of 1302 insects collected, 1204 are included here (foliage: 629; epiphytes: 532; trunks, branches: 43); the remaining insects came from five trees not mentioned or were impossible to classify

	Insect category													
	*Total population**			*Foliage population*[+]						*Trunk, branches population*		*Epiphytes population*		*Egg-laying type of total population*
Trees (n)	*Q*	*%*	*Qm/T*	*Qm/T*	*Nsp*	*Sp*	*%Sp*	*Stsp*	*Csp*	*Qm/T*	*Nsp*	*Qm/T*	*%*	
Lauraceae (5)	303	23.3	60.6	42.4	13	3	85	3	2	2	6	16.2	6.2	Ep
Combretaceae (7)	263	20.2	37.4	25	16	1	67.5	0		0.7	3	11.7	6.3	S
Myristicaceae (4)	102	7.8	25.5	16	12	1	57.8	1		1	3	8.5	2.6	Ep
Lecythidaceae (15)	150	11.5	10	4.8	12					0.8	3	4.4	5.1	S
Sapotaceae (6)	138	10.6	23	0.2	1					0.8	3	22	10.1	Ep
Sterculiaceae (2)	58	4.5	29	9	7					1.5	3	18.5	2.8	S/Ep
Vochysiaceae (4)	42	3.2	10.5	7	14	1		1		0.25	1	3.2	0.9	S
Caryocaraceae (2)	43	3.3	21.5	13	6					1	2	7.5	1.2	Ep
Celestraceae (2)	42	3.2	21	6	6					0.5	1	14.5	2.2	S
Leguminoseae (P + C + M = 6)	46	3.6	7.6	2.5	8							5.1	2.4	S
Moraceae (1)	15	1.2	15	3	3							12	0.9	Ep
Apocynaceae (1)	2											2		

* Q, total population; %, proportion of the total sample; Qm/T, mean population per tree

[+] Nsp, number of species; Sp, number of specialized species; %Sp, % of specialists in the foliage population; Stsp, number of strictly specialized species; Csp, number of coexisting specialized species; Ep, epiphyte; S, edaphic.

(Romaleidae), and *Dendrophilacris lorenzi* (Acrididae), forming 85% of the foliage population. At least two of these three species coexisted in important populations on all the suitable trees. The most common pairing was *A. contumax* and *D. lorenzi*, on *Beilschmieda*. *A. limbatipes*, although rare in the present samples, formed important colonies with *A. contumax* in the absence of *D. lorenzi* (on ?*Aniba*, Table 14.1). Considering both foliage and epiphytes, the major egg-laying behaviour of grasshoppers on Lauraceae is epiphylle (except perhaps *Dendrophilacris* for which the oviposition site could not be established: ovipositor valves not regressed but nymphs found only in the canopy).

Myristicaceae (*Osteophloeum*) appeared similar to Lauraceae, but with a single specialist, *Ophthalmolampis truculenta* (Romaleidae). Combretaceae (three species of *Terminalia*) were inhabited by very important colonies of *Trybliophorus octomaculatus* (almost 70% of the population). Due to the fact that *Trybliophorus* oviposits in the soil (where the litter serves as a refuge when disturbed), the most important egg-laying mode observed on Combretaceae was edaphic. Young nymphs have, nevertheless, always been found rapidly to reach the crowns. Epiphytes and lianas supported by Combretaceae appeared to harbour the normal fauna, specialized or not (cf. guilds).

Lecythidaceae, a dominant family (Prance, 1990), appeared to differ markedly from other tree families in having a low abundance of grasshoppers on the foliage and epiphytes. Lecythidaceae supported a greater number of polyphagous species (see Table 14.2, for main canopy Romaleidae). In another sample, a small colony of *Apophylacris incondita* was found on this tree family. The occupation of Lecythidaceae by more mobile polyphagous species, which are also less gregarious, combined with a poor epiphyte population, is expressed by a predominantly edaphic egg-laying behaviour.

For Sapotaceae and Moraceae the presence of latex compromises habitation by grasshoppers and these tree families support a medium-density population, mainly on epiphytes, and do not seem to harbour specialists. Other families, such as Celastraceae and Sterculiaceae, were not found in sufficient numbers in this ecosystem to provide significant data relative to their foliage grasshopper populations. However, examination of other samples comprising these families have shown similar results to Sapotaceae and Moraceae. Caryocaraceae, which was generally rare, was inhabited by *Calohippus* sp. feeding on foliage, whereas this species had previously been found to feed only on various epiphytes. The type of occupation of this set of families, mostly by generalists of the canopy, is expressed by a mixed edaphic and epiphylle egg-laying mode.

Leguminoseae had depauperate grasshopper assemblages. Typical grasshopper groups for Mimosaceae included Eumastacids, Proscopids and possibly some large Romaleinae, such as different species of the genus *Titanacris* (adults and nymphal bands found on the genus *Inga*), although not as permanent residents (C. Amedegnato, personal observation, see also Descamps and Carbonell, 1985).

Other tree families that were not inhabited (inedible and without important epiphytes) included Apocynaceae.

General data on late restoration canopies

Canopies in the process of restoration appeared to be very different from undisturbed forest canopies (data from the same region, Denevan *et al.*, 1984; Flores-Paitan, 1984). A few restoration grasshopper species are also found in the forest; however, the restoration period is characterized by the succession of typical secondary formation species (Amedegnato and Descamps, 1980a,b; Amedegnato and Poulain, 1987, also unpublished data), which are nearly absent in the forest. Exceptions to this general rule include two species of *Ophthalmolampis* from 20-year-old restorations on Annonaceae or Flacourtiaceae and which were weakly present in the forest (see Table 14.1) and some species of *Poecilocloeus*, one of which, *P. ferus*, is more typical of the undisturbed forest.

Grasshopper density could be estimated using secondary forest samples (500 m^2):

- 15 years (five surfaces, Ampiyacu): 0.1–0.5/m^2
- 20 years (four surfaces, Ampiyacu): 0.2–0.4/m^2; Mamepo, Yubineto: 0.7/m^2
- 30–40 years (seven surfaces, Ampiyacu): 0.1–0.2/m^2

Therefore, even considering the very important differences in sampling methods for forests and for successions, the values obtained were found to vary within the same limits as in undisturbed forest. The densities, which also depend on the composition of the vegetation, are clearly comparable to forest densities and not as high as previously interpreted (Amedegnato and Descamps, 1980b; Riede, 1993).

An outline of guilds

At the level of an individual forest tree, the different modes of participation in the canopy fauna allow all families to be represented. The distribution of adaptive characteristics within these guilds is relatively independent of the species phyletic relationships.

The main life-forms of canopy grasshoppers can be defined by the stratum they inhabit in the canopy and by the type of resources they exploit. From a resource point of view, foliage-mass exploitation was found to be shared as follows: one-third of acridids feed exclusively on

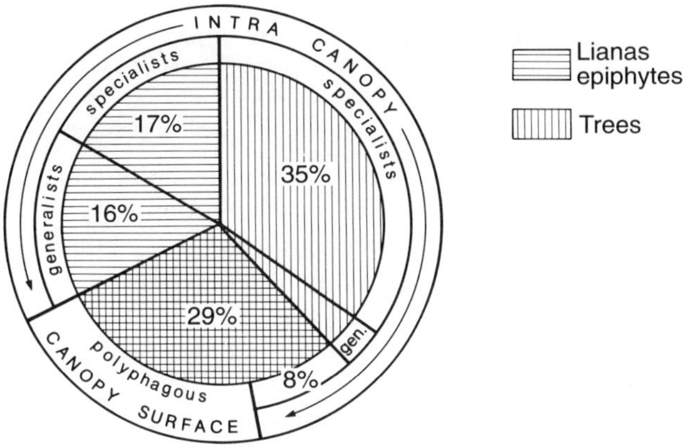

Figure 14.1 The canopy acridid community: quantitative partitioning of the phyllophagous fauna, in relation to their life zone and diet (except corticolous species).

the leaves of trees, one-third exploit exclusively the foliage of epiphytes (most often lianas) and one-third are apparently indifferently exploiting these two principal resource types (Figure 14.1).

From a spatial viewpoint the intermediate stratum, except for edges or gaps, is weakly or not occupied. Consequently, the stratification of canopy acridids (deduced from the exploitation of host plants and their spatial location) seemed roughly reduced to two levels: inside and outside the canopy. The canopy surface forms an interface between the most heliophilous intracanopy species, those species exclusive to the surface, and supracanopy acridids. The intracanopy level appeared to be divided into two well-differentiated units: trunks and branches on the one hand, with a rather sciaphilous population, and, on the other hand, terminal branches and leaves with a heliophilous population.

Combining these sets of criteria gives rise to the following classification (which, however, one must be wary of considering as strict):

1 Palm feeders
2 Intracanopy tree foliage feeders
3 Intracanopy liana feeders
4 Corticolous intracanopy species
5 Intracanopy epiphyte specialists
6 Surface-active canopy species
7 Supracanopy species

Most of these guilds are formed of both generalists and specialists, in variable proportions (Table 14.1 and Figure 14.1), and the majority of forms included were previously listed in the more general categories 'dendrorhabdophiles' and 'dendrophiles' (Descamps, 1976).

Palm feeders

The common characteristics of palm feeders consist of a cylindrical and elongated morphology of graminicolous type, but with the typical arboricolous tarsal adaptation and a special palmicolous adaptation of the mouth parts (Amedegnato and Poulain, 1986).

Intracanopy tree foliage feeders

Even if only species that are strictly limited to tree foliage are taken into account, this guild appeared quantitatively the most important (38% of the population). From a trophic perspective it partly overlapped with the widespread phyllophagous group, feeding on both trees and lianas (Figure 14.1). In the narrow sense, the guild of intracanopy tree foliage feeders appeared to be distributed into 35% of specialists and only 3% of generalists, the weakness of this latter percentage probably resulting from overlap with the widespread phyllophagous feeders (Figure 14.1).

This guild comprises mainly Romaleidae without an acoustic communication mechanism (Ophthalmolampae and *Trybliophorus*), as well as Proctolabinae. The species have a common set of characteristics: gregarious adults, cryptic homochrome chromatic type, an adaptive morphology (lengthened tarsi) and an egg-laying type that is mostly epiphylle (although perhaps edaphic or similar, such as egg-laying in suspended soil or decaying bark, as occurs in Proctolabinae, for example). Their reproductive potential varies from weak for Opthalmolampae (few eggs per pod, ovarioles 2 × 4/5) to medium for Proctolabinae (ovarioles, *Dendrophilacris*: 2 × 6, *Witotacris*: 2 × 8/10).

The Bactrophorini (polyphagous) are provisionally included in this guild. Their morphology is extremely adapted to life on branches (modification of all femora, hind tibiae and tarsi, thoracic sternal concavity) as well as to edaphic egg-laying (except *Hyleacris*), indicative of a special and not yet well understood type of behaviour.

Intracanopy liana feeders

This guild comprises 32% of the population and is equally composed of generalists and specialists well adapted to canopy life (lengthened tarsi, epiphylle egg-laying type). Almost all have a low reproductive potential (ovarioles: 2 × 2 to 2 × 4) and an epiphylle egg-laying type. Specialists, all Romaleidae, including two dominant species *Euprepacris piperacipes* and *Pseudonautia maculipes*, are characterized by a disruptive coloration (white to fluorescent spots and stripes), a gregarious sedentary behaviour inside lianas and a loss of the stridulatory mechanism, presumably due to the development of visual communication (Riede, 1987).

In contrast, generalists, including three dominant species *Habrolampis bicolor, Saltonacris avellinoi* and *Calohippus arboreus* (one in each Acridoidea group), are characterized by a cryptic, non-disruptive chromatic type. *Saltonacris* (Proctolabinae) is mimetic of *Trybliophorus* (Romaleidae). The way these species are distributed in the vegetation seems to show a far greater mobility, notably for *Habrolampis* and *Calohippus*.

Corticolous intracanopy species

This category (3.4%) is formed by a provisional grouping of species with a complete absence of typical arboricolous morphological modifications. The chromatic type is always cryptic, homochromatic with tree branches. Egg pods are edaphic and medium in abundance (ovarioles: $2 \times 7–8/9$), although weaker for the two genera which appeared morphologically more specialized (*Agrotacris*: $2 \times 4/5$, *Hysterotettix*: 2×3).

Vilernae living on moss-covered trunks are sciaphilous, brachypterous to apterous and endowed with a very powerful jump (thamnophilous morphological type). In contrast, Oulenotacrae are of a thamnogeophilous appearance (winged and flying insects, that are more heliophilous).

Intracanopy epiphyte specialists

Outside of the polyphagous category, grazing on many small epiphytes as well as on lianas, some winged insects with an endophytic egg-laying mode (ovarioles: 2×5) have adapted to unknown plants with very thick leaves and seem to constitute a particular guild. Belonging to Ommatolampinae, they have evolved from divergent lineages (known genera appearing quite disparate). Morphologically, they are characterized by few typical canopy adaptations (normal tarsi), and a cryptic homochromatic coloration. Abundance is generally low.

Surface-active canopy species

Most of the widespread polyphagous feeders on tree and liana foliage are non-flying or weakly flying species. This important category is formed by elements of all groups, but mainly of Eumastacoidea (*Apioscelis* sp. being a dominant) and *Hisychius nigrispinus* (Romaleidae). Although poor fliers or non-fliers, Acridids and Eumastacids have an extremely powerful jump. This is not the case for Proscopids. All seem to oviposit in the soil, particularly in gaps. Their egg pods seem to be medium (ovarioles: 2 × 6 for Eumastacids and Proscopids) to large in size (2 × 16 for *Hisychius*).

Two forms, *Coscineuta* and *Adelotettix* (a seasonal genus) are good flyers and seem to be intermediate between surface-active species and supracanopy species. They have a higher fecundity, *Coscineuta* ovarioles: 2 × 9/10, *Adelotettix* ovarioles: 2 × 57–63. These insects are all of a cryptic, homochromatic type and, unlike intracanopy species, are not gregarious and sedentary, although one of the sibling species, *C. pulchripes*, shows an important tendency to swarm (C. Amedegnato, personal observation), like other species of the genus (Popov *et al.*, 1994).

Supracanopy species

In the adult stages supracanopy species are large, solitary and territorial winged forms, with powerful stridulatory and visual communication (stridulation displaying brightly coloured wings). They are cryptic and homochromatic. All are Romaleidae, with abundant to very abundant egg pods (ovarioles 2 × 125–145 or more) laid singly or multiply (Carbonell, 1986), in gaps or at edges, giving rise to important gregarious nymphal bands comprising 20–30 individuals of different ages which are often characterized by a disruptive coloration.

DISCUSSION

For insects in tropical forest canopies the problem is still more one of evaluation of the incredible diversity rather than its functional meaning. Except for some recent attempts (Stork, 1987), the approach taken toward understanding canopy systems in terms of functional organization and the determinants of diversity is still hampered. Linked to this problem is the question of the density of life in the canopy. The only coherent reply to these questions, at this stage of the analysis, is to study one particular group, here the acridids. They are a sort of macroguild, composed of insects that are basically phyllophagous, and having roughly the same behaviour in the nymphal and adult stages.

Although no strictly comparative data that provide a measure of the relative importance of the group are available, the results presented here may contribute to the understanding of these canopy systems. The data show several kinds of patterns for the canopy acridid community:

1. A pattern of vegetation occupation: canopy vegetation, as in other ecosystems, is divided into more or less separate units, taxonomic and/or structural, that provide different possibilities for grasshopper diversification.

2. A pattern of vegetation exploitation by each taxonomic insect group, which is more or less divided to occupy the main habitat, according to behavioural adaptations, resulting in the formation of guilds.
3. Taxonomic groups of grasshoppers do not occupy all canopy parts in the same manner. The main characteristics linked to these differences appear to be behavioural, with a strong historical component involved as well. This pattern expresses the phyletic and historic components of the origin of acridid diversity.

Each pattern, whether (1) as an extrinsic driving force, (2) as realized groups for functional habitat exploitation, or (3) as a pool for biological diversification, is linked in one manner or another to canopy grasshopper diversity. I do not intend to discuss here the greater problem of the basis of phytophagous insect diversity, but only to focus on questions more restricted to acridids.

Validity domain of the results

Can the results obtained from a specific location be generally applied? That depends both on the situation of the samples in the Amazon ecological and biogeographical context, and also on the degree of knowledge concerning the rest of the region. This study, as well as those that frame it, was undertaken in the lower part of western Amazonia, a region that is least likely to be affected by the Pleistocene coolings (Colinvaux, 1987, 1993). This zone is also the most highly diversified site in Amazonia. It seems to be the same for the vegetation (Gentry, 1982). Moreover, general data obtained elsewhere indicate, in most cases, a density well below that obtained here. This, as well as other factors (essentially phyletic and biogeographic ones), means that the present samples can be considered as a central sample from which comparisons can be more easily and more meaningfully made.

With respect to its botanical composition, as far as it could be judged by the most frequent families, the forest sampled seemed to be similar to Central Amazonian formations, where the same families rank in the top 10 (Prance *et al.*, 1976; Gentry, 1990, 1993; Prance, 1990). Nevertheless, the acridid density there is a great deal less than in this part of north-western Amazonia. One canopy acridofauna sample that was taken north of Manaus comprised only 279 specimens (Descamps, 1981), which is extremely poor (see also Fittkau and Klinge, 1973). Roughly speaking, a valid domain for these results extends over at least north-western Amazonia and for the main lines (except Proctolabinae) over the entire Amazon basin, with vicariant genera or species according to different kinds of subhabitats.

Possible determinants of vegetation occupation and associated grasshopper diversity

The pattern observed is constituted by trees or lianas appearing as central for acridid populations, surrounded by more or less inedible plants. In terms of spatial occupation by insects, non-edible trees (Sapotaceae, for example) often have numerous lianas and epiphytes (mean populations per tree are similar). Acridid density is directly related to the spatial structure of the vegetation, to differences in chemical characteristics (specialists) and to a tree's capacity to support epiphytes. Therefore, the main determinant of acridid diversity and abundance is vegetational heterogeneity. This strongly suggests forest dynamics as a causative factor. These considerations mainly concern the higher stratum, as populations of the other strata were found to be very weak, if not absent, except at edges and gaps. Heliophily of most of the grasshoppers could be the main determinant of this difference, but the absence of any evolutionary radiation there suggests that a true second strata does not really exist.

Comparing grasshopper distribution in the canopy with the model provided by the eco-unit concept (surfaces characterized by one set of trees with the same beginning in development time, see Oldeman, 1983, 1989), is particularly instructive. One could consider, for example, that in many cases palm trees with their specialized fauna that move from palms to surrounding trees would constitute the closing of a gap (Kahn and de Granville, 1994). An eco-unit nucleus in a mature forest, with tree varieties in close proximity being generally different (Hubbell and Foster, 1990) may provide the best conditions for acridid diversity. Acridid density likely varies between the extreme values obtained. The mean density resulting appears to be similar to that recorded for other tropical habitats favourable to grasshoppers.

The case of the apparently weakly inhabited family Lecythidaceae is still unresolved. It does not seem to be of seasonal origin or to be strictly local. Moreover, Lecythidaceae is one of the most abundant families, without apparent toxicity and consumed by many polyphagous feeders which are more or less excluded from tree families supporting important colonies of specialists. Lecythidaceae could constitute a sort of trophic reserve. Trees of this family seem to be characteristic of the mature forest (Mori, 1989) and the weakness of their epiphytes and liana population could be related to forest dynamics; it could also be the same for its acridid population.

Similarly, the coexistence of several species of Nautiae on Bignoniaceae lianas could also be linked to sylvigenesis. Grasshopper species could succeed according to changing environmental factors (luminosity for example), notably at the level of some old eco-unit boundaries. The relative importance of Nautiae in decaying formations (a 70-year-old restoration, late successional stages: 27/30 years) favours this hypothesis. In the sample studied, the young edge fauna is severely lacking. This is especially revealing of some structural characters of the forest sampled.

Besides the determinants linked to the structure of undisturbed mature forest, another dimension involved in acridid/botanical diversity is clearly shown by the different faunas corresponding to forest heterogeneity resulting from different local ecological conditions. This importance of spatial heterogeneity would tend to confirm hypotheses that emphasize multiple possibilities for survival (Colinvaux, 1987, 1993) for plant species and associated animal species (such as Vochysiaceae and *Witotacris*) in the origin of diversity.

Behaviour, spatial occupation and the origin of diversity

The second pattern observed, which is the differential occupation of the vegetation by acridid taxonomic groups, was mainly linked to behaviour. Canopy life more or less implies behavioural and morphological preadaptations, which give rise, through evolution, to more and more adapted forms, particularly for mobility and the ability to stay in the canopy (e.g. mechanisms of attachment and modes of egg-laying). The formation of guilds with a general taxonomic predominance expresses this process, they are evolving units where diversification is enhanced:

1. With regard to mobility. One of the most commonly circulating ideas, at least for grasshoppers, is that the forest (and the canopy, by extension) would be populated by apterous or very brachypterous species (Rowell, 1978, 1987; Farrow, 1990) which are specialized and have reduced mobility. This may result in speciation due to isolation and may be one of the main causes of diversity. The link between a higher speciation of taxa and their specialization is obvious, but the link with an effect of isolation due to a lack of mobility on host-plant 'islands' cannot be supported.

In this study a good part of the fauna appeared to be relatively specialized at various levels (Tables 14.2 and 14.3; Figure 14.1). It was also noted that even apparently strict specialists are occasionally capable of feeding on other hosts. This mechanism, well known in low vegetation and temperate biotas (Bernays and Graham, 1988; Chapman, 1990), is probably a vital necessity, especially when linked to frequent population disturbances (falling insects, perturbations of colonies by predators, etc.). As only the most abundant plant families harbour specialist populations, inter-individual distances are insufficient to prevent intermixing populations. Therefore, there can be no confusion between plant families or species considered by particular sets of characteristics where grasshopper evolution takes root (maybe as theoretical islands?) and individual plants isolated from conspecifics (rare species as physical islands). This concept implies real possibilities for displacement. Some often-reported data involving lack of mobility in Central American forests (Rowell, 1987) refer to thamnophilous gap species, whose usual morphology is reduced wings. In fact, the present results show that few species are apterous or micropterous (17 species). Also, most of these are endowed with an extremely powerful jump and are generally polyphagous and widespread. This is especially true for species dominating in gaps and secondary formations, as well as for canopy surface species. The possibility of mobility within or at the surface of the canopy is the rule for almost all species. The most mobile, outside of the supracanopy species, being the canopy surface-active species and those inhabiting the trunks and branches (resource limitation, c.f. guilds). In no case can weak mobility be evoked in the origin of species diversification in the canopy, but gregarious behaviour is probably a reinforcing or accelerating factor for the Ophthalmolampae.

The high level of speciation among trunk and branch insects does not seem to be linked to trophic specialization, compared with the Ophthalmolampae for which specialization is at least a strong part of the origin of the species.

2. With regard to reproductive modes. Another important adaptive characteristic rarely encountered elsewhere, but especially developed in canopy acridids, is the epiphylle egg-laying mode. By examining the distribution of this behaviour in guilds and plant families (Tables 14.1 and 14.3), it can be observed that it is quite general for specialists as well as for liana- and epiphyte-generalists. On the contrary, the edaphic egg-laying behaviour is widespread among supracanopy and canopy surface generalists, as well as among some global polyphagous intracanopy species and among trunk and branch insects. Therefore, narrow arboricolous adaptation largely includes the epiphylle egg-laying mode and, like the presence of colonies of specialists on some families, this behaviour could be considered to be a result of long-term parallel evolution of plants and insects in forests.

Another aspect of this parallel evolution is expressed by the linear relation of egg-laying potential (at least for the egg pod) and the behaviour of the insects. Indeed, egg-laying potential roughly decreases from supracanopy generalists (average > 200) and canopy surface grasshoppers (average 23) to intracanopy generalists (average 14) and intracanopy specialists (average 7.6). The seemingly least exposed species (e.g. Nautiae inside compact masses of Bignoniaceae lianas) has the weakest potential (4). The former are affected by seasonal changes, whereas the latter show continuous reproduction. This partly corresponds to the classic r/K demographic strategy.

The combination of mobility, gregarious habit or not, egg-laying behaviour and whether or not the grasshoppers have a specialized diet, has produced the observed pattern of occupation and diversity at the tree level: general representation of the main guilds, non-equivalence of different trees, trees with moving populations of cohabiting species (some being part of colonies whereas others are transitory individuals).

Functional organization, phyletic origin and historical components of diversity

Potentially, at the level of any forest tree, all taxonomic groups could largely coexist. In general, one tree's acridid community is composed of about 20 common species, with one-third to one-half of species being more or less randomly present, surrounding a stable core of polyphagous species which are always present (cf. guilds) and of specialists according to the precise tree or liana. The stable nuclei are, therefore, variable in composition. However, relative dominances of the different higher taxa and guild composition express different representation in the ecosystem's functional organization. The main structuring force behind these patterns is the phyletic background. It may, to a certain extent, be the result of parallel evolution between acridid phyla and the botanical ecosystem. The phyletic relationships of canopy Acridomorpha are illustrated in Table 14.1.

For the dominant group Romaleidae, largely represented by Ophthalmolampini, it has been seen that the generic structure of Ophthalmolampae and Nautiae largely parallels the functional structure. *Nothonautia* (Ophthalmolampae mimicking a Nautiae) could represent a mimetic adaptation or a true intermediate form on the way toward specialization on Bignoniaceae, but it belongs entirely to the Ophthalmolampae's adaptative radiation for environment exploitation. The third group Lagarolampae, except for *Habrolampis bicolor*, is more characteristic of mountain forests of north-western Andean slopes and is especially found in successions of the west-Amazonian dispersal centre (*sensu* Amedegnato and Descamps, 1982; Amazonian centre of Müller, 1973) (Amedegnato, 1977; Descamps, 1983). Lagarolampae, therefore, typically represent a different ecological adaptation, part of the adaptative radiation in the tribe Ophthalmolampini. Besides the Ophthalmolampini group are found *Trybliophorus, Eurostacris* and *Pseudeurostacris*. These genera, according to their very large distribution, weak speciation and their less pronounced adaptations, would seem to belong to an anterior radiation.

The other Romaleidae, far more diverse in their evolutionary origin and their functional integration, are clearly less intimately linked to the canopy ecosystem. They in no way represent a single evolutionary radiation. Some, like Hisychiini, are in the same case as Lagarolampae (Amedegnato and Poulain, 1994), but for most of the others, including Leguini (palm specialists), their evolution largely exceeds the Amazon basin and stands at the continental level. It is notable that Romaleidae not only represent the predominant group in canopy, but also occupy all the forest strata from the litter to the supracanopy zone, except trunks and branches. Outside Lagarolampae and Hisychiini, the entire population structure of the Romaleidae is similar throughout the Amazon basin. This is probably a sign of the antiquity of the evolution of this endemic group in relation to present or past forests.

Concerning Acrididae, the only group presenting important morphological adaptations to canopy life is Proctolabini (Proctolabinae), whose main forest forms are intracanopy species. However, in contrast to most Ophthalmolampini, the entire evolution of the group cannot be detected in the undisturbed forest habitat. Besides the forest genera mentioned here, the majority of forms are more characteristic of secondary formations and Eucephalacrae are of cerrado (shrubland) origin. Proctolabinae could have represented the arboricolous fauna of the forest's cold periods (more thamnophilous and adapted to a dry season). Their relationship with Nearctic Melanoplinae (Amedegnato, 1977) would favour this hypothesis. Acrididae specialists of palm trees (Copiocerinae, Copiocerae) are very close relatives of this subfamily.

The other large group of Acrididae, Ommatolampinae along with two close subfamilies, represents the widest neotropical assemblage (140 genera) and is equal in importance to Romaleidae. Caloscirtae and Syntomacrae are true representatives of the arboricolous Amazon fauna, of which *Calohippus* is one of the most typical elements. In contrast, Syntomacrae, an important group at the lower level of the vegetation, are more marginal in the forest (edge canopies, neither morphological nor egg-laying mode adaptations). Therefore, the main representation of the subfamily in the undisturbed forest is on trunks and branches, e.g. two closely related phyletic groups, Vilernae and Oulenotacrae, without arboricolous morphological adaptations. The moss-feeders, Vilernae, inhabiting trunks and big branches are clearly derived from the rest of the group (thamnophilous insects of low bushy biotopes), particularly from forest understorey genera like *Aptoceras*. Oulenotacrae are close to Vilernae, from which they are derived (Amedegnato, 1977; Descamps and Amedegnato, 1989). Some (at edges, gaps, or in open forests) feed on foliage. The more specialized species living on branches and highly characteristic of the undisturbed forest clearly derive from the former and have undergone an important speciation whose origins are unknown. It is clear, therefore, that the adaptation of Ommatolampinae to forest crowns is secondary, recent, and concerns niches left vacant by other groups (adaptative radiation of Vilernae and Oulenotacrae).

Eumastacids and Proscopids, canopy surface groups, are largely spread over the entire South American continent. While the world-wide-ranging Eumastacids are more diversified in forest and in the sub-Andean region, their South American endemic vicariant, the Proscopids (Amedegnato, 1993), have a southern centre of diversification outside the Amazon basin. Equally present at lower forest levels, in open habitats and in secondary formations, the two groups join and juxtapose, in some way, at the surface of the Amazon canopy where they do not seem to present adaptative radiations similar to those of Acridoidea.

Finally, of the three main neotropical endemic groups of the true acridid fauna, only one, Romaleidae (the most ancient endemic), is a fundamental component of the canopy fauna (Amedegnato, 1990). The second, Proctolabinae, although arboricolous, adapts to the present rainforest by means of secondary formations, while the third, Ommatolampinae, despite the large number of forest genera and species, clearly occupies narrow interstitial niches between the larger fundamental compartments of the canopy (Amedegnato, 1990). Acrididae, Proctolabinae, Ommatolampinae and Copiocerinae appear relatively complementary. The family Acrididae, ranging world-wide, seems of recent origin in South America (with a probable Cenozoic-to-present development; Amedegnato, 1977, 1993).

Although the main part of the arboreal acridid community appears very recent, the place occupied by the different higher-level taxonomic (and phyletic) components is clearly related to the long-term historical development of the faunas (and ecosystems) on the continent.

In conclusion: a structured ecosystem and community?

Is this a structured ecosystem with interdependent species, or a randomly organized ecosystem made up of independent species? The canopy acridofauna is neither chaotic nor absolutely coherent, with only complementary species. Results obtained show that both the vegetation and their acridid communities are organized in stable cores that are strongly related, with the remainder of the vegetal or animal ecosystem seeming to be in a more diffuse state of interrelations. The community is partitioned in three parts: strict liana and foliage feeders and the remaining polyphagous population reflecting an intermediate stage, probably resulting from a relation between the abundance of resources and the abundance of insects exploiting these categories. It is likely to be the same for the weak relative abundance of trunk and branch species (which have a more limited habitat). Other aspects of the complementarity of species for the exploitation of the forest mosaic are the existence of gap-, succession- and undisturbed forest-species and also, more clearly, the complementary exploitation of the dominant tree families by different grasshopper species. This gives rise to about two-thirds of the community forming stable nuclei, each one having a centre adapted to the botanical circumstance, surrounded by opportunists, and the whole adapted to an ecological environment (upland/lowland forests, for example) inhabited by vicariant species. As pointed out by Gentry (1986), despite its richness and spatial heterogeneity, the composition of the Amazon flora is highly structured and predictable, probably reflecting an ecological equilibrium. It is, therefore, likely to be the same for the acridid population.

Regarding specific relations within the acridid community, despite an overview of guilds being outlined, most of their internal structure still remains to be defined. A nearly coherent organization can be found only in very recent radiations, probably contemporary with the last climatic warming and, therefore, in the recent to present extension of the present forest type (i.e. at the very most 10 000 years; Colinvaux, 1993, e.g. Ophthalmolampae, Vilernae and Oulenotacrae radiations). Radiations in others types of forest (e.g. Proctolabinae, Copiocerinae), or more ancient radiations (without the accompanying process of explosive speciation, e.g. *Trybliophorus*, *Eurostacris* and *Pseudeurostacris*) remain perceptible. Most ancient groups apparently also persist in the present canopy, without clear organization.

Thus, one of the main determinants of diversity appears to originate in the partitioning of resources among species of the recent fauna (functional organization), associated with persistence superimposed on more ancient faunas of various origins and adaptations (dry forest, mountain forests).

As indicated by the relationships between behaviour and types of vegetal support, the specialization really seems to have developed largely at the behavioural level, as suggested by Futuyma (1983), and on the most abundant families, indicative of the importance of resource abundance in the process (Price, 1984). Yet, in what measure is resource partitioning the result of evolutionary forces? Trees supporting specialist populations can give some clues. Community density, excluding generalists and non-tree foliage-feeders, is variable, with a mean maximum density of 30–40 individuals. In the case of two coexisting specialists, the maximum populations observed were of 60–80 individuals. These populations seem stable for long enough periods, so it is probable that saturation could be reached on individual trees. In what measure can there be saturation at a larger scale? Host-trees that are poorly inhabited are rare. It is therefore probable that through colony development the possibility of saturation does really exist, at least on a small scale. Species such as *Trybliophorus octomaculatus*, which is extremely abundant and potentially a relatively polyphagous species (perhaps a consequence of its locally high abundance?), could develop in these circumstances, with populations adapting to a support other than the Combretaceae which they usually occupy, and thus enter into a radiative phase (polytypic species). However, tree families which are apparently uninhabited by specialists still exist and may, therefore, still provide diversification possibilities for canopy grasshopper groups.

Finally, it would be somewhat disappointing to find in what is presumably so rich a biotope such a simple organization, with a density and richness that does not seem extraordinary, especially compared with the environment. Nevertheless, looking more carefully, it is rather reassuring and a sign that the comprehension of this ecosystem, even if it is difficult and time and perseverance demanding to study, is not at all inaccessible for cause of exuberance. On the contrary, the coherent, recent and historic aspects of evolutionary radiations linked to the exploitation of some types of resources corresponds to the possibility for observation of evolution in action, with the biogeographical component implied, in one of the last intact ecosystems on Earth.

Acknowledgements

I would like to thank particularly my collaborator, Simon Poulain, who carried out most of the sampling work and without whom this study would not exist, as well as the Indians of Estiron for their knowledge of the forest and their enthusiastic comprehension of our work. I am particularly indebted to my colleague, anthropologist Jürg Gasche for his great efficiency. His field experience was infinitely important to us. Botanical sample determinations are due to the late A. Gentry, S. Barrier, F. Encarnacion, C. Sastre and T. Schröder, whom I thank for their help. This work was realized through the collaboration of the IIAP (Instituto de Investigaciones de la Amazonia Peruana), whose Director and personnel are particularly acknowledged.

REFERENCES

Adis J., Lubin, Y.B. and Montgomery, G.G. (1984) Arthropods from the canopy of inundated and terra firme forests near Manaus, Brasil, with critical considerations on the pyrethrum-fogging technique. *Studies on Neotropical Fauna and Environment*, **19**(4), 233–46.

Amedegnato, C. (1977) *Etude des Acridoidea centre et sud-américains (Catantopinae sensu lato), anatomie des genitalia, classification, répartition, phylogénie*. Thèse de Doctorat d'Etat, Paris.

Amedegnato, C. (1985) Descriptions d'Ommatolampinae inféodés aux troncs et branches maîtresses (Orthoptera, Acrididae). *Annales de la Société Entomologique de France (N.S.)*, **21**(1), 51–68.

Amedegnato, C. (1990) Organization phylétique, spatiale et trophique des peuplements d'Acridiens forestiers amazoniens. (abstract). Actas de la 5° Reunion Internacional de la Sociedad de Orthopterologos. 17–20 Juillet 1989,

Valsain (Ségovie) , Espagne. *Boletin de Sanidad Vegetal,* Fuera de Serie no. 20, 403.

Amedegnato, C. (1993) African–American relations in the Acridians (Insects, Orthoptera), in *The Africa–South America Connection* (eds W. George and R. Lavocat), Oxford Monographs on Biogeography no. 7, Oxford Science Publications, Clarendon Press, Oxford University Press, pp. 59–75.

Amedegnato, C. and Descamps, M. (1980a) Evolution des populations d'Orthoptères d'Amazonie du nord-ouest dans les cultures et les formations secondaires d'origine anthropique. *Acrida,* **9**, 1–33.

Amedegnato, C. and Descamps, M. (1980b) Etude comparative de quelques peuplements acridiens de la forêt néotropicale. *Acrida,* **9**, 171–216.

Amedegnato, C. and Descamps, M. (1982) Dispersal centres of the Amazonian acridids. *Acta Amazonica,* **12**(1), 155–65.

Amedegnato, C. and Poulain, S. (1986) Diagnoses et signalisations de Romaleidae arboricoles amazoniens (Orthoptera, Acridoidea). *Annales de la Société Entomologique de France (N.S.),* **22**, 423–55.

Amedegnato, C. and Poulain, S. (1987) Les acridiens néotropicaux I: Proctolabinae Amazoniens (Orthoptera, Acrididae). *Annales de la Société Entomologique de France (N.S.),* **23**(4), 399–434.

Amedegnato, C. and Poulain, S. (1994) Nouvelles données sur les peuplements acridiens nord andéens et nord-ouest amazoniens: la famille des Romaleidae. *Annales de la Société Entomologique de France (N.S.),* **30**(1), 1–24.

Bernays, E.A. and Graham, M. (1988) On the evolution of host range in phytophagous insects. *Ecology,* **69**, 886–92.

Bolivar, I. (1881) Nota entomologicas V. Nuevas especies de ortopteros sud-americanos del Viaje al Pacifico. *Anales de la Sociedad Española de historia Natural,* **10**, 31–53.

Bolivar, I. (1890) Diagnosis de ortopteros nuevos. *Anales de la Sociedad Española de Historia Natural,* **19**, 299–334.

Bruner, L. (1910) Report on an interesting collection of locusts from Peru. *Horae Societatis Entomologicae Rossicae,* **39**, 464–88.

Carbonell, C.S. (1986) Revision of the neotropical genus *Tropidacris* (Orthoptera, Acridoidea, Romaleidae, Romaleinae). *Proceedings of the Academy of Natural Sciences of Philadelphia,* **138**(2), 366–402.

Chapman, R.F. (1990) Food selection, in *Biology of Grasshoppers* (eds R.F. Chapman and A. Joern), John Wiley & Sons, Chichester, pp. 39–72.

Colinvaux, P. (1987) Amazon diversity in light of the palaeoecological record. *Quaternary Science Review,* **6**, 93–114.

Colinvaux, P. (1993) Pleistocene biogeography and biological relationships between Africa and South America, in *Biological Relationships between Africa and South America* (ed. P. Goldblatt), Yale University, pp. 473–99.

De Geer, C. (1773) Onzième mémoire, Des criquets, in *Mémoires pour servir à l'histoire des insectes.* Tome 3, 460–504.

Denevan, W.M., Treacy, J.M., Acorn, J., Padoch, C., Demslow, J. and Flores-Paitan, S. (1984) Indigenous agroforestry in the Peruvian Amazon: Bora Indian management of swidden fallows. *Interciencia,* **9**, 346–57.

Descamps, M. (1971) Les Eumastacidae de Colombie. Révision des Paramastacinae et Eumastacinae (Acridomorpha, Eumastacoidae). *Caldasia (Bogota),* **11**(51), 99–192.

Descamps, M. (1972) Geographical regions and taxonomic groups of Acridomorpha in need of study. *Proceedings of the International Study Conference on Current and Future Problems of Acridiology,* London 1970, pp. 9–20.

Descamps, M. (1976a) Le peuplement acridien d'un périmètre d'Amazonie colombienne. *Annales de la Société Entomologique de France (N.S.)*, **12**(2) 291–318.

Descamps, M. (1976b) La faune dendrophile néotropicale. I. Revue des Proctolabinae (Orth. Acrididae). *Acrida*, **5**, 63–167.

Descamps, M. (1977) Etude des écosystéme guyanais. I. Eumastacoidea et Acridoidea dendrophiles (Orthoptera). *Annales de la Société Entomologique de France (N.S.)*, **13**(2), 193–236.

Descamps, M. (1978) La faune dendrophile néotropicale, II. Revue des Taeniophorini et Ophthalmolampini (Orth. Romaleidae). Bulletin du Muséum national d'Histoire Naturelle, Paris, 3ème sér., no. 517. *Zoologie*, **355**, 371–476.

Descamps, M. (1979a) La faune dendrophile néotropicale, IV. Le groupe des Oulenotacrae (Orthoptera, Ommatolampinae) (première partie). *Revue Française d'Entomologie (N.S.)*, **1**(3), 117–31.

Descamps, M. (1979b) Eustamacoidae néotropicaux. Diagnoses, signalisations, notes biologiques. *Annales de la Société Entomologique de France (N.S.)*, **15**(1), 117–55.

Descamps, M. (1979c) La faune dendrophile néotropicale. IV. Le groupe des Oulenotacrae (Orthoptera, Ommatolampinae) (deuxieme partie). *Revue Française d'Entomologie (N.S.)*, **1**(4), 150–9.

Descamps, M. (1980) La faune dendrophile néotropicale. V. Seconde revue des Proctolabinae amazoniens et guyanais (Orthoptères, Acrididae). *Annales de la Société Entomologique de France (NS)*, **16**(1), 19–47; **16**(2), 161–95.

Descamps, M. (1981) La faune dendrophile néotropicale. VI. Diagnoses génériques et spécifiques d'Acridoidea de la région de Manaus (Orthoptera). *Annales de la Société Entomologique de France (N.S.)*, **17**(3), 311–30.

Descamps, M. (1982) Eumastacoidea néotropicaux, diagnoses, signalisations, notes biologiques II (Orth.). *Bulletin de la Société Entomologique de France (NS)*, **87**(5–6), 141–80.

Descamps, M. (1983a) La faune dendrophile néotropicale. VII. Seconde revue des Ophthalmolampini. Le groupe des Lagarolampae (Orthoptera, Romaleidae). *Annales de la Société Entomologique de France (N.S.)*, **19**, 17–35.

Descamps, M. (1983b) La faune dendrophile néotropicale. IX. Seconde revue des Ophthalmolampae (Orthoptera, Romaleidae). *Annales de la Société Entomologique de France (NS)*, **19**(4), 367–404.

Descamps, M. (1984) Revue préliminaire de la tribu des Copiocerini (Orth. Acrididae). Mémoires du Muséum National d'Histoire Naturelle, Paris, Sér. A, *Zoologie*, **130**, 1–72.

Descamps, M. and Amedegnato, C. (1971a) Contribution à la faune des Acridoidea de Colombie (Missions M. Descamps). I. Les genres *Syntomacris* Walker, 1870 et *Pseudococama*, nov. *Annales de la Societé Entomologique de France (NS)*, **7**(1), 95–113.

Descamps, M. and Amedegnato, C. (1971b) Contribution à la faune des Acridoidea de Colombie (Missions M. Descamps). II. Les genres *Taeniophora* Stal, 1873 et *Megacephalacris*, nov. *Annales de la Societé Entomologique de France (NS)* **7**(1), 115–46.

Descamps, M. and Amedegnato, C. (1972) Contribution à la faune des Acridoidea de Colombie (Missions M. Descamps). III. Diagnoses de Catantopinae (sensu lato). *Annales de la Societé Entomologique de France (NS)*, **8**(3), 505–59.

Descamps, M. and Amedegnato, C. (1989) Les genres *Vilerna*, *Locheuma* et *Pseudovilerna* nov. I. Le genre *Vilerna* Stal, 1873 (Orthoptera, Acrididae, Ommatolampinae). *Revue Française d'Entomologie (N.S.)*, **11**(1), 17–28.

Descamps, M. and Carbonell, C.S. (1985) Revision of the neotropical arboreal genus *Titanacris* (Orthoptera, Acridoidea, Romaleidae). *Annales de la Société Entomologique de France (N.S.)*, **21**(3), 259–85.

Descamps, M. and Wintrebert, D. (1966) Pyrgomorphidae et Acrididae de Madagascar; Observations Biologiques et Diagnoses. *Revista Española de Entomología*, 42 (1–2), Madrid.

Dirsh, V.M. (1962) The Acridoidea (Orthoptera) of Madagascar I. Acrididae (except Acridinae). *Bulletin of the British Museum (Natural History) Entomology*, **12**(6), 276–350.

Dirsh, V.M. and Descamps, M. (1968) *Faune de Madagascar XXVI. Insectes Orthoptères Acridoidea, Pyrgomorphidae et Acrididae*. ORSTOM–CNRS.

Farrow, R.A. (1990) Flight and migration in Acridoids, in *Biology of Grasshoppers* (eds R.F. Chapman and A. Joern), John Wiley & Sons, Chichester, pp. 227–314.

Fittkau, E.J. and Klinge, H. (1973) On biomass and trophic structure of the central Amazonian rain forest ecosystem. *Biotropica*, **5**(1), 2–14.

Flores-Paitan, S. (1984) Analysis of old Bora swidden fallows, in *Indigenous Agroforestry in the Northeast Peruvian Amazon*, Report of the University of Wisconsin–Universidad Nacional de la Amazonia Peruana, Man and the Biosphere Project 1981–1983, pp. 111–223.

Foster, R.B. (1990) The floristic composition of the Rio Manu floodplain, in *Four Neotropical Rainforests* (ed. A.H. Gentry), Yale University Press, pp. 99–111.

Futuyma, D.J. (1983) Selective factors in the evolution of host choice by phytophagous insects, in *Herbivorous Insects: Host Seeking Behavior and Mechanisms* (ed. S. Ahman), Academic Press, pp. 227–44.

Gentry, A.H. (1982) Neotropical floristic diversity: phytogeographical connection between Central and South America, Pleistocene climatic fluctuations, or an accident of the Andean Orogeny? *Annals of the Missouri Botanical Gardens*, **69**, 557–93.

Gentry, A.H. (1986) Quelques modèles phytogéographiques du domaine néotropical et leurs implications pour la conservation en Equateur, in *Colloques et Séminaires Equateur 1986, vol. 1, Quito 7–12 Juillet 1986*, Editions de l'ORSTOM, pp. 145–61.

Gentry, A.H. (1990) Floristic similarities and differences between Southern Central America and Upper and Central Amazonia, in *Four Neotropical Rainforests* (ed. A.H. Gentry), Yale University Press, pp. 141–57.

Gentry, A.H. (1993) Diversity and floristic biological relationships between Africa and South America, in *Biological Relationships between Africa and South America* (ed. P. Goldblatt), Yale University, pp. 500–47.

Gerstaecker, A. (1889) Carakteristil einer Reihe bemerkenswerther Orthopteren. Mittheilungen aus dem naturwissenschaftlichen *Verein für Neu-Vorpommern und Rugen in Greifswald* **20**, 1–58.

Gunther, K (1940) Uber die Verbreitung eininger Insekten im Gebiete des Amazonenstromes und die Frage eines solumbischen Faunendistriktes in der brasilianischen Subregion. *Archiv für Naturgeschichte, Leipzig, (NF)*, **9**(4), 450–500.

Hammond, P.M. (1990) Insect abundance and diversity in the Dumoga-Bone National Park, N. Sulawesi, with special reference to the beetle fauna of lowland rain forest in the Toraut region, in *Insects and the Rain Forests of South East Asia (Wallacia)* (eds W.J. Knight and J.D. Holloway), Royal Entomological Society, London, pp. 197–254.

Hubbell, S.P. and Foster, R.B. (1990) Structure, dynamics, and equilibrium status of old-growth forest on Barro Colorado Island, in *Four Neotropical Rainforests* (ed. A.H. Gentry), Yale University Press, pp. 522–41.

Kahn, F. and de Granville, J.J. (1994) *Palms in Forest Ecosystems of Amazonia*, Ecological Studies 95, Springer-Verlag, Berlin, Heidelberg.

Linnaeus, C. (1758) Systema Naturae per regna tria Naturae, secundum Classes, Ordines, Genera, Species, cum characteribus, differentiis, synonymis, locis. Editio decima reformata, Holmiae, tonus I, 824 pp.

Mori, S.A. (1989) Diversity of Lecythidaceae in the Guianas, in *Tropical Forests Botanical Dynamics, Speciation and Diversity* (eds L.B. Holm-Nielsen, I.C. Nielsen and H. Balslev), Academic Press, pp. 319–32.

Müller, P. (1973) *The Dispersal Centres of Terrestrial Vertebrates in the Neotropical Realm. A Study in the Evolution of Neotropical Biota and its Native Landscapes*, Biogeographica, Vol. 2, Dr W. Junk, The Hague.

Oldeman, R.A.A. (1983) The design of ecologically sound agroforests, in *Plant Research and Agroforestry* (ed. P. Huxley), ICRAF, Nairobi.

Oldeman, R.A.A. (1989) Dynamics in tropical rain forests, in *Tropical Forests Botanical Dynamics, Speciation and Diversity* (eds L.B. Holm-Nielsen, I.C. Nielsen and H. Balslev), Academic Press, pp. 3–21.

Popov, G.B., McComie, L.D. and Launois-Luong, M.H. (1994) The Moruga grasshopper in Trinidad *Coscineuta virens* (Thunberg 1815) (Acrididae: Proctolabinae). *Journal of Orthoptera Research*, **2**, 49–60.

Prance, G.T. (1990) The floristic composition of the forests of central Amazonian Brazil, in *Four Neotropical Rainforests* (ed. A.H. Gentry), Yale University Press, pp. 112–40.

Prance, G.T., Rodriguez, W. and da Silva, M.F. (1976) Inventario florestal de um hectare de mata de terra firme Km 30 da Estrada Manaus Itacoatiara. *Acta Amazonica*, **6**(1), 9–35.

Price, P.W. (1984) *Insect Ecology*, John Wiley & Sons, Chichester, UK.

Rehn, J. A. G. and Rehn, J. W. H. (1942) A review of the New World Eumastacinae (Orthoptera, Acrididae). Part II. *Proceedings of the Academy of Natural Sciences of Philadelphia*, **94**, 1–88, pl. 1–2.

Riede, K. (1987) A comparative study of mating behaviour in some neotropical grasshoppers (Acridoidea). *Ethology*, **76**, 265–96.

Riede, K. (1993) Abundance differences of phytophagous tropical insects: the case of forest grasshoppers (Orthoptera, Caelifera), in *Animal–Plant Interactions in Tropical Environments* (eds W. Barthlott, C.M. Naumann, K. Schmidt-Loske and K.L. Schumann), Results of the annual meeting of the German Society for Tropical Ecology, Bonn, February 13–16, 1992. Bonn, Zoologisches Forschungsinstitut und Museum Alexander Koenig, pp. 211–18.

Roberts, H.R. (1973) Arboreal Orthoptera in the rain forests of Costa Rica collected with insecticide: a report on the grasshoppers (Acrididae), including new species. *Proceedings of the Academy of Natural Sciences of Philadelphia*, **125**(3), 49–66.

Roberts, H.R. (1980) *New taxa of the subfamily Leptysminae from Brazil (Acrididae: Orthoptera).* Notula Naturae, Academy of Natural Sciences of Philadelphia, No. 458.

Roberts, H. R. and Carbonell, C. S. (1979) A revision of the genera *Stenopola* and *Cornops* (Orthoptera, Acrididae, Leptysminae). *Proceedings of the Academy of Natural Sciences of Philadelphia*, **131**, 104–30.

Rowell, C.H.F. (1978) Food plant specificity in neotropical rain-forest acridids. *Entomologia Experimentalis et Applicata*, **24**, 651–62.

Rowell, C.H.F. (1987) The biogeography of Costa Rican acridid grasshoppers, in relation to their putative phylogenetic origins and ecology, in *Evolutionary Biology of Orthopteroid Insects* (ed. B. Baccetti), John Wiley & Sons, Chichester, UK, pp. 470–82.

Schnell, R. (1987) *La Flore et la Végétation de l'Amérique Tropicale*, I. Masson, Paris.

Scudder, S. H. (1869) A study of the gigantic lobe-crested grasshoppers of South and Central America. *Proceedings of the Boston Society of Natural History*, **12**, 345–55.

Serville, J.G. (Audinet-Serville) (1831) Revue méthodique des insectes de l'Ordre des Orthoptères. *Annales des Sciences Naturelles, Paris*, **22**, 28–65, 134–67, 262–92.

Stal, C. (1878) Systema acridiodeorum. Essai d'une systématisation des acridoidées. *Bihang till Kongliga Svenska Vetenskaps-akademiens Handlingar*, **5**(4), 1–100.

Stoll, C. (1813) Naturlijke en naar leven nauwerkeurige gekleurde afbeeldingen en beschrijvingen der spoken, wandelende bladen, zabelspringhanen, krekels, trekspringhanen en kakkerlaken, in alle vier deelen del wereld Europa, Asia, Afrika en Amerika. Représentation exactement colorée d'apres nature des spectres ou phasmes, des mantes, des sauterelles, des grillons, des criquets et des blattes qui se trouvent dans les quatre parties du monde, l'Europe, l'Asie, l'Afrique et l'Amerique. Amsterdam, J.C. Sepp et fils, vol. 1, 79 pp., 25 pl.; vol. 2, 28 + 42 pp., 13 + 23 pl.

Stork, N.E. (1987) Guild structure of arthropods from Bornean rain forest trees. *Ecological Entomology*, **12**, 69–80.

Thunberg, C. P. (1842) Grylli monographia, illustrata. *Mémoires de l'Académie Impériale des Sciences de St. Pétersbourg*, **3**, 390–430, pl. 14.

Walker, F. (1870) Catalogue of the specimens of Dermaptera Saltatoria in the collection of the British Museum. British Museum (Natural History) London, **3**, 425–604; **4**, 605–809; **5**, 811–50.

Williams, C.A., Harborne, J.B. and Clifford, H.T. (1973) Negatively charged flavones and tricin as chemosystematic markers in the Palmae. *Phytochemistry*, **12**, 2417–30.

Appendix 14A List of canopy grasshopper species mentioned

ACRIDOIDEA: ROMALEIDAE
OPTHALMOLAMPINI

Ophthalmolampae

Adrolampis contumax Descamps, 1983
Adrolampis limbatipes Descamps, 1983
Apophylacris incondita Descamps, 1983
Nothonautia splendens Descamps, 1978
Ophthalmolampis brunneiceps Descamps, 1983
Ophthalmolampis condita Descamps, 1983
Ophthalmolampis elaborata Descamps, 1983
Ophthalmolampis praeferox Descamps, 1983
Ophthalmolampis sigillata (Descamps, 1978)
Ophthalmolampis truculenta Descamps, 1978
Peruviacris cerciata Descamps, 1983

Nautiae

Euprepacris mutilipennis Descamps, 1983
Euprepacris piperacipes Descamps, 1983
Pseudonautia maculipes Descamps, 1978
Pseudonautia seducta Descamps, 1983

Lagarolampae

Habrolampis bicolor Descamps, 1978
Helolampis coloniana Descamps, 1983

Appendix 14A continued

Lagarolampis amazonica Descamps, 1978
Helicopacrae
Helicopacris viridans Descamps, 1978

BACTROPHORINI
Bactrophora sp.
Bora nemoralis Amedegnato and Descamps, 1979
Hyleacris rubrogranulata Amedegnato and Descamps, 1979
Mezentia sp.

TAENIOPHORINI
Taeniophora caqueta Descamps and Amedegnato, 1971

HISYCHIINI
Hisychius nigrispinus Stall, 1878
Pseudhisychius nigroornatus Amedegnato and Poulain, 1986

LEGUINI
Ampiacris insolita Amedegnato and Poulain, 1986
Legua crenulata Stoll, (1813)

EUROSTACRINI
Eurostacris puncticrus Descamps, 1978
Pseudeurostacris valida Descamps, 1978

Other tribes
Aprionacris coerulescens (Bolivar, 1890)
Prionacris cantrix Descamps, 1981
Titanacris albipes (De Geer, 1773)
Titanacris humboldtii (Scudder, 1869)
Titanacris picticrus marginalis Descamps and Carbonell, 1985
Tropidacris cristata (Linnaeus, 1758)
Trybliophorus octomaculatus Serville, 1831
Xomacris nuptialis (Gerstaecker, 1889)

ACRIDOIDEA: ACRIDIDAE
COPIOCERINAE
Copiocera austera Gerstaecker, 1889
Copiocera nigricans (Walker, 1870)
Copiocera rubricrus Descamps, 1984
Eumecacris colombiana Descamps, 1978
Eumecacris pygmaea Descamps, 1984

PROCTOLABINAE
COSCINEUTINI
Coscineuta pulchripes (Gerstaecker, 1889)
Coscineuta cicatricosa Bolivar, 1890

Appendix 14A continued

PROCTOLABINI
Proctolabae
Adelotettix obscurus (Bruner, 1910)
Cercoceracris grandicula Descamps, 1980
Dendrophilacris bora Descamps, 1980
Dendrophilacris lorenzi Amedegnato and Poulain, 1987
Dendrophilacris secoya Descamps, 1980
Halticacris spinifer Descamps, 1976
Poecilocloeus ferus Descamps, 1976
Poecilocloeus estironana Amedegnato and Poulain, 1987
Poecilocloeus uncinatus Descamps, 1980
Witotacris concinna Descamps, 1976
Witotacris annulicrus Amedegnato and Poulain, 1987
Saltonacrae
Loretacris fascipes Amedegnato and Poulain, 1987
Saltonacris avellinoi Descamps, 1976
Saltonacris lucicola Descamps, 1976
Eucerotettix ludificator Descamps, 1980
Eucephalacrae
Eucephalacris miguelangeli Descamps, 1980
Pareucephalacris luridicrus Descamps, 1980

RHYTIDOCHROTINAE
Galidacris eckardtae (Gunther, 1940)
Galidacris purmae in litt.

LEPTYSMINAE
Stenopola laticerca Roberts, 1980*
Stenopola bicoloripes (Descamps and Amedegnato, 1972)

OMMATOLAMPINAE
SYNTOMACRINI
Caloscirtae
Anoptotettix levis Amedegnato and Descamps, 1979
Calohippus sp.
Machigengacris nigrovittata in litt.
Occiotettix lubricus Amedegnato and Descamps, 1979
Rhyphoscirtus stragulatus Amedegnato and Descamps, 1979
Stigacris rubropicta Descamps, 1977
Syntomacrae
Syntomacris sp.
N. gen. (*S. balachowskii*) Descamps and Amedegnato, 1971

OMMATOLAMPINI
Oulenotacrae
Agrotacris corticicolor Descamps, 1979
Agrotacris nubilosa Descamps, 1979
Agrotacris witotae Amedegnato, 1985

Appendix 14A continued

Agrotacris sp.
Anablysis gaschei Descamps, 1979
Anablysis guyoti Descamps, 1979
Eurybiacris luteoguttata Descamps, 1979
Hysterotettix nigricoxa Descamps, 1979
Vilernae
Bryophilacris muscicolor Descamps, 1976
Hypsipages dives Gerstaecker, 1889
Rhabdophilacris sylvatica Amedegnato, 1985
Other groups
Agenacris subbrevis Amedegnato and Descamps, 1979
Sciponacris amazonica Amedegnato, 1985

EUMASTACOIDEA
PROSCOPIIDAE
Apioscelis sp.
Proscopia sp. 1, sp. 2

EUMASTACIDAE
Eumastax becharai Descamps, 1971
Eumastax megacephala Descamps, 1982
Eumastax rubriventris Descamps, 1982
Eumastax vittata napoana Descamps, 1982
Eumastax zumuniana Descamps, 1982
Eumastacops caligo Rehn and Rehn, 1942
Pareumastacops witotae Descamps, 1979
Pseudeumastacops militaris (Gerstaecker, 1889)
Pseudomastax personata (Bolivar, 1881)

---

* *Stenopola nigricans* Roberts and Carbonell 1979, very similar to *Stenopola* (*Innusia*) *dorsalis* (Thunberg, 1824), but a strictly arboricolous forest species (compared with this last), comprises, according to Roberts, two subspecies: *S.n. nigricans*, Guyanese, and *S.n. laticerca*, only known from northwest Amazonia. The two forms are from our samples and it is considered that *S. laticerca* is a distinct species.

15

Dipteran tree-crown assemblages in a diverse southern temperate rainforest

R.K. Didham

ABSTRACT

The abundance, richness and diversity of canopy-dwelling Diptera in a New Zealand rainforest varied significantly with habitat type. Diptera in southern beech tree-crowns (*Nothofagus solandri*, Fagaceae) were more abundant and diverse than the fauna in the crowns of three species of Podocarpaceae or in the crowns of *Beilschmiedia tawa* (Lauraceae). The *B. tawa* tree fauna was particularly depauperate.

Isolating the significant determinants of dipteran abundance and diversity on trees was confounded by the coincidence of *N. solandri* sites at the forest edge due to site microenvironmental preferences, and the influence of honeydew secretions as an attractant for arthropods. Close proximity to forest edge and low percentage canopy cover were important determinants of high abundance and diversity.

Patterns of family representation and species dominance varied markedly between habitat types and to a lesser degree between tree species. While many species were specific to different habitat types, few were host-specific. Non-random space utilization by Diptera in the forest canopy, combined with the swarming habit of many species, indicate that Diptera play a much more important role in arboreal community interactions than implied by their traditional designation as 'tourists' in the canopy.

INTRODUCTION

The taxonomic composition and abundance of arthropods on trees is broadly determined by a variety of deterministic and stochastic factors which influence community assembly (Wiens *et al.*, 1991). These include: the size and composition of the regional arthropod species pool

Canopy Arthropods. Edited by N.E. Stork, J. Adis and R.K. Didham. Published in 1997 by Chapman & Hall, London. ISBN 0 412 74900 9

(Southwood, 1960, 1961a,b, 1973; Opler, 1974; Strong *et al.*, 1984; Kennedy and Southwood, 1984; and review in Cornell and Lawton, 1992), dispersal and colonization potentials of particular species (Southwood, 1961b; Southwood and Kennedy, 1983), population sizes of prey and predators (Cohen, 1977), the diversity of plant species within the community (Murdoch *et al.*, 1972; Witkowski, 1973; Southwood *et al.*, 1979; Brown and Southwood, 1983), host-species preferences (Beaver, 1979; Futuyma and Gould, 1979; Mitter *et al.*, 1991; Basset, 1992; Gaston, 1993), and chance variations in weather and colonization (Strong *et al.*, 1984; Krebs, 1985; Wiens *et al.*, 1991). The factor or factors which are of prime importance to a given community are clearly context-sensitive, depending on such variables as locality, history or season, and may vary through time. The relative importance of these determining factors will also depend on the arthropod taxon under investigation. For example, one focus of explanations for the composition of arthropods on plants is the role of primary nutrition and secondary metabolites of plants and the concept of plant–herbivore coevolution (Ehrlich and Raven, 1964; Feeny, 1976; Rhoades and Cates, 1976; Rosenthal and Janzen, 1979; Coley *et al.*, 1985). This may be true for some phytophages, but such complex interactions are difficult to study at the community level and often have little bearing on associated arthropod trophic levels (but see Price *et al.*, 1980; Schultz, 1983). In this paper I examine the abundance and species composition of Diptera in the canopies of five tree species from three different habitat types in a southern temperate rainforest, New Zealand, and investigate local biotic and abiotic factors structuring the observed patterns.

Diptera have received relatively little attention in community diversity studies. An unfortunate development of many studies has been the classification of the majority of dipteran families as 'tourists' (Moran and Southwood, 1982; West, 1986; Stork, 1987, 1991). The tourist guild (*sensu* Root, 1967, 1973) was defined by Moran and Southwood (1982), with respect to the canopy fauna, as those species having no lasting or intimate association with the plant, but which may be subject to local predators and so form part of the trophic web of the arboreal community. Whether Diptera should be considered more as tourists than as residents is debatable. Some authors consider that a handful of dipteran families contribute integrally to various trophic guilds in the canopy (*viz.* predators: Empididae, Dolichopodidae, Ceratopogonidae, Rhagionidae; parasitoids: Tachinidae, Pipunculidae; scavengers/fungal feeders: Mycetophilidae, Sciaridae; phytophages: Cecidomyiidae, Tephritidae; Moran and Southwood, 1982; West, 1986; Stork, 1987, 1991), but that most adult Diptera do not actively feed in the canopy and hence should be classed merely as tourists (Stork, 1991). These species may be using the canopy for such purposes as mating, avoiding predators or unsuitable microclimatic conditions, for resting sites, or as a path to other habitats.

The designation 'tourist' holds specific connotations with respect to the ephemeral nature and non-specificity of the interaction between the species in question and its 'host' community. These premises are naturally very difficult to test, even in relatively simple communities. It is arguable, though, that even ephemeral dipteran associations with the 'resident' arboreal community are made all the more significant by the swarming habit of many taxa. These swarms must provide a vast resource for canopy predators, and by their very nature represent an integral component of arboreal communities. This importance may to some extent be negated if tourist activity is patchy and non-specific in nature, but again it is not clear whether Diptera are using the canopy in a random or non-random manner. If space utilization is non-random, i.e. if species tend to be habitat-specific or host tree-specific to some degree, then Diptera may be predictable and regular components of the canopy community.

METHODS

Study site and sample collection

Arthropod samples were collected using flight intercept traps (FITs), consisting of a clear 0.25-m^2 PVC sheet and suspended collecting tray (500 × 300 mm) (Didham, 1992). FITs were placed in the canopies of 14 trees in the Blue Duck Scientific Reserve, South Island, New Zealand (Figure 15.1). The forest is a diverse mixed podocarp–broadleaf stand of approximately 85 ha, interspersed with isolates of mountain beech (Wardle, 1961; Williams, 1980). The reserved area represents one of the last significant stands of temperate rainforest surviving along the east coast of South Island (D.A. Norton, personal communication). The five tree species sampled were: (i) three mountain beech trees (*Nothofagus solandri* var. *cliffortioides*, Fagaceae); (ii) three matai trees (*Prumnopitys taxifolia*, Podocarpaceae); (iii) three rimu trees (*Dacrydium cupressinum*, Podocarpaceae); (iv) three totara trees (*Podocarpus totara*, Podocarpaceae); and (v) two tawa trees (*Beilschmiedia tawa*, Lauraceae).

Site characteristics are listed in Table 15.1. A full description of study sites, including plant composition and tree characteristics, is given by Didham (1992). Matai, rimu and totara were the dominant podocarp trees. Average canopy height was 20–25 m and subcanopy and shrub layers were extremely diverse. In contrast, tawa, which is at the southern limit of its distribution in the reserve, formed a pure, dense, low-canopy (15–20 m) stand with very sparse subcanopy and shrub layers. Beech formed small monocultural habitats with a low (15–20 m), open canopy.

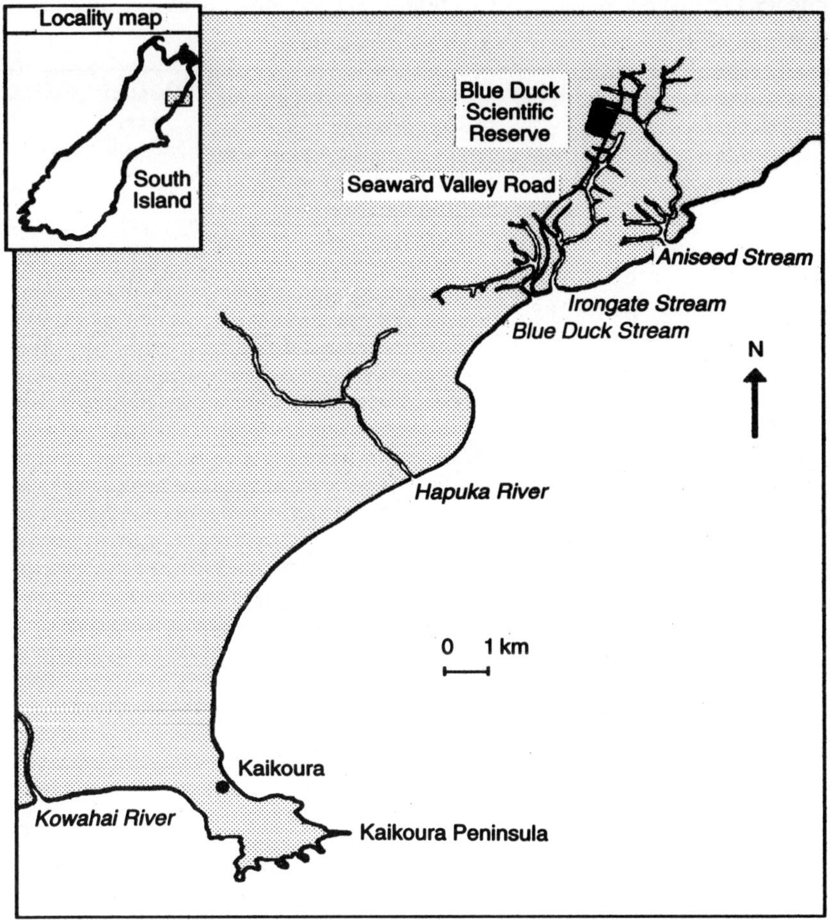

Figure 15.1 Locality map showing the study site, the Blue Duck Scientific Reserve, 30 km north of Kaikoura on the east coast of South Island, New Zealand.

A unique feature of some *Nothofagus* species in New Zealand is that they support populations of sooty beech scale, *Ultracoelostoma assimile* and *U. brittini* (Hemiptera, Margarodidae). The scale insects produce honeydew secretions which support profuse growths of sooty mould fungi (*Capnodium, Trichopeltheca* and *Capnocybe* spp., Hughes, 1972) and provide a huge resource base for forest animals (Gaze and Clout, 1983; Boyd, 1987; Moller and Tilley, 1987, 1989). The influence of honeydew on the beech tree arthropod community was investigated by Didham (1993).

Table 15.1 Selected measures of habitat structure and canopy architecture. See text for further explanation

	Taxon no.	*Canopy depth (m)*	*Canopy cover (%)*	*Tree height (m)*	*Trap height (m)*	*Vegetational biomass (basal area m²/plot)**	*Distance from edge (m)*	*Epiphyte load (0–3)⁺*
Beech–1	1	8.0	30	20.0	15.0	1.94	30	0.0
Beech–2	1	8.0	50	18.0	13.5	3.28	50	0.0
Beech–3	1	8.0	50	21.0	15.0	2.09	80	0.0
Matai–1	2	6.0	85	22.0	20.0	9.34	300	0.3
Matai–2	2	7.0	80	23.0	19.0	7.75	300	0.3
Matai–3	2	5.0	75	22.0	17.0	5.13	150	1.3
Rimu–1	3	8.0	35	23.5	20.0	6.12	300	1.7
Rimu–2	3	7.0	20	21.0	15.0	3.98	270	1.0
Rimu–3	3	11.0	60	22.0	13.0	3.96	120	2.0
Totara–1	4	9.0	80	23.0	21.0	6.74	280	1.0
Totara–2	4	4.0	65	20.0	17.0	7.34	280	1.3
Totara–3	4	10.0	85	19.0	12.0	6.68	160	3.0
Tawa–1	5	6.0	85	16.0	12.0	4.50	240	0.0
Tawa–2	5	6.0	95	18.5	13.5	4.16	240	0.7

* Basal area of trees larger than 10.0 cm diameter at breast height (DBH), circular plot, radius 15 m, centred on the study tree.

⁺ Subjective scale 0–3 indicating increasing epiphyte abundance. Values are averaged scores from orchid, lilly and vine loads.

FITs were placed 13–20 m above the ground and sampled trees were 25–800 m apart. Six complete series of weekly samples collected between November 1990 and February 1991 were analysed. Adult Diptera (excluding Cecidomyiidae and Sciaridae) were sorted to recognizable taxonomic units, hereafter referred to as species.

Habitat structure and canopy architecture

Eight measures of forest structure (Table 15.1) were correlated with Dipteran abundance and species richness in tree-crowns. Percentage canopy cover was used in preference to leaf shape or leaf area which have been used in previous studies (Moran and Southwood, 1982; Southwood *et al.*, 1982a; Kennedy and Southwood, 1984; Leather, 1986). In some of these studies leaf length was found to be an important determinant of insect species richness on trees. However, leaf length does not take into account related factors such as leaf density, total leaf area per tree, or distance between adjacent trees, all of which may affect the composition of the insect community at a site. Consequently a 'composite' measure, percentage canopy cover, was taken to represent the product of leaf area, density and tree spacing at each site.

Environmental variables were correlated with dipteran abundance and species richness using Pearson's product moment correlation coefficient, *r*, and stepwise multiple regression on the MINITAB statistical package. Variables were entered and removed from multiple regression models at a probability level of 0.05.

Not all habitat variables were statistically independent. Trap height was significantly correlated with tree height ($r = 0.773$, $P < 0.001$), and tree species, vegetational biomass and trap height were significantly correlated with distance from forest edge ($r = 0.611$, $r = 0.792$ and $r = 0.583$, respectively, $P < 0.05$). Beech sites provided pivotal weight in each of the latter cases, as beech sites were characterized by close proximity to the forest edge (where conditions favour this tree species), low vegetational biomass due to open stand structure and small average tree diameter at breast height (DBH), and low trap height as a consequence of structurally small tree-crowns. In general, then, beech habitat was structurally very different to the other forest types in the area.

RESULTS

The fauna

Diptera dominated FIT samples with 166 210 specimens (91.1% of the total fauna). From six (of 10) weekly sample sets analysed, 94 299 Diptera, representing 34 families and 373 species (species count excludes Sciaridae and Cecidomyiidae) were recorded from 14 tree-crowns (Figure 15.2). Mycetophilidae dominated, with 72% of individuals and 26% of species. Sciaridae, Ceratopogonidae, Cecidomyiidae and Dolichopodidae were also abundant families in canopy samples. In terms of species richness, however, Tipulidae and Tachinidae were particularly important.

Seven of the 10 most abundant species were Mycetophilidae and the four most common Diptera were all species of the genus *Tetragoneura* (Table 15.2).

Patterns of abundance and species richness

Diptera abundance varied by an order of magnitude between tree-crowns. Beech sites had high abundances, as did a number of podocarp sites. Tawa tree canopies were comparatively depauperate (Figure 15.3). In a forward stepwise multiple regression analysis, percentage canopy cover was the single best predictor of variation in abundance between sites ($r^2 = 0.35$, Table 15.3); more open canopy sites had a higher dipteran

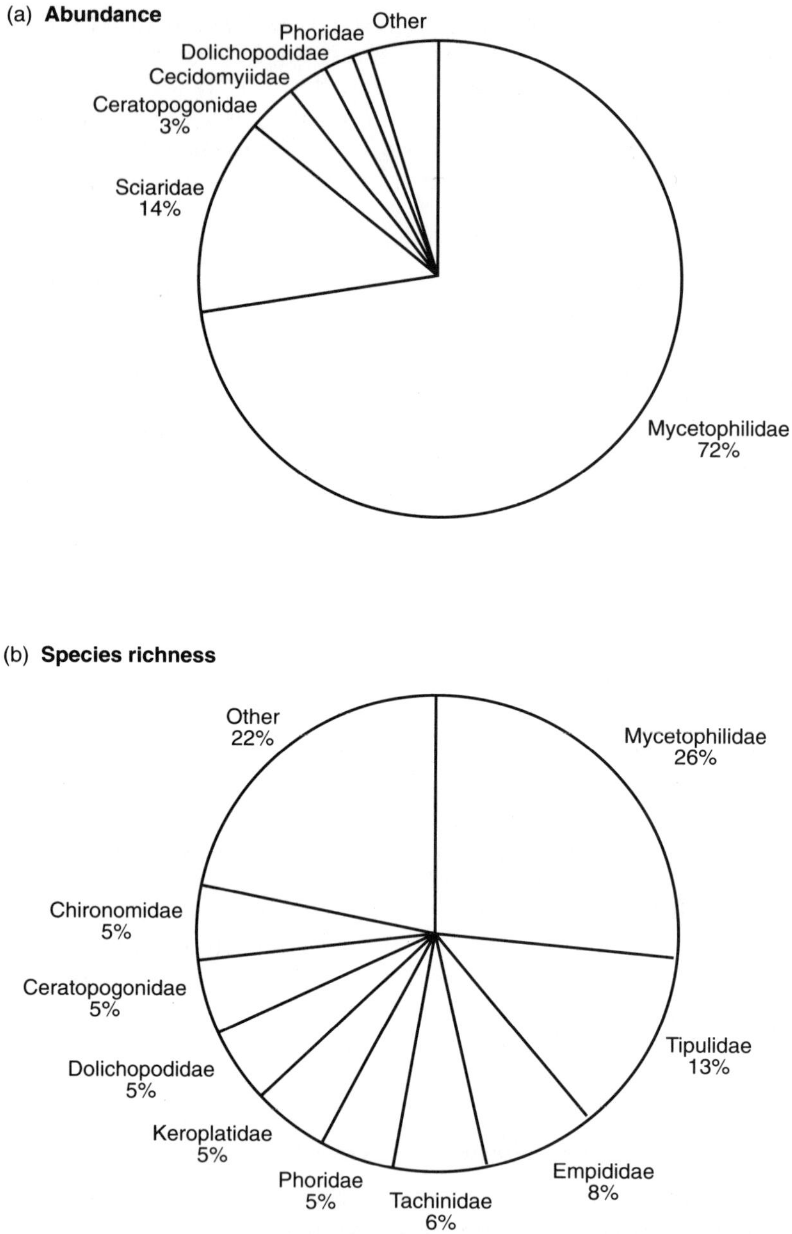

Figure 15.2 (a) Abundance and (b) species richness of dominant dipteran families captured in the forest canopy. Total abundance = 94 299, total number of species = 373.

Table 15.2 The 10 most abundant Diptera captured in the forest canopy. All Mycetophilidae except as indicated

Rank dominance	*Taxa*	*Abundance*
1	*Tetragoneura spinipes*	36 786
2	*Tetragoneura* sp.1	24 524
3	*Tetragoneura* sp.4	1 463
4	*Tetragoneura* sp.3	1 225
5	*Mycetophila fagi*	965
6	Ceratopogonidae sp.6	920
7	Ceratopogonidae sp.4	878
8	Dolichopodidae sp.12	636
9	*Cycloneura* sp.2	635
10	*Anomalomyia guttata*	624

abundance. Some of the variation in Dipteran abundance was due to the super-abundance of two species of Mycetophilidae, *Tetragoneura* sp.1 and *T. spinipes*, as reflected in a strong correlation of site abundance with the Berger–Parker Dominance Index, BPI (r = 0.723, P <0.01) (Berger and Parker, 1970).

Beech tree-crowns also had greatest numbers of species and greatest species diversity, as measured by William's α (Fisher *et al.*, 1943) (Figure 15.3). Williams α-diversity was calculated without the two dominant species, *Tetragoneura* sp.1 and *T. spinipes*, because of the strong influence of the abundance of these species on diversity values (correlation between William's α and BPI, r = –0.575, P <0.05). Stepwise multiple regression analyses identified distance from forest edge and percentage canopy cover as the best predictors of variation in both species richness (total r^2 = 0.75) and species diversity (total r^2 = 0.66) (Table 15.3).

Species composition

In terms of abundance, dipteran family representation varied considerably between tree species (Figure 15.4). Beech, rimu and totara trees were similar in having a 70–80% mean representation of Mycetophilidae, 5–12% mean representation of Sciaridae, and important contributions from Cecidomyiidae, Ceratopogonidae and Dolichopodidae. Matai tree-crowns had a considerably higher mean representation of Sciaridae. In tawa tree-crowns, Sciaridae dominated and Ceratopogonidae and Cecidomyiidae were better represented, but Mycetophilidae were considerably less abundant (Figure 15.4).

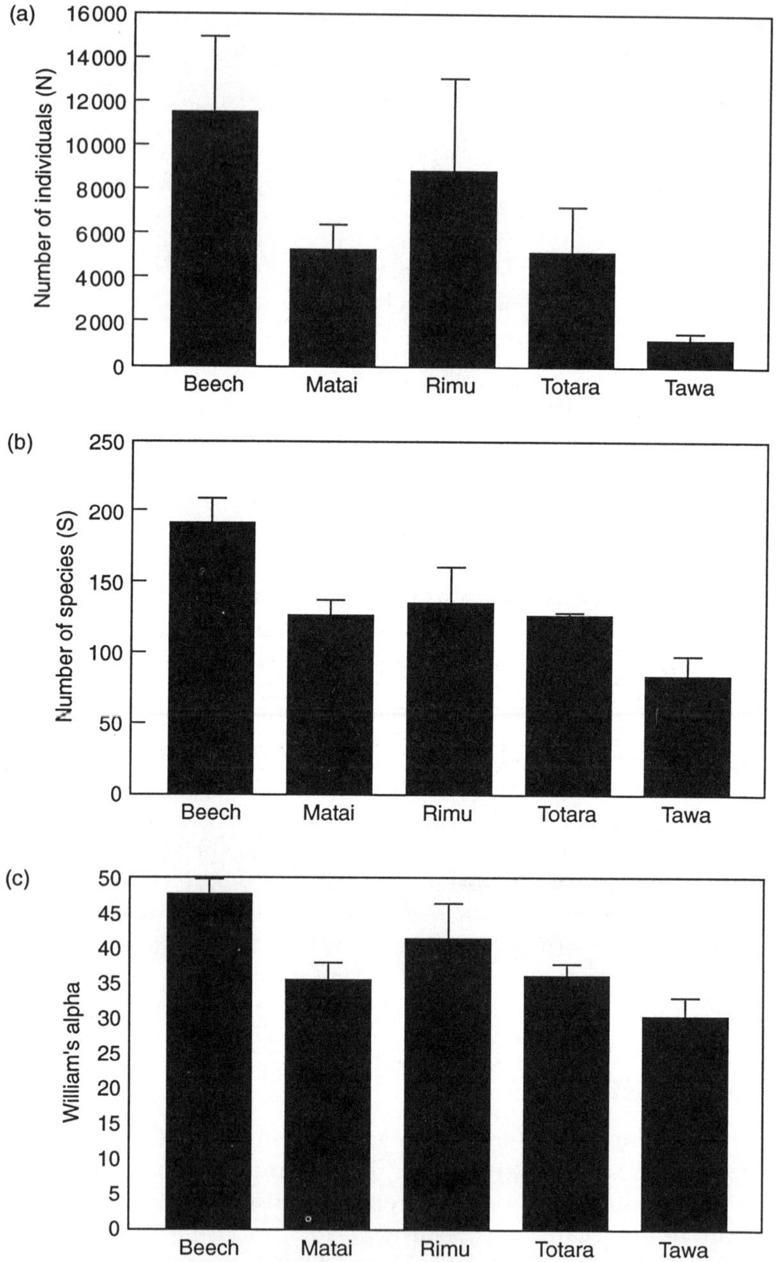

Figure 15.3 Mean (± S.E.M.) (a) abundance, (b) species richness and (c) species diversity of diptera in the canopies of different tree species (n = 3 trees, except tawa, n = 2). William's α-diversity is calculated excluding *Tetragoneura* sp.1 and *T. spinipes*.

Table 15.3 Stepwise multiple regression analysis to identify significant environmental predictors of dipteran abundance, species richness and species diversity in tree-crowns. α(exc.) is William's α-diversity calculated from samples after the exclusion of the two dominant species, *Tetragoneura* sp.1 and *T. spinipes* (see Results)

Variable entered	*Parameter estimate*	*Partial r^2*	*Model r^2*
Abundance			
Intercept	14163.00		
Percentage canopy cover	–138.00	0.35	0.35
Species richness			
Intercept	226.00		
Distance from forest edge	–0.25	0.58	0.58
Percentage canopy cover	–0.70	0.17	0.75
William's α(exc.)			
Intercept	59.17		
Distance from forest edge	–0.052	0.52	0.52
Percentage canopy cover	–0.147	0.14	0.66

In terms of species richness, mean family representation was very much more similar between tree species (Figure 15.5). Beech, matai and rimu trees, in particular, were practically identical in mean family representation. In totara and tawa tree-crowns, Dolichopodidae, Phoridae and Ceratopogonidae were better represented, while Empididae and to some extent Tipulidae were less well represented (Figure 15.5).

Table 15.4 shows the rank dominance of the five most abundant dipteran species in each tree-crown. The concordance of species ranking among sites was low (Kendall's coefficient of concordance $W = 0.451$, for 18 dominant species of Diptera). Beech sites were dominated by *Tetragoneura* sp. A, *T. spinipes*, Ceratopogonidae sp.6, *Anomalomyia guttata* and *Cycloneura* sp.2. Podocarp sites were characteristically dominated by *T. spinipes*, *T.* sp.1, *T.* sp.3, *T.* sp.4, and Ceratopogonidae sp.4 and sp.5. Tawa sites were dominated instead by Ceratopogonidae sp.4 and sp.5, reinforcing the lesser importance of Mycetophilidae in tawa tree-crowns. Pair-wise Kendall's rank correlation coefficients (Table 15.5) show clearly that beech canopies were dominated by different species to those in podocarp and tawa canopies. No differentiation in species dominance could be made between different podocarp tree species. Only Rimu-1 was aberrant in species dominance (see Table 15.4), but the differences between Rimu-1 and other podocarps were not apparent in the pair-wise rank correlations given in Table 15.5.

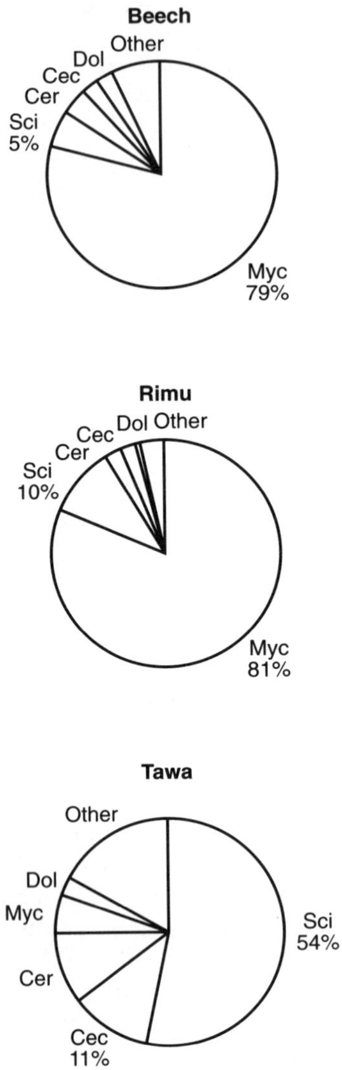

Figure 15.4 Mean representation, in terms of abundance, of the five dominant dipteran families in the canopies of five tree species (mean of $n = 3$ trees, except tawa, $n = 2$). Myc, Mycetophilidae; Sci, Sciaridae; Cer, Ceratopogonidae; Cec, Cecidomyiidae; Dol, Dolichopodidae.

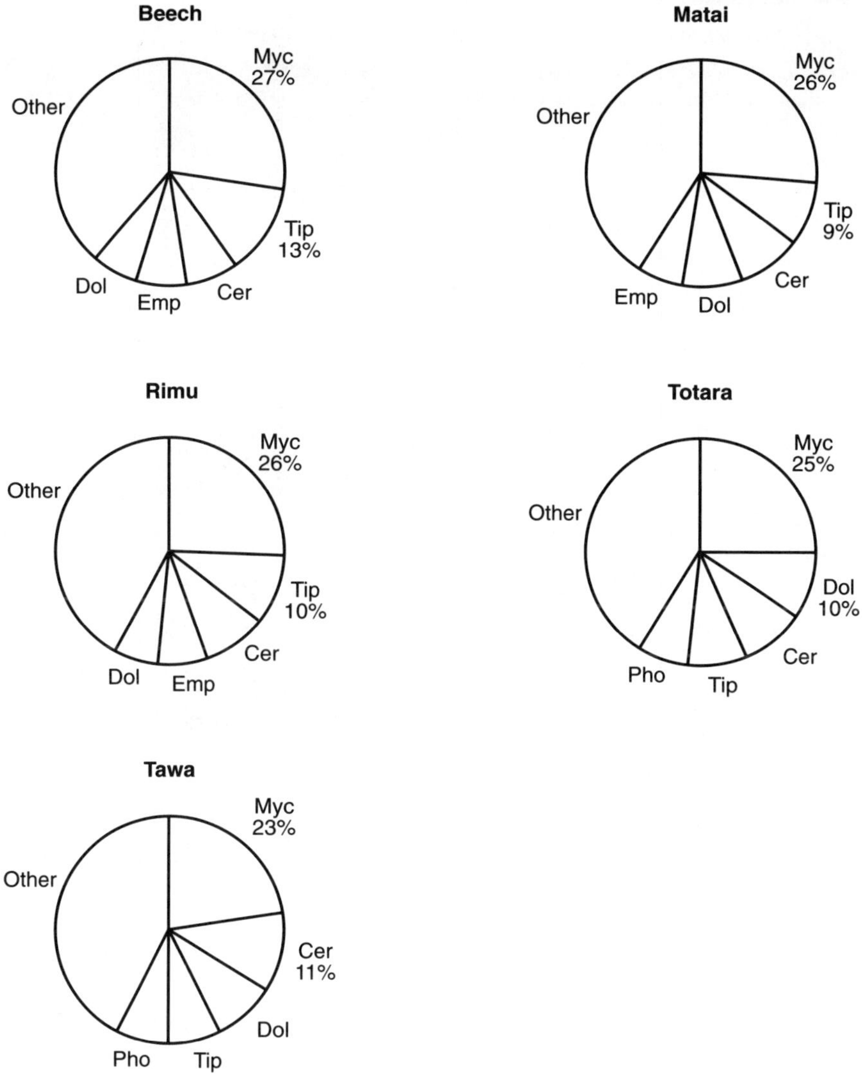

Figure 15.5 Mean representation, in terms of species richness, of the five dominant dipteran families in the canopies of five tree species (mean of $n = 3$ trees, except tawa, $n = 2$). Emp, Empididae; Tip, Tipulidae; Pho, Phoridae. Other abbreviations as in Figure 15.4. Note that Sciaridae and Cecidomyiidae were not sorted to species.

Table 15.4 The five most abundant species captured in 14 tree-crowns. *T.* = *Tetragoneura*, *A.* = *Allophylopsis*

	Rank dominance of species				
Site	*1*	*2*	*3*	*4*	*5*
Beech 1	*T.* sp.1	*T. spinipes*	Dolichopodidae sp.1	*Anomalomyia guttata*	*Cycloneura* sp.2
Beech 2	*T.* sp.1	*T. spinipes*	Ceratopogonidae sp.6	*Cycloneura* sp.2	*Anomalomyia guttata*
Beech 3	*T.* sp.1	*T. spinipes*	Ceratopogonidae sp.6	*Mycetophila fagi*	*T. rufipes*
Matai 1	*T. spinipes*	*T.* sp.3	*T.* sp.4	Ceratopogonidae sp.5	Dolichopodidae sp.12
Matai 2	*T. spinipes*	*T.* sp.1	Ceratopogonidae sp.4	*T.* sp.4	*T.* sp.3
Matai 3	*T. spinipes*	*T.* sp.3	*T.* sp.4	Ceratopogonidae sp.5	*T.* sp.1
Rimu 1	Ceratopogonidae sp.5	*T.* sp.3	Undetermined sp.1	*T.* sp.4	Dolichopodidae sp.12
Rimu 2	*T. spinipes*	*T.* sp.1	Ceratopogonidae sp.4	*T.* sp.4	*T.* opaca
Rimu 3	*T. spinipes*	*T.* sp.1	*T.* sp.4	*T.* sp.3	Ceratopogonidae sp.4
Totara 1	*T. spinipes*	*T.* sp.1	*T.* sp.4	Dolichopodidae sp.1	Undetermined sp.2
Totara 2	*T. spinipes*	*T.* sp.1	Ceratopogonidae sp.4	*T.* sp.4	*T.* sp.3
Totara 3	*T. spinipes*	*T.* sp.4	*T.* sp.3	Dolichopodidae sp.12	*T.* sp.1
Tawa 1	Ceratopogonidae sp.4	Ceratopogonidae sp.5	Undetermined sp.1	*T.* sp.3	*T. spinipes*
Tawa 2	Ceratopogonidae sp.5	Ceratopogonidae sp.4	Dolichopodidae sp.12	*Limonia sponsa*	*A. scutellata*

Table 15.5 Significant (P <0.01) Kendall's rank correlation coefficients for pair-wise comparisons of species dominance. Rank correlation of 18 species of Diptera occurring 'dominantly' (i.e. in the top five most abundant species) in at least one of the 14 sites. – indicates correlations at P >0.01

	B-1	*B-2*	*B-3*	*M-1*	*M-2*	*M-3*	*R-1*	*R-2*	*R-3*	*T-1*	*T-2*	*T-3*	*Ta-1*
Beech-2	0.66												
Beech-3	0.62	0.80											
Matai-1	–	–											
Matai-2	–	–	–	–									
Matai-3	–	–	–	0.81	0.80								
Rimu-1	–	–	–	0.82	–	0.82							
Rimu-2	–	–	–	0.59	0.86	0.71	–						
Rimu-3	–	0.69	–	0.60	0.76	0.69	–	0.77					
Totara-1	–	–	–	–	0.66	0.60	–	–	–				
Totara-2	–	–	–	–	0.84	–	–	0.69	0.60	0.78			
Totara-3	–	–	–	0.83	0.76	0.89	0.78	0.82	0.67	–	0.62		
Tawa-1	–	–	–	0.76	–	0.64	0.66	–	–	–	–		
Tawa-2	–	–	–	0.67	–	–	0.59	–	–	–	–	–	0.64

Habitat- and host tree-specificity

Of 189 species of Diptera classed as 'common' for the purposes of analysis (i.e. $n \geq 5$), 51 species (27%) were specific to one habitat type (Table 15.6): 29 species to beech sites and 22 species to podocarp sites. No Diptera were specific to tawa habitat. As well as a greater total number of species specific to beech habitat, on a 'per trap' basis beech sites had approximately four times as many habitat-specific species as podocarp sites.

Most species were widely distributed within their respective habitats, showing low affinities with tree species or sites, although it should be noted that it was not possible to discern whether beech-specific species were restricted to beech trees as a unique habitat type or unique host-tree. The lack of host tree-specificity amongst podocarp Diptera suggests that beech-dwelling Diptera may be habitat-specific also. Only Dolichopodidae sp.9 was host tree-specific to totara.

Diptera specific to beech habitat were dominated by Empididae, Mycetophilidae and Tipulidae (Table 15.6). Important genera were *Tetragoneura* and *Mycetophila* (Mycetophilidae), and *Amphineurus* (Tipulidae). The inclusion of species of Ceratopogonidae, Chironomidae and Ephydridae as beech-specific indicators and the relative development of Tipulidae may indicate the close proximity of some beech sites

Table 15.6 Habitat- and host tree-specificity of 'common' arboreal Diptera, defined as the capture of 75% or more of all individuals of a species in a single habitat or single host tree. For the purposes of analysis, 'common' was classed as $n \geq 5$ individuals

Beech	*Podocarp*	*Tawa*
Ceratopogonidae		
Ceratopogonidae sp.11	Dolichopodidae	
Chironomidae	Dolichopodidae sp.7	
Chironomidae sp.1	Dolichopodidae sp.9	
Dolichopodidae	(totara)	
Dolichopodidae sp.17	Dolichopodidae sp.13	
Dolichopodidae sp.18	Dolichopodidae sp.14	
Drosophilidae	Dolichopodidae sp.16	
Drosophila sp.	Keroplatidae	
Empididae	'*Platyura*' sp.2	
Empididae sp.3	'*Platyura*' sp.3	
Empididae sp.8	Mycetophilidae	
Empididae sp.12	*Mycetophila opaca*	
Ephydridae	*Tetragoneura spinipes*	
Ephydridae sp.1	*Tetragoneura* sp.3	
Ephydridae sp.2	*Tetragoneura* sp.4	
Helosciomyzidae	*Tetragoneura* sp.10	
Helosciomyza spinicosta	*Tetragoneura* sp.13	
Mycetophilidae	Phoridae	
Cycloneura flava	Phoridae sp.14	
Cycloneura sp.2	Psychodidae	
Mycetophila dilatata	*Nemapalpus zelandiae*	
Mycetophila integra	*Psychoda* sp.4	
Mycetophila sp.11	Sciomyzidae	
Mycetophila sp.14	*Neolimnia* sp.	
Parvicellula gracilis	Tachinidae	
Tetragoneura sp.6	Tachinidae sp.26	
Tetragoneura sp.7	Tipulidae	
Tetragoneura sp.8	*Amphineurus horni*	
Phoridae	*Gynoplistia magnifica*	
Phoridae sp.7	*Limonia seducta*	
Psychodidae	Undetermined	
Pericoma sp.2	Undetermined sp.9	
Tipulidae		
Amphineurus sp.AMA		
Amphineurus hudsoni		
Amphineurus perdecorus		
Limnophila sp.LIA		
Limonia tristigmata		
Molophilus sp.MOP		

to small forest streams. Other features, such as the relative importance of Empididae over Dolichopodidae (both highly active predatory groups) in podocarp habitat are more difficult to assess, but may represent true faunal differentiation between habitats. Tipulidae and Mycetophilidae were similarly important in podocarp habitat.

Tree-crown assemblages

DECORANA multivariate ordination separated sites into three principal groups based on their associated dipteran faunas (Figure 15.6). These groupings corresponded to beech, podocarp and tawa habitats. The close association of podocarp tree species and their broad within-group variability (Figure 15.6) are indicative of a high degree of dipteran species overlap between podocarp tree species. DECORANA separated tawa sites from beech and podocarp sites despite there being no Diptera specific to this tree species.

Analysis of site ordering along the DECORANA axes showed that only a small amount of the total variation in the data was accounted for

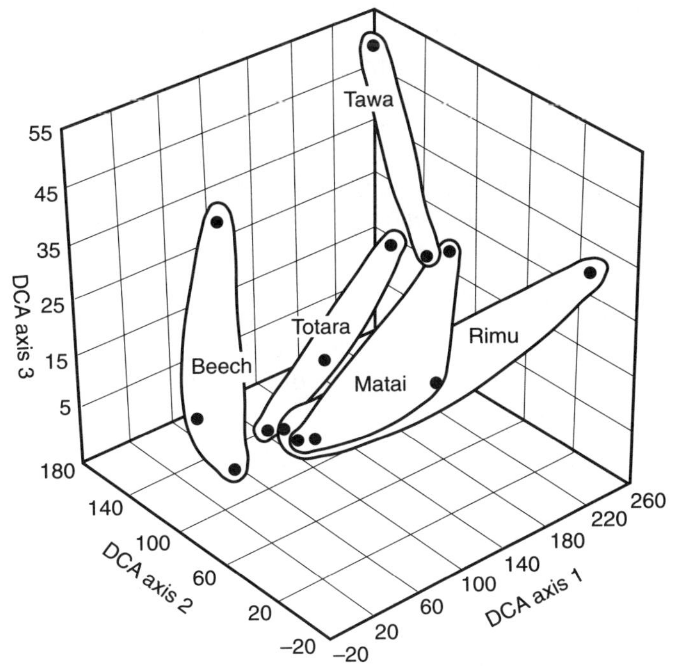

Figure 15.6 Three-dimensional representation of site clustering in DECORANA principal axes space. Eigenvalues for the DCA axes are (axes 1 to 3 respectively): 0.310, 0.046 and 0.019.

Table 15.7 Significant environmental predictors of dipteran site ordering along DECORANA community ordination axes, using stepwise multiple regression

Variable entered	*Parameter estimate*	*Partial r^2*	*Model r^2*
DECORANA axis 1			
Intercept	5.08		
Percentage canopy cover	1.87	0.25	0.25
DECORANA axis 2			
Intercept	93.30		
Tree species	21.60	0.32	0.32
Epiphyte load	–27.10	0.30	0.62
DECORANA axis 3			
Intercept	269.10		
Tree height	–8.70	0.31	0.31

(eigenvalues for axes 1–3, respectively): 0.310, 0.046 and 0.019. Axis 1 was not significantly correlated with any of the selected site structural variables, but in a stepwise regression analysis percentage canopy cover was the best single predictor of site ordering, although it explained little variation (r^2 = 0.25, Table 15.7). Examining Figure 15.6 qualitatively, sites along axis 1 appear to fall broadly into three categories corresponding to different habitat types: beech, podocarp and tawa. Qualitatively, then, axis 1 may be regarded as a habitat separation axis.

Axis 2 was significantly correlated with tree taxa (r = –0.670, P <0.05), accounting for 32% of variation in site ordering within perceived habitat types (Table 15.7). The principal importance of axis 2 was in the separation of the three podocarp tree species, although component sites overlapped considerably. Axis 3 further separated podocarp tree-crown communities into non-overlapping groups. Axis 3 ordering was (poorly) related to tree height (r^2 = 0.31, Table 15.7).

DISCUSSION AND CONCLUSIONS

Beech tree-crowns supported greater numbers of individuals, greater numbers of species and had higher species diversity than either podocarp or tawa tree-crowns. Tawa tree-crowns, in particular, were depauperate in terms of both abundance and species richness.

Explanations for the local richness of insects on plants have been derived from direct correlations with plant diversity (Murdoch *et al.*, 1972; Witkowski, 1973; Southwood *et al.*, 1979; Farrell and Erwin, 1988; Mitter *et al.*, 1991), the influence of plant architectural complexity

(Lawton and Schröder, 1977; Lawton, 1978; Lawton and Price, 1979; Strong *et al.*, 1984) and host-plant affinities (Beaver, 1979; Futuyma and Gould, 1979; Mitter *et al.*, 1991; Basset, 1992). None of these factors appeared to provide significant explanation for the observed variation in dipteran abundance and species richness in this study. On the contrary, the more abundant and species-rich beech sites had lower plant diversity (both canopy and understorey; Wardle, 1961; Didham, 1992) and lower architectural complexity (beech trees were less highly branched, had smaller crowns, lower tree height and lower tree biomass). Host-tree taxonomic status was also a non-significant predictor of abundance and species richness, suggesting that tree species was not an important factor structuring community diversity. Rather, distance from forest edge and percentage canopy cover were consistently the best predictors of richness and abundance.

It was not possible to resolve the microenvironmental site preferences of beech trees (coincident with edges in this forest reserve) from intrinsic tree properties, particularly the presence of honeydew. Arthropod abundance and richness are known to increase with both proximity to forest edge (Hopkins and Webb, 1984; Helle and Muona, 1985; Duelli *et al.*, 1990; Greatorex-Davies, 1991; Malcolm, 1991, 1997, Chapter 25, this volume; Buse and Good, 1993; Báldi and Kisbenedek, 1994; Didham, in press [a]; Ozanne *et al.*, 1997, Chapter 26, this volume; reviewed in Didham, in press [b]) and presence of honeydew on beech trees (Didham, 1993). The results of Didham (1993) for beech trees with and without honeydew, at identical distances from the forest edge, suggest that honeydew is important in promoting arthropod abundance, but that it is not clear whether species richness may also be influenced by edge effects. On the other hand, the lack of an apparent influence of edge effects on podocarp communities would suggest that patterns of abundance and richness may be related broadly to habitat type.

In tawa tree-crowns, low abundance and species richness correlated well with high percentage canopy cover. Moran and Southwood (1982) and Southwood *et al.* (1982a,b) found similar results for the morphologically similar, narrow-leaved willows, *Salix alba* and *S. capensis*. These trees supported fewer arthropod families, species and individuals than broadleaved counterparts. Claridge and Wilson (1981) also noted that narrow-leaved *Populus* spp. had the lowest species richness of mesophyll-feeding leafhoppers (Typhlocybinae) on British trees. The suggested explanation for the observed impoverishment of arboreal faunas on narrow-leaved trees was that flexible stems and branches, with leaves held closely together, move readily in the wind and rub one against another. As such, narrow-leaved trees do not offer an inviting refuge for insects because attachment and movement within the tree-crowns is extremely limited by mechanical abrasion and habitat instability (Moran and Southwood, 1982). I would also suggest that the high leaf density and smooth bark of tawa limit the size and diversity of the flying arthropod fauna by reducing clear flight space, promoting the risk of mechanical damage and offering few suitable retreats for sheltering individuals.

While abundance, species richness and diversity characterized habitats to some degree, the high variability between individual tree-crowns meant that there was relatively little predictive power provided by sample descriptors. A more profitable approach was to investigate differences from a biological standpoint, looking at host-tree preferences of different species and at tree-crown species-assemblages. Diptera show a relatively high degree of habitat affiliation (27% of species), but apparently low host-tree specificity. Interestingly, no Diptera were specific to tawa trees. In a DECORANA ordination, different habitat types had markedly different species associations, but in addition it was apparent that tree species could also be broadly classified based on their associated dipteran fauna. Given the high vagility of most Diptera, this finding is significant. It shows that even though very few Diptera may be restricted to a single tree species, each tree species supports a fairly characteristic arthropod species assemblage, much like a unique 'fingerprint' for that tree. Clearly, though, the within-group variability of podocarp tree species was in some cases higher than the between-group variation, notably for rimu tree-crowns. I speculate that under exceptional conditions, characteristic species assemblages may break down or become modified under external influence. For example, the fauna at Rimu-1 may have been more strongly influenced by the adjacent tree-fall gap and sparse canopy cover, than by intrinsic tree properties. Other factors which may exert an influence on arthropod composition are natural and man-induced edge effects, introduced species, epiphyte abundance, tree health and environmental stochasticity.

The mechanisms underlying patterns of habitat use by adult Diptera in the canopy are not as clear as for some other taxa. Chrysomelid leaf beetles, for example, obviously feed directly on foliage in the canopy and are often restricted to a single host plant (Farrell and Erwin, 1988). Most adult Diptera, though, probably take no more than small quantities of nectar and water to sustain their short adult life. By comparison with the length of larval life and the importance of larvae in predator, scavenger and fungal-feeding trophic groups, the role of adults in sequestering resources in the canopy is negligible. Consequently, most adult Diptera would not be expected to be at all closely linked to a particular tree species (although some do have arboreal larvae). What, then, accounts for the association of particular Diptera with particular tree species? Larval preferences for different habitats or microenvironmental site conditions may account for patterns of habitat use by some species (Kon, 1989), but there is no necessity for adults to remain in the same habitat as the larvae. A number of authors have found that adult Diptera swarm in response to very specific marker cues, such as vegetation type, gaps, different heights or colours (Cooter, 1989; Blackwell *et al.*, 1992; Svensson and Petersson, 1992; Yuval and Bouskila, 1993; Jones and Schreiber, 1994). Some species also have extremely clearly defined resting heights on trees (Memmott, 1991, 1992). The reasons for this are much argued over, but provide a likely mechanism accounting for non-random patterns of space utilization by adult Diptera.

The specificity of space utilization raises questions about the classification of adult Diptera into trophic groups and their role in the canopy food web. Trophic group classifications made in previous studies (see Introduction) are largely based on larval biology, but are curiously ambiguous in their classification of some families into trophic groups (as larvae, not adults), while the majority of families are simply termed tourists (as adults, not larvae). A consistent guild classification based on larval biology is a valid and valuable method for the relative characterization of sites, but it bears little relation to the importance of adults in the canopy. Most adult Diptera in fact do not feed and, superficially at least, fulfil the criteria for Moran and Southwood's (1982) classification as tourists. However, the non-random space utilization of adult Diptera found in this study and the swarming habits of many species, are not indicative of a fauna that is an ephemeral and random component of the canopy community. Nor are Diptera insignificant in terms of species interactions in the canopy, as the 'tourist' designation implies. Many adult dipteran species are integral components of canopy communities and must significantly influence predator–prey and competitive interactions. The 'tourist' guild is wholly inadequate in describing these interactions.

There is a basic need for research into the role of adult Diptera in the forest canopy, from both autecological and synecological standpoints. Perhaps the greatest hinderance to advances in the study of canopy arthropod communities is the dearth of knowledge of the biology of the organisms under investigation, none more so than for Diptera. Drawing inferences from the known biologies of a handful of 'representative' taxa and the generalist, community-orientated approach to canopy entomology will, in the long run, do little for the advancement of the science.

Acknowledgements

I thank N.E. Stork, J.H. Lawton and P.M. Johns for useful discussion and comments on the manuscript. D.A. Norton, I.R. Didham and M.D. Tocher provided invaluable assistance in the field. The research was funded by the University of Canterbury, New Zealand, and preparation of the manuscript was supported by the Commonwealth Scholarship Commission and The British Council, UK, The Natural History Museum, London, the NERC Centre for Population Biology, Silwood Park, UK, and a University of Canterbury Doctoral Scholarship award, New Zealand.

REFERENCES

Báldi, A. and Kisbenedek, T. (1994) Comparative analysis of edge effects on bird and beetle communities. *Acta Zoologica Academiae Scientarum Hungaricae*, **40**, 1–14.

Basset, Y. (1992) Host specificity of arboreal and free-living insect herbivores in rain forests. *Biological Journal of the Linnean Society*, **47**, 115–33.

Beaver, R.A. (1979) Host specificity of temperate and tropical animals. *Nature*, **281**, 139–41.

Berger, W.H. and Parker, F.L. (1970) Diversity of planktonic Foraminifera in deep-sea sediments. *Science*, **168**, 1345–7.

Blackwell, A., Mordue, A.J., Young, M.R. and Mordue, W. (1992) The swarming behaviour of the Scottish biting midge, *Culicoides impunctatus* (Diptera: Ceratopogonidae). *Ecological Entomology*, **17**(4), 319–25.

Boyd, S.A. (1987) *Patterns of Use of Honeydew by Birds and Insects*, Unpublished MSc Thesis, University of Auckland, New Zealand.

Brown, V.K. and Southwood, T.R.E. (1983) Trophic diversity, niche breadth and generation times of exopterygote insects in a secondary succession. *Oecologia*, **56**, 220–5.

Buse, A. and Good, J.E.G. (1993) The effects of conifer forest design and management on abundance and diversity of rove beetles (Coleoptera: Staphylinidae): implications for conservation. *Biological Conservation*, **64**, 67–76.

Claridge, M.F. and Wilson, M.R. (1981) Host plant associations, diversity and species-area relationships of mesophyll-feeding leafhoppers of trees and shrubs in Britain. *Ecological Entomology*, **6**, 217–38.

Cohen, J.E. (1977) Ratio of prey to predators in community food webs. *Nature*, **270**, 165–7.

Coley, P., Bryant, J. and Chapin, F. (1985) Resource availability and plant anti-herbivore defense. *Science*, **230**, 895–9.

Cooter, R.J. (1989) Swarm flight behaviour in flies and locusts, in *Insect Flight* (eds G.J. Goldsworthy and C.H.J. Wheeler), CRC Press, Inc., Boca Raton, Florida, USA, pp. 165–203

Cornell, H.V. and Lawton, J.H. (1992) Species interactions, local and regional processes, and limits to the richness of ecological communities: a theoretical perspective. *Journal of Animal Ecology*, **61**, 1–12.

Didham, R.K. (1992) *Faunal Composition, Diversity and Spatial Heterogeneity of the Arboreal Arthropod Community in a Temperate Rainforest Canopy, New Zealand*. Unpublished MSc Thesis, Zoology Department, University of Canterbury, New Zealand.

Didham, R.K. (1993) The influence of honeydew on arthropods associated with beech trees in New Zealand. *New Zealand Natural Sciences*, **20**, 47–53.

Didham, R.K. The influence of edge effects and forest fragmentation on leaf litter invertebrates in Central Amazonia, in *Tropical Forest Remnants: Ecology, Management and Conservation of Fragmented Communities* (eds W.F. Laurance and R.O. Bierregaard, Jr), University of Chicago Press, Chicago (in press, a).

Didham, R.K. An overview of invertebrate responses to forest fragmentation, in *Forest and Insects* (eds A. Watt, N.E. Stork and M. Hunter), The 18th Symposium of the Royal Entomological Society of London, Chapman & Hall, London (in press, b).

Duelli, P., Studer, M., Marchand, I. and Jakob, S. (1990) Population movements of arthropods between natural and cultivated areas. *Biological Conservation*, **54**, 193–207.

Ehrlich, P.R. and Raven, P.H. (1964) Butterflies and plants: a study in coevolution. *Evolution*, **18**, 586–608.

Farrell, B.D. and Erwin, T.L. (1988) Leaf-beetle community structure in an amazonian rainforest canopy, in *Biology of Chrysomelidae* (eds P. Jolivet, E. Petitpierre and T.H. Hsiao), Kluwer Academic Publishers, Dordrecht, pp. 73–90.

Feeny, P.P. (1976) Plant apparency and chemical defence. *Recent Advances in Phytochemistry*, **10**, 1–40.

Fisher, R.A., Corbet, A.S. and Williams, C.B. (1943) The relation between the number of species and the number of individuals in a random sample of an animal population. *Journal of Animal Ecology*, **12**, 42–58.

Futuyma, D.J. and Gould, F. (1979) Associations of plants and insects in a deciduous forest. *Ecological Monographs*, **49**(1), 33–50.

Gaston, K.J. (1993) Herbivory at the limits. *Trends in Ecology and Evolution*, **8**(6), 193–4.

Gaze, P.D. and Clout, M.N. (1983) Honeydew and its importance to birds in beech forests of the South Island, New Zealand. *New Zealand Journal of Ecology*, **6**, 33–7.

Greatorex-Davies, J.N. (1991) Edge management in woodlands. *Forestry Commission Occasional Paper*, **28**, 25–30.

Helle, P. and Muona, J. (1985) Invertebrate numbers in edges between clear-fellings and mature forests in northern Finland. *Silva Fennica*, **19**, 281–94.

Hopkins, P.J. and Webb, N.R. (1984) The composition of the beetle and spider faunas on fragmented heathlands. *Journal of Applied Ecology*, **21**, 935–46.

Hughes, S.J. (1972) New Zealand fungi 17. Pleomorphism in Euantennariaceae and Metacapnodiaceae, two new families of sooty moulds. *New Zealand Journal of Botany*, **10**, 225–42.

Jones, C.J. and Schreiber, E.T. (1994) Color and height affects oviposition site preferences of *Toxorhynchites splendens* and *Toxorhynchites rutilus rutilus* (Diptera: Culicidae) in the laboratory. *Environmental Entomology*, **23**(1), 130–5.

Kennedy, C.E.J. and Southwood, T.R.E. (1984) The number of species of insects associated with British trees: a re-analysis. *Journal of Animal Ecology*, **53**, 455–78.

Kon, M. (1989) Swarming and mating behaviour of *Chironomus flaviplumus* (Diptera: Chironomidae), compared with a sympatric congeneric species, *C. yoshimatsui*. *Journal of Ethology*, **7**(2), 125–31.

Krebs, C.J. (1985) *Ecology: The Experimental Analysis of Distribution and Abundance*, 3rd edn, Harper and Row Publishers, New York.

Lawton, J.H. (1978) Host-plant influences on insect diversity: the effects of space and time, in *Diversity of Insect Faunas* (eds L.A. Mound and W.A. Waloff), Symposium of the Royal Entomological Society of London, Vol. 9, Blackwell Scientific Publications, Oxford, pp. 105–25.

Lawton, J.H. and Price, P.W. (1979) Species richness of parasites on hosts: agromyzid flies on the British Umbelliferae. *Journal of Animal Ecology*, **48**, 619–37.

Lawton, J.H. and Schröder, D. (1977) Effects of plant type, size of geographic range and taxonomic isolation on number of insect species associated with British plants. *Nature*, **265**, 137–40.

Leather, S.R. (1986) Insect species richness of the British Rosaceae: the importance of host range, plant architecture, age of establishment, taxonomic isolation and species–area relationships. *Journal of Animal Ecology*, **55**, 841–60.

Malcolm, J.R. (1991) *The Small Mammals of Amazonian Forest Fragments: Pattern and Process*, Unpublished PhD Thesis, University of Florida.

Malcolm, J.R. (1997) Insect biomass in Amazonian forest fragments, in *Canopy Arthropods* (eds N.E. Stork, J. Adis and R.K. Didham), Chapman & Hall, London, pp. 510–33 (this volume).

Memmott, J. (1991) Sandfly distribution and abundance in a tropical rainforest. *Medical and Veterinary Entomology*, **5**(4), 403–11.

Memmott, J. (1992) Patterns of sandfly distribution in tropical forest: a causal hypothesis. *Medical and Veterinary Entomology*, **6**(3), 188–94.

Mitter, C., Farrell, B. and Futuyma, D.J. (1991) Phylogenetic studies of insect–plant interactions: insights into the genesis of diversity. *Trends in Ecology and Evolution*, **6**(9), 290–3.

Moller, H. and Tilley, J.A.V. (1987) *Honeydew, a South Island Beech Forest Resource*, Alpha 58, DSIR Publishing, Wellington, New Zealand.

Moller, H. and Tilley, J.A.V. (1989) Beech honeydew: seasonal variation and use by wasps, honeybees and other insects. *New Zealand Journal of Zoology*, **16**, 289–302.

Moran, V.C. and Southwood, T.R.E. (1982) The guild composition of arthropod communities on trees. *Journal of Animal Ecology*, **51**, 289–306.

Murdoch, W.W., Evans, F.C. and Peterson, C.H. (1972) Diversity and pattern in plants and insects. *Ecology*, **53**, 819–29.

Opler, P.A. (1974) Oaks as evolutionary islands for leaf-mining insects. *American Scientist*, **62**, 67–73.

Ozanne, C.M.P., Hambler, C., Foggo, A. and Speight, M.R. (1997) The significance of edge effects in the management of forests for invertebrate biodiversity, in *Canopy Arthropods* (eds N.E. Stork, J. Adis and R.K. Didham), Chapman & Hall, London, pp. 534–550 (this volume).

Price, P.W., Bouton, C.E., Gross, P., McPheron, B.A., Thompson, J.N. and Weis, A.E. (1980) Interactions among three trophic levels: influence of plants on interactions between insect herbivores and natural enemies. *Annual Review of Ecology and Systematics*, **11**, 41–65.

Rhoades, D.F. and Cates, R.G. (1976) Toward a general theory of plant antiherbivore chemistry. *Recent Advances in Phytochemistry*, **10**, 168–213.

Root, R.B. (1967) The niche exploitation pattern of the Blue-Gray Gnatcatcher. *Ecological Monographs*, **37**(4), 317–50.

Root, R.B. (1973) Organization of a plant–arthropod association in simple and diverse habitats: the fauna of collards (*Brassica oleracea*). *Ecological Monographs*, **43**(1), 95–124.

Rosenthal, G. and Janzen, D. (1979) *Herbivores: Their Interactions with Secondary Metabolites*, Academic Press, New York.

Schultz, J.C. (1983) Impact of variable plant chemistry defenses on insect susceptibility to parasites, predators, and diseases. *Symposia of the American Chemical Society*, **208**, 37–55.

Southwood, T.R.E. (1960) The abundance of the Hawaiian trees and the number of their associated insect species. *Proceedings of the Hawaiian Entomological Society*, **17**, 299–303.

Southwood, T.R.E. (1961a) The evolution of the insect–host tree relationship – a new approach. *Proceedings of the XIth International Congress of Entomology*, **1**, 651–5.

Southwood, T.R.E. (1961b) The number of species of insects associated with various trees. *Journal of Animal Ecology*, **30**, 1–8.

Southwood, T.R.E. (1973) The insect/plant relationship – an evolutionary perspective, in *Insect/Plant Relationships* (ed. H.F. Van Emden), Symposium of the Royal Entomological Society, London, Vol. 6, Blackwell Scientific Publications, Oxford, pp. 3–30.

Southwood, T.R.E. and Kennedy, C.E.J. (1983) Trees as islands. *Oikos*, **41**, 359–71.

Southwood, T.R.E., Brown, V.K. and Reader, P.M. (1979) The relationship of plant and insect diversities in succession. *Biological Journal of the Linnean Society*, **12**, 327–48.

Southwood, T.R.E., Moran, V.C. and Kennedy, C.E.J. (1982a) The richness, abundance and biomass of the arthropod communities on trees. *Journal of Animal Ecology*, **51**, 635–49.

Southwood, T.R.E., Moran, V.C. and Kennedy, C.E.J. (1982b) The assessment of arboreal insect faunas: comparisons of knockdown sampling and faunal lists. *Ecological Entomology*, **7**, 331–40.

Stork, N.E. (1987) Guild structure of arthropods from Bornean rainforest trees. *Ecological Entomology*, **12**, 69–80.

Stork, N.E. (1991) The composition of the arthropod fauna of Bornean lowland rain forest trees. *Journal of Tropical Ecology*, **7**, 161–80.

Strong, D.R., Lawton, J.H. and Southwood, T.R.E. (1984) *Insects on Plants: Community Patterns and Mechanisms*, Blackwell Scientific Publications, Oxford.

Svensson, B.G. and Petersson, E. (1992) Why insects swarm: testing the models for lek mating systems on swarming *Empis borealis* females. *Behavioural Ecology and Sociobiology*, **31**(4), 253–61.

Wardle, P. (1961) The forest of Aniseed Stream near Kaikoura. *New Zealand Journal of Forestry*, **8**, 477–81.

West, C. (1986) Insect communities in tree canopies, in *Kora: an Ecological Inventory of the Kora National Reserve* (eds M. Coe and N.M. Collins), Kenya. Royal Geographic Society, London, pp. 209–22.

Wiens, J.A., Cates, R.G., Rotenberry, J.T., Cobb, N., Van Horne, B. and Redak, R.A. (1991) Arthropod dynamics on sagebrush (*Artemisia tridentata*): effects of plant chemistry and avian predation. *Ecological Monographs*, **61**(3), 299–321.

Williams, P.A. (1980) *Blue Duck Scientific Reserve. Reserves of Marlborough*, Department of Lands and Survey, New Zealand.

Witkowski, Z. (1973) Species diversity, stability and succession. Studies on weevils (Curculionidae, Coleoptera) and their host plants during the succession of mowed meadows under the influence of drainage. *Bulletin de l'Academie Polonaise des Sciences*, **21**, 223–8.

Yuval, B. and Bouskila, A. (1993) Temporal dynamics of mating and predation in mosquito swarms. *Oecologia*, **95**(1), 65–9.

16

Diversity and recolonization dynamics of selected arthropod groups on different tree species in a lowland rainforest in Sabah, Malaysia with special reference to Formicidae

A. Floren and K.E. Linsenmair

ABSTRACT

To obtain a better idea of the mechanisms maintaining tropical diversity, arboreal arthropod communities of a lowland rainforest in Borneo, Sabah, Malaysia, were sampled using an improved method of canopy fogging. For the first time this selective fogging allowed comparison between alpha- and beta-diversities of single tree-crowns and analysis of recolonization dynamics.

Controls were adopted to check the efficiency of the fogging method. The results demonstrate that it is methodologically possible to sample tree-crown inhabiting arthropod communities selectively and quantitatively.

At a high taxonomic level arthropod communities were composed very regularly, independent of the number of arthropods found in a single tree-crown. However, this was not true at the morphospecies level, as was shown exemplarily for the Formicidae and was evident after sorting most other orders. On the contrary, ant communities were always very heterogeneous in composition and no convergent structuring mechanism could be detected. This was also the case where tree-crowns were dominated by the same ant species.

The communities of those trees that were re-fogged after 7 months were equally heterogeneous in comparison with first-fogged trees. High beta-diversity and variability of alpha-diversity and rarefaction curves, unpredictable species appearance and

Canopy Arthropods. Edited by N.E. Stork, J. Adis and R.K. Didham. Published in 1997 by Chapman & Hall, London. ISBN 0 412 74900 9

disappearance, and the obvious lack of successional stages or climax equilibrium, suggested that community composition was unpredictable and highly influenced by stochastic events.

Formicidae were the only group consistently dominant in the trees investigated. We do not know what many dominant ant species feed on in the canopy and what the consequences of feeding mode are with regard to community structure. Observations and baiting experiments indicated that all ants were generalists and thus most probably generate a high predation pressure. The predominance of mobile arthropod taxa in the canopy is perhaps a function of this.

INTRODUCTION

Since the first investigations of tropical forests the high floristic and faunistic diversity has been recognized as one of the characteristics of these ecosystems (Wallace, 1869). A central challenge is to understand the mechanisms responsible for the maintenance of species coexistence. In the past decade studies have questioned whether deterministic equilibrium models are a necessary and sufficient explanation for community structure, or whether non-equilibrium models, in which chance plays a major role, are more important (Wiens, 1984; Cornell and Lawton, 1992; Huston, 1994).

The model of the ecological niche forms the basis for deterministic equilibrium theory. Each organism maintains a defined position in its environment, which can be imagined as a n-dimensional niche-space (Sugihara, 1980; Giller, 1984). Driven by competition, the system runs through defined successional stages and ends up with a structurally predictable climax equilibrium. In contrast, stochastic non-equilibrium models assume that competition might be of local importance but may not be the crucial form of interaction determining final community structure. The presence of a species at the most favourable time is regarded as more important and may lead to an advantage, for example in growth, that ensures dominance against all species that arrive later. As a result, neither successional stages nor a climax community can emerge. Necessarily, free sites are created by a disturbance regime that produces a many-fold pattern of different habitats (Pickett and White, 1985; Chesson and Case, 1986; Huston, 1994).

Until recently investigation of these competing models has been mainly botanical, and empirical evidence has demonstrated both deterministic and stochastic structuring mechanisms (Hubbell and Foster, 1986; Ashton, 1992; Huston and DeAngelis, 1994; Tilman, 1994). Zoologically, our knowledge is much more incomplete, especially for the highly diverse arthropod communities of tropical ecosystems. Because of the immense taxonomic problems and difficulty of access to tree-crowns, systematic investigation of canopy arthropods only started relatively few years ago (Stork, 1987; Basset and Kitching, 1991; Basset and Arthington, 1992; Stork and Blackburn, 1993). Tree-crowns are thought to harbour an enormous number of species (Erwin, 1982, 1988; Stork, 1988, 1993) and most investigations tend to be descriptive. Comparable studies of the mechanisms structuring arthropod communities in the tropics and in temperate ecosystems are almost completely lacking.

In order to assess the importance of deterministic versus stochastic processes in determining species coexistence we investigated arthropod communities of different tree-crowns in a lowland rainforest of Sabah, Malaysia, using canopy fogging (Erwin, 1982; Stork, 1987). Contrary to other studies, this approach required sampling tree-crowns both quantitatively and selectively. For that purpose a suitable method had to be developed. As well as establishing an inventory of the arthropod fauna we asked the following questions:

1. How efficient is the fogging method and is it a useful tool to investigate the dominant mechanisms structuring arboreal arthropod communities?
2. How are arthropod communities composed on a high taxonomic level and especially within the Formicidae?
3. What conclusions can be drawn from the analysis of recolonization seven months after initial fogging and can the results be interpreted with respect to deterministic or stochastic processes?
4. To what extent do ants influence arthropod community structure?

METHODS

Selective fogging was used in the lowland forest of Kinabalu Park in Sabah, Malaysia (6°5'N 160°33'E), concentrating on relatively small trees of the lower canopy layer. A 100-m^2 cotton roof was stretched horizontally above the crown of the study tree in order to prevent arthropods living in taller neighbouring trees from falling into the trays beneath the fogged tree. This can be done without disturbing the study tree since there is usually free space of more than 10–15 m between the upper and lower canopy. At least 10 weeks before conducting the fogging, all lianas and epiphytes were removed from the tree-crown in order to sample arthropods associated only with the study tree. Baits were used to sample abundant ant species in order to map their nesting sites and to facilitate species identification.

The fogging experiment was conducted early in the morning (06:00–07:00 h) when there was little wind drift. Natural pyrethrum with a technical oil (Essobayol 82) as a carrier was applied to the trees for 10 minutes using an insecticide fogging machine. A concentration of 2% active ingredient ensured a high knockdown capacity. During the fogging procedure roughly 1.5 litres of insecticide were sprayed. The insecticide only affects invertebrates, and breaks down in minutes in direct sunlight. To do the fogging as effectively as possible, the tree was climbed and fogged immediately, ensuring that the whole tree was fogged evenly. Almost the entire crown projection (80–90%) was covered by funnels arranged 1 m above the forest floor and all arthropods that dropped into the funnels for 2 hours following fogging were collected and stored in 80% alcohol. The efficacy of fogging was improved by the roof, since the rising fog was held back in the tree-crown. On the other hand, both installing the roof and climbing the tree were disturbances and as a result sensitive species may have disappeared. Thus, this technique might result in a reduced measure of diversity.

The efficiency of the fogging method

Several controls were adopted to check the efficiency of the fogging method used: (i) observations in the tree-crowns before and after the fogging experiment; (ii) approximately 3 hours after fogging the tree was vigorously shaken to dislodge any remaining arthropods into the collecting funnels, which could be then evaluated separately; and (iii) instead of shaking, two study trees were fogged again in an identical manner 3 hours after the first fog.

Tree species fogged

Three tree species were selected, two species of Euphorbiaceae (*Aporusa lagenocarpa* and *A. subcaudata*) and one Polygalaceae (*Xanthophyllum affine*). All three are relatively small tree species of the lower canopy of primary forest, with an average crown diameter of 8–9 m and with a maximum height of 28 m. While *A. lagenocarpa* had been found in two large clusters of up to 50 trees, separated by 3 km, *A. subcaudata* was found frequently, mostly as individual trees. *X. affine* tends to appear as isolated trees. In order to keep intraspecific variability as low as possible, only those trees which corresponded highly in their biotic and abiotic location parameters, as for example altitude, exposure, edaphic factors, epiphyte covering and so on, were chosen for fogging.

During four field visits, 40 fogging experiments on a total of 10 individuals of *A. lagenocarpa*, five *A. subcaudata* and four *X. affine* were carried out. In the context of this paper the results of 19 trees fogged for the first time and of 10 trees re-fogged after time intervals of at least 7 months are presented.

The morphospecies concept

Since most arthropods of the canopy of tropical forests have not been taxonomically classified, identification of the groups investigated

Table 16.1 Alpha-diversities arranged in the Hill-series

	Index	*Hill-transformation*
May:	$J = \dfrac{\min(n_i)}{N}$	1/J
Species number:	S	S
Shannon–Weiner:	$H = -\sum_{i=1}^{S} \pi_i \ln \pi_i$	exp. H
Simpson:	$D = \sum_{i=1}^{S} \dfrac{n_i(n_i - 1)}{N(N - 1)}$	1/D
Berger–Parker:	$d = \dfrac{\max(n_i)}{N}$	1/D

n_i = individuals of the i-th of S species
N = number of total individuals
π = relative abundance of the i-th species
S = total number of species

(Coleoptera and Formicidae) was carried out by distinguishing morphospecies, which discriminates in a manner similar to a 'formal' determination. Since most mistakes by this method should be conservative, it is more likely that real diversities are underestimated.

Alpha-diversities

To analyse the structure of ant communities the fitting of species-abundance distribution models was checked for each community. Alpha-diversities, arranged in the Hill-series, were computed to consider the increasing influence of dominant species (Wolda, 1983). In their transformed reciprocal or exponential forms these indices have been shown to be mathematically related, resulting in direct comparability (Hill, 1973) (Table 16.1).

The rarefaction method

This method allows a direct comparison of two communities after standardizing abundances (Sanders, 1968; Hurlbert, 1971). The number of expected species (ES) is computed from a sample of n individuals selected at random from all collected species and individuals (without replacement). With increasing sample size (m) the result is a characteristic species accumulation curve from which information about community structure can be derived, provided the same sampling technique is used. A high degree of evenness would cause a curve to flatten out. If one or a few species dominate the community, the curve is straighter. At low sample sizes (*m*) similarity is determined by dominant species, while rare species are more influential in large sample sizes. As a rule of thumb all samples should be standardized at least to the smallest sample. Since rarefaction curves illustrate temporal variation, they are a useful tool to analyse arthropod community structure after different time intervals of recolonization.

$$ES\ (\mathrm{m}) = S - \sum_{i=1}^{S} \frac{\binom{N-n_i}{m}}{\binom{N}{m}}$$

where $ES(m)$ is the expected number of species in subsample m; N is the total number of individuals; n_i are the individuals of the i-th of S species; and S is the total number of species.

The rarefaction method can also be applied to quantitative data (Shinozaki, 1963; Smith *et al*, 1979). The result is a species accumulation curve which can be interpreted as a measure of beta-diversity. The slope gives information about the quality (completeness) of sampling effort (Achtziger *et al.*, 1992). If the slope is zero, further sampling will probably not result in additional species. If it is larger than zero, sampling effort was not sufficient to give a complete picture of the community.

$$ES\ (q) = S - \sum_{i=1}^{S} \frac{\binom{K-k_i}{q}}{\binom{K}{q}}$$

where $ES(q)$ is the expected number of species in q samples; q is the number of discrete samples; K is the total number of samples; k_i is the number of species of the i-th of K species; and S is the total number of species.

Beta-diversity

To measure differences between two communities the Sörensen and NESS indices were computed.

Sörensen index

$$C_s = \frac{2S_{1,2}}{S_i + S_2}$$

where $S_{1,2}$ is the number of similar species in samples 1 and 2.

NESS index

An extension of the rarefaction method to two different samples is the NESS index (Normalized Expected Species Shared) (Grassle and Smith, 1976). It computes the expected number of species in two random samples from all collected species and individuals (without replacement).

$$ESS\ (N_1, N_2, m) = \sum_{i=1}^{S} \left(1 - \frac{\binom{N_1 - n_{1i}}{m}}{\binom{N_1}{m}}\right)\left(1 - \frac{\binom{N_2 - n_{2i}}{m}}{\binom{N_2}{m}}\right)$$

where m is the subsample size; S is the total number of species of both samples; N_1 the individuals of sample 1; and N_2 the individuals of sample 2.

RESULTS

The efficiency of the fogging method

In 15 of the 19 trees fogged for the first time, observations of several hours were conducted before the fogging experiment. In all cases ants were highly dominant and continuously present. In contrast, other arthropods were seen only sporadically, despite the fact that most tree-crowns could be surveyed easily and searched intensively. Directly after fogging, no arthropods could be seen on the leaves or branches. At 3 hours after fogging the tree was climbed again and shaken vigorously in order to dislodge any remaining arthropods into the collecting funnels. This usually resulted in an increase of 5% of the total number of arthropods from the study trees ($n = 12$). Again, 40–50% of these were ants.

The subsequent control fog knocked down an additional 16% of total arthropod abundance. Although Formicidae were still dominant, other highly mobile groups of Diptera, Coleoptera, other Hymenoptera and arachnids were found in larger relative abundances in these samples, while all other arthropods were represented only by few individuals.

The roof was highly efficient in preventing the sampling of arthropods from neighbouring trees, as evident from the many dead arthropods lying on top of the roof.

Taxonomic composition

The combined total number of arthropods from all 40 fogging experiments was 155 000, of which 88 502 (57.1%) were collected from the first fogged trees and 30 335 (19.6%) from those trees that were fogged again after

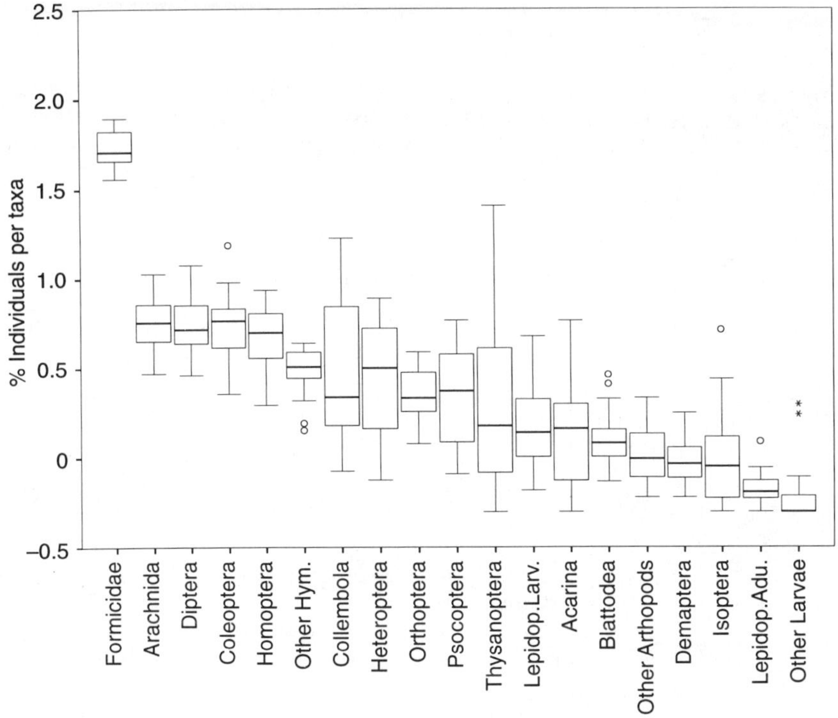

Figure 16.1 Mean relative abundances of individuals (as box-plots) of the main taxa of all trees fogged for the first time (y-axis on a log scale). A circle indicates extreme values between one and three times the box length. An asterisk indicates extreme values larger than three times the box length.

7 months. The remaining arthropods were collected during further foggings not mentioned in the context of this paper. As shown in Table 16.2 the ranking of the major orders was remarkably constant, irrespective of the total number of all arthropods found in a single tree-crown. This is also reflected in the low coefficient of variation. A graphical illustration of the means of all orders and tree-crowns is shown in Figure 16.1.

A one-way ANOVA confirms highly significant differences between the insect orders on all first-fogged trees ($F = 40.71$, $P < 0.001$). A multiple comparison with the Scheffes-test results in three distinguishable groups of taxa ($\alpha = 0.05$). First, Formicidae dominated all communities and contributed on average 54% of all individuals of a community. Second, Arachnida, Diptera, Coleoptera and Homoptera make up approximately 5% of each community. Third, Blattodea, Dermaptera, adult Lepidoptera and other arthropods do not exceed 1%, on average.

Table 16.2 Numbers and percentages of arthropods grouped by major taxa collected from 19 trees. 'Other Hymenoptera' includes all Hymenoptera other than Formicidae. 'Other' taxa includes Mantodea, Phasmida, Neuroptera, Ephemeroptera, Plecoptera and Trichoptera which were found regularly but at low abundances per tree. All further groups were found frequently but as single individuals. 'Other larvae' includes all larvae not already represented by those taxa. (From a carton nest of termites found on Al. #52, 13 400 individuals were sampled but not considered for the computation of relative abundances). Al. 5–74 = *Aporusa lagenocarpa* trees, As. 8–51 = *Aporusa subcaudata* trees, Xa. = *Xanthophyllum affine* trees. Vk = coefficient of variation, Ind./m² = number of arthropods caught per collecting funnel m²

Aporusa lagenocarpa

		Al. 5	*Al. 15*	*Al. 52*	*Al. 57*	*Al. 62*	*Al. 70*	*Al. 71*	*Al. 72*	*Al. 73*	*Al. 74*	*Total*	*Mean%*	*Vk*
Formicidae	N	**957**	**524**	**3833**	**761**	**1028**	**3052**	**760**	**3052**	**5010**	**3527**	**22504**		0.27
	%	65.77	38.00	41.34	77.73	41.07	51.75	35.8	46.09	47.05	46.61		49.12	
Arachnida		**76**	**104**	**655**	**29**	**177**	**339**	**217**	**398**	**696**	**360**	**3051**		0.30
		5.22	7.54	7.06	2.96	7.07	5.75	10.22	6.01	6.54	4.76		6.31	
Diptera		**92**	**142**	**361**	**40**	**128**	**261**	**109**	**283**	**533**	**608**	**2557**		0.36
		6.32	10.3	3.89	4.09	5.11	4.43	5.13	4.27	5.01	8.03		5.66	
Coleoptera		**59**	**203**	**373**	**23**	**148**	**314**	**144**	**346**	**326**	**286**	**2222**		0.63
		4.05	14.72	4.02	2.35	5.91	5.32	6.78	5.23	3.06	378		5.52	
Homoptera		**73**	**92**	**390**	**19**	**146**	**351**	**47**	**391**	**836**	**328**	**2673**		0.37
		5.02	6.67	4.21	1.94	5.83	5.95	2.21	5.90	7.85	4.33		4.99	
Other Hymenoptera		**23**	**52**	**232**	**9**	**84**	**122**	**81**	**257**	**390**	**215**	**1465**		0.37
		1.58	3.77	2.50	0.92	3.36	2.07	3.82	3.88	3.66	2.84		2.84	
Heteroptera		**39**	**19**	**470**	**9**	**124**	**278**	**108**	**206**	**402**	**529**	**2184**		0.48
		2.68	1.38	5.07	0.92	4.95	4.71	5.09	3.11	3.78	6.99		3.87	
Orthoptera		**24**	**36**	**141**	**7**	**52**	**87**	**64**	**83**	**203**	**171**	**868**		0.36
		1.65	2.61	1.52	0.72	2.08	1.48	3.01	1.25	1.91	2.26		1.85	
Psocoptera		**27**	**50**	**29**	**23**	**43**	**215**	**87**	**133**	**180**	**391**	**1178**		0.55
		1.86	3.63	0.31	2.35	1.72	3.65	4.10	2.01	1.69	5.17		2.65	

Table 16.2 continued

	Al. 5	*Al. 15*	*Al. 52*	*Al. 57*	*Al. 62*	*Al. 70*	*Al. 71*	*Al. 72*	*Al. 73*	*Al. 74*	*Total*	*Mean%*	*Vk*
Thysanoptera	**4**	**12**	**2340**	**12**	**324**	**59**	**8**	**144**	**151**	**113**	**3167**		1.73
	0.27	0.87	25.24	1.23	12.94	1.00	0.38	2.17	1.42	1.49		4.70	
Collembola	**14**	**55**	**157**	**11**	**42**	**463**	**348**	**1071**	**1400**	**389**	**3950**		0.92
	0.96	3.99	1.69	1.12	1.68	7.85	16.39	16.17	13.15	5.14		6.81	
Lepidoptera larvae	**22**	**36**	**31**	**3**	**39**	**90**	**46**	**32**	**67**	**64**	**430**		0.67
	1.51	2.61	0.33	0.31	1.56	1.53	2.17	0.48	0.63	0.85		1.20	
Lepidoptera adults	**2**	**4**	**13**	**0**	**1**	**15**	**5**	**5**	**29**	**28**	**102**		0.67
	0.14	0.29	0.14	0.00	0.04	0.25	0.24	0.08	0.27	0.37		0.18	
Blattodea	**8**	**6**	**103**	**6**	**15**	**37**	**5**	**24**	**53**	**57**	**314**		0.41
	0.55	0.44	1.11	0.61	0.60	0.63	0.24	0.36	0.50	0.75		0.58	
Dermaptera	**7**	**5**	**17**	**1**	**7**	**26**	**27**	**43**	**46**	**23**	**202**		0.73
	0.48	0.36	0.18	0.10	0.28	0.44	1.27	0.65	0.43	0.30		0.45	
Other taxa	**24**	**9**	**15**	**1**	**10**	**54**	**28**	**32**	**22**	**70**	**265**		0.76
	1.65	0.65	0.16	0.10	0.40	0.92	1.32	0.48	0.21	0.93		0.68	
Other larvae	**4**	**0**	**0**	**0**	**1**	**0**	**0**	**0**	**0**	**0**	**5**		2.75
	0.27	0.00	0.00	0.00	0.04	0.00	0.00	0.00	0.00	0.00		0.03	
Isoptera	**0**	**17**	**0**	**22**	**0**	**42**	**0**	**26**	**62**	**102**	**271**		1.16
	0.00	1.23	0.00	2.25	0.00	0.71	0.00	0.39	0.58	1.35		0.65	
Acarina	**0**	**13**	**112**	**3**	**134**	**93**	**39**	**96**	**242**	**306**	**1038**		0.87
	0.00	0.94	1.21	0.31	5.35	1.58	1.84	1.45	2.27	4.04		1.90	
Total	**1455**	**1379**	**9272**	**979**	**2503**	**5898**	**2123**	**6622**	**10648**	**7567**	**48446**		
Ind./m^2	63.3	81.1	403.1	42.6	108.8	226.8	96.5	315.3	507.0	360.3			

Table 16.2 continued

Aporusa subcaudata

	As. 8	*As. 9*	*As. 10*	*As. 50*	*As. 51*	*Total*	*Mean%*	*Vk*
Formicidae N	**3370**	**2837**	**1066**	**3168**	**3530**	**13971**		0.20
%	76.71	50.12	74.49	63.60	50.65		63.12	
Arachnida	**108**	**361**	**62**	**149**	**539**	**1219**		0.47
	2.46	6.38	4.33	2.99	7.73		4.78	
Diptera	**169**	**395**	**44**	**119**	**372**	**1099**		0.43
	3.85	6.98	3.07	2.39	5.34		4.33	
Coleoptera	**129**	**513**	**82**	**168**	**531**	**1423**		0.46
	2.94	9.06	5.73	3.37	7.62		5.74	
Homoptera	**135**	**419**	**45**	**213**	**369**	**1181**		0.39
	3.07	7.40	3.14	4.28	5.29		4.64	
Other	**120**	**142**	**15**	**141**	**185**	**603**		0.31
Hymenoptera	2.73	2.51	1.05	2.83	2.65		2.35	
Heteroptera	**77**	**100**	**14**	**244**	**207**	**642**		0.62
	1.75	1.77	0.98	4.90	2.97		2.47	
Orthoptera	**31**	**134**	**20**	**50**	**238**	**473**		0.62
	0.71	2.37	1.40	1.00	3.42		1.78	
Psocoptera	**32**	**111**	**9**	**25**	**50**	**227**		0.66
	0.73	1.96	0.63	0.50	0.72		0.91	
Thysanoptera	**41**	**9**	**0**	**374**	**9**	**433**		1.86
	0.93	0.16	0.00	7.51	0.13		1.75	
Collembola	**15**	**247**	**21**	**18**	**598**	**899**		1.16
	0.34	4.36	1.47	0.36	8.58		3.02	
Lepidop. larvae	**7**	**150**	**24**	**213**	**46**	**440**		0.87
	0.16	2.65	1.68	4.28	0.66		1.88	
Lepidop. adults	**3**	**13**	**2**	**6**	**13**	**37**		0.42
	0.07	0.23	0.14	0.12	0.19		0.15	
Blattodea	**36**	**92**	**4**	**35**	**74**	**241**		0.55
	0.82	1.63	0.28	0.70	1.06		0.90	
Dermaptera	**28**	**36**	**6**	**10**	**18**	**98**		0.48
	0.64	0.64	0.42	0.20	0.26		0.43	
Other taxa	**20**	**45**	**9**	**40**	**17**	**131**		0.41
	0.46	0.80	0.63	0.80	0.24		0.59	
Other larvae	**0**	**11**	**4**	**0**	**103**	**118**		1.59
	0.00	0.19	0.28	0.00	1.48		0.39	
Isoptera	**10**	**0**	**1**	**8**	**15**	**34**		0.73
	0.23	0.00	0.07	0.16	0.22		0.13	
Acarina	**62**	**45**	**3**	**0**	**55**	**165**		0.87
	1.41	0.80	0.21	0.00	0.79		0.64	
Total	**4393**	**5660**	**1431**	**4981**	**6969**	**23434**		
Ind./m²	191.0	246.1	65.0	216.6	316.8			

Table 16.2 continued

Xanthophyllum affine

	Xa. 5	*Xa. 5*	*Xa. 11*	*Xa. 12*	*Total*	*Mean%*	*Vk*
Formicidae N	**1188**	**4201**	**2817**	**2433**	**10639**		0.15
%	50.25	66.18	72.70	59.99		62.28	
Arachnida	**117**	**163**	**180**	**148**	**608**		0.27
	4.95	2.57	4.65	3.65		3.95	
Diptera	**241**	**726**	**119**	**124**	**1210**		0.65
	10.19	11.44	3.07	3.06		6.94	
Coleoptera	**189**	**112**	**209**	**239**	**749**		0.49
	7.99	1.76	5.39	5.89		5.26	
Homoptera	**193**	**93**	**77**	**158**	**521**		0.78
	8.16	1.47	1.99	3.90		3.88	
Other Hymenoptera	**85**	**127**	**97**	**133**	**442**		0.26
	3.60	2.00	2.50	3.28		2.84	
Heteroptera	**8**	**16**	**19**	**23**	**66**		0.35
	0.34	0.25	0.49	0.57		0.41	
Orthoptera	**75**	**86**	**27**	**124**	**312**		0.60
	3.17	1.35	0.70	3.06		2.07	
Psocoptera	**81**	**201**	**21**	**35**	**338**		0.75
	3.43	3.17	0.54	0.86		2.00	
Thysanoptera	**17**	**295**	**107**	**235**	**654**		0.64
	0.72	4.65	2.76	1.79		3.48	
Collembola	**25**	**93**	**35**	**16**	**169**		0.46
	1.06	1.47	0.90	0.39		0.96	
Lepidop. larvae	**19**	**15**	**21**	**59**	**114**		0.68
	0.80	0.24	0.54	1.45		0.76	
Lepidop. adults	**17**	**7**	**1**	**6**	**31**		1.26
	0.72	0.11	0.03	0.15		0.25	
Blattodea	**49**	**45**	**28**	**97**	**219**		0.60
	207	0.71	0.72	2.39		1.47	
Dermaptera	**18**	**16**	**30**	**25**	**89**		0.40
	0.76	0.25	0.77	0.62		0.60	
Other taxa	**30**	**27**	**12**	**5**	**74**		0.95
	1.27	0.43	0.31	0.12		0.53	
Other larvae	**0**	**78**	**0**	**0**	**78**		2.00
	0.00	1.23	0.00	0.00		0.31	
Isoptera	**10**	**29**	**35**	**188**	**262**		1.27
	0.42	10.46	0.90	4.64		1.60	
Acarina	**2**	**18**	**40**	**8**	**68**		1.09
	0.08	0.28	1.03	0.20		0.40	
Total	**2364**	**6348**	**3875**	**4056**	**16643**		
Ind./m^2	102.8	276.0	203.9	176.3			

In all cases other Hymenoptera (2.9%) and Orthoptera (1.9%) were not significantly different between trees. While the variability of Heteroptera and Isoptera is explainable (see below), the highly fluctuating abundances of Collembola, Acarina, Thysanoptera and Psocoptera, which were separated as a fourth group, is not yet understood.

Some remarks on individual taxa

Except for the Formicidae and Coleoptera evaluation of all other orders has been confined until now to a high taxonomic level. The Coleoptera certainly represent one of the most diverse orders and are still being evaluated. Few groups were found with relatively constant abundance in all tree-crowns. On all trees that were fogged for the first time, Chrysomelidae were dominant, representing on average 23.9% of all individuals. Besides Chrysomelidae, only Curculionidae (11.1%) and Staphylinidae (10.9%) were captured regularly, while all other families fluctuated greatly in abundance between tree-crowns.

Heteroptera were collected in higher abundances only on the Euphorbiaceae with on average 4% of total arthropod abundance, while they only represented 0.4% on Polygalaceae. The large abundances found on both *Aporusa* tree species are explained by one tingid bug (*Physatocheila*, Tingidae) which was the only non-social arthropod species that attained abundances of a few hundred individuals per tree. Four further tingid species (all singletons) were captured during all foggings (these belonged to the genera *Perissonemia, Penottus* and *Cromerus*, J. Pericart, personal communication).

The Orthoptera were dominated by the Ensifera, especially by Tettigoniidae and Gryllidae. Omnivorous crickets of the genus *Ornebius* were remarkably abundant in tree-crowns, accounting for one-third of all Ensifera. In contrast, Caelifera were extremely rare, contributing less than 1% of all Orthoptera. Of all Orthoptera, 70–80% were nymphs. A more detailed analysis of the Orthoptera is in preparation.

Lepidoptera were divided into larvae and adults. Adults were found regularly, but as single individuals, while caterpillars were more abundant but showed larger fluctuations between the trees. One arctiid caterpillar species occurred in the crown of *A. subcaudata* #50 with 213 individuals, the highest density found.

Isoptera were recorded from only five tree-crowns and constituted the dominant arthropod group in only a single case. On *A. lagenocarpa* #52 a large carton nest of *Lacessititermes* was found. Although the termites were not active during the fogging procedure, 13 400 individuals were caught. Since an influence of termites on the composition of the arthropod communities was not visible, they were not taken into account for the computation of the relative abundances of all taxa (this does not mean

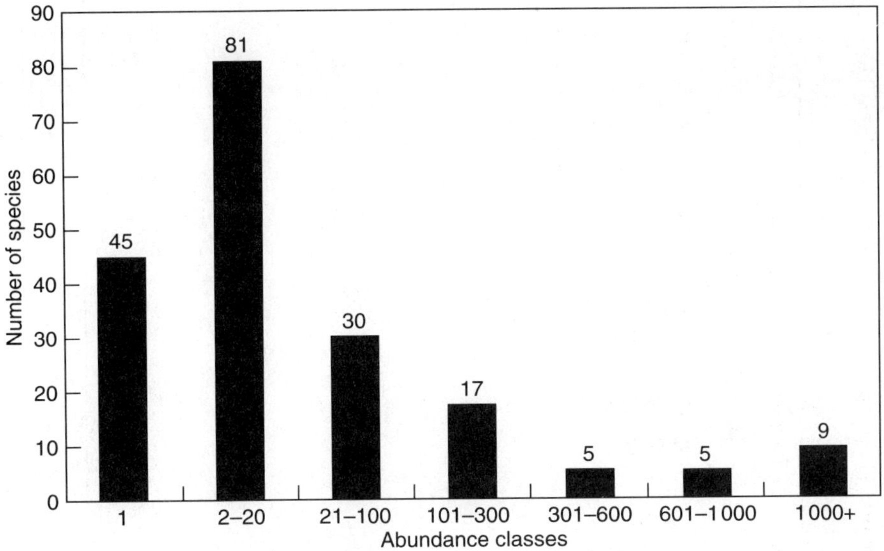

Figure 16.2 Ant species of all first-time fogged trees distributed by different abundance classes.

that termites do not have any effect on the composition of arthropod communities, but simply that there was no visible effect). Including termites would have masked the underlying relative abundances. Similarly, all individuals of a newly established nest found after the second fog of *A. lagenocarpa* #5 were not considered in this evaluation.

The structure of Formicidae communities

From all fogging experiments, 218 ant species from seven subfamilies were identified (Table 16.3), 192 from the first-fogged trees and 35 more species from all re-fogging experiments. At the generic level the Myrmicinae were the most diverse subfamily, while the Formicinae (with just the genera *Camponotus* and *Polyrhachis*) were the most species rich. However, the Dolichoderinae, represented by only four genera and 22 species, contributed most of the dominant species. The Ponerinae and the Cerapachyinae were found regularly in the samples with the genera listed in Table 16.3. The single species of Aenictinae almost certainly got into the collecting funnels accidentally, as this species usually forages in large groups. The Pseudomyrmicinae, represented by the genus *Tetraponera,* provided the only taxa which are considered to be strictly arboreal.

At most, 56 species of ants were found in a single tree-crown. Usually,

Table 16.3 Formicidae genera and species found during all fogging experiments. The systematic classification follows Urbani *et al.* (1992)

Formicinae		*Myrmicinae*		*Dolichoderinae*		*Pseudomyrmicinae*		*Ponerinae*		*Cerapachyinae*		*Aenictinae*	
Genera	*Sp.*	*Genera*	*Sp.*	*Genera*	*Sp.*	*Genera*	*Sp.*	*Genera*	*Sp.*	*Genera*	*Sp.*	*Genera*	*Sp.*
Campanotus	47	*Crematogaster*	16	*Tapinoma*	8	*Tetraponera*	9	*Hypoponera*	4	*Cerapachys*	4	*Aenictus*	1
Polyrhachis	33	*Tetramorium*	9	*Dolichoderus*	7			*Diacamma*	2				
Paratrechina	6	*Vollenhovia*	8	*Technomyrmex*	6			*Odontoponera*	1				
Echinopla	6	*Oligomyrmex*	6	*Philidris*	1								
Prenolepis	2	*Pheidole*	5										
Gesomyrmex	2	*Monomorium*	5										
Plagiolepis	2	*Cataulacus*	4										
Anoplolepis	1	*Dilobocondyla*	3										
Acantholepis	1	*Rhophalomastix*	3										
Foreophilus	1	*Vombisidris*	2										
		Cardiocondyla	2										
		Meranoplus	2										
		Strumigenys	2										
		Pheidologeton	2										
		Leptothorax	1										
		Myrmicaria	1										
		Paratopula	1										
		Smithistroma	1										
		Mayriella	1										
Total													
10	101	19	74	4	22	1	9	3	7	1	4	1	1

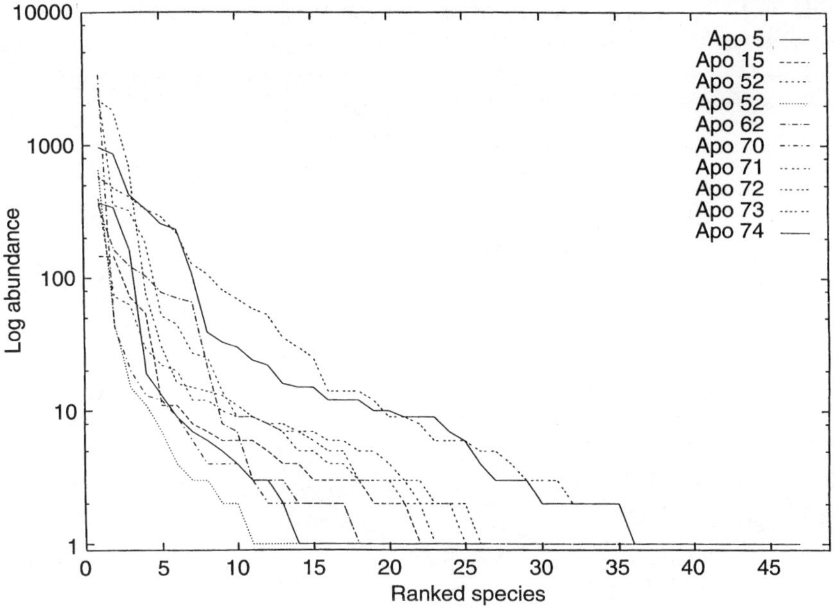

Figure 16.3 Rank-abundance plot of the Formicidae found on first-fogged trees, shown for all ten tree individuals of *Aporusa lagenocarpa*. Abundances on a log scale.

one can expect between 30 and 40 species, of which two to three may be dominant. All communities are characterized by a high percentage of rare species. This is shown for all first-time fogged trees in Figure 16.2.

An average of 40.9% (S.D. 12.7) of all species appeared as singletons and 79.8% (S.D. 7.5) of species were represented by less than 20 individuals. This very general finding is shown in the rank-abundance plot in Figure 16.3.

In order to separate rare species from those living permanently in tree-crowns, we counted as members of the actual community only those species which had more than 20 individuals. This preliminary definition was a consequence of the observation of many small ant-nests with a similar low number of individuals.

Alpha-diversity

The best-fit distribution model (broken-stick, geometric series, log-series or log-normal distribution) for all communities of first-time fogged trees was determined by a Goodness of Fit Test (G-Test) (Sokal and Rohlf, 1995) (Table 16.4). The broken-stick distribution and the geometric series

Table 16.4 Distribution models of first-fogged trees showing G-test values (G), degrees of freedom (d.f.) and significance values (*P*).

	Log-series			*log-normal distribution*		
	G	*d.f.*	*P*	*G*	*d.f.*	*P*
Aporusa lagenocarpa						
5	14.2	3	<0.001	6.4	2	<0.05
15	17.5	5	<0.05	3.6	3	<0.3
52	13.9	6	<0.05	6.9	3	<0.7
57	14.76	1	<0.001	Not computable*		
62	15.6	3	<0.05	8.4	3	<0.05
70	20.3	2	<0.001	3.1	2	<0.3
71	11.2	5	<0.05	10.2	3	<0.025
72	51.0	7	<0.7	3.3	6	<0.08
73	25.4	7	<0.001	10.1	3	<0.025
74	19.7	7	<0.05	11.2	5	<0.05
Aporusa subcaudata						
8	18.5	8	<0.025	2.9	5	<0.8
9	21.9	7	<0.05	6.1	4	<0.2
10	31.3	4	<0.001	Not computable*		
50	13.0	6	<0.05	3.2	3	<0.5
51	13.5	7	<0.1	1.5	5	<0.95
Xanthophyllum affine						
4	26.0	5	<0.001	2.4	2	<0.95
5	10.5	3	<0.025	0.7	2	<0.975
11	14.9	6	<0.025	7.8	3	<0.05
12	21.1	6	<0.05	4.9	4	<0.3
Regional	43.54	11	<0.001	58.87	6	<0.001

* A distribution is not computable when too few abundance classes are available, resulting in d.f. ≤0.

showed significant deviations in all cases (*P* <0.001). In addition the log-series has to be rejected except for the crown of *A. lagenocarpa* #72 for which a 70% probability fit was found. The log-normal distribution, however, cannot be rejected with certainty in 10 cases.

Table 16.5 shows computed alpha-diversities arranged in the Hill-series. Alpha-diversities are highly variable between trees, independent of whether crowns of the same tree species are considered or even those trees which were dominated by the same ant species. Table 16.6 shows that *Dolichoderus* #23 (which belongs to the *Dolichoderus thoracicus* group), found in seven tree-crowns, was the most dominant ant species. Consequently, this species is particularly suitable to assess the influence

Table 16.5 Alpha-diversities arranged in the 'Hill-series' for all ant communities on tree-crowns fogged for the first time. In addition, the α of the log-series and rarefaction values for sample size m = 200 were computed. Communities dominated by the most common dominant ant species (*Dolichoderus* #23) are shaded

Tree no.	*(1/J)*	*(S)*	*(exp H)*	*(1/D)*	*(1/d)*	α	*RAF m = 200*
Aporusa lagenocarpa							
5	1063	25	4.43	3.27	2.58	4.70	12.59
15	3125	41	9.55	5.41	3.58	10.42	25.96
52	523	35	3.20	1.91	1.41	5.32	11.54
57	1754	21	2.00	1.33	1.16	4.0	10.56
62	609	26	7.64	5.36	2.87	4.85	14.19
70	826	27	1.31	1.08	1.04	3.97	6.34
71	3571	31	6.81	3.10	1.82	6.58	22.04
72	724	43	12.63	9.01	5.25	7.09	21.69
73	3030	38	3.68	2.86	2.27	5.59	9.85
74	5000	48	8.84	5.87	3.64	7.86	19.19
Aporusa subcaudata							
8	129	52	7.06	4.73	3.22	8.72	17.39
9	900	44	4.22	2.24	1.54	7.42	15.91
10	2777	28	2.27	1.56	1.27	5.27	9.40
50	316	34	3.81	2.30	1.60	5.32	13.16
51	2857	56	8.67	5.11	2.87	9.90	21.08
Xanthophyllum affine							
4	1149	36	2.56	1.45	1.21	7.01	14.59
5	4166	26	2.54	1.59	1.27	3.71	11.06
11	2439	34	3.91	2.43	1.68	5.45	12.67
12	434	36	4.67	2.72	1.73	5.99	13.56
Regional	50000	192	17.38	9.52	5.03	25.57	

of dominants. As the Simpson (1/D) and Berger–Parker index (1/d) show, there are tree-crowns which are strongly influenced by dominant ant species (e.g. *A. subcaudata* #9 and #10) and communities which show no influence (e.g. *A. lagenocarpa* #72 and #74). Further comparisons reflect highly variable conditions for all tree-crowns. Remarkably, Dolichoderinae were found to be dominant in 16 tree-crowns. Two species, *Dolichoderus* #22 and #23 (both *Dolichoderus thoracicus* group), were dominant in 10 crowns. *Dolichoderus sulcaticeps* #100 dominated two tree-crowns and *Technomyrmex* #29 four crowns. The Myrmicinae contributed four dominant species, all belonging to the genus *Crematogaster*. Likewise the Formicinae contributed four species, one

Table 16.6 Dominant ant species found on all first-time fogged trees. Communities dominated by the most dominant ant species (*Dolichoderus* #23) are indicated by shading. Species in the second and third columns were found in slightly lower abundances, but were counted as dominant species

Tree no.			*Dominant ant species*			
Aporusa lagenocarpa						
5	*Crematogaster*	#155	*Dolichoderus*	#22		
15	*Dolichoderus*	#23	*Camponotus*	#74		
52	*Technomyrmex*	#29				
57	*Dolichoderus*	#100				
62	*Camponotus*	#115				
70	*Plagiolepis*	#154				
71	*Camponotus*	#115				
72	*Dolichoderus*	#23	*Plagiolepis*	#154	*Pheidole*	#56
73	*Dolichoderus*	#23	*Plagiolepis*	#154		
74	*Crematogaster*	#115	*Dolichoderus*	#23		
Aporusa subcaudata						
8	*Dolichoderus*	#23	*Technomyrmex*	#29	*Dolichoderus*	#22
9	*Dolichoderus*	#23				
10	*Dolichoderus*	#23				
50	*Dolichoderus*	#100				
51	*Dolichoderus*	#22				
Xanthophyllum affine						
4	*Crematogaster*	#44				
5	*Technomyrmex*	#29				
11	*Technomyrmex*	#29				
12	*Crematogaster*	#155				

Plagiolepis and three *Camponotus*, which dominated six crown communities. In total, 27 of all 192 species collected by fogging were dominant in at least one tree-crown, but only nine species had more than 1000 individuals (seven Dolichoderinae, one Myrmicinae and one Formicinae). No preference of dominants for a single tree species could be detected.

The similar results computed for rarefaction sample sizes of $m = 200$ demonstrates that the observed patterns are independent of sample size. In addition, the heterogeneity of all rarefaction curves confirm that ant communities in different tree-crowns are structurally different, as is shown for all first-fogged *A. lagenocarpa* trees (Figure 16.4). Even communities with very similar rarefaction curves, for instance *A. lagenocarpa* #52 and #72, have clearly different underlying dominance patterns, as is confirmed by comparison with the species-abundance matrices.

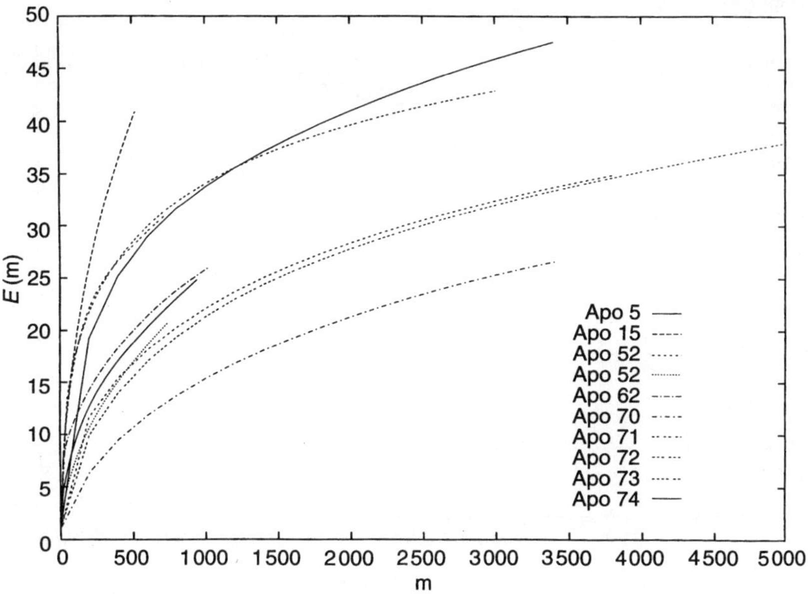

Figure 16.4 Rarefaction curves of all first-fogged *Aporusa lagenocarpa* trees.

Regional diversity

If the rarefaction curve is computed for all canopy samples it can be interpreted as a measure of regional diversity. As the Shinozaki curve shows, beta-diversity is high and a multitude of new ant species can still be expected in the canopy. This can be explained mainly by the large number of rare species, as is indicated by the bending of the rarefaction curve (Figure 16.5).

Re-fogging experiments after 7 months

Table 16.7 shows that the taxonomic composition of those communities fogged again after 7 months almost completely approximated the composition of first-fog samples. In only a few cases did we observe fluctuations of relative abundances of taxa despite the fact that six out of ten communities had a distinctly lower number of arthropods, largely as a result of a strong decrease in the abundance of Formicidae (see Table 16.8). For instance, 13% Diptera on *A. lagenocarpa* #62 instead of the average 5% found after the first fog. Likewise, the colonization of an unknown *Bulbitermes* sp. termite was recorded in the crown of *A. lagenocarpa* #5. Furthermore, in some crowns taxonomic groups that were rather

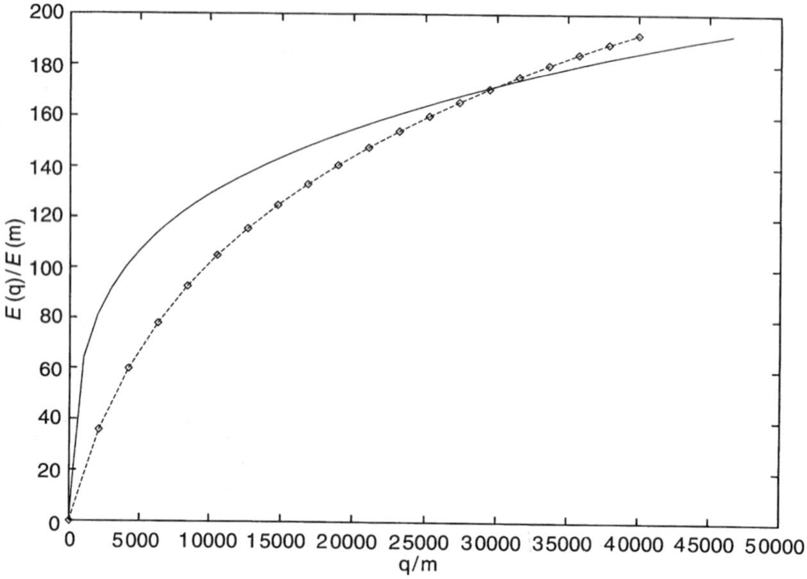

Figure 16.5 Rarefaction (–) and Shinozaki (◊) curves for all ant communities on all trees fogged for the first time. *E*(q) = Expected number of species in q samples, q = number of discrete samples. *E*(m) = Expected number of species in m subsamples, m = size of the subsample.

inconspicuous during former foggings were sampled in greater abundances, such as Dermaptera on *A. subcaudata* #51 (from 0.3% to 5.2%) and Heteroptera on *X. affine* #4 (from 0.3% to 5%, mainly nymphs, but no tingid bugs as were found abundantly on Euphorbiaceae). In general, such changes were the exception to the rule.

Formicidae communities

As in first-fog communities, none of the species-abundance distribution models could be fitted. Since a correspondingly high variability of all computed diversity indices was found, these results are not presented here. When community structure of the undisturbed tree-crowns is compared with that of trees re-fogged after 7 months the impression is very heterogeneous (Table 16.8). In all cases dominance hierarchies changed drastically and community structure of first-fogged and re-fogged trees was obviously distinct. Newly immigrated species became dominant on four trees and in one further tree (*A. lagenocarpa* #57) a newly established species was second dominant in rank. Newly

Table 16.7 Absolute numbers and relative percentages of main arthropod taxa in re-fog samples after time intervals of more than 7 months. Some 2527 termites were found on *Aporusa lagenocarpa* #5 (fog 2) which were not considered in the calculation of relative abundances

	Aporusa lagenocarpa									
	Al.5 (F2)		*Al.52 (F2)*		*Al.57 (F2)*		*Al.62 (F2)*		*Al.73 (F2)*	
	N	%	N	%	N	%	N	%	N	%
Formicidae	3222	74.70	1855	46.51	1212	64.26	819	33.18	4986	69.86
Arachnida	121	2.81	248	6.22	115	6.10	186	7.54	260	3.64
Diptera	197	4.57	321	8.05	129	6.84	327	13.25	351	4.92
Coleoptera	107	2.48	188	4.71	79	4.19	148	6.00	273	3.83
Homoptera	76	1.76	224	5.62	65	3.45	122	4.94	287	4.02
Other Hymenoptera	89	2.06	104	2.61	49	2.60	106	4.29	132	1.85
Heteroptera	49	1.14	81	2.03	15	0.80	59	2.39	122	1.71
Orthoptera	68	1.58	61	1.53	27	1.43	65	2.63	198	2.77
Psocoptera	214	4.96	242	6.07	58	3.08	57	2.31	166	2.33
Thysanoptera	36	0.83	45	1.13	1	0.05	14	0.57	55	0.77
Collembola	46	1.07	405	10.16	64	3.39	438	17.75	88	1.23
Lepidoptera (larvae)	0	0.00	63	1.58	44	2.33	53	2.15	94	1.32
Lepidoptera (adults)	1	0.02	5	0.13	0	0.00	4	0.16	11	0.15
Blattodea	11	0.26	20	0.50	5	0.27	26	1.05	16	0.22
Dermaptera	14	0.32	10	0.25	1	0.05	11	0.45	9	0.13
Other taxa	8	0.19	15	0.38	4	0.21	13	0.53	23	0.32
Other larvae	14	0.32	2	0.05	0	0.00	0	0.00	0	0.00
Isoptera	0	0.00	26	0.65	10	0.53	1	0.04	48	0.67
Acarina	40	0.93	73	1.83	8	0.42	19	0.77	18	0.25
Total individuals per m^2	4313		3988		1886		2468		7137	

	Aporusa subcaudata						*Xanthophyllum affine*			
	As.8 (F2)		*As.50 (F2)*		*As.51 (F2)*		*Xa4 (F2)*		*Xa11 (F2)*	
	N	%	N	%	N	%	N	%	N	%
Formicidae	918	57.02	1220	49.17	497	36.30	2208	65.91	1251	72.19
Arachnida	95	5.90	197	7.94	135	9.86	147	4.39	53	3.06
Diptera	102	6.34	129	5.20	60	4.38	135	4.03	87	5.02
Coleoptera	66	4.10	96	3.87	113	8.25	207	6.18	79	4.56
Homoptera	55	3.42	125	5.04	56	4.09	66	1.97	11	0.63
Other Hymenoptera	28	1.74	52	2.10	87	6.36	31	0.93	43	2.48

Table 16.7 continued

	Aporusa subcaudata						*Xanthophyllum affine*			
	As.8 (F2)		*As.50 (F2)*		*As.51 (F2)*		*Xa4 (F2)*		*Xa11 (F2)*	
	N	%	N	%	N	%	N	%	N	%
Heteroptera	60	3.73	149	6.01	33	2.41	167	4.99	4	0.23
Orthoptera	15	0.93	36	1.45	44	3.21	56	1.67	18	1.04
Psocoptera	81	5.03	38	1.53	153	11.18	50	1.49	21	1.21
Thysanoptera	1	0.06	5	0.20	4	0.29	2	0.06	15	0.87
Collembola	102	6.34	238	9.59	9	0.66	152	4.54	33	1.90
Lepidoptera (larvae)	29	1.80	80	3.22	61	4.46	63	1.88	47	2.71
Lepidoptera (adults)	1	0.06	1	0.04	1	0.07	0	0.00	0	0.00
Blattodea	14	0.87	28	1.13	22	1.61	14	0.42	15	0.87
Dermaptera	4	0.25	24	0.97	71	5.19	11	0.33	28	1.62
Other taxa	11	0.68	49	1.98	3	0.22	6	0.18	6	0.35
Other larvae	0	0.00	0	0.00	1	0.07	0	0.00	0	0.00
Isoptera	5	0.31	14	0.56	9	0.66	19	0.57	12	0.69
Acarina	23	1.43	0	0.00	10	0.73	16	0.48	10	0.58
Total individuals per m^2	1610		2481		1369		3350		1733	

colonized species were very important for the reorganization of the ant communities in 50% of all cases. Although high beta-diversities were mainly a result of the presence of many rare species (which have not been classified as an integral part of the crown communities), the variability remained high if only the more abundant nesting species were considered. Communities structured by newly immigrated species were no more or less variable than those in which former dominants were still in a dominant position. The differences observed are illustrated in rarefaction curves which can be classified into three types depending on their shape; four examples are given (one type A-, one type B-, and two type C curves) (Figure 16.6):

Type A: the tree *A. subcaudata* #8 is characterized by non-diverging rarefaction curves, with the second fogging showing lower species diversity. It is characterized by a loss of species and abundance of 54.2% and 73.0%, respectively. No new species were established; however, the three dominant Dolichoderinae species #22, #23 and #29 (each with 700–1000 individuals) that were dominant initially were replaced by *Crematogaster* #155 (600 individuals, represented originally by 300 specimens). Of all former dominant species only *Dolichoderus* #23 could be found with 15%

Table 16.8 Changes within ant communities in those trees re-fogged after 7 months. Evaluation considers only species represented by more than 20 individuals. The NESS index was calculated with a subsample m that was standardized with regard to the smaller of the two communities being compared

	Aporusa lagenocarpa					*Aporusa subcaudata*			*Xanthophyllum affine*	
	Al.5	Al.52	Al.57	Al.62	Al.73	As.8	As.50	As.51	Xa.4	Xa.11
Abundance Fog 1	957	3833	759	1028	5010	3370	3168	3530	1188	2817
Abundance Fog 2	3222	1855	1221	819	4986	918	1220	497	2208	1251
% Change	+336.7	–48.4	+160.9	–20.1	–0.5	–72.7	–61.5	–85.9	+185.9	–55.6
Species number Fog 1	23	33	20	28	39	48	32	56	36	32
Species number Fog 2	38	31	18	39	32	26	27	25	17	17
Newly immigrated	4	3	1	3	–	–	3	–	2	1
Immigrated dominants	No	Yes	No	Yes	No	No	No	No	Yes	Yes
Disappeared species	–	3	2	5	2	5	5	8	4	4
Changes in dominance structure	Yes	Yes	Yes	Yes	Yes	Yes	Yes	Yes	Yes	Yes
NESS index	0.65	0.75	0.61	0.47	0.77	0.75	0.48	0.82	0.50	0.43
Sörensen index	0.44	0.58	0.50	0.35	0.50	0.46	0.55	0.52	0.38	0.28

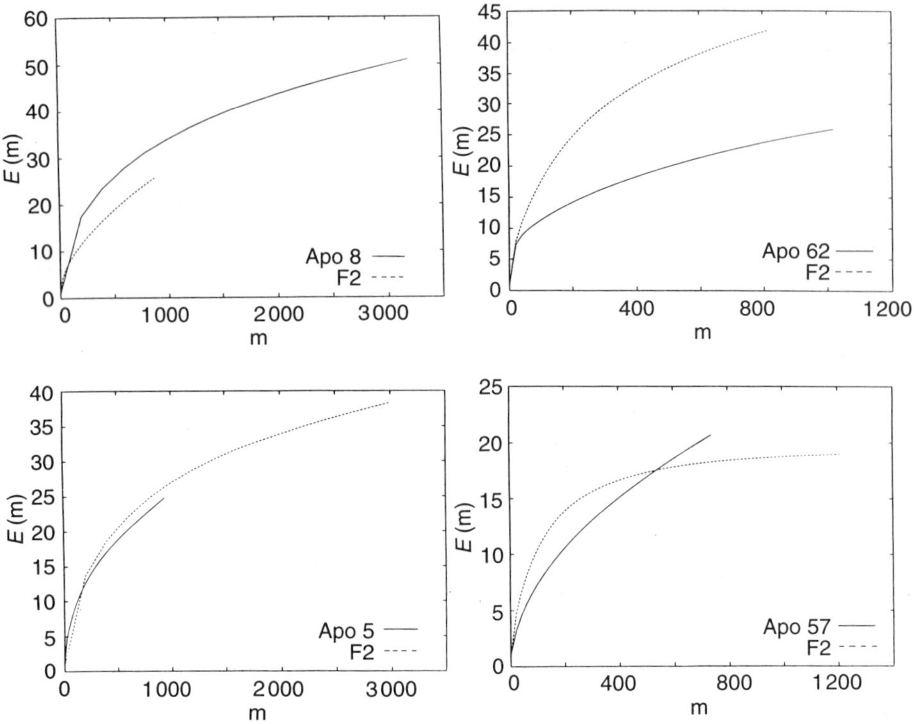

Figure 16.6 Examples of typical rarefaction curves, comparing first-fogged (–) and re-fogged (....) communities.

of its original abundance. Type A communities include trees *A. subcaudata* #8, #50 and #51, and *A. lagenocarpa* #52 and #73.

Type B: *A. lagenocarpa* #5 is characterized by non-diverging rarefaction curves, but here considerably more species and individuals were collected after the second fog. Of two dominant species, *Crematogaster* #155 and *Dolichoderus* #22, found at the beginning in equal numbers, *Crematogaster* had almost tripled in abundance in re-fog samples. However, even though *Dolichoderus* was represented by more specimens it was exceeded in total abundance by *Technomyrmex* #27, which was originally represented in much lower numbers. *Pheidole* #56, represented by a single individual in first-fog samples, had more than 300 specimens in the re-fog.

Type C: *A. lagenocarpa* #62 is characterized by strongly divergent rarefaction curves. Considerably more species were found after the second fogging, but total abundances were lower. The former dominant *Camponotus* (*Colebopsis*) #115 (with 358 individuals) was second dominant behind the newly immigrated *Dolichoderus* #23 (with 317 individuals) after the second fog. All six original dominant species were strongly reduced in their abundances or even totally missing in the second fog. *Technomyrmex* #27 and *Echinopla* #90 were newly established with 100 and 45 specimens, respectively. Ant communities of this type comprise *A. lagenocarpa* #57 and #62, and *X. affine* #4 and #11.

Type C: The last example is *A. lagenocarpa* #57 another Type C curve. This was the tree with the lowest number of species and individuals. Although the dominant *Dolichoderus* #100 held its dominant position, *Technomyrmex* #27 colonized successfully and was second dominant with 253 individuals. *Camponotus* #91 has been sampled from both crowns with approximately 40 specimens.

A comprehensive summary of all species-abundance fluctuations is given in Table 16.9. As already mentioned, dominant ant species were the group most strongly affected by the fogging. During recolonization they benefited in only three tree-crowns. Appearance and disappearance of species seemed to be unpredictable. Three Dolichoderinae and one Myrmicinae species established new nests and became dominant. *Crematogaster* #155 and *Technomyrmex* #29 became dominant by improving their rank position. Of all 65 species considered (n >20 individuals) 45% were not found in their original community after 7 months. Disappearance affected both dominant species and species of all following rank orders or taxonomic position (Table 16.9).

DISCUSSION

The efficiency of the selective fogging method

The investigations of the efficiency of the fogging method showed that arboreal arthropod communities can be sampled both quantitatively and selectively. This was made possible by the special roof construction and the high knockdown capacity of the pyrethrum, which attacks the nervous system of arthropods immediately after application (Soderlund and Bloomquist, 1989; Schulz *et al.*, 1993), killing even highly mobile groups, as can be seen by their large quantities in all samples. Of course not all arthropod groups can be sampled with equal efficiency.

Ants nesting within tree trunks are only eliminated by fogging if their queen(s) have been caught or if colony abundance is reduced to such a degree that the species will be replaced by a better competitor (hints to both possibilities can be derived from the analysis of the re-fog communities after 7 months). Similarly, species building carton nests were more affected by the pyrethrum application than those nesting in the tree trunk, but without the carton-nesting species being more successful colonists. The results of the two control-fogs conducted 3 h after the first

Table 16.9 Changes in the composition of ant communities after the second fogging; included are all species with more than 20 individuals. Dominant species of first-time fogged trees are indicated in bold. Shading represents new dominant species of ants on re-fogged trees

	Newly immigrated species	Disappeared species	Advantaged species	Disadvantaged species
Aporusa lagenocarpa				
Al.5	#36 *Cataulacus*		#27 *Technomyrmex*	
	#38 *Monomorium*		**#155 *Crematogaster***	
	#56 *Pheidole*		**#22 *Dolichoderus***	
	#58 *Monomorium*			
Al.52	#22 *Dolichoderus* (shaded)	#29 ***Technomyrmex***	#27 *Technomyrmex*	
	#28 *Technomyrmex*	#107 *Paratrechina*		
	#72 *Echinopla*	#132 *Camponotus*		
Al.57	#27 *Technomyrmex*	#23 *Dolichoderus*	**#100 *Dolichoderus***	
Al.62	#23 *Dolichoderus* (shaded)	#72 *Echinopla*		#65 *Oligomyrmex*
	#27 *Technomyrmex*	#88 *Polyrhachis*		#112 *Polyrhachis*
	#90 *Echinopla*	#97 *Dolichoderus*		**#115 *Camponotus***
		#105 *Crematogaster*		
		#116 *Crematogaster*		
Al.73		#6 *Polyrhachis*	**#23 *Dolichoderus***	#6 *Polyrhachis*
		#173 *Polyrhachis*	#92 *Tetraponera*	#29 *Technomyrmex*
				#154 *Plagiolepis*
Aporusa subcaudata				
As.8		#6 *Polyrhachis*	#155 *Crematogaster* (shaded)	**#23 *Dolichoderus***
		#22 *Dolichoderus*		#29 *Technomyrmex*
		#38 *Monomorium*		#56 *Pheidole*
		#60 *Tapinoma*		
		#100 *Dolichoderus*		

Table 16.9 continued

	Newly immigrated species	*Disappeared species*	*Advantaged species*	*Disadvantaged species*
As.50	#17 *Camponotus*	#10 *Polyrhachis*	#15 *Echinopla*	#22 *Dolichoderus*
	#74 *Camponotus*	#29 *Technomyrmex*	#72 *Echinopla*	**#100 *Dolichoderus***
	#137 *Pheidologeton*	#38 *Monomorium*		
		#65 *Oligomyrmex*		
		#73 *Camponotus*		
As.51		#1 *Camponotus*	#46 *Crematogaster*	**#22 *Dolichoderus***
		#2 *Camponotus*		#155 *Crematogaster*
		#24 *Camponotus*		
		#27 *Technomyrmex*		
		#29 *Technomyrmex*		
		#43 *Crematogaster*		
		#68 *Dolichoderus*		
Xanthophyllum affine				
Xa.4	#21 *Acantholepis*	#27 *Technomyrmex*	#29 *Technomyrmex*	
	#94 *Philidris*	**#44 *Crematogaster***		
		#88 *Polyrhachis*		
		#11 *Camponotus*		
Xa.11	#223 *Crematogaster*	#23 *Dolichoderus*		#29 ***Technomyrmex***
		#24 *Camponotus*		
		#27 *Technomyrmex*		
		#60 *Tapinoma*		
		#62 *Tapinoma*		
		#65 *Oligomyrmex*		
		#115 *Camponotus*		
		#155 *Crematogaster*		

fog, indicate that a strong recolonization dynamic exists. This can be seen in the distinctly larger samples from second fogs (16% of the abundance of arthropods in the first fog) in comparison with the shaking-samples (5–6%). The percentage of arthropods that newly immigrated during this time-span was therefore about 10% and was composed mainly of species of highly mobile groups such as Diptera, Coleoptera, other Hymenoptera, and surprisingly always arachnids. Since the cotton roof that stretched above the study tree hampered the recolonization dynamic, one has to expect recolonization to be much higher under natural conditions.

The cotton roof proved to be an effective barrier to arthropods falling from the higher canopy stratum since many dead arthropods were found on top of the roof, and were thus successfully prevented from falling into the collecting funnels beneath the study tree. This study utilizes for the first time a methodology that allows true selective sampling of the arthropods of a single tree-crown and therefore provides the precondition for the investigation of structuring mechanisms.

The composition of arthropod communities in the tree-crowns

The constancy of rank ordering of the major arthropod taxa was described by Stork (1991) after fogging 10 trees in a lowland rainforest in Brunei. In all tree-crowns, Formicidae were found to be dominant. Besides species of the eusocial Formicidae and Isoptera, no other species were found in high abundances. Most frequently, a tingid bug (Heteroptera, Tingidae) was found with a few hundred individuals per tree.

While Riede (1993) found mainly grasshoppers (Acridoidea) among the Orthoptera in a lowland rainforest in north-west Amazonia, our investigations showed very different results. Almost exclusively we found species of Ensifera, mainly Tettigoniidae and Gryllidae, while Caelifera were poorly represented (less than 1% of total abundance). The reason for these differences are not yet understood. Possibly they result from different crown structure and number of suitable food plants (Riede, 1993), but perhaps historical reasons are responsible.

Termites were dominant in only one tree-crown. Although many species build sheltered trail systems, observations in the tree-crowns excluded the possibility that they were abundant but overlooked. Another study in Kinabalu Park, conducted by Leideritz (1993), confirms that only 12% of nests can be found in the crown region. Similar observations were described by Collins (1983), working in Mulu National Park in Sarawak. On the other hand, arboreal termites play a much more important role in other rainforests, for example in the flooded forest of the Amazon, where termites are described as abundant and important structuring keystone species (Martius, 1994).

The highly fluctuating abundances of Collembola, Acarina, Thysanoptera and Psocoptera can not be fully interpreted as yet. It is known that Collembola migrate vertically along tree trunks (Bowden *et al.*, 1976); thus, their high abundances in some samples might result from random sampling of these migrations. In the case of mites, the presence of leaf domatia might be the reason that only a relatively small proportion were caught (O'Dowd and Wilson, 1991). The higher occurrence of Psocoptera and Thysanoptera in some tree-crowns might be correlated with the flowering of these trees, but this has not been found in the majority of the cases investigated.

Further investigations are necessary before more detailed conclusions can be reached.

The composition of Formicidae communities

With 19 trees fogged (for the first time) and 192 species collected this is the most comprehensive study of the canopy-inhabiting Formicidae fauna of a lowland rainforest (compare Wilson, 1987; Majer, 1990; Harada and Adis, 1997, Chapter 17, this volume). Characteristically, the many rare species in each community prevented the fitting of any species-abundance distribution model. For most ant communities the log-normal distribution fits best, even if large uncertainties are found in most cases. This is most likely a direct consequence of the central limits theorem as discussed by May (1975) and therefore biological interpretation does not allow more than the inference of a multifunctional influence on community structure (Ludwig and Reynolds, 1988; Colwell and Coddington, 1994). Alpha-diversity also gives no hint as to a homogeneous structuring mechanism. Theoretically, those tree-crowns dominated by the same ant species should show the highest conformity, if (as may be expected) these dominant species are important biotic structuring elements. In seven of the tree-crowns sampled the dominant species was *Dolichoderus* sp. #23, but highly variable alpha-diversities clearly showed its different influence in the respective crown communities. In addition, the lack of a convergent structuring mechanism is indicated by the highly variable rarefaction curves, confirming also the independence of the results from sample size.

In this context it is remarkable that dolichoderines contributed 60% of all dominant ant species in the tree-crowns and that 56% of these species belonged to the *Dolichoderus thoracicus* group. These species are polydom and polygynous and therefore able to occupy new nesting sites with considerable speed. The dominance of dolichoderines in tree-crowns is a well-known phenomenon and has previously been documented in the Neotropics (Adis *et al.*, 1984; Wilson, 1987; Tobin, 1991), in Australia (Majer, 1990) and in Malaysia (Way and Khoo, 1991). These findings are clearly distinct from the epigeic ant fauna, in which dolichoderines play a minor role (K. Rosciszewski, personal communication). This is largely due to the primarily arboreal evolution of this subfamily, which depends on trophobiotic relations that occur only in the canopy (U. Maschwitz, personal communication). Oddly, such mutualism could not be found in the trees under study. Because of the large numbers of mutualists needed to nourish thousands of workers it seems unlikely that they have been overlooked during the observations in the trees, even in the case of inconspicuous scale insects (Coccoidea). Only in carton nests of *Dolichoderus* and *Monomorium* spp., arranged along the underside mid-vein of leaves of Euphorbiaceae were scale insects detected as symbiotic partners. Currently, information regarding how much of the energy supply can be provided by homopterans is lacking. Irrespective of this question, it is not known how ants cover their nitrogen requirement. In comparison with other south-east Asian lowland rainforests, trophobiosis is relatively rare in the canopy in Kinabalu Park, as subsequently confirmed for the shrub layer (Aug, 1995). Since extra-floral nectaries are a limited food resource found only on the newly flushed leaves of *X. affine*, and arthropods, furthermore, were not important prey items (Götzke, 1993), the question arises as to what dolichoderines and other ants feed on in tree-crowns. Long-term observations and bait samples indicate that all arboreal ant species found so far are generalists that opportunistically exploit all potential prey items. Baits were monopolized very quickly and taken away, and in some cases aggressive interactions were observed. Possible feeding strategies might be the use of the leaf microfauna, of pollen, or nematodes, among others, but this is speculative at this stage.

The re-fogging experiments

Besides the comparison of the arthropod communities of all trees that were fogged for the first time, the analysis of recolonization dynamics should give information about mechanisms structuring communities. In particular, it is interesting to consider to what extent community reorganization at a high taxonomic level, as well as at the morphospecies level, shows similarities to the conditions of the first fog. If there are definite similarities, does the establishment of communities follow certain patterns? As the results show, the taxonomic composition at a high level almost completely approximated the composition of the first fog, but this is not true for species composition, as was shown for Formicidae. In most cases tree-crowns had a lower number of individuals. It seems as if the communities had not yet recovered from the first fogging procedure.

As can be concluded from rarefaction curves and beta-diversity calculated for the ant community, the fogging procedure caused such strong disturbance that in all cases dominance hierarchies changed drastically. In no case was the structure of the reorganized communities predictable, independent of whether the former dominant species were still in a dominant position or whether dominant species had changed. It does not seem very likely that the structural heterogeneity of the newly established communities was the result of a successional pattern which was merely not detected, since initial community composition was accordingly heterogeneous. Crucial to the structure of a community might be the arrival of a species at the most favoured time, resulting in a growth advantage and consequently in the ability to outcompete potential competitors. Such a situation would resemble a chaotic system depending mainly on its starting constraints (Hastings *et al.*, 1993). This would also explain the unpredictability of the development of ant communities (in this context see Logan and Allen, 1992). Competition might be of immediate importance, for instance when species originally separated by dominance hierarchies are freed from competition as a result of a major disturbance. These interactions, however, would not result in a predictable climax equilibrium. In addition, our results demonstrate that ant mosaics, as they have been described very generally

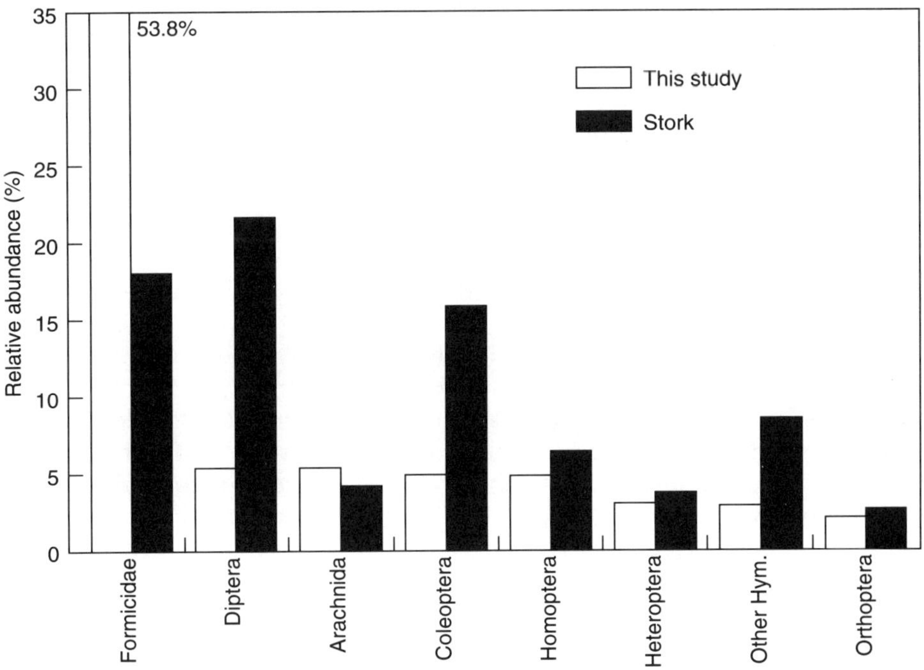

Figure 16.7 The most important arthropod taxa in tree-crowns fogged for the first time, compared with the results of Stork (1987, 1991).

in the literature (Leston, 1978; Jackson, 1984; Majer and Camer-Pesci, 1991; Majer, 1994), do not exist in this form in highly complex lowland rainforests.

Comparison with other fogging studies

There are some striking differences between this study and the results of the fogging experiments conducted by Stork (1987, 1991), particularly in the lower arthropod abundance per tree (Figure 16.7), although Stork investigated much taller trees. The number of collecting funnels was comparable. In particular, Formicidae were represented by a conspicuously lower proportion in each crown in Stork's (1987, 1991) study. In total, Stork caught 4429 ants sorted to 98 species. In contrast to these results we found 47 568 individuals from the first-fogged trees alone, belonging to 192 species. Even from the crowns of the four Polygalaceae trees more than twice as many individuals and 82 species were collected. Averaged over all trees, Stork's investigations showed Diptera as the largest group with 5046 individuals, ranking ahead of Formicidae. In four tree-crowns he found even more beetles than ants, a result never found in our fogging experiments, where total abundance of Formicidae (with one exception) always exceeded 1000 individuals.

These differences might indicate that the two habitats are not directly comparable because of structural differences. Stork mentioned periodical flooding of the forest investigated and furthermore he fogged trees of the upper canopy stratum which might in itself support differently structured arthropod communities, merely by differences in abiotic factors. Similarly, the results of Stork and Brendell (1990) after fogging tree-crowns in Sulawesi are difficult to interpret. Here, Formicidae represented just 7% of total abundance, while Diptera dominated all samples. These results might be due to possible methodological biases. First, fogging from the ground started by radio-control is certainly not as effective as when conducted inside relatively small trees of the lower canopy. Second, by using a cotton roof the fog is held back for a longer time in the tree-crown and sampling of arthropods in surrounding taller trees is prevented. Third, fogging smaller trees guarantees that wind drift cannot bias sampling. Control observations and control collections on the canopy walk-way installed in Kinabalu Park confirm the dominance of Formicidae in the crowns of the higher canopy. The dominance of Formicidae has been described for most comparable ecosystems in the Neotropics (Erwin, 1983; Adis *et al.*, 1984; Harada and Adis, 1997, Chapter 17, this volume) as well as in Africa (Basset *et al.*, 1992; T. Wagner, personal communication). Seasonal effects, which might explain these differences, cannot be derived from our data.

The structural importance of Formicidae for arthropod communities

The correct interpretation of the main differences between Stork's and our work is important, since one might glean information about different structuring factors in the crowns of the different canopy strata. For Formicidae, a structural influence on the composition of the associated arthropod community has been exemplified manifold, as a few examples should illustrate. For mutualistic ant–plant relationships, ants have been shown to remove arthropods as well as lianas from their host plant (Fiala *et al.*, 1989, 1994). Investigations on bamboo in a dry forest in Costa Rica showed that leaf mining chrysomelids have a significantly higher probability of hatching if ants are excluded from their host plants (Memmott *et al.*, 1993) and Majer (1993) found that arthropod community structure in the ant mosaics of tropical plantations is determined by dominant ant species. In tree-crowns of northern European forests, ants are also more important than expected. This is illustrated by studies on birch trees, which show that there are distinct differences in the composition of arthropod communities where ants are highly abundant compared with those trees not visited by ants (Fowler and Macgarvin, 1985; Mahdi and Whittaker, 1993). Corresponding effects have been described in northern Japanese oak forests (Ito and Higashi, 1991). In north Swedish forests it was shown that the permanent presence of highly abundant ants (*Formica aquilonia*) in crowns of *Picea abies* increased prey search-time of insectivorous bird species relative to trees without ants (Haemig, 1994).

In a similar way one can assume that ants are an important structuring factor in the trees investigated here. Thus, observations in the tree-crowns showed that few other sedentary arthropods could be detected and fogging samples consisted mainly of highly mobile arthropod groups that were able to avoid the ants. In contrast to the tropics, most trees in temperate forests are not dominated by ants. Here one usually finds characteristically greater numbers of arthropod groups such as Lepidoptera, tenthredinids or Coleoptera larvae and aphids. While it has been argued that the lack of aphids in tropical systems is caused by the rareness of their host plants (Dixon *et al.*, 1987), the rareness of other arthropod groups may be due to ant predation pressure (see also Wagner, 1997, Chapter 9, this volume). If this is the case, arthropods must necessarily be highly mobile (as was found for most groups), well camouflaged, or possess specific adaptations protecting them from ant attacks. These defensive mechanisms might be of mechanical, chemical or ethological character. Even if only very few data are available for such comparisons, the significantly higher proportion of obligate ant mutualism found among the lycaenid butterflies might be interpreted as an adaptation to constant high predation pressure (K. Fiedler, personal communication). In any case, the Lepidoptera as a highly diverse taxon obviously possesses very successful strategies to avoid Formicidae predation.

An understanding of the functional relations of these interactions is only possible if similar investigations are extended to more tree species.

Acknowledgements

We are grateful to the Director of Sabah Parks, Datuk Ali Lamri, for permission to work in Kinabalu National Park. Stefan Messner supported the evaluation by providing his BIODIV computer program for calculating the indices and rarefaction values. We thank Krzysztof Rosciszewski, who supplied help for morphospecies-level identification of the ants and Dr Balgooy and Anne Schot, Botanical Garden, Leiden, for identifying the tree species. We are grateful to ESSO for providing the technical oil used for the fogging experiments. Johannes Henschel, Ulmar Grafe and Christiane Brandt gave valuable suggestions and improved the English. The study was financially supported by the Deutsche Forschungsgemeinschaft (DFG) SPP/Li 150/13–2 and in part by the Stifterverband für die Deutsche Wissenschaft.

REFERENCES

Achtziger, R., Nigmann, U. and Zwölfer, H. (1992) Rarefaction-Methoden und ihre Einsatzmöglichkeiten bei der zooökologischen Zustandsanalyse und Bewertung von Biotopen. *Zeitschrift für Ökologie und Naturschutz*, **1**, 89–105.

Adis, J., Lubin, Y.D. and Montgomery, G.G. (1984) Arthropods from the canopy of inundated and terra firme forests near Manaus, Brazil, with critical considerations on the pyrethrum-fogging technique. *Studies on Neotropical Fauna and Environment*, **4**, 223–36.

Ashton, P.S. (1992) The structure and dynamics of tropical rain forest in relation to tree species richness, in *The Ecology and Silviculture of Mixed-Species Forests* (ed. M.J. Kelty), Kluwer Academic Publishers, Dordrecht, pp. 53–64.

Aug, E. (1995) *Tropische Ameisenzönosen – Ressourcennutzung, Aktivität und Diversität. Ein Vergleich verschiedener Habitate und Höhenstufen.*, Msc-thesis, Julius-Maximilians-Universität, Würzburg.

Basset, Y. and Arthington, A.H. (1992) The arthropod community of an Australian rainforest tree: abundance and component taxa, species richness and guild structure. *Australian Journal of Ecology*, **17**, 89–98.

Basset, Y. and Kitching, R.L. (1991) Species number, species abundance and body length of arboreal arthropods associated with an Australian rainforest tree. *Ecological Entomology*, **16**, 391–402.

Basset, Y., Aberlenc, H.-P. and Delvare, G. (1992) Abundance and stratification of foliage arthropods in a lowland rain forest of Cameroon. *Ecological Entomology*, **17**, 310–18.

Bowden, J., Haines, I.H. and Mercer, D. (1976) Climbing collembola. *Pedobiologia*, **16**, 298–312.

Chesson, P.L. and Case, T.J. (1986) Overview: nonequilibrium community theories: chance, variability, history, and coexistence, in *Community Ecology* (eds J. Diamond and T.J. Case), Harper and Row Publishers, NY, pp. 229–39.

Collins, N.M. (1983) Termite populations and their role in litter removal in Malaysian rain forests, in *Tropical Rain Forest: Ecology and Management* (eds S.L. Sutton, T.C. Whitmore and A.C. Chadwick), Blackwell Scientific Publications, Oxford, pp. 311–25.

Colwell, R.K. and Coddington, J.A. (1994) Estimating terrestrial biodiversity through extrapolation. *Philosophical Transactions of the Royal Society of London, Series B*, **345**, 101–18.

Cornell, H.V. and Lawton, J.H. (1992) Species interactions, local and regional processes, and limits to the richness of ecological communities: a theoretical perspective. *Journal of Animal Ecology*, **61**, 1–12.

Dixon, A.F.G., Kindlmann, P., Leps, J. and Holman, J. (1987) Why there are so few species of Aphids, especially in the tropics. *The American Naturalist*, **129**(4), 580–92.

Erwin, T.L. (1982) Tropical forests: their richness in Coleoptera and other arthropod species. *The Coleopterist's Bulletin*, **36**(1), 74–5.

Erwin, T.L. (1983) Beetles and other insects of tropical forest canopies at Manaus, Brazil, sampled by insecticidal fogging, in *Tropical Rain Forest: Ecology and Management* (eds S.L. Sutton, T.C. Whitmore and A.C. Chadwick), Blackwell Scientific Publications, Oxford, pp. 59–75.

Erwin, T.L. (1988) The tropical forest canopy: the heart of biotic diversity, in *Biodiversity* (eds E.O. Wilson and F.M. Peter), National Academy Press, Washington DC, pp. 123–9.

Fiala, B., Maschwitz, U., Pong, T.Y. and Helbig, A.J. (1989) Studies of South East Asian ant–plant association: protection of *Macaranga* trees by *Crematogaster borneensis*. *Oecologia*, **79**, 463–70.

Fiala, B., Grunsky, H., Maschwitz, U. and Linsenmair, K.E. (1994) Diversity of ant–plant interactions: protective efficiency in *Macaranga* species with different degrees of an ant association. *Oecologia*, **97**, 186–92.

Fowler, S.V. and Macgarvin, M. (1985) The impact of hairy wood ants, *Formica lugubris*, on the guild structure of herbivorous insects on birch, *Betula pubescens*. *Journal of Animal Ecology*, **54**, 847–55.

Giller, P.S. (1984) *Community Structure and the Niche*, Chapman & Hall, New York.

Götzke, A. (1993) *Ameisenzönosen ausgewählter tropischer Baumkronen: Struktur, Diversität und Ressourcennutzung der Gemeinschaft.*, Diplomarbeit, Julius-Maximilians-Universität, Würzburg.

Grassle, J.F. and Smith, W. (1976) A similarity measure sensitive to the contribution of rare species and its use in investigation of variation in marine benthic communities. *Oecologia*, **25**, 13–22.

Haemig, P.D. (1994) Effects of ants on the foraging of birds in spruce trees. *Oecologia*, **97**, 35–40.

Harada, A.Y. and Adis, J. (1997) The ant fauna of tree canopies in central Amazonia: a first assessment. in *Canopy Arthropods* (eds N.E. Stork, J. Adis and R.K. Didham), Chapman & Hall, London, pp. 382–400 (this volume).

Hastings, A., Hom, C.L., Ellner, S., Tirchin, P. and Godfray, H.C.J. (1993) Chaos in ecology: is mother nature a strange attractor? *Annual Review of Ecology and Systematics*, **24**, 1–33.

Hill, M.O. (1973) Diversity and evenness: a unifying notation and its consequences. *Ecology*, **54**(2), 427–32.

Hubbell, S.P. and Foster, R.B. (1986) *Biology, Chance, and History and the Structure of Tropical Rain Forest Communities*, Harper and Row Publishers, New York.

Hurlbert, S.H. (1971) The nonconcept of species diversity: a critique and alternative parameters. *Ecology*, **52**, 577–86.

Huston, M. A. (1994) *Biological Diversity*, Cambridge University Press, Cambridge.

Huston, M.A. and DeAngelis, D.L. (1994) Competition and coexistence: the effects of resource transport and supply rates. *The American Naturalist*, **144**(6), 954–77.

Ito, F. and Higashi, S. (1991) Variance of ant effects on the different life forms of moth caterpillars. *Journal of Animal Ecology*, **60**, 327–34.

Jackson, D.A. (1984) Ant distribution patterns in a Cameroonian cocoa plantation: investigation of the ant mosaic hypothesis. *Oecologia*, **62**, 318–24.

Leideritz, B. (1993) *Untersuchungen zur Höhenzonierung der Termitenzönose am Mount Kinabalu (Sabah/Malaysia)*, Diplomarbeit, Freie Universität Berlin.

Leston, D. (1978) A neotropical ant mosaic. *Annals of the Entomological Society of America*, **71**, 649–53.

Logan, J.A. and Allen, J.C. (1992) Nonlinear dynamics and chaos in insect populations. *Annual Review of Entomology*, **37**, 455–77.

Ludwig, J.A. and Reynolds, J.F. (1988) *Statistical Ecology, a Primer on Methods and Computing*, John Wiley & Sons, New York.

Mahdi, T. and Whittaker, J.B. (1993) Do birch trees (*Betula pendula*) grow better if foraged by wood ants? *Journal of Animal Ecology*, **62**, 101–16.

Majer, J.D. (1990) The abundance and diversity of arboreal ants in Northern Australia. *Biotropica*, **22**(2), 191–9.

Majer, J.D. (1993) Comparison of the arboreal ant mosaic in Ghana, Brazil, Papua New Guinea and Australia – its structure and influence on arthropod diversity, in *Hymenoptera and Biodiversity* (eds J. LaSalle and I.D. Gauld), CAB International, Bristol.

Majer, J.D. (1994) Arboreal ant community patterns in Brazilian cocoa farms. *Biotropica*, **26**(1), 73–83.

Majer, J.D. and Camer-Pesci, P. (1991) Ant species in Australian tree crops and native ecosystems – is there a mosaic? *Biotropica*, **23**(2), 173–81.

Martius, C. (1994) Termite nests as structural elements of the Amazon floodplain forest. *Andrias*, **13**, 137–50.

May, R.M. (1975) Patterns of species abundance and diversity, in *Ecology and Evolution of Communities* (eds M.L. Cody and J. Diamond), Belknap Press of Harvard University, Cambridge, Massachusetts, pp. 81–120.

Memmott, J., Godfray, H.C.J. and Bolton, B. (1993) Predation and parasitism in a tropical herbivore community. *Ecological Entomology*, **18**, 348–52.

O'Dowd, D.J. and Willson, M.F. (1991) Associations between mites and leaf domatia. *Trends in Ecology and Evolution*, **6**(6), 179–82.

Pickett, S.T.A. and White, P.S. (1985) *The Ecology of Natural Disturbance and Patch Dynamics*, Academic Press, Orlando, Florida.

Riede, K. (1993) Abundance differences of phytophagous tropical insects: the case of forest grasshoppers (Orthoptera, Caelifera), in *Animal–Plant Interactions in Tropical Environments* (eds W. Barthlott, C.M. Naumann, K. Schmidt-Loske and K.L. Schumann), Results of the annual meeting of the German Society for Tropical Ecology, Bonn, February 13–16, 1992. Bonn, Zoologisches Forschungsinstitut und Museum Alexander Koenig, pp. 211–18.

Sanders, H.L. (1968) Marine benthic diversity: a comparative study. *The American Naturalist*, **102**, 243–82.

Schulz, J., Schmoldt, A. and Schultz, M. (1993) Phyrethroide: Chemie und Toxikologie einer Insektizidgruppe. *Pharmazeutische Zeitung*, **15**, 1141–56.

Shinozaki, K. (1963) Note on the species–area curve, in Proceedings of the 10th Annual Meeting of the Ecological Society of Japan, vol. 5, Tokyo. Aus dem Japaischen von S. Kobayashi.

Smith, W., Grassle, J.F. and Kravitz, D. (1979) Measures of diversity with unbiased estimates, in *Ecological Diversity in Theory and Practice* (eds J.F. Grassle, G.P. Patil, W. Smith and C. Taillie), International Cooperative Publishing House, Fairland, Maryland, pp. 177–91.

Soderlund, D. M. and Bloomquist, J.R. (1989) Neurotoxic actions of pyrethroid insecticides. *Annual Review of Entomology*, **34**, 77–96.

Sokal, R.R. and Rohlf, F.J. (1995) *Biometry*, W.H. Freeman & Co., New York.

Stork, N.E. (1987) Arthropod faunal similarity of Bornean rain forest trees. *Ecological Entomology*, **12**, 219–26.

Stork, N.E. (1988) Insect diversity: facts, fiction and speculation. *Biological Journal of the Linnaean Society*, **35**, 321–37.

Stork, N.E. (1991) The composition of the arthropod fauna of Bornean lowland rain forest trees. *Journal of Tropical Ecology*, **7**, 161–80.

Stork, N.E. (1993) How many species are there? *Biodiversity and Conservation*, **2**, 215–32.

Stork, N.E. and Blackburn, T.M. (1993) Abundance, body size and biomass of arthropods in tropical forest. *Oikos*, **67**, 483–9.

Stork, N.E. and Brendell, M.J.D. (1990) Variation in the insect fauna of Sulawesi trees with season, altitude and forest type, in *Insects and the Rain Forest of South East Asia (Wallacea)* (eds W.J. Knight and J.D. Holloway), Royal Entomological Society of London. pp. 173–90.

Sugihara, G. (1980) Minimal community structure: an explanation of species abundance patterns. *The American Naturalist*, **116**(6), 770–87.

Tilman, D. (1994) Competition and biodiversity in spatially structured habitats. *Ecology*, **75**(1), 2–16.

Tobin, J.E. (1991) A neotropical rain forest canopy ant community: some ecological considerations, in *Ant–Plant Interactions* (eds C.R. Huxley and D.F. Cutler), Oxford University Press, Oxford. pp. 536–8.

Urbani, C.B., Bolton, B. and Ward, P.S. (1992) The internal phylogeny of ants (Hymenoptera: Formicidae). *Systematic Entomology*, **17**, 301–29.

Wagner, T. (1997) The beetle fauna of different tree species in forests of Rwanda and East Zaire, in *Canopy Arthropods* (eds N.E. Stork, J. Adis and R.K. Didham), Chapman & Hall, London, pp. 169–183 (this volume).

Wallace, A.R. (1869) *The Malay Archipelago*, Tyron Press, Scotland.

Way, M.J. and Khoo, K.C. (1991) Colony dispersion and nesting habits of the ants, *Dolichoderus thoracicus* and *Oecophylla smaragdina* (Hymenoptera: Formicidae), in relation to their success as biological control agents on cocoa. *Bulletin of Entomological Research*, **81**, 341–50.

Wiens, J.A. (1984) On understanding a non-equilibrium world: myth and reality in community patterns and processes, in *Ecological Communities. Conceptual Issues and the Evidence* (eds D.R. Strong, D. Simberloff, L.G. Abele and A. Thistle), Princeton, New Jersey. pp. 439–57.

Wilson, E.O. (1987) The arboreal ant fauna of Peruvian Amazon forests: a first assessment. *Biotropica*, **19**(3), 245–51.

Wolda, H. (1983) Diversity, diversity indices and tropical cockroaches. *Oecologia*, **56**, 290–8.

17

The ant fauna of tree canopies in Central Amazonia: a first assessment

A.Y. Harada and J. Adis

ABSTRACT

Two canopies of the widely distributed Amazonian tree species *Goupia glabra* (Celastraceae) were fogged during different seasons in a primary upland forest at Adolpho Ducke Forest Reserve, Manaus, Brazil.

The 2613 ants collected represented 100 species, 21 genera and five subfamilies. Up to 82 ant species occurred on a single tree. The two trees had 28 species in common. Most ant species were foraging in the canopy: at least 60% of all morphospecies collected were arboricolous, but only two to six nests were obtained per tree crown.

Foraging activity of ants was higher during the dry season as compared with the wet season. An interval of 6 months between fogging events was apparently not long enough for ants to re-establish their nests.

INTRODUCTION

Ants of Amazonian forests represent in most cases the dominant taxa of trunk and canopy arthropods with respect to abundance and biomass (Erwin, 1983, 1989; Adis and Schubart, 1985; Adis *et al.*, 1984). Preliminary data also indicate that their diversity is high. Adis (1981) caught 71 species representing 30 genera on three adjacent tree trunks in a blackwater inundation forest in Central Amazonia during a 6-month period. Wilson (1987) reported 43 species representing 21 genera from a single tree of a secondary floodplain forest in Western Amazonia. Since 1991, the German and Brazilian Research Foundations have supported

Canopy Arthropods. Edited by N.E. Stork, J. Adis and R.K. Didham. Published in 1997 by Chapman & Hall, London. ISBN 0 412 74900 9

a research programme on 'Mechanisms which maintain tropical diversity' (Linsenmair, 1990). Studies concentrate on the question of whether these mechanisms are of stochastic or deterministic nature (Floren and Linsenmair, 1997, Chapter 16, this volume). On the assumption that each tree represents an 'island' in the neotropical forest, our objective was to test how fast recolonization is observed after emptying the canopy of arthropods by using the fogging technique, and which mechanisms are found to occur at the morphospecies level in different orders. This paper gives preliminary results for the ant fauna.

STUDY SITE AND METHODS

Sampling was conducted in two specimens of the widely distributed Amazonian tree species *Goupia glabra* (Celastraceae, common name 'Cupiuba') in a Central Amazonian primary upland forest of the Adolpho Ducke Forest Reserve. This reserve is located about 26 km north-east of Manaus (2°55'S; 59°59'W) and belongs to the National Institute for Amazonian Research (INPA) in Manaus, Brazil. It represents one of the most intensively studied upland forest sites in Central Amazonia (Adis *et al.*, 1997, Chapter 4, this volume).

The first tree sampled, Cupiuba 59 (height 45 m, crown diameter approximately 15 m), was fogged four times in August 1991 (dry season; cf. Ribeiro and Adis 1984) on two consecutive days. As knockdown agent, a 1% solution of natural pyrethrum (without synergist) diluted in diesel oil was used for the first three fogging events and synthetic pyrethrum Baythroid 0.15% for the fourth fogging. Arthropods were intercepted in 26 funnel-shaped trays near the ground. Eighteen of the trays were placed directly under the canopy and two trays were mounted in each of the four cardinal points, the first at 10 m and the second at 20 m distance from the tree trunk, to monitor possible wind drift of falling arthropods. Further details of study site and methodology are given in Adis *et al.* (1997, Chapter 4, this volume).

Cupiuba 59 was refogged with 1% natural pyrethrum (without synergist) during the wet season (February, 1992) and the dry season (August, 1992), together with a second tree specimen, Cupiuba 64 (height 38 m, crown diameter about 14 m).

RESULTS AND DISCUSSION

Knockdown efficiency of natural pyrethrum

Given a drop time of 120 minutes and monitoring five selected funnels directly under the canopy of Cupiuba 59, approximately 69% of the falling ants were collected within the first hour after the first fogging

(a)

(b)

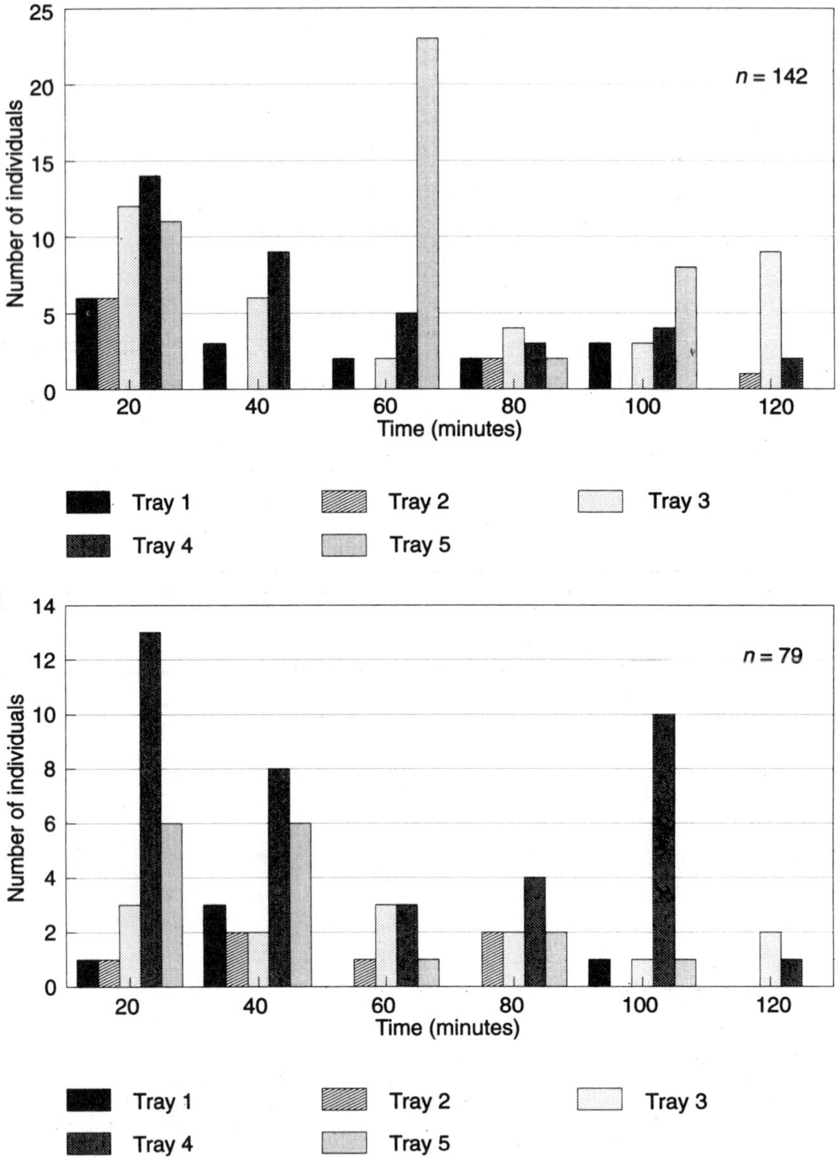

Figure 17.1 (a) Number of ants collected every 20 min from five selected trays after fogging the canopy of 'Cupiuba 59' (*Goupia glabra* Aubl.) on 21 August, 1991 (06:00 h, dry season) at Adolpho Ducke Forest Reserve, Manaus, Brazil (total drop time 120 min). (b) Number of ants collected every 20 min from five selected trays after re-fogging the canopy of 'Cupiuba 59' on the same day (08:00 h) (total drop time 120 min).

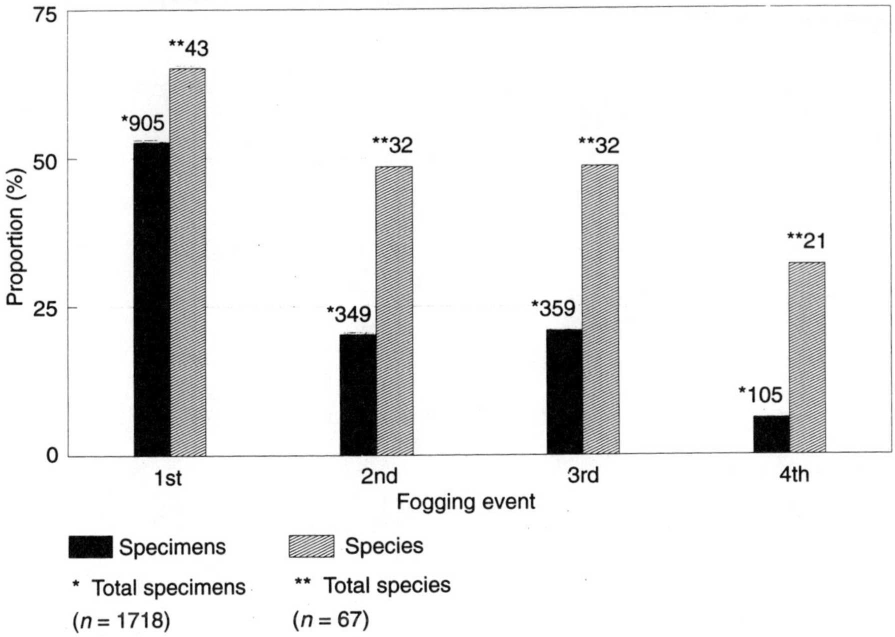

Figure 17.2 Proportion (%) of ant individuals and species obtained from 18 trays after four sequential fogging events in the canopy of 'Cupiuba 59' (*Goupia glabra* Aubl.) at Adolpho Ducke Forest Reserve (Manaus, Brazil). Total drop-time 120 min for each fog, using natural pyrethrum (without synergist) or the synthetic pyrethrum Baythroid, as follows: 1st fog – 21 August, 1991 at 06:00 h, natural pyrethrum 1.0%; 2nd fog – 21 August, 1991 at 08:00 h, natural pyrethrum 1.0%; 3rd fog – 22 August, 1991 at 06:00 h, natural pyrethrum 1.0%; 4th fog – 22 August, 1991 at 08:00 h, synthetic pyrethrum 0.15%.

event and 67% within the first hour after the second fogging. However, capture amounts varied between funnels (Figure 17.1). The total number of intercepted ants decreased from 905 specimens and 43 species after the first fogging event to 105 specimens and 21 species after the fourth fogging (Figure 17.2), if data from all 18 funnels under the canopy are evaluated. After the fourth fogging, the canopy was considered free of foraging ants and at least three out of six nest sites were eradicated (Table 17.1: *Dolichoderus diversus, Cephalotes atratus, Pheidole* sp. 7, versus *D. lutosus, Crematogaster* sp.2 and *C.* sp.8). There was evidence that foraging ants immigrated within 24 hours (from neighbouring trees with overlapping canopies, from the lower trunk region and from the forest floor), as the total number of specimens in the second and third foggings

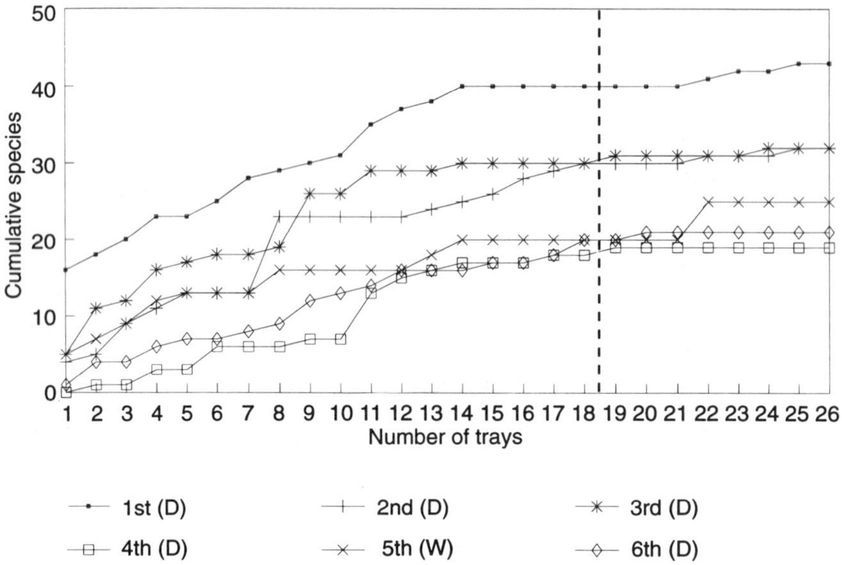

Figure 17.3 Cumulative number of ant species obtained from 26 trays after six fogging events in the canopy of 'Cupiuba 59' (*Goupia glabra* Aubl.) during the dry (D) and wet (W) seasons 1991/1992 at Adolpho Ducke Forest Reserve, Manaus, Brazil. Trays 1–18 were placed directly below the canopy, trays 19–26 at 10-m or 20-m distances from the tree trunk.

were not significantly different (Figure 17.3). In addition, there was a higher similarity of the less abundant ant species found to occur after the third fogging event (Table 17.2, Renkonen's coefficient 71.58 and 71.59, respectively). This is also supported by the fact that highest similarity of the total ant fauna was found between the first and second fogging events, i.e. on the same day (Table 17.2, Sörensen's quotient, 61.33).

Eighteen funnels installed directly under the canopy were considered sufficient to collect the ant species of both trees (Figures 17.3 and 17.4). Up to 2.9% of the total ants collected in all trays per fogging event were found in the trays 20 m distant from the two tree trunks. In some cases these specimens contributed to an increase in cumulative number of species (Figures 17.3 and 17.4). In part they may represent species which lived in close vicinity to the fogged trees (for example *Discothyrea* sp.1 and *Pachycondyla* sp.16 on 23.2.92; Table 17.1).

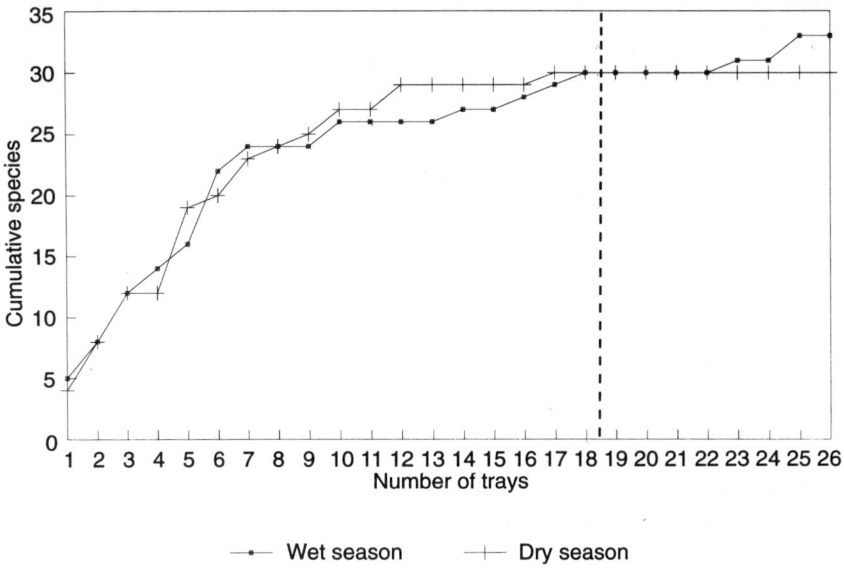

Figure 17.4 Cumulative number of ant species obtained from 26 trays after two fogging events in the canopy of 'Cupiuba 64' (*Goupia glabra* Aubl.) during the dry (D) and wet (W) seasons 1992 at Adolpho Ducke Forest Reserve, Manaus, Brazil. Trays 1–18 were placed directly below the canopy, trays 19–26 at 10-m or 20-m distances from the tree trunk.

Relative abundance, species richness, taxonomic composition and recolonization

In Cupiuba 59 ants represented 13–45% of the total arthropods collected in the three 6-monthly fogging samples. They comprised 82 species (six with nests) and 21 genera (Figure 17.5; Table 17.1). This is the highest number of ant species ever recorded on a single tree. Longino and Nadkarni (1990) collected 21 species of ants from the canopies of *Clusia*, *Didymopanax*, *Quercus*, *Pouteria* and *Ficus* trees in a cloud forest in Costa Rica. Wilson (1987) reported 43 species and 26 genera from an unidentified leguminous tree fogged by Erwin and collaborators in a secondary floodplain forest of the Tambopata Reserve, located in Peruvian Amazonia (Erwin, 1989). Floren and Linsenmair (1994, 1997, Chapter 16, this volume) collected up to 61 species on single trees of *Aporusa lageno-carpa* and *A. subcaudata* (Euphorbiaceae) in the Kota Kinabalu Park, located in the Malaysian part of North Borneo. Stork (1991) obtained up to 32 ant species per tree when fogging the canopies of 10 single trees in a lowland floodplain forest in Brunei, Borneo.

Table 17.1 Distribution of ant species in the canopy of two trees of 'Cupubia'(*Goupia glabra* Aubl.) at Adolpho Ducke Forest Reserve, Manaus, Brazil. Data for fogging events in August 1991 and 1992 represent the dry season (D) and data for fogging events in February 1992 the wet season (W)

	Cupiuba 59						*Cupiuba 64*						
	D (21.8.91)		*D* (22.8.91)		*W* (23.2.92)	*D* (26.8.92)	*W* (23.2.92)	*D* (26.8.92)			*Habitat*		
	1st	*2nd*	*3rd*	*4th*	*5th*	*6th*	*1st*	*2nd*	Σ	X	*a*	*t*	*?*
Subf. Dolichoderinae													
Azteca sp. 2	–	–	–	–	–	–	26	17	43	2.4	x		
Azteca sp. 5	–	–	–	–	–	1	–	–	1	2.2	x		
Azteca sp. 6	1+	–	–	–	–	–	–	–	1	1.9	x		
Azteca sp. 7	–	–	–	–	–	–	–	25	25	2.3	x		
Azteca sp. 8	–	–	–	–	–	–	32	–	32	2.3	x		
Azteca sp. 9	–	–	–	–	–	–	1	–	1	2.3	x		
Dolichoderus diversus	27	4	54	7	1	–	–	1	94	5.1	x	x	
Dolichoderus lamellosus	21	5	9	1	–	–	2	3	41	3.6	x		
Dolichoderus lutosus	58*	3	7	5	2	2*	–	–	77	3.4	x		
Linepithema sp. 6	1	–	–	–	–	–	–	–	1	1.9	x		
Tapinoma sp. 4	2	–	–	–	–	–	1	1	4	1.3			x
Tapinoma sp. 5	–	–	–	–	–	–	–	1	1	1.6			x
Tapinoma sp. 6	–	–	–	–	–	–	11	–	11	1.9			x
Tapinoma sp. 7	–	–	–	–	–	–	1	–	1	1.8			x
Tapinoma sp. 8	–	–	–	–	3	–	–	–	3	1.4			x
Subf. Formicinae													
Camponotus crassus	–	1*	–	–	–	15	–	–	16	3.7	x?		
Camponotus godmani	–	–	–	–	–	–	130	21	151	3.2	x?		
Camponotus latangulus	–	–	–	–	2	–	–	–	2	4.2	x?		

Table 17.1 continued

	Cupiuba 59						Cupiuba 64						
	D (21.8.91)		*D* (22.8.91)		*W* (23.2.92)	*D* (26.8.92)	*W* (23.2.92)	*D* (26.8.92)			*Habitat*		
	1st	*2nd*	*3rd*	*4th*	*5th*	*6th*	*1st*	*2nd*	Σ	X	*a*	*t*	?
Camponotus nr. bidens	1	2	8	–	–	2	17	1	31	3.0		x?	
Camponotus rapax	–	–	–	–	–	1	2	2	5	7.9		x?	
Camponotus trapezoideus	5	3	3	1	–	–	8	–	20	2.9			x
Camponotus sp. 1	3	2	2	–	6	–	–	–	13	4.0			x
Camponotus sp. 3	–	–	1	–	1	–	–	–	2	4.0			x
Camponotus sp. 4	7	–	–	–	1	1	–	–	9	4.8			x
Camponotus sp. 5	1	–	–	–	–	–	–	–	1	4.5			x
Camponotus sp. 8	1	–	–	–	–	2	–	–	3	6.3			x
Camponotus sp. 23	–	–	–	–	–	–	7	–	7	6.5			x
Camponotus sp. 30	–	–	–	–	–	–	–	2	2	3.7			x
Camponotus sp. 31	–	–	–	–	–	–	–	4	4	7.3			x
Camponotus sp. 32	–	–	–	–	–	–	1	–	1	3.9			x
Camponotus sp. 33	–	–	1+	–	–	–	–	–	1	5.3			x
Camponotus sp. 34	–	–	–	–	–	1	–	1	2	2.1			x
Campanotus sp. 35	1+	–	–	–	–	–	–	–	1	3.7			x
Dendromyrmex sp.4	–	–	–	–	–	1	–	–	1	5.7			x
Myrmelachista sp.3	2	–	1	–	8	–	1	1	13	1.3	x?		x
Myrmelachista sp.5	–	1	–	1	–	–	–	3	5	1.2	x?		x
Subf. Pseudomyrmecinae													
Pseudomyrmex nr. *euryblema*	–	–	–	–	–	–	1	–	1	6.9	x		

Table 17.1 continued

	Cupiuba 59						*Cupiuba 64*						
	D (21.8.91)		*D* (22.8.91)		*W* (23.2.92)	*D* (26.8.92)	*W* (23.2.92)	*D* (26.8.92)			*Habitat*		
	1st	*2nd*	*3rd*	*4th*	*5th*	*6th*	*1st*	*2nd*	Σ	X	*a*	*t*	*?*
Pseudomyrmex nr.fiebridgi	–	–	–	1	–	–	–	–	1	2.7	x		
Pseudomyrmex filiformis	–	1	–	–	–	–	–	–	1	2.3	x		
Pseudomyrmex flavidulus	1	–	–	–	–	–	–	–	1	3.6	x		
Pseudomyrmex gracilis	4	4	–	2	–	1	2	–	13	8.2	x		
Pseudomyrmex gr. gracilis	–	–	–	–	–	1	–	–	1	7.0	x		
Pseudomyrmex gr. oculatus	–	–	1	–	–	–	–	–	1	2.2	x		
Pseudomyrmex gr. pallidus	1*	–	–	–	–	-	–	–	1	6.1	x		
Pseudomyrmex pupa	–	1	–	–	–	–	4	–	5	6.5	x		
Pseudomyrmex simplex	3	1	1	3	1	–	–	–	9	4.9	x		
Pseudomyrmex aff. spiculus	–	1	–	–	–	–	1	–	2	2.5	x		
Pseudomyrmex venustus	2	–	–	1	–	–	–	–	3	6.8	x		
Pseudomyrmex sp. 13	–	–	1	–	–	–	–	–	1	3.3	x		
Pseudomyrmex sp. 23	–	–	–	-	–	–	1	–	1	5.6	x		
Pseudomyrmex sp. 24	–	–	–	–	–	–	1	–	1	7.0	x		
Subf. Myrmicinae													
Cephalotes atratus	50	51*	40	12	–	–	–	2	155	10.1	x		
Cephalotes serraticeps	3	1	–	–	1	1	–	12	18	7.5	x?		
Eucryptocerus placidus	–	–	–	–	–	–	–	8	8	6.3			x
Procryptocerus goeldii	16	–	6	1	–	2	12	1	38	4.0	x		
Procryptocerus marginatus	16	1	5	–	4	2	12	6	46	5.8	x		

Table 17.1 continued

	Cupiuba 59						*Cupiuba 64*						
	D (*21.8.91*)		*D* (*22.8.91*)		*W* (*23.2.92*)	*D* (*26.8.92*)	*W* (*23.2.92*)	*D* (*26.8.92*)			*Habitat*		
	1st	*2nd*	*3rd*	*4th*	*5th*	*6th*	*1st*	*2nd*	Σ	X	*a*	*t*	*?*
Procryptocerus scabriusculus	2	2	–	1	1	3	4	–	13	3.8	x	x	
Zacryptocerus cordatus	–	–	1	–	–	–	–	–	1	4.1	x		
Zacryptocerus duckei	3	1	2	2	–	–	–	–	8	3.0	x		
Zacryptocerus grandinosus	–	3	–	1	1	–	1	1	7	3.1	x		
Zacryptocerus minutus	–	2	–	–	–	–	–	–	2	2.9	x		
Zacryptocerus simillimus	–	–	–	–	–	–	8	1	9	3.0	x		
Zacryptocerus sp. 4	–	1	–	–	–	–	–	–	1	2.9	x		
Zacryptocerus sp. 13	6	–	–	–	–	1	–	–	7	3.0		x	
Crematogaster sp. 1	–	–	–	1	–	–	1	3	5	2.0	x?		
Crematogaster sp. 2	370	193	152	51	200*	44	5	13*	1028	2.2	x?		
Crematogaster sp. 3	25	–	1	–	5	2	9	11	53	2.2	x?		
Crematogaster limata (sp. 5)	7	–	–	–	–	–	19	–	26	2.7	x?		
Crematogaster sp. 7	–	–	–	–	3	–	1	–	4	2.6	x?		
Crematogaster sp. 8	163	35	21	–	82	–	–	–	301	2.5	x?		
Crematogaster sp. 9	11	4	3	–	–	–	–	–	18	2.0	x?		
Crematogaster sp. 10	3	1	–	2	–	–	–	–	6	1.9	x?		
Crematogaster sp. 12	–	–	–	–	1	–	–	–	1	3.6	x?		
Leptothorax (*Nesomyrmex*) sp. 2	1	–	–	–	–	–	3	1	5	1.6	x?		x

Table 17.1 continued

	Cupiuba 59						Cupiuba 64				Habitat		
	D (21.8.91)		D (22.8.91)		W (23.2.92)	D (26.8.92)	W (23.2.92)	D (26.8.92)					
	1st	2nd	3rd	4th	5th	6th	1st	2nd	Σ	X	a	t	?
Leptothorax (*Nesomyrmex*) sp. 3	–	–	1	–	–	–	–	–	1	2.5	x?		x
Leptothorax (*Macromischa*) sp. 4	–	–	1*	–	–	–	–	–	1	3.7	x?		x
Monomorium pharaonis	–	–	–	–	4	–	–	–	4	1.9	x		
Pheidole sp. 7	41	8	12	–	–	–	–	–	61	2.1			x
Pheidole sp. 28	–	–	–	1	–	–	–	–	1	1.2			x
Pheidole sp. 31	1	–	–	–	–	–	–	–	1	3.3			x
Pheidole sp. 36		–	1	–	–	–	–	–	1	1.8		x	
Pheidole sp. 47	–	–	–	–	–	1	–	–	1	2.0		x	
Solenopsis (*Diplorhophtrum*) sp. 8	6	2	–	–	–	–	–	2	10	1.1	x?		
Solenopsis (D.) sp. 9	1	2	–	–	–	–	–	–	3	1.1	x?		
Solenopsis (D.) sp. 10	–	–	2	3	–	–	2	–	7	1.3	x?		
Solenopsis (D.) sp. 12	–	–	1	–	–	–	–	–	1	1.1	x?		
Solenopsis (D.) sp. 17	7	3	1	–	–	–	–	2	13	1.1	x?		
Solenopsis (D.) sp. 18	4	3	5	–	–	–	–	2	14	1.2	x?		
Solenopsis (D.) sp. 19	6	6	9	6	–	–	–	–	27	1.3	x?		
Solenopsis (D.) sp. 22	–	–	–	–	–	–	1	*	–	1	2.0	x?	
Solenopsis (D.) sp. 23	–	–	–	–	1	–	–	–	1	1.8	x?		

Table 17.1 continued

	Cupiuba 59						Cupiuba 64				Habitat		
	D (21.8.91)		D (22.8.91)		W (23.2.92)	D (26.8.92)	W (23.2.92)	D (26.8.92)					
	1st	*2nd*	*3rd*	*4th*	*5th*	*6th*	*1st*	*2nd*	Σ	X	*a*	*t*	?
Cyphomyrmex costatus	–	1*	–	–	–	–	–	–	1	1.8		x	
Cyphomyrmex sp. 6	–	–	–	2*	–	–	–	–	2	2.1		x	
Subf. Ponerinae													
Discothyrea sp. 1	–	–	–	–	1	*	–	–	–	1	1.3		
x													
Ectatomma sp. 6	–	–	–	–	1	–	–	–	1	8.7	x?		
Pachycondyla crenata	7	–	–	–	2	–	–	–	9	6.9	x?	x	
Pachycondyla striatinodis	12	–	6	–	1	1	–	–	20	7.7	x		
Pachycondyla villosa	2	–	–	–	–	–	–	–	2	11.7	x?		
Pachycondyla sp. 16	–	–	–	–	1	–	–	–	1	4.8		x	
Total no. of ants	905	349	359	105	334	85	328	148	2613				
Total no. of species	43	32	32	21	25	20	33	28	100				
Total no. of genera	15	11	12	11	14	11	12	12	21				

Σ, Total no. collected; X, mean body length (mm).
– no records; a, arboricolous; t, terricolous; ?, habitat uncertain; *, queen; +, male.

Table 17.2 Indices of similarity of less abundant ant species (Renkonen's coefficient) and of total ant species (Sörensen's quotient) obtained by fogging the canopies of two trees of 'Cupiuba' (*Goupia glabra* Aubl.) during the dry (D) and wet (W) seasons 1991/1992 at Adolpho Ducke Forest Reserve, Manaus, Brazil

	RENKONEN *Cupiuba 59*						*Cupiuba 64*	
	21.8.91		*22.8.91*		*23.2.92*	*26.8.92*	*23.2.92*	*26.8.92*
	1st	*2nd*	*3rd*	*4th*	*5th*	*6th*	*1st*	*2nd*
Cupiuba 59								
1st (D)		68.97	71.58	59.13	63.56	52.94	–	–
2nd (D)	61.33		71.59	69.28	69.77	54.59	–	–
3rd (D)	53.33	53.13		68.76	52.27	49.66	–	–
4th (D)	43.75	52.83	41.51		50.36	54.00	–	–
5th (W)	41.18	35.09	38.60	26.09		55.09	–	–
6th (D)	41.27	30.77	26.92	24.39	35.57		–	–
Cupiuba 64								
1st (W)	–	–	–	–	–	–		34.99
2nd (D)	–	–	–	–	–	–	49.18	
	SÖRENSEN							

In Cupiuba 64, which was 1 km distant, ants represented 12–21% of the total number of arthropods collected during the two fogging events. On this tree, only 46 species, 13 genera and two ant nests were sampled (Figure 17.5; Table 17.1).

On both trees a total of 2613 ants were collected. They represented 100 species (at least 31 of them described) and 21 genera of five subfamilies. Abundance and species richness was highest in the Dolichoderinae, Formicinae and Myrmecinae. The 2537 specimens in these latter families belonged to 17 genera, with the highest number of species being recorded for *Dolichoderus* (three species), *Camponotus* (19 species) and *Crematogaster* (nine species), respectively (Table 17.1). All genera collected from the canopy were known from habitats less than 2 m above the ground and at least 80% of the named species had previously been recorded from the Manaus area. At least 60% of all morphospecies collected were considered arboricolous (Table 17.1).

In Cupiuba 59, the number of species decreased from 67 in August 1991 to 20 species in August 1992 (Table 17.2; Figure 17.5). *Crematogaster* sp. 2 dominated (cf. Tables 17.1 and 17.3). Species turnover between the first and second fogging studies was 36% (Table 17.1; Aug. 91 vs. Feb 92).

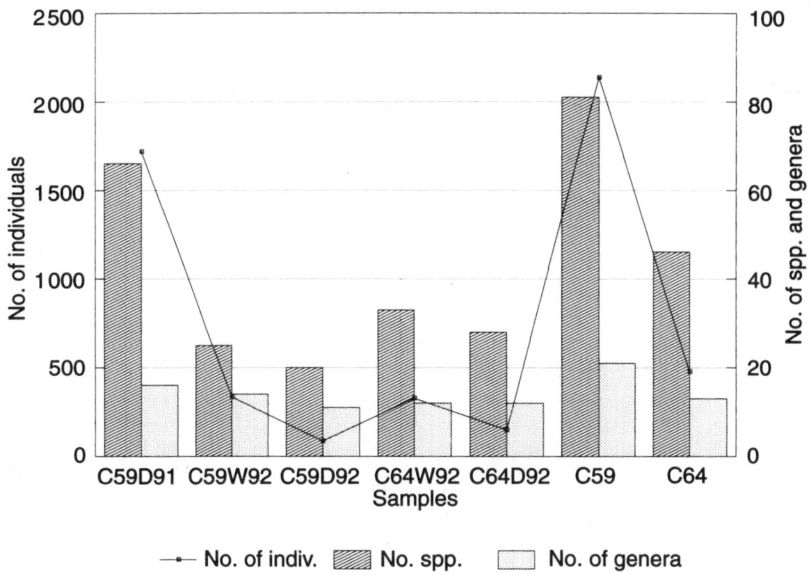

Figure 17.5 Comparison of the ant fauna from two *Goupia glabra* Aubl. ('Cupiuba 59 and 64') obtained by fogging at Adolpho Ducke Forest Reserve (Manaus, Brazil). Total numbers of individuals, species and genera are given for each fogging event during the dry (D) and wet (W) seasons 1991/1992 per tree (C59, C64) as well as for the total fogging events per tree.

Species turnover between the first and second fogging studies (Aug 91; combined) and the third fogging study (Aug 92) was 30%. Only rare species were affected in both cases; out of 15 newly recorded species, 11 species were represented by one specimen (singletons) and one species by two specimens, independent of body length (1.8–8.7 cm; Table 17.1). In Cupiuba 64, species number decreased from 33 in February 1992 to 2 8 species in August 1992 (Tables 17.1 and 17.3). Species turnover was 50%. Of 18 newly recorded species in the second fogging study, four species were singletons and five species had two specimens (Table 17.1). Our results are in contrast with those obtained by Floren and Linsenmair (1997, Chapter 16, this volume) in Northern Borneo where the number of ant species on *Aporusa lagenocarpa* trees increased between the first and second foggings (natural pyrethrum without synergist applied to 10 trees, sampling interval 7 months), species turnover was 94% and both abundant and rare species were affected. This may be attributed not only to the different tropical fauna and flora in Borneo (cf. Stork, 1988) but also to the fact that understorey trees were investigated.

Table 17.3 Species richness and diversity indices for ants obtained by fogging the canopies of two trees of 'Cupiuba' (*Goupia glabra* Aubl.) during the dry (D) and wet (W) seasons 1991/1992 at Adolpho Ducke Forest Reserve, Manaus, Brazil

Sample/fog no.	*S*	*N*	*A*	*D*	*Hs*	*E*	$H_{(max)}$*(logs)*	*Dominant species*	*Density (ind/m²)*	*Total*
Cupiuba 59										
August 1991 (D)	67	1718	13.4	0.23	3.29	0.55	4.17	*Crematogaster #2*	29.46	66.04
								Crematogaster #8	8.42	
								Cephalotes atratus	5.88	
1st	43	905	9.4	0.21	3.27	0.60	3.76	*Crematogaster #2*	14.23	34.81
2nd	32	349	8.6	0.34	2.59	0.52	3.47	*Crematogaster #2*	7.42	13.42
3rd	32	359	8.4	0.22	2.94	0.67	3.47	*Crematogaster #2*	5.85	13.81
4th	21	105	7.9	0.26	2.94	0.67	3.04	*Crematogaster #2*	1.96	4.04
February 1992 (W)										
5th	25	334	6.3	0.42	2.00	0.43	3.22	*Crematogaster #2*	7.69	12.85
August 1992 (D)										
6th	20	85	8.9	0.31	2.70	0.62	3.00	*Crematogaster #2*	1.65	3.27
Cupiuba 64										
February 1992 (W)	33	328	9.1	0.19	3.46	0.69	3.50	*Camponotus godmani*	5.00	12.62
August 1992 (D)	28	148	10.2	0.09	3.96	0.81	3.40	*Azteca #7*	0.96	5.77

S = number of species, N = total number of individuals, A = alpha index, D = Diversity (1 – 1pi² (Simpson); Hs = Shannon–Wiener (H (log_2))
E = eveness (H/H_{max} (Pieloŭ)

There was evidence for nest establishment of only one species, *Camponotus crassus*, in one of the two trees (Cupiuba 59) during the 6-month interval between foggings. Whether ants need a longer period for nest re-establishment will be investigated using data obtained from another fogging study of Cupiuba 59 in July 1995, representing a sampling interval of 2 years.

Diversity and similarity of ants between trees

Of the 100 species of ants obtained from the canopies of the two Cupiuba trees, 54 species only occurred on Cupiuba 59 and 18 species solely on Cupiuba 64 (Table 17.1). The two trees had 28 species in common. Thirty-seven species (37%) were represented by singletons, 42 species by up to 10 specimens, and only three species by more than 100 specimens. In Cupiuba 59, 43% (35 species) of the ants were represented by one specimen, in most part species which forage in the canopy. In Cupiuba 64, singletons amounted to 26% (12 species). In Whittaker plots of species-abundance data, all samples are well described by the logarithmic series (Figures 17.6 and 17.7). Also, when comparing the first and second fogging study, the highest similarity of less abundant ant species

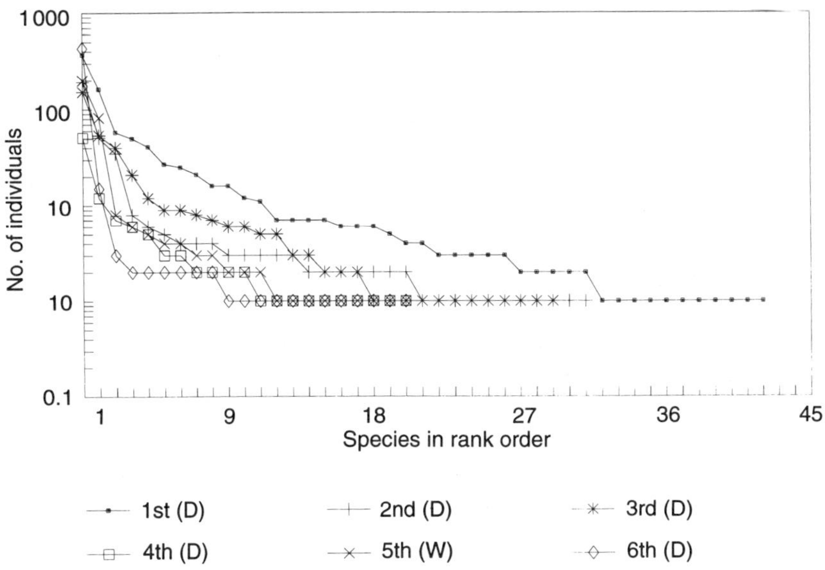

Figure 17.6 Rank-abundance plots (Whittaker plots, log scale) of ant species, based on specimens collected from 26 trays after six fogging events in the canopy of 'Cupiuba 59' (*Goupia glabra* Aubl.) during the dry (D) and wet (W) seasons 1991/1992 at Adolpho Ducke Forest Reserve, Manaus, Brazil.

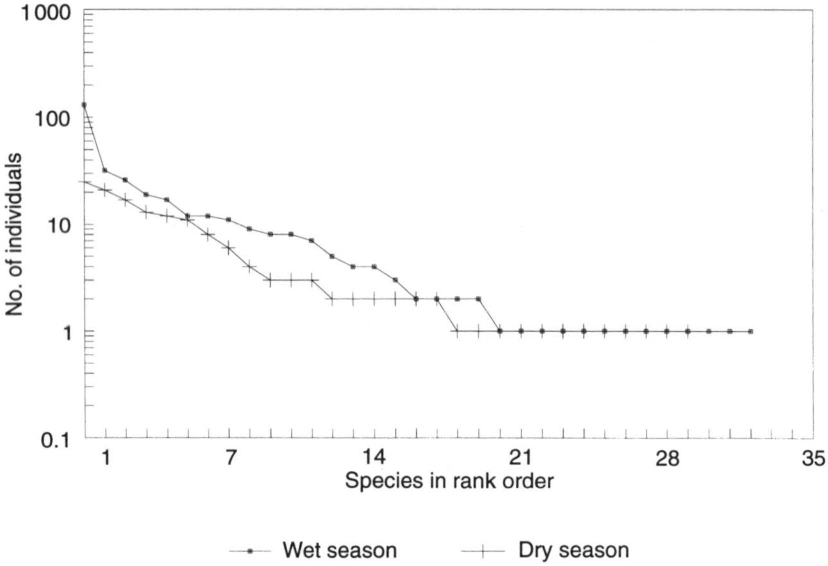

Figure 17.7 Rank-abundance plots (Whittaker plots, log scale) of ant species, based on specimens collected from 26 trays after two fogging events in the canopy of 'Cupiuba 64' (*Goupia glabra* Aubl.) during the dry (D) and wet (W) seasons 1992 at Adolpho Ducke Forest Reserve, Manaus, Brazil.

(Renkonen's quotient) was found on Cupiuba 59 (Table 17.2; 63.56 in C59 vs. 34.99 in C64). These data indicate that the ant fauna is to a large extent distinct in each Cupiuba tree. However, only longer fogging intervals (≥2 years) may show if recolonization of the canopy leads to the re-establishment of the original ant fauna and to what extent this is influenced by deterministic or stochastic processes. Data also indicate that multiple foggings over a short time interval can have long-term consequences in depressing ant diversity (Figure 17.3), whereas a single fog has no discernible effect 6 months later (Figure 17.4).

The lowest dominance of ants found on both trees occurred during the wet season (Table 17.4). Alpha-values for species diversity (Table 17.3) are higher during the dry season. Both results indicate a higher foraging activity of ants during the dry season, which was previously reported from primary and secondary upland forests as well as from inundation forests in the Central Amazon (Figure 17.5; Adis and Schubart, 1984; Adis, 1992).

Table 17.4 Indices of minor dominance among the ant fauna from two trees of 'Cupiuba' (*Goupia glabra* Aubl.) obtained by fogging the canopy during the dry and wet seasons 1991/1992 at Adolpho Ducke Forest Reserve, Manaus, Brazil

	Cupiuba 64	
	23.2.92 (Wet) *1st*	*26.8.92 (Dry)* *2nd*
Cupiuba 59		
21.8.92		
1st	15.06	19.96
2nd	10.02	16.90
22.8.91		
3rd	12.26	16.91
4th	12.04	14.86
23.2.92		
5th	6.05	12.97
26.8.92		
6th	9.79	15.94

Acknowledgements

This study is part of a 6-year project on 'Mechanisms which maintain tropical diversity', funded by the German Research Foundation, the German Agency of Technical Cooperation (GTZ: project 85.2522.2–06100) and the Brazilian Research Foundation (CNPq: project CNPq/MPG 91.0304/90–4) since 1991. Additional funding was received from the Tropical Ecology Working Group of the Max-Planck-Institute for Limnology in Plön, Germany, and Manaus, Brazil, the German Academic Exchange Service (DAAD) and the National Institute for Amazonian Research (INPA) in Manaus, Brazil. We thank Dr Nigel Stork of The Natural History Museum in London, UK, for his participation and help during our first canopy fogging in August 1991. We are especially grateful to PD Dr Wolfgang J. Junk, Dr Maria Teresa Fernandez Piedade and Dr Jörg Ohly for logistical support in Manaus via the 'Projéto INPA/Max-Planck'. We heartily thank all scientists of INPA, especially Dr Claudio Ruy V. da Fonseca, the technical staff of INPA as well as the participants of the postgraduate course 'Entomological Field Ecology' of INPA/Univ. Amazonas (February 1992) who joined the canopy fogging studies. Paulo Moutinho of the University of Pará in Belém, Brazil, kindly helped with the statistical analyses. Jack Longino, OTS, La

Selva, Costa Rica, gave valuable comments which helped to improve the manuscript. Miriam Ribeiro and João Claudino Neto of the University of Amazonas, Manaus, Brazil, are thanked for pinning the ants.

REFERENCES

Adis, J. (1981) Comparative ecological studies of the terrestrial arthropod fauna in Central Amazonian inundation-forests. *Amazoniana*, **7**(2), 87–173.

Adis, J. (1992) Überlebensstrategien terrestrischer Invertebraten in Überschwemmungswäldern Zentralamazoniens. *Verhandlungen des Naturwissenschaftlichen Vereins in Hamburg*, (NF) **33**, 21–114.

Adis, J. and Schubart, H.O.R. (1985) Ecological research on arthropods in Central Amazonian forest ecosystems with recommendations for study procedures, in *Trends in Ecological Research for the 1980s* (eds J.H. Cooley and F.B. Golley), NATO Conference Series I: Ecology, Plenum Press, New York, pp. 111–44.

Adis, J., Lubin, Y.D. and Montgomery, G.G. (1984) Arthropods from the canopy of inundated and terra firme forests near Manaus, Brazil, with critical considerations on the pyrethrum-fogging technique. *Studies on Neotropical Fauna and Environment*, **19**, 223–36.

Adis, J., Paarmann, W., da Fonseca C.R.V. and Rafael, J.A. (1997) Knockdown efficiency of natural pyrethrum and survival rate of arthropods obtained by canopy fogging in Central Amazonia, in *Canopy Arthropods* (eds N.E. Stork, J. Adis and R.K. Didham), Chapman & Hall, London, pp. 67–79 (this volume).

Erwin, T.L. (1983) Beetles and other insects of tropical forest canopies at Manaus, Brazil, sampled by insecticidal fogging, in *Tropical Rain Forest: Ecology and Management* (eds S.L. Sutton, T.C. Whitmore and A.C. Chadwick), Blackwell Scientific Publishers, Oxford, pp. 59–75.

Erwin, T.L. (1989) Canopy arthropod biodiversity: a chronology of sampling techniques and results. *Revista peruana de Entomologia*, **32**, 71–7.

Floren, A. and Linsenmair, K.E. (1994) Zur Diversität und Wiederbesiedelungsdynamik von Arthropoden auf drei Baumarten in einem Regenwald in Sabah, Malaysia. *Andrias*, **13**, 23–8.

Floren, A. and Linsenmair, K.E. (1997) Diversity and recolonization dynamics of selected arthropod groups on different tree species in a lowland rainforest in Sabah, Malaysia, in *Canopy Arthropods* (eds N.E. Stork, J. Adis and R.K. Didham), Chapman & Hall, London, pp. 344–81 (this volume).

Linsenmair, K.E. (1990) Tropische Biodiversität: Befunde und offene Probleme. *Verhandlungen der Deutschen Zoologischen Gesellschaft*, **83**, 245–61.

Longino, J.T. and Nadkarni, N.M. (1990) A comparison of ground and canopy leaf litter ants (Hymenoptera: Formicidae) in a neotropical montane forest. *Psyche*, **97**(1/2), 81–93.

Ribeiro, M. de N.G. and Adis, J. (1984) Local rainfall variability – a potential bias for ecological studies in the Central Amazon. *Acta Amazonica*, **14**, 159–74.

Stork, N.E. (1988) Insect diversity: facts, fiction and speculation. *Biological Journal of the Linnean Society*, **35**, 321–37.

Stork, N.E. (1991) The composition of the arthropod fauna of Bornean lowland rain forest trees. *Journal of Tropical Ecology*, **7**, 161–80.

Wilson, E.O. (1987) The arboreal ant fauna of Peruvian Amazon forests: a first assessment. *Biotropica*, **19**(3), 245–51.

18

Estimation of arboreal and terrestrial arthropod densities in the forest canopy as measured by insecticide smoking

H. Watanabe

ABSTRACT

Among arboreal arthropods collected by insecticide 'smoking', Acari (mainly Oribatei) and Collembola are usually the most numerous, accounting for over 60% of total arthropod abundance. Densities of arboreal arthropods are usually much higher in coniferous forests than in deciduous or evergreen broad-leaved forests. Even though most Acari and Collembola are found on tree trunks or in the canopy, they usually migrate from the soil substratum.

There are no marked differences in the species composition of Collembola and Acari among different tree species or forest types. This may be due to the fact that Oribatei and Collembola feed on plant debris and are associated with epiphyte habitats.

INTRODUCTION

At present, the only method which can effectively collect 'whole' arboreal arthropod communities in the forest canopy, including minute springtails (Collembola) and mites (Acari), is insecticide smoking (fogging). Except for a few groups of arthropods such as wood-borers which penetrate into the inner wood, or scale insects which stick to leaves, virtually all arthropods in the canopy can be collected by insecticide smoking. For sampling the arthropod faunas of large trees or the whole forest canopy, insecticide smoking is undoubtedly the most effective method.

Canopy Arthropods. Edited by N.E. Stork, J. Adis and R.K. Didham. Published in 1997 by Chapman & Hall, London. ISBN 0 412 74900 9

Aerial application of insecticides by aircraft or helicopter for controlling outbreaks of insect pests also provides an ideal opportunity for research.

One important problem is the evaluation of the rate of individuals dropping, relative to the total number killed. While there is no doubt that smoking is very effective, susceptibility to particular insecticides is distinctly different amongst arthropod species. It must also be understood correctly and emphasized that, except for animals collected by trays positioned below the trees, all other animals are killed without collection, usually when fog drifts across a large area (unless the insecticide used has high knockdown but low kill properties, see Stork and Hammond, 1997, Chapter 1, this volume; Adis *et al.*, 1997, Chapter 4, this volume, Paarmann and Kerck, 1997, Chapter 3, this volume). In addition, one disadvantage of this method is that repeated sampling of the same sites, for example examining seasonal changes, is difficult.

ESTIMATION OF ARBOREAL ARTHROPOD DENSITY AND ARTHROPODS MIGRATING FROM THE GROUND TO THE CANOPY

In order to control pine-wilt disease caused by the pine-wood nematode (*Bursaphelenchus xylophilus*) and the Japanese pine sawyer (*Monochamus alternatus*), aerial spraying of mixed and unmixed emulsions of insecticides has been practised on a large scale since the early 1970s, mainly in western Japan. Using these mass spraying opportunities, several research projects were undertaken to estimate the density of arboreal arthropods and to examine the effects of spraying on other arthropods that inhabit the forest. Results of a case study in Nara Park, Nara, Japan, in a mature, mixed stand of Japanese red pine (*Pinus densiflora*) and Japanese black pine (*P. thumbergii*), with a tree density of 1575/ha above 5 cm diameter at breast height (DBH), were as follows. Emulsions of insecticides MEP (Fenitrothion) *O,O*-dimethyl *O*-4-nitro-*m*-tolyl phosphorothioate ($C_6H_{12}NO_5PS$) and EDB (ethylene dibromide) 1,2-dibromethane ($C_2H_{14}Br_2$) or MEP alone were applied by helicopter onto the pine stand. On each occasion, 10–12 funnel-shaped trays, each 1 m^2, were suspended beneath the trees to catch all falling arthropods. Collections were made on three to five consecutive days. The applications were repeated twice a year for eight successive years (Watanabe, 1983). As shown in Table 18.1, the density of arboreal arthropods was estimated at 680.5/m^2. Araeopidae and Jassidae (Hemiptera) (29.2%) and Formicidae (Hymenoptera) (25.5%) were the most abundant taxa, followed by Collembola (9.5%), Psocoptera (5.7%) and Diptera (3.7%).

In the results obtained in a mixed pine forest, Collembola was estimated at 64.5/m^2, and among them the genera *Xenylla, Uzelia, Hypogastrura, Lophognanthella* and *Entomobrya* were identified from the

Table 18.1 Density (numbers/m^2) and biomass (wet weight, mg/m^2) of invertebrates collected following the first application of insecticide to a mixed pine stand in Nara, Japan. (After Watanabe, 1983.)

	Density	*Biomass*
Mollusca	0.2	0.7
Phalangida	0.8	16.4
Acari	7.4	1.7
Araneae	23.8	216.5
Diplopoda	0.2	3.6
Chilopoda	0.1	0.1
Collembola	64.5	18.1
Plecoptera	0.1	3.7
Orthoptera	6.8	119.3
Blattodea	1.8	13.3
Mantodea	0.4	10.4
Phasmida	0.5	25.4
Psocoptera	38.5	48.3
Thysanoptera	16.6	1.2
Hemiptera	256.0	186.8
Araeopidae, Jassidae	(198.8)	(100.3)
Aphididae	(36.6)	(19.8)
Neuroptera	10.6	33.6
Trichoptera	0.1	0.6
Lepidoptera	6.8	307.1
Diptera	41.7	48.9
Cecidomyiidae	(25.4)	(14.8)
Hymenoptera	189.9	343.3
Ichneumonoidea	(14.8)	(5.9)
Formicidae	(173.3)	(215.1)
Coleoptera	11.4	144.9
Others	2.4	6.4
Total	680.5	1546.6

first-fog samples. The former three genera are known to be arboreal, while the latter two are often found on the soil surface. After the second application of insecticide, however, arboreal groups were replaced by the soil-dwelling species of *Entomobrya* and *Lophognanthella*. It is believed that after arboreal Collembola (i.e. *Xenylla* sp. and *Uzelia* sp.) had become locally extinct due to the repeated application of insecticides, soil-dwelling species migrated to the tree-crown and replaced the original resident species (see also Prinzing, 1997, Chapter 22, this volume).

Table 18.2 Densities of arboreal arthropods in different pine stands. (Revised from Watanabe, 1983.)

Forest type	*Insecticide applied*	*Densities(number/m^2)*	*Percentage of*		*Reference*
			Acari	*Collembola*	
Seashore black pine (Kochi)	BHC emulsion 120 l/ha, sprayed by helicopter	179.6	1.8	56.5	Ochi *et al.* (1968)
Black pine mixed with evergreen oak (Kochi)	BHC emulsion 120 l/ha, sprayed by helicopter	32.6	0.1	18.2	Ochi *et al.* (1968)
Red pine (25 years) (Miyazaki)	BHC fogging	438.3			Ono *et al.* (1968)
Red pine (35 years) (Kyoto)	BHC 40% fogging	600.0		60.0	Kikuzawa and Shidei (1967)
Red pine (Natural)	Lindane, DDVP	160.0			Yamashita and Ishii (1976)
(Kagoshima)	fogging	103.0	13.1	52.2	
Red pine (11–18 years) (Mie)	NAC emulsion (1/50) 5000 l/ha, by spray gun	1467.0	25.7	2.6	Society of Natural Science of Mie Prefecture (1978)
Red pine (14 years)	Permethrin	145.2	66.9	8.0	Society of Nature
(Tokushima)	fogging	93.6	50.6	3.4	Conservation (1980)
Mixed red pine and black pine, Nara	MEP (40%), EDB (20%) 1/38, 90 l/ha sprayed by helicopter	680.5	1.1	9.5	Watanabe (1983)
Pinus reginosa Ontario, Canada	Pyrethrin, piperonyl butoxide, fogging		18.6–44.0	9.5–10.6	Martin (1966)

Similar samples have been collected from pine stands at several localities in Japan, with densities varying considerably from 32.6 to 1467/m^2 (Table 18.2). This may be due to differences in stand structure, such as tree density, stand age, canopy tree species composition and understorey tree species composition. In addition, differences in the mode of application and the insecticides used, amounts of insecticides applied and local weather conditions (e.g. wind velocity and rainfall) also cause differences in arthropod density and species composition. In general, Acari (mainly Oribatei) and Collembola are dominant and account for 60–70% of total arthropod abundance. In Japan, an extensive study of arboreal arthropods was carried out by Yamashita and Ishii (1976, 1977) using permethrin smoking in various forest types (Table 18.3). In that study, densities of arboreal arthropods in coniferous forests were usually much higher than those in deciduous broad-leaved forests. This is probably due to leaf-loss, and hence lack of food and shelter for arthropods, on deciduous trees in winter. There is a similar high percentage of Acari and Collembola in the total abundance. The data obtained from evergreen broad-leaved forest are insufficient for comparisons.

Regarding tropical forests, only a few studies have dealt with the density of arboreal arthropods. In order to understand the structure and density of the arthropod community of tree-crowns in tropical forests, research was carried out in a dry evergreen, seasonal rainforest at Nam Phrom in Chaiyaphoom Province, located about 140 km west of Khon Kaen in north-eastern Thailand (Watanabe and Ruaysoongnern, 1989). In this forest, the overstorey is a mixture of both deciduous trees, such as *Lagerstroemia* sp., *Terminalia* sp., *Dalbergia cochinchinensis*, and some evergreen trees, such as *Diospyros castanea, Xylia kerrii* and *Pterocarpus macrocarpus*, which form a closed canopy 20–25 m in height. The density of trees with DBH larger than 4.5 cm is about 1219/ha and basal area is 35.2 m^2/ha. The above-ground biomass of trees is about 350 ton/ha.

At Nam Phrom annual rainfall is about 1500 mm, but rainfall distribution is uneven, with very little or no rain in the dry season (November to March) and heavy precipitation in the rainy season (April to October). Mean temperature (23.1°C) fluctuates little throughout the year. On each sampling occasion, two cans of insecticide (Fuji Sumijet, containing DDVP (Dichlorvos) 6%, MEP (Fenithrothion) 6%, total weight 1 kg) were released in the early morning. One can was on the ground and another in the lower canopy. Smoking was conducted in December 1979 and January 1980 during the dry season, and in June 1980 and September 1981 during the rainy season. According to the manufacturer's directions, three cans per hectare are sufficient to control an outbreak of insect pests. Falling arboreal arthropods were collected at 6, 24, 48 and 72 hours after smoking. Total densities of arboreal arthropods were estimated at 256.4 and 140.4/m^2 in the dry season, and 195.2 and 123.1/m^2 in the

Table 18.3 Densities of arboreal arthropods in various kinds of forest

Forest type	*Insecticide applied*	*Densities (number/m^2)*	*Percentage of*		*Reference*
			Acari	*Collembola*	
Evergreen broad-leaved trees					
Quercus acuta	Lindane, DDVP smoking	1221.0	68.8	19.3	Yamashita and Ishii (1976)
(Kirishima, Japan)					
Evergreen and	DDVP, MEP smoking	256.4	2.7	22.5	Watanabe and
deciduous tree		140.4	6.8	14.9	Ruaysoongnern (1989)
mixed forest					
(Chaiyaphoom,		195.2	1.9	66.9	
N.E. Thailand)		123.0	2.1	29.2	
Tropical rainforest	Pyrethroid fogging	117.4			Stork (1991)
(Brunei)					
(Sulawesi, Indonesia)	Pyrethroid fogging	63.9–461.4			Stork and Brendell (1990)
Mangrove	Pyrethroid fogging	92.0			Stork and Brendell (1990)
Agricultural trees		258.7			
Pole forest		235.7			
Swamp		204.3			
Coffee plantation					
(Sulawesi, Indonesia)		442.5			
Deciduous broad-leaved trees					
Fagus crenata					
(Odaigahara, Japan)	Lindane, DDVP smoking	261.0	6.5	44.3	Yamashita and Ishii (1976)
(Ishizuchi, Japan)		101.0	11.2	23.0	

Table 18.3 continued

Forest type	*Insecticide applied*	*Densities (number/m²)*	*Percentage of*		*Reference*
			Acari	*Collembola*	
(Hakkoda, Japan)		44.0	3.0	9.1	
(Kyoto, Japan)	BHC 40% smoking	87.5			Kikuzawa and Shidei (1966)
Alnus maximowiczii					
(Hakkoda, Japan)	Lindane, DDVP smoking	1501.0	77.0	1.4	Yamashita and Ishii (1976)
Betula ermani					
(Daisetu, Japan)	Lindane, DDVP smoking	259.0	2.5	4.1	Yamashita and Ishii
(Ontake)		291.0	6.6	41.6	(1976)
Buddleia salvoifolia					
(Cape, South Africa)	Pyrethrum spraying	85.5			Southwood *et al.* (1982)
Quercus robur					
(Ascot, UK)	Pyrethrum spraying	591.3			Southwood *et al.* (1982)
(Cape, South Africa)		41.4			
Robinia pseudoacacia					
(UK)	Pyrethrum spraying	42.3			Southwood *et al.* (1982)
Salix cinerea					
(UK)	Pyrethrum spraying	139.6			Southwood *et al.* (1982)
Evergreen conifers					
Cryptomeria japonica					
(Inabu, Japan), 15 years	Permethrin smoking	3754.8	16.6	64.1	Hijii (1986)
(Inabu, Japan)	Permethrin smoking	1898.1			Hijii (1989)
		3537.3			
Chamaecyparis obtusa					
(Inabu, Japan)	Permethrin smoking	1227.7			Hijii (1983, 1984)

Table 18.3 continued

Forest type	*Insecticide applied*	*Densities (number/m²)*	*Percentage of*		*Reference*
			Acari	*Collembola*	
(Inabu, Japan)	Permethrin smoking	2011.6	81.7	10.8	Terakawa and Ohsawa (1981)
(Ontake, Japan)	Lindane, DDVP smoking	265.0	9.9	24.0	Yamashita and Ishii (1976)
Picea jezoensis					
(Hokkaido, Japan)	Lindane, DDVP smoking	842.0	7.1	25.8	Yamashita and Ishii (1976)
(Ontake)		1057.0	3.6	41.1	
(Odaigahara)		950.0	2.3	82.9	
Picea glehhnii					
(Hokkaido, Japan)	Lindane, DDVP smoking	2333.0	11.5	58.2	Yamashita and Ishii (1976)
Abies veitchi					
(Ontake, Japan)	Lindane, DDVP smoking	2578.0	7.3	19.2	Yamashita and Ishii (1976)
(Ishizuchi)		2997.0	7.6	40.3	
Deciduous conifers					
Larix leptolepis					
(Inabu, Japan)	Permethrin smoking	957.8	74.3		Terakawa (1982)
(Inabu, Japan)	Permethrin smoking	969.5	76.3	6.2	Terakawa and Ohsawa (1981)

Table 18.4 Densities (numbers/m^2) of arboreal invertebrates in a dry evergreen (seasonal) forest in northeastern Thailand. (After Watanabe and Ruaysoongnern, 1989.)

	Dry season		*Rainy season*	
	December	*January*	*July*	*September*
Mollusca	0.1	–	0.6	0.3
Phalangida	–	–	0.1	–
Acari	6.8	9.5	3.8	2.6
Araneae	12.9	5.2	2.3	1.1
Diplopoda	–	–	0.3	–
Chilopoda	–	–	–	0.1
Collembola	57.6	20.8	130.6	36.0
Orthoptera	1.8	0.5	0.2	0.1
Blattodea	0.5	0.1	0.3	–
Mantodea	–	–	–	0.1
Phasmida	0.1	–	0.1	–
Psocoptera	5.3	9.7	3.1	47.0
Thysanoptera	79.0	42.3	10.6	1.7
Hemiptera	13.9	18.5	5.5	5.2
Lepidoptera	4.3	3.3	1.8	1.5
Diptera	14.6	9.4	18.6	4.2
Hymenoptera	51.9	17.4	16.6	20.5
Coleoptera	7.5	3.8	0.9	2.7
Others	0.2	–	–	–
Total	256.4	140.0	195.2	123.1

rainy season (Table 18.4). There were no marked differences in density and faunal structure between the dry and the rainy seasons except for the markedly increased abundance of Thysanoptera in the dry season samples. Densities of arboreal arthropods in this seasonal forest in Thailand, however, are lower than those in temperate forests (see Table 18.3).

In a tropical rainforest in Brunei, Borneo, a similar low arthropod density (117.4/m^2) was recorded (Stork, 1991) as was the case in Sulawesi, Indonesia (63.9–461.4/m^2; Stork and Brendell, 1990). The density of arboreal arthropods in tropical forest canopies appears to be considerably lower than values obtained from temperate forests. In Thailand, the most abundant arthropod groups in tree-crowns were Collembola, Thysanoptera, Hymenoptera (mainly Formicidae) and Hemiptera. However, Stork and Brendell (1990) found Diptera to be the most common arthropod group, while Formicidae and Collembola were surprisingly poorly represented in lowland rainforest in Sulawesi.

COLLEMBOLA AND ACARI AS MAJOR COMPONENTS IN THE CANOPY

There have been a number of studies on Collembola and Acari living in tree canopies and/or tree trunks (Trave, 1963; Bowden *et al.*, 1976). As shown in Tables 18.1 and 18.2, Collembola and Acari (Oribatei) are often the most abundant orders on temperate trees and some tropical trees, followed by Hemiptera and Diptera.

Uchida and Kojima (1966) examined Collembola obtained from insecticide application in a natural black pine (*Pinus thumbergii*) forest to prevent pine-wilt disease. *Xenylla brevispina* was extremely abundant (approximately 98% of total abundance), while *Sphyrotheca multifasciata, Pseudisotoma monochaeta, Homidia nigrocephala, Salina celebensis*, and *Tomocerus minutus* were also collected. Except for *S. celebensis*, which usually inhabits trees and was collected by the beating method, all other species were normally observed on the soil surface and were collected by funnel extraction.

Similarly, Suma (1993) has reported that many Collembola were observed on tree trunks of alder (*Alnus japonica* var. *arguta*), including the dominant species *Folsomia pusillus* (83.3%) and *Anurophorus larisis* (11.5% of total arthropod abundance). Collembola have been further classified into two types: (i) true arboreal Collembola, such as *F. pusillus, A. laricis, Entomobrya aino* and *Xenyllu acauda*, which are collected from the upper trunk or in the canopy; and (ii) semi-arboreal Collembola such as *Xenylla brevispina, X. subcavernarum* and *Tomocerus aokii*, species of Sminthuridae and Entomobryidae which are observed on the lower trunk.

M. Igarashi (unpublished data) attempted to collect all arthropods ascending to the canopy from the lower trunk of beech trees (*Fagus crenata*) using a barrier made of vinyl hose and tyre tube about 1 m above the ground. Numbers totalled approximately 15 000 arthropods, including 6850 Collembola from a single beech tree of 40–50 cm in DBH. It is unlikely that all individuals ascend to the canopy, however, with some stopping at the lower or upper trunk. Recently (Itoh, 1991), clarification of the life-cycle and seasonal migration of a common species of arboreal Collembola, *Xenylla brevispina*, has revealed that they lay eggs in the leaf litter at the soil surface where the juveniles develop. In June, juveniles begin to climb trees toward the canopy where they mature by November. The adults descend from the canopy at the onset of winter and remain in the soil to breed in the following March and April (Itoh, 1991). *X. brevispina*, previously believed to be a truly arboreal Collembola, is actually a migrant between soil surface and tree canopy.

Regarding arboreal Acari, Aoki (1970, 1971, 1974) identified about 40 species of oribatid mites in samples collected in various forest types by insecticide smoking (Yamashita and Ishii, 1976, 1977). He noted that

Scapheremaeus yamashitai, Dendrozetes caudatus and *Megeremaeus expansus* are 'true arboreal forms', whereas *Camisia spinifer, Trhypochthonius japonicus, Cepheus latus, Ceratoppia bipilis,* and *Eupelops acromios* are 'wandering forms', migrants between soil surface and tree-crowns. In addition, Aoki described arboreal oribatid mites as more tolerant of desiccation and therefore adapted to life in the canopy.

It is interesting to note that there are no marked differences in species composition of Collembola and Acari among the different tree species studied. This is one important matter yet to be clarified.

DISCUSSION

In samples from a Brazilian inundation rainforest, Formicidae comprised more than 40% of the arthropods. Other major groups included Coleoptera, Hemiptera (Homoptera) and Diptera (Adis and Schubart, 1985). The low density of Collembola is probably due to seasonal flooding on the forest floor.

Some arthropods are specific to particular species of plants. This means that high floral diversity may give rise to high faunal diversity. This fact must be examined in tropical forests composed of various tree species. Stork's (1987a,b) study of insects on Bornean trees suggested that for many insect groups there is less host-specificity in the tropics than in temperate regions, and the similarity of different trees' epiphyte loads has a greater effect on faunal similarities than the taxonomic similarity of the trees themselves. This is supported by the fact that, in temperate forests, there is no distinct difference between arboreal and ground-dwelling oribatid mites and Collembola, which depend mainly on dead plant debris. In addition, many arthropods feed on epiphytes or are associated with epiphyte habitats (vines, ferns, bromeliads, mosses, lichens, algae, etc.).

Also noteworthy is the fact that in tropical rainforests many leaves and twigs are interrupted in their fall towards the ground, dangling or getting caught on the lower branches. In this situation Collembola and Acari are playing distinct and important roles as decomposers in the canopy.

Given that: (i) oribatid mites and Collembola are usually the most abundant of arboreal arthropods; and (ii) the dominant species are detritus and litter feeders, and not specific to a particular plant, it can be concluded that the number of species in the world may not be as great as previously calculated, even in tropical forests. It should be emphasized that there is a need for more research in tropical forests consisting of diverse tree species to confirm these conclusions.

Lastly, promotion of taxonomy of arthropods in tropical regions and more detailed studies on arboreal arthropods at the species level are needed for clarification of guild structure, niche structure and diversity. Research must be carried out in collaboration with taxonomists. While some such results have been reported (Erwin, 1983a,b; Hijii, 1983, 1984; Paarmann and Stork, 1987; Stork, 1987a,b, 1991), usually only one order of insects has been examined in detail because of the problems of sorting specimens to species (Roberts, 1973; Erwin and Scott, 1980).

REFERENCES

Adis, J. and Schubart, H.O.R. (1985) Ecological research on arthropods in central Amazonian forest ecosystems with recommendations for study procedures, in *Trends in Ecological Research for the 1980s* (eds J.H. Cooley and F.B. Golley), NATO Conference Series No.1, Ecology, Plenum Press, pp.111–14.

Adis, J., Paarmann, W., da Fonseca C.R.V. and Rafael, J.A. (1997) Knockdown efficiency of natural pyrethrum and survival rate of arthropods obtained by canopy fogging in Central Amazonia, in *Canopy Arthropods* (eds N.E. Stork, J. Adis and R.K. Didham), Chapman & Hall, London, pp. 67–79 (this volume).

Aoki, J. (1970) Descriptions of Oribatid mites collected by smoking of trees with insecticides I. Mt. Ishizuchi and Mt. Odaigahara. *Bulletin of the National Science Museum Tokyo*, **13**(4), 585–602.

Aoki, J. (1971) Soil mites (Oribatids) climbing trees. *Proceedings of the 3rd International Congress of Acarology*, Prague, pp. 59–65.

Aoki, J. (1974) Descriptions of Oribatid mites collected by smoking of trees with insecticides. II. A new subspecies of the genus *Ommatocepheus* from Mt. Odaigajhara. *Bulletin of the National Science Museum Tokyo*, **17**(1), 53–5.

Bowden J., Haines, I.H. and Mercer, D. (1976) Climbing Collembola. *Pedobiologia*, **16**, 298–321.

Erwin, T.L. (1983a) Tropical forest canopies: the last biotic frontier. *Bulletin of the Entomological Society of America*, **30**, 14–19.

Erwin, T.L. (1983b) Beetles and other insects of tropical forest canopies at Manaus, Brazil, sampled by insecticidal fogging, in *Tropical Rainforest: Ecology and Management* (eds S.L. Sutton, T.C. Whitmore and A.C. Chadwick), Blackwell Scientific Publications, Oxford, pp. 59–75.

Erwin, T.L. and Scott, J.C. (1980) Seasonal and size patterns, trophic structure, and richness of Coleoptera in the tropical arboreal ecosystem: the fauna of the tree *Luehea seemannii* Triana and Planch in the canal zone of Panama. *The Coleopterist's Bulletin*, **34**, 305–22.

Hijii, N. (1983) Arboreal arthropod fauna in a forest. I. Preliminary observation on seasonal fluctuations in density, biomass, and faunal composition in a *Chamaecyparis obtusa* plantation. *Japanese Journal of Ecology*, **33**, 435–44.

Hijii, N. (1984) Arboreal arthropod fauna in a forest. II. Presumed community structures based on biomass and number of arthropods in a *Chamaecyparis obtusa* plantation. *Japanese Journal of Ecology*, **34**, 187–93.

Hijii, N. (1986) Density, biomass, and guild structure of arboreal arthropods as related to their inhabited tree size in a *Cryptomeria japonica* plantation. *Ecological Research*, **1**, 97–118.

Hijii, N. (1989) Arthropod communities in a Japanese cedar (*Cryptomeria japonica* D.Don) plantation: abundance, biomass and properties. *Ecological Research*, **4**, 243–60.

Itoh, R. (1991) Growth and life cycle of an arboreal Collembola, *Xenylla brevispina*

Kinoshita, with special reference to its seasonal migration between tree and forest floor. *Edaphologia*, **45**, 33–48.

Kikuzawa, K. and Shidei, T. (1966) On sampling techniques to estimate the density and the biomass of forest arthropods. *Japanese Journal of Ecology*, **16**, 24–8.

Kikuzawa, K. and Shidei, T. (1967) On the biomass of arthropods of the Japanese red pine forest in the vicinity of Kyoto. *Bulletin of Kyoto University Forestry*, **39**, 1–8.

Martin, J. L. (1966) The insect ecology of red pine plantations in central Ontario. IV. The crown fauna. *Canadian Entomologist*, **98**, 10–27.

Ochi, K., Katagiri, K. and Kojima, K. (1968) Studies on the composition of the fauna of invertebrate animals in the canopy-strata of pine forests. I. On the method of investigation and simple description of the fauna. *Bulletin of the Government Forest Experiment Station, Meguro Tokyo*, **217**, 1–48.

Ono, Y., Higashi, K. and Hamada, H. (1977) Estimation of population density and biomass of invertebrates living in the forest, in *Studies on Methods of Estimating Density, Biomass and Productivity in Terrestrial Animals* (ed. M. Morisita), University of Tokyo Press, Tokyo, *JIBP Synthesis*, **17**, 63–84.

Paarmann, W. and Kerck, K. (1997) Advances in using the canopy fogging technique to collect living arthropods from tree-crowns, in *Canopy Arthropods* (eds N.E. Stork, J. Adis and R.K. Didham), Chapman & Hall, London, pp. 53–66 (this volume).

Paarmann, W. and Stork, N.E. (1987) Canopy fogging, a method of collecting living insects for investigations of life history strategies. *Journal of Natural History*, **21**, 563–6.

Prinzing, A. (1997) Spatial and temporal use of microhabitats as a key strategy for the colonization of tree bark by *Entomobrya nivalis* (Collembola: Entomobryidae), in *Canopy Arthropods* (eds N.E. Stork, J. Adis and R.K. Didham), Chapman & Hall, London, pp. 453–76 (this volume).

Roberts, H.R. (1973) Arboreal Orthoptera in the rainforests of Costa Rica collected with insecticide. A report on the grasshoppers (Acrididae) including new species. *Proceedings of the Academy of Natural Sciences Philadelphia*, **125**, 46–66.

Society of Natural Science of Mie Prefecture (1978) *Report on Effect of Insecticide Application for Preventing Pine Wilt Disease*, Tsu.

Society of Nature Conservation of Tokushima (1980) *Research of the Effect of Aerial Application of Insecticide to Control Pine Wilt Disease*, Tokushima.

Southwood, T.R.E., Moran, V.C. and Kennedy, C.E.J. (1982) The richness, abundance and biomass of the arthropod communities on trees. *Journal of Animal Ecology*, **51**, 635–49.

Stork, N.E. (1987a) Guild structure of arthropods from Bornean rainforest trees. *Ecological Entomology*, **12**, 69–80.

Stork, N.E. (1987b) Arthropod faunal similarity of Bornean rainforest trees. *Ecological Entomology*, **12**, 219–26.

Stork, N.E. (1991) The composition of the arthropod fauna of Bornean lowland rainforest trees. *Journal of Tropical Ecology*, **7**, 161–80.

Stork, N.E. and Hammond, P.M. (1997) Sampling arthropods from tree-crowns by fogging with knockdown insecticides: lessons from studies of oak tree beetle assemblages in Richmond Park (UK), in *Canopy Arthropods* (eds N.E. Stork, J. Adis and R.K. Didham), Chapman & Hall, London, pp. 3–26 (this volume).

Stork, N.E. and Brendell, M.J.D. (1990) Variation in the insect fauna of Sulawesi trees with season, altitude and forest type, in *Insects and the Rainforests of South East Asia (Wallacea)* (eds W.J. Knight and J.D. Holloway), Royal Entomological

Society of London, pp. 173–90.

Suma, Y. (1993) Studies on the arboreal Collembola in Kushiro marsh. Investigation 1990. *Jezoensis*, **20**, 175–8.

Terakawa, N. (1982) *Analysis of Community Structure of the Arthropod Fauna in a Japanese Larch Plantation*, Unpublished MSc Thesis, Nagoya University, Japan.

Terakawa, N. and Ohsawa, W. (1981) The arboreal arthropod fauna in Inabu experiment forest of Nagoya University II. The relation between body length and number of individuals. *Transactions of the 91st Meeting of the Japanese Forest Society*, pp. 349–50.

Trave, J. (1963) Ecologie et biologie des Oribates (Acariens) saxicoles et arboricoles. *Vie et Milieu, Supplement*, **14**, 1–267.

Uchida, H. and Kojima, K. (1966) On the Collembola obtained by the aircraft chemical application. *Kontyu*, **34**, 317–26.

Watanabe, H. (1983) Effects of repeated aerial applications of insecticides for pine-wilt disease on arboreal arthropods in a pine stand. *Journal of the Japanese Forestry Society*, **65**(8), 282–7.

Watanabe, H. and Ruaysoongnern, S. (1989) Estimation of arboreal arthropod density in a dry evergreen forest in northeastern Thailand. *Journal of Tropical Ecology*, **5**, 151–8.

Yamashita, Z. and Ishii, T. (1976) Basic structure of the arboreal arthropod fauna in the natural forest of Japan. Ecological studies of the arboreal arthropod fauna. 1. *Report Environmental Science, Mie University* No. 1, pp. 81–111.

Yamashita, Z. and T. Ishii (1977) Basic structure of the arboreal arthropod fauna in the natural forest of Japan. Ecological studies of the arboreal arthropod fauna. 2. *Report Environmental Science, Mie University* No. 2, pp. 69–4.

Part Four

The Biology of Canopy Arthropods

19

The ecology and behaviour of arboreal dung beetles in Borneo

A.J. Davis, J. Huijbregts, A.H. Kirk-Spriggs, J. Krikken and S.L. Sutton

ABSTRACT

Dung beetles foraging in the rainforest canopy have been recorded on every subcontinent. Work at the Danum Valley Field Centre in Sabah, Malaysian Borneo, revealed a guild of arboreal dung beetles found only metres above the forest floor up to the high forest canopy, and that feed on primate dung caught on canopy vegetation. The flight activity, spatial distribution and feeding ecology of this previously unrecorded guild of arboreal beetles is examined, and related to the ecology, behaviour and gut morphology of the primate species resident in the same forest.

The dung-relocation behaviour of the Bornean dung beetles is atypical for the genus in which they are found, and breaks with the existing dichotomy of dung beetle tribes into dung rollers and non-dung rollers. The implications of this discovery are discussed.

This work gives a very real indication of how new light can be shed on the ecology of tropical insect communities by studying the strato-orientation of insects in response to conditions in the rainforest canopy, and the importance of canopy work in forming comprehensive descriptions of rainforest arthropod communities.

INTRODUCTION

The distribution of insects in different strata of architecturally rich habitats is a little-studied area of ecology (Sutton and Hudson, 1980; Roubik, 1993; Peng *et al.*, 1992), with most work concentrating on uniform agricultural systems in temperate latitudes (Peng *et al.*, 1992). Few studies have looked at the distribution of insects in relation to the strata of a tropical rainforest (Sutton and Hudson, 1980; Sutton, 1983, 1989). The rainforest canopy is now known to be a major source of arthropod

Canopy Arthropods. Edited by N.E. Stork, J. Adis and R.K. Didham. Published in 1997 by Chapman & Hall, London. ISBN 0 412 74900 9

diversity (Erwin and Scott, 1980; Stork, 1988, 1993), and many of the species found here do not exist in any other habitat. The forest floor and the upper canopy represent two ends of a microclimatic spectrum: a transition from low-light regimes to conditions of high incident solar radiation. Selection pressures on insect populations lie between these extremes, and provide a vertical axis along which potentially competing species can differentiate spatially (i.e. 'strato-orientation', *sensu* Roubik, 1993, can occur). The canopy also provides a third dimension in which species can partition resources, in addition to the two represented by the ground surface. Given this potential for strato-orientation in tropical rainforests, such patterns may play a role in the processes that underlie and maintain regional species diversity, and enhance the spatial and temporal complexity of insect communities in such ecosystems.

In this paper, we examine the ecology and behaviour of a previously unrecorded guild of arboreal dung beetle from Borneo, and the role of strato-orientation in structuring these communities, and those of arboreal mammals on whose dung they feed. We also review for the first time the available literature on arboreal dung beetles, and show how the Bornean arboreal beetles diverge from the classic taxonomic distinction of the Scarabaeid dung beetles into dung-rolling and tunnelling species, thereby highlighting the importance of canopy work in forming complete descriptions of rainforest arthropod communities.

ARBOREAL DUNG BEETLES

Most species of dung beetle (family Scarabaeidae) forage within close proximity to the ground. For communities that live in open habitats, resources can only be found in two dimensions, and so ground-based foraging behaviour is the sole option available to them. Utilization of a third dimension only becomes possible when dung and carrion are available in the arboreal environment in sufficient quantities to enable any arboreal species to remain reproductively viable. In tropical rainforests, a significant portion of the vertebrate fauna is arboreal (MacKinnon, 1972; Malcolm, 1997, Chapter 25, this volume). A large proportion of the dung produced by these animals never reaches the forest floor (A.J. Davis, personal observation) as much of it is caught on branches and leaves on the way down. This paper concerns those dung beetles that forage in dung caught in the upper forest canopy and lower vegetational layers.

Dung beetles foraging in the rainforest canopy have been recorded on every sub-continent. Globally, the average (± S.E.) number of dung beetle species recorded from rainforest sites is 57.9 ± 7.041 (based on data from 11 studies; Davis, 1993). In North Sulawesi, at least two arboreal species from a total of 46 dung and carrion beetles species recorded (Hanski and Krikken, 1991) were collected on the Royal Entomological Society of

London expedition to the Dumoga-Bone National Park in 1985 ('Project Wallace'). One of these species, *Onthophagus magnipygus* Boucomont, was rarely caught in ground traps, but regularly caught in traps at 5–20 m above the ground. The second species, *Phaeochrous emarginatus* Castelnau, a carrion specialist, was found to be active at all levels from the ground to the upper canopy (Hanski and Krikken, 1991). *O. magnipygus* may well be an arboreal specialist, whereas *P. emarginatus* is ubiquitous throughout the forest. Several canopy-feeding species have been recorded in the Neotropics. Two South American species – *Canthon angustatus* Harold and *C. subhyalinus* Harold – were observed on leaves making and rolling balls from howler monkey dung, and then falling to earth with dung balls tucked between their hind legs (Howden and Young, 1981). Other South American beetles, belonging to the genera *Trichillum* Harold, *Pedaridium* Harold and *Uroxys* Westwood, live in the fur of sloths, and some species of the genus *Glaphyrocanthon* Martínez likewise live on monkeys and tapirs (Halffter and Matthews, 1966; Howden and Young, 1981). The greatest number of arboreal dung beetle species to be recorded from one site were collected in the Makokou Reserve, Gabon, West Africa (Walter, 1984), where collections comprised four species of the genus *Onthophagus* (*O. laeviceps* d'Orbigny, *O. ahenomicans* d'Orbigny, *O. possoi* Walter and *O. mpassa* Walter) and one species of the genus *Sisyphus* (*S. arboreus* Walter). Only *S. arboreus* showed ball-rolling behaviour, but was not observed on the forest floor and was presumed to exist entirely in the forest canopy. Collections from Madagascar have only revealed one arboreal dung beetle species to date: *Arachnodes goudoti* Castelnau. This small canthonine roller was found to be common in traps set in small trees, as low as 50 cm above the ground (Vadon, 1947), and was observed demonstrating the same behaviour as the South American dung rollers (*C. angustatus* and *C. subhyalinus*).

MATERIALS AND METHODS

Research was carried out at the Danum Valley Field Centre, Sabah, Malaysian Borneo (5°01'N, 117°47'E). Investigation of the existence of arboreal dung beetles, and if present, of their species composition, was approached by trapping and direct observation. Traps were positioned within the Danum Valley Conservation Area and in the surrounding forest within the Ulu Segama Reserve, complementing a programme of ground pitfall trapping in the area (Davis, 1993; see also Holloway *et al.*, 1992). Traps were suspended under trees within a variety of forested areas (Davis, 1993). Much of the Danum Valley Conservation Area is composed of lowland, evergreen dipterocarp forest (<760 m above sea level), where the Dipterocarpaceae make up approximately 88% of the total volume of large trees (Newbery *et al.*, 1992). Arboreal traps were of the same basic design as pitfall traps used to collect beetles from the forest floor

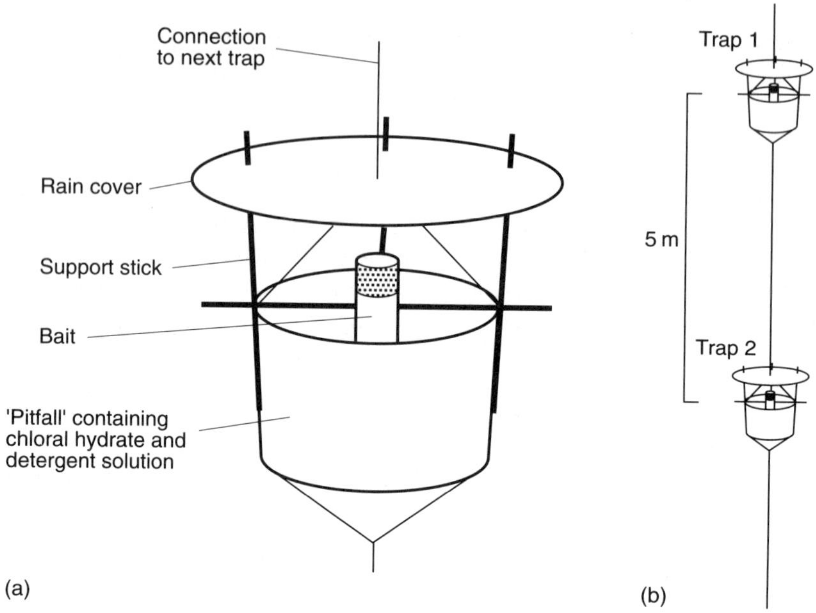

Figure 19.1 (a) An aerial 'pitfall' trap. (b) Diagram illustrating the arrangement of traps placed vertically from a tree platform. Eight aerial traps were used from 5 to 40 m in height, placed at 5-m intervals.

(Hammond, 1990; Davis, 1993), but with certain modifications. Plastic containers (12 cm diameter × 13 cm depth) were suspended by string and attached to longer pieces of rope. A plastic plate was held over the container by three sticks, to keep rain from entering the trap. The bait was held in a plastic tube, aerated at the top end, and suspended over the pitfall (Figure 19.1). Traps were baited between 08:00 h and 10:00 h, and collected 72 hours later. Pitfalls from the main collecting programme were placed at 5 m or 10 m above ground. Studies of the community ecology of dung beetle populations in Western Africa used similar 'aerial pitfall traps' (Walter, 1984), but pitfalls had platforms attached to their rim, to enable beetles to alight on a surface before approaching the bait. Initial tests at Danum Valley showed that such modifications were not necessary for this study (see below).

In addition to their use within the main pitfall trapping programme, a series of traps were set up from a canopy platform on the Danum Valley Nature Trail (an area of primary forest), along a single vertical transect. The platform was positioned on a tree that formed part of the upper canopy, adjacent to a mature strangling fig. The understorey (5–15 m in height) formed a separate vegetational layer beneath the upper canopy. Eight aerial pitfall traps were positioned at intervals of 5 m, joined together by rope, to a height of 40 m (Figure 19.1). The transect included one pitfall on the forest floor, and was baited four times (three times with human faeces and once with carrion).

A NEW ARBOREAL FAUNA

Two aerial pitfall traps were hung from an observation tower on the Nature Trail at Danum Valley between 21st and 24th August, 1990 (at 5 m and 10 m), during a preliminary study on the existence, or absence, of an arboreal dung beetle fauna. This was the first time that such traps had been used in Borneo. After 72 hours the traps were emptied, and the contents examined. Remarkably, of a total 2378 beetles present in these samples (780 at 10 m and 1598 at 5 m), only one specimen (*Onthophagus pavidus* Harold) was recognized as representing a species present in collections made from ground traps in primary forest. Subsequent collections made from aerial traps within the main pitfall trapping programme also showed that, while the arboreal species were dominant in arboreal traps, they were scarce or entirely absent in ground traps (Davis, 1993). To find such high numbers of beetles so close to the ground, and yet of such a different nature to that of the ground fauna, implies that the species present in aerial traps were arboreal in habit, foraging in the upper canopy and understorey rather than the ground. Subsequent behavioural studies confirmed this hypothesis (see below).

Taxonomy of the arboreal fauna

The classification of dung beetles used in this paper follows Balthasar (1963) and Hanski and Cambefort (1991). A paper fully describing the taxonomy of the group is currently in preparation (J. Krikken and J. Huijbregts, in preparation).

The arboreal beetles belong to the genus *Onthophagus* Latreille, tribe Onthophagini, a genus that is abundant on the forest floor. On average, they are smaller than equivalent ground-dwelling species (being only ca. 4–5 mm in length, compared with 6–8 mm for the ground fauna). Individuals can be split into two distinct morphological groups. The first (Arboreal #1) has an elongate curved hind metatarsus, and the second (Arboreal #2), a tibial spur on the back legs (see Davis, 1993). Subsequent analysis showed that the Arboreal #2 group consists of only one species, *Onthophagus* sp.2. Arboreal #1 is dominated by one species closely related to *O. deliensis* Lansberge, 1885, although a second species has been found in small numbers (closely related to *O. falculatus* Boucomont, 1914). The taxonomy of the arboreal species is currently under revision, and will be dealt with in a future paper (see also Davis, 1993).

A third arboreal group (Arboreal #3) was found in very low numbers. This third group is represented by only one species, *Onthophagus* sp.3 (J. Krikken and J. Huijbregts, in preparation). Most were collected from one site within primary forest (11 of the 16 individuals), which suggests that this species is highly clumped. Although *Onthophagus* sp.3 shows a propensity towards foraging higher than most ground-dwelling species (56% of all specimens were collected in aerial traps), it cannot be said to be an arboreal species in the same way that *Onthophagus* sp.2 and *O. deliensis* complex (and related forms) are.

The reasons for Arboreal #1 and Arboreal #2 (*Onthophagus* sp.2) forms meriting the distinction of being truly arboreal in habit, and their position within the superfamily Scarabaeoidea, both in taxonomic and functional terms, are examined below.

Dung beetle functional groups

Dung beetles can be divided into four functional groups: tunnellers, rollers, dwellers, and kleptoparasites (Halffter and Matthews, 1966; Hammond, 1976; Klemperer, 1983). The ball rollers and tunnellers form the dominant functional groups in tropical latitudes, comprising the family Scarabaeidae. The Scarabaeidae is split into two subfamilies, the Scarabaeinae, which comprise the functional group of ball rollers, and the Coprinae, which are tunnellers (Table 19.1). The division of the Scarabaeidae into the Scarabaeinae and the Coprinae is seen as reflecting fundamental behavioural and taxonomic differences between species that roll balls of dung and those that relocate dung by tunnelling beneath into the soil.

Table 19.1 The division of the Family Scarabaeidae into subfamilies and tribes. (After Balthasar, 1963.)

Subfamily	*Tribe*
Coprinae (tunnellers)	Coprini
	Dichotomiini
	Oniticellini
	Onitini
	Onthophagini
	Phanaeini
Scarabaeinae (ball-rollers)	Canthonini
	Eucraniini
	Eurysternini
	Gymnopleurini
	Scarabaeini
	Sisyphini

ECOLOGY AND BEHAVIOUR OF THE ARBOREAL BEETLES

Feeding specializations

No arboreal beetles were attracted to traps baited with carrion. Neither were they attracted to bat guano or common palm civet (*Paradoxurus hermaphroditus*) dung. This, along with the observation that both Arboreal #1 and Arboreal #2 are diurnal (as are the South American arboreal species), suggests that these beetles specialize in the dung of frugivores and folivores, and specifically on the dung of monkeys and apes. This hypothesis is supported by the fact that beetles belonging to Arboreal #1 were collected from freshly deposited orang-utan (*Pongo pygmaeus*) dung within the Danum Valley Conservation Area, and were also collected from orang-utan dung from within the Sepilok Forest Reserve (near Sandakan, in eastern Sabah).

Diel flight activity

Data on the diel flight activity patterns were obtained by trapping at regular intervals during the day and night. Five such activity studies were carried out, each over a 34-hour period. Traps were emptied at 2-hourly intervals during the day and 4-hourly intervals at night. One aerial trap was used for this purpose, located on the Nature Trail close to the Danum Valley Field Centre. Similar activity studies were also carried out on the ground fauna (Davis, 1993).

Figure 19.2 shows the flight activity of the two groups of arboreal beetles, Arboreal #1 and Arboreal #2. Figures given are average numbers of beetle per pitfall (± S.E). Hours of darkness were between 18:30 h and 05:00 h. Arboreal species #2 shows peak activity at dawn and dusk, with no activity towards mid-day, whereas Arboreal species #1 peaks between 12:00 h and 14:00 h. Neither group is active at night. These findings correspond with direct behavioural observations (below).

In terms of arboreal dung beetle activity, diurnal primate incidence and abundance, and in particular the time of dung deposition, can be expected to play a major role, for reasons outlined above (see also Davis, 1993). All 10 species of primate found in eastern Sabah are found in the conservation area, including the orang-utan and proboscis monkey (*Nasalis larvatus*). Several of these species are seen on a regular basis, most notably the Bornean gibbon (*Hylobates muelleri*), the red langur (*Presbytis rubicunda*), and two species of macaque (the long-tailed macaque, *Macaca fascicularis,* and the pig-tailed macaque, *Macaca nemestrina*) (A.J. Davis, personal observation).

The old-world monkeys of the subfamily Colobinae (which include the leaf monkeys, or langurs, of south-east Asia) differ from other primates

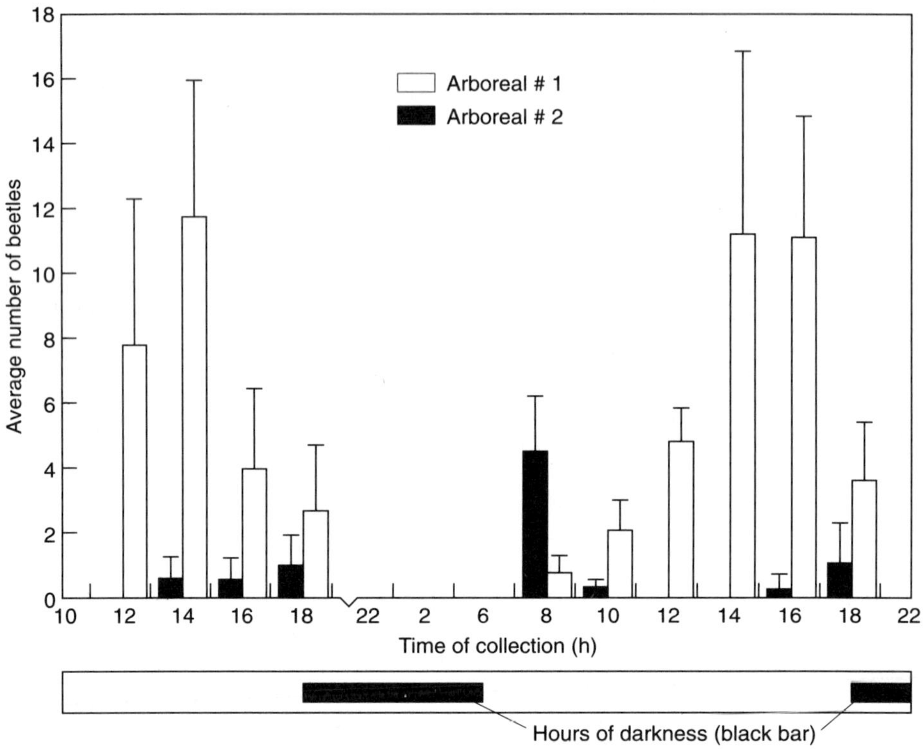

Figure 19.2 Flight activity of two groups of arboreal dung beetles (Arboreal #1 and Arboreal #2). Nocturnal hours are compressed along the x-axis (traps emptied every 4 hours, compared with every 2 hours during day-time).

in that they have a large and multi-chambered stomach which supports bacterial microflora with cellulose-digesting abilities, as in ruminant artiodactyl ungulates (Bauchop and Martucci, 1968; Janis, 1976; Kay *et al.*, 1976; Davies, 1991; Chivers, 1994; Kay and Davies, 1994; Oates and Davies, 1994). Other primates, in contrast, have only a simple stomach. The howler monkeys (*Alouatta* spp.) of the Neotropics, while depending on leaves for a significant portion of their diet (Estrada and Coates-Estrada, 1985), lack an enlarged forestomach (Kay and Davies, 1994).

Stomach morphology has a profound influence on ecology (Oates and Davies, 1994), and has led to a dichotomy in the Bornean primate fauna, between the leaf- and seed-eating colobines and fruit-eating orang-utan, gibbon and both species of macaque (Rodman, 1978; Leighton and Leighton, 1983; Caldecott, 1986; Davies, 1991). It has been noted by workers in Africa (Struhsaker, 1975; E.D. Starin, personal communication) that terrestrial dung-rolling species appear to follow red colobus monkey troops (*Colobus badius tephrosceles*). Colobus monkeys were observed being shadowed by *Gymnopleurus crenulatus* in the Kibale Forest Reserve in Western Uganda by Struhsaker, and by *Onthophagus tridens* in the Abuko Nature Reserve in The Gambia by Starin. In both cases beetles appear to specialize in the dung of these monkeys rather than that of other primates. Feeding specialization of dung beetles on herbivore dung has been recorded elsewhere (Kingston, 1977), and thus the specialized gut morphology of forestomach fermenters may give rise to the specializations described above. It is possible that such behaviour is also exhibited by the arboreal beetles in Borneo. The early morning activity of *Onthophagus* sp.2 coincides with a period of defecation by primates as they move from their night positions (Ellefson, 1974; MacKinnon, 1974; N. Ghaffar, personal communication). The majority of this dung would be produced by predominantly fruit-eating primates (gibbons, macaques and orang-utans). Primates generally rest through the mid-day period (MacKinnon, 1974; Caldecott, 1986), when Arboreal group #1 is at its peak of activity. Differences in food retention time between foregut fermenters and monogastric primates (Kay and Davies, 1994) may result in a second period of defaecation by leaf-monkeys towards mid-day. This, combined with differences in dung consistency between primates with different digestive morphology, may well account for the temporal differentiation seen in the Bornean arboreal dung beetles.

Spatial distribution

Figure 19.3 shows the vertical distribution of Arboreal #1 and Arboreal #2, from the ground (trap 9) to a height of 40 m (trap 1). The graph demonstrates the greater abundance of Arboreal #1 compared with Arboreal #2 (see also Davis, 1993) and the significantly higher canopy (>5 m) abundance of Arboreal #1. Arboreal #1 was most abundant between 5 and 20 m above the forest floor, declining in abundance from 20 to 40 m. Arboreal #2 was collected between 5 and 20 m above the ground, with only one specimen collected above 20 m (Figure 19.3). This suggests differential spatial aggregation of the two dominant arboreal groups (Arboreal #3 was not present in these samples). The sudden reduction in abundance from 5 m to the ground is striking, and highlights the truly arboreal nature of these species. The high arboreal beetle abundance between 5 and 20 m may reflect the tendency of these beetles to forage in the understorey, where most arboreal sources of dung could be expected to accumulate.

The high standard errors in Figure 19.3 illustrate the great variability in abundance between transects. This heterogeneity may reflect variability in the spatial distribution of monkey troops, and therefore

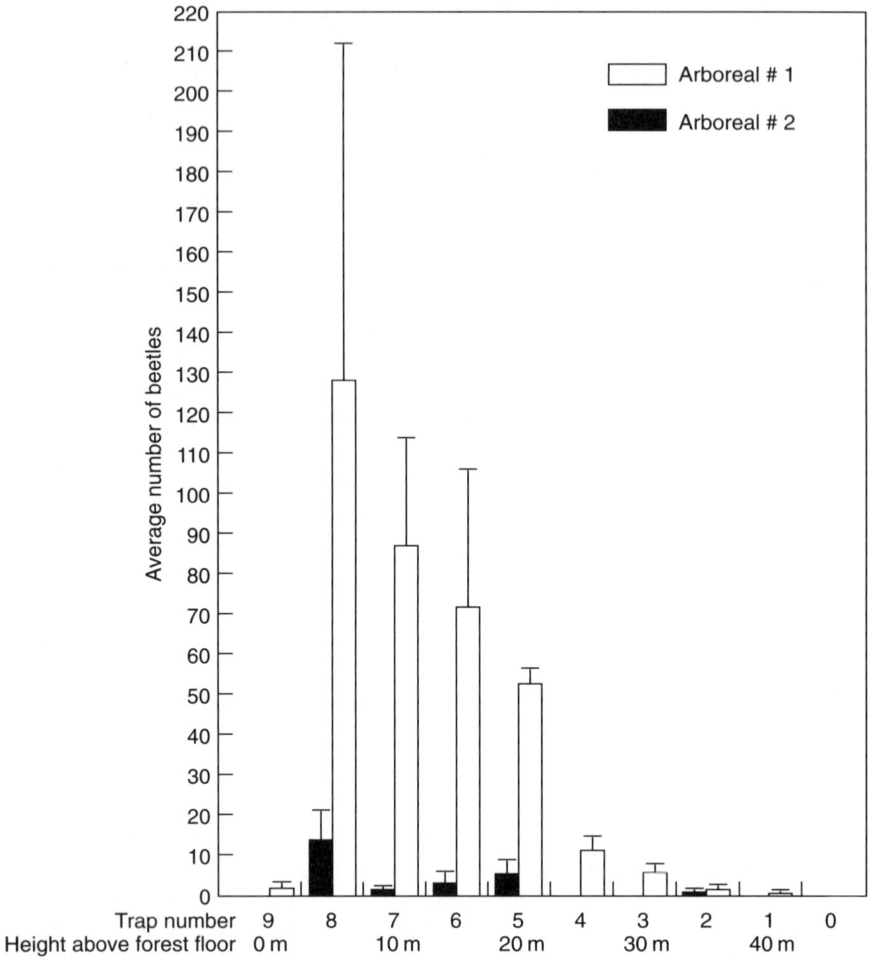

Figure 19.3 Vertical distribution of two groups of arboreal dung beetles (Arboreal #1 and Arboreal #2). Mean (± S.E.) number of beetles per pitfall trap at eight heights above the ground and one ground trap, averaged over three transects at one site.

variability in dung deposition, and/or changes in environmental conditions. Arboreal mammals differ in body size, group size, territoriality, population densities and activity cycles (Estrada and Coates-Estrada, 1985), all of which will influence dung deposition. Distribution and abundance of primate groups will depend on the spatial pattern of fruits (for orang-utan, gibbon and macaques) and of edible leaves (for leaf monkeys) (Davies, 1991).

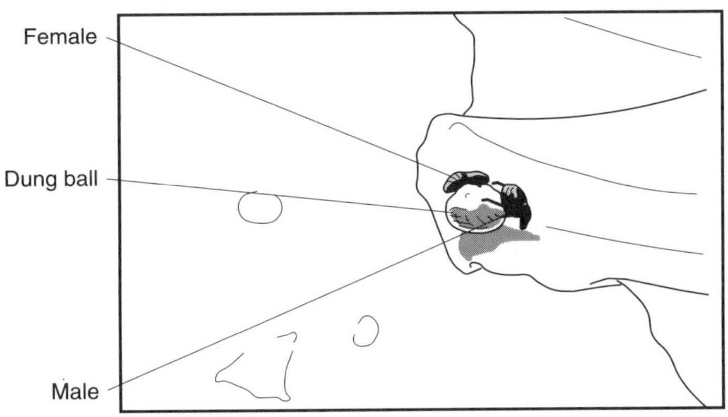

Figure 19.4 A male arboreal dung beetle (Arboreal #1) on the verge of rolling a dung ball (and female) over the edge of a leaf.

Direct observation

Some 10 hours were spent in the field observing the most abundant arboreal species (Arboreal #1 and *Onthophagus* sp.2) *in situ* in the understorey. A full description of the methods employed and of the behaviour demonstrated by the arboreal beetles have been discussed elsewhere (Davis, 1993) and only a brief review is given here.

Arboreal dung beetles were only active during the hours of daylight. Both Arboreal #1 and Arboreal #2 displayed ball-rolling behaviour, mirroring behaviour displayed by ground-dwelling ball-rollers (Halffter and Matthews, 1966). This was true of both ball-forming activity and copulatory behaviour (Davis, 1993). The two groups worked separately, with little interference. Intraspecific ball stealing and interference competition were both observed. Leaf surfaces provide only limited surface area for dung-rolling activity. Upon reaching the periphery of a leaf, individuals or male–female pairs would continue rolling at the same speed, propelling the ball over the edge of the leaf (Figure 19.4). If the dung ball was caught on an intermediate obstacle, the beetles would continue dung-rolling activity until the ground was reached.

The relationship between beetle morphology and rolling behaviour

Canopy dung beetle behaviour can be related to the morphology of the hind legs. Modifications correspond to analogous adaptations seen in ground dung-rollers. The curved metatarsi of Arboreal #1 beetles act like callipers, into which the dung ball fits. Large ground-based rollers have a variation on this adaptation, with curved tibiae rather than curved metatarsi (Halffter and Matthews, 1966). It has been demonstrated that the calliper-like nature of the legs not only assists in rolling, but also in the formation of the ball by allowing the beetle to estimate and change the nature of the ball in a way that consolidates its shape. It is possible that the curved segment in Arboreal #1 beetles also fulfils this function. The tibial spur of Arboreal #2 can be seen as an independently derived adaptation to achieve the same ends. Instead of elongate tarsal segments this species has only slightly curved metatarsi, with the spur compensating for the shorter length by allowing for an alternative source of purchase on the dung ball. In both cases, the ability to form a ball of dung, and maintain a grip on it even while falling to the ground, are the factors that make it possible for these two groups to be truly arboreal in habit.

DISCUSSION

The discovery of a separate guild of beetles that are abundant only 5 m above the ground, while rare or entirely absent at ground-level, poses questions as to how they developed this remarkable niche. Arboreal beetles may have evolved their arboreal lifestyles as a response to competition with the ground fauna, developing behaviours that led to them foraging and orienting themselves upwards towards the lighter conditions of the upper canopy. The different microclimatic conditions found in the rainforest canopy compared with those at ground level are well documented (Shuttleworth, 1984), and the development of vertical stratification within the dung beetles may have resulted from differential responses to microclimatic conditions.

Differences were found in both the temporal and spatial distributions of arboreal groups #1 and #2. These differences may be the outcome of competition for a common resource that is both limited and ephemeral, and may help explain the coexistence of closely related species. Local dung beetle species richness, as a combination of both arboreal and ground elements, is therefore influenced not only by the availability of arboreal sources of dung, but also by the temporal and spatial differentiation of the species that utilize this resource.

This is the first time that ball-rolling behaviour has been observed in the genus *Onthophagus*. This discovery represents a break with the classical division of the Scarabaeidae into rollers (Scarabaeinae) and tunnellers (Coprinae), as the genus *Onthophagus* (tribe Onthophagini) lies within the family Coprinae (Table 19.1). Some species of *Onthophagus* demonstrate facultative rolling behaviour, pushing and pulling lumps of dung as and when the need arises (Halffter and Edmonds, 1982), but none has been seen to portray the morphological adaptations and classical ball-rolling behaviour demonstrated by the Bornean arboreal beetles. It is remarkable that two groups within the same tropical forests have developed the same novel behaviour through different morphological adaptations. The fact that such behaviour has gone unnoticed for so long is surprising, as the genus *Onthophagus* is the largest within the family Scarabaeidae. This would suggest that such behaviour may well be limited to species of the rainforest canopy. A morphologically different but very close species to *Onthophagus* sp.2 exists in Thailand, and taxa with possible arboreal habits belonging to the *deliensis* and *falculatus* complexes are known from Peninsular Malaysia, Sumatra, Java, Kangean, Bali, Flores and Adonara (J. Krikken and J. Huijbregts, unpublished data), which suggests that arboreal dung beetles may be widely distributed throughout the Sunda shelf region (J. Krikken and J. Huijbregts, in preparation).

The activity of the arboreal Bornean beetles, and their method of dung relocation, is the same as the arboreal species of *Canthon* seen in South America (G. Halffter, personal communication). This convergence in community structure presents a striking example of two geographically distant sets of species, with taxonomically distinct origins, evolving to fill the same niche hyperspace by developing similar morphological adaptations. The discovery of a new guild of dung-rolling arboreal dung beetles in Borneo, in a species-rich genus not previously known for dung-rolling activity, gives a very real indication of how new light can be shed on the ecology of tropical insect communities by studying the strato-orientation of insects in response to conditions in the rainforest canopy.

Acknowledgements

This paper is based on material collected while A.J.D. was a participant in the Royal Society's South-east Asia Rainforest Research Programme (Programme Publication No. A/123), and this work formed part of a PhD study funded by the Natural Environment Research Council, UK. We wish to thank N.E. Stork for his comments on an earlier version of this paper, and to N. Ghaffar and E.D. Starin for adding invaluable detail to the discussion on the primatological aspects of this study.

REFERENCES

Balthasar, V. (1963) *Monographie der Scarabaeidae und Aphodiidae der Palaearktischen und Orientalischen Region (Coleoptera: Lamellicornia)*, vols 1–3, Verlag der Tschechoslowakischen Akademie der Wissenschaften, Prague.

Bauchop, T. and Martucci, R.W. (1968) Ruminant-like digestion of the langur monkey. *Science*, **161**, 698–700.

Caldecott, J.O. (1986) An ecological and behavioural study of the pig-tailed macaque. *Contributions to Primatology*, **21**, 1–259.

Chivers, D.J. (1994) Functional anatomy of the gastrointestinal tract, in *Colobine Monkeys: Their Ecology, Behaviour, and Evolution* (eds A.G. Davies and J.F. Oates), Cambridge University Press, pp. 205–28.

Davis, A.J. (1993) *The Ecology and Behaviour of Rainforest Dung Beetles in Northern Borneo*, PhD Thesis, University of Leeds, UK.

Davies, A.G. (1991) Seed-eating by red-leaf monkeys (*Presbytis rubicunda*) in dipterocarp forest in northern Borneo. *International Journal of Primatology*, **12**, 119–44.

Ellefson, J.O. (1974) A natural history of white-handed gibbons in the Malayan Peninsular, in *Gibbon and Siamang, Vol.3: Natural History, Social Behaviour, Reproduction, Vocalization, Prehension* (ed. D.M. Rumbaugh), S. Karger, Basel, Switzerland, pp. 1–209.

Erwin, T.L. and Scott, J.C. (1980) Seasonal and size patterns, trophic structure, and richness of Coleoptera in the tropical arboreal ecosystem: the fauna of the tree *Luehea seemannii* in the Canal Zone of Panama. *The Coleopterist's Bulletin*, **34**, 305–55.

Estrada, A. and Coates-Estrada, R. (1985) A preliminary study of resource overlap between howling monkeys (*Alouatta palliata*) and other arboreal mammals in the tropical rain forest of Los Tuxtlas, Mexico. *American Journal of Primatology*, **9**, 27–37.

Halffter, G. and Edmonds, W.D. (1982) *The Nesting Behavior of Dung Beetles (Scarabaeinae): an Ecological and Evolutive Approach*, Instituto de Ecología, México, D.F.

Halffter, G. and Matthews, E.G. (1966) The natural history of dung beetles of the subfamily Scarabaeinae (Coleoptera: Scarabaeidae). *Folia Entomológica Mexicana*, **12–14**, 1–312.

Hammond, P.M. (1976) Kleptoparasitic behaviour of *Onthophagus suturalis* Peringuey (Coleoptera: Scarabaeidae) and other dung-beetles. *The Coleopterists' Bulletin*, **30**, 245–9.

Hammond, P.M. (1990) Insect abundance and diversity in the Dumoga-Bone National Park, N. Sulawesi, with special reference to the beetle fauna of lowland rain forest in the Toraut region, in *Insects and the Rain Forests of South-East Asia (Wallacea)* (eds W.J. Knight and J.D. Holloway), The Royal Entomological Society of London, pp. 197–254.

Hanski, I. and Cambefort, Y. (eds) (1991) *Dung Beetle Ecology*, Princeton University Press.

Hanski, I. and Krikken, J. (1991) Dung beetles in tropical forests in South-East Asia, in *Dung Beetle Ecology* (eds I. Hanski and Y. Cambefort), Princeton University Press, pp. 179–97.

Holloway, J.D., Kirk-Spriggs, A.H. and Chey Vun Khen (1992) The response of some rain forest insect groups to logging and conversion to plantation. *Philosophical Transactions of the Royal Society of London, Series B*, **335**, 425–36.

Howden, H.F. and Young, O.P. (1981) Panamanian Scarabaeinae: taxonomy, distribution, and habits (Coleoptera, Scarabaeidae). *Contributions to the American Entomological Institute* (Ann Arbor, Michigan), **18**, 1–204.

Janis, C. (1976) The evolutionary strategy of the Equidae and the origins of rumen and cecal digestion. *Evolution*, **30**, 757–74.

Kay, R.N.B. and Davies, A.G. (1994) Digestive physiology, in *Colobine Monkeys: Their Ecology, Behaviour, and Evolution* (eds A.G. Davies and J.F. Oates), Cambridge University Press, pp. 229–84.

Kay, R.N.B., Hoppe, P. and Moloiy, G.M.O. (1976) Fermentative digestion of food in the colobus monkey, *Colobus polykomos*. *Experientia*, **32**, 485–7.

Kingston, T.J. (1977) *Natural Manuring by Elephants in Tsavo National Park, Kenya*, DPhil Thesis, University of Oxford, UK.

Klemperer, H.G. (1983) The evolution of parental behaviour in Scarabaeinae (Coleoptera: Scarabaeidae): an experimental approach. *Ecological Entomology*, **8**, 49–59.

Krikken, J. and Huijbregts, J. Notes on the taxonomy of arboreal dung beetles in the Sunda region (in preparation).

Leighton, M. and Leighton, D.R. (1983) Vertebrate responses to fruiting seasonality within a Bornean rain forest, in *Tropical Rain Forests: Ecology and Management* (eds S.L. Sutton, T.C. Whitmore and A.C. Chadwick), Blackwell Scientific Publications, Oxford, pp. 181–96.

MacKinnon, J.R. (1972) *The Behaviour and Ecology of Wild Orang-utan, Pongo pygmaeus*, DPhil Thesis, Oxford University, UK.

MacKinnon, J.R. (1974) The behaviour and ecology of wild orang-utan (*Pongo pygmaeus*). *Animal Behaviour*, **22**, 3–74.

Malcolm, J.R. (1997) Insect biomass in Amazonian forest fragments, in *Canopy Arthropods* (eds N.E. Stork, J. Adis and R.K. Didham), Chapman & Hall, London, pp. 510–33 (this volume).

Newbery, D.McC., Campbell, E.J.F., Lee, Y.F., Ridsdale, C.E. and Still, M.J. (1992)

Primary lowland dipterocarp forest at Danum Valley, Sabah, Malaysia: structure, relative abundance and family composition. *Philosophical Transactions of the Royal Society of London, Series B*, **335**, 341–56.

Oates, J.F. and Davies, A.G. (1994) What are the colobines ? in *Colobine Monkeys: Their Ecology, Behaviour, and Evolution* (eds A.G. Davies and J.F. Oates), Cambridge University Press, pp. 1–10.

Peng, R.K., Sutton, S.L. and Fletcher, C.R. (1992) Spatial and temporal distribution patterns of flying Diptera. *Journal of Zoology*, **228**, 329–40.

Rodman, P.S. (1978) Diets, densities, and distributions of Bornean primates, in *The Ecology of Arboreal Folivores* (ed. G.G. Montgomery), Smithsonian Institution Press, Washington DC, pp. 465–78.

Roubik, D.W. (1993) Tropical pollinators in the canopy and understorey: field data and theory for stratum 'preferences'. *Journal of Insect Behaviour*, **6**, 659–73.

Shuttleworth, W.J. (1984) Observations of radiation exchange above and below Amazonian forest. *Quarterly Journal of the Royal Meteorological Society*, **110**, 1163–9.

Stork, N.E. (1988) Insect diversity: facts, fiction and speculation. *Biological Journal of the Linnean Society*, **35**, 321–37.

Stork, N.E. (1993) How many species are there ? *Biodiversity and Conservation*, **2**, 215–32.

Struhsaker, T.T. (1975) *The Red Colobus Monkey*. The University of Chicago Press, Chicago.

Sutton, S.L. (1983) The vertical distribution of flying insects in lowland rain forests of Panama, Papua New Guinea and Brunei. *Zoological Journal of the Linnean Society*, **78**, 287–97.

Sutton, S.L. (1989) The spatial distribution of flying insects, in *Tropical Rain Forest Ecosystems* (eds H. Leith and M.J.A. Werner), Elsevier Science Publisher, Amsterdam, pp. 427–36

Sutton, S.L. and Hudson, P.J. (1980) The vertical distribution of small flying insects in the lowland rain forest of Zaire. *Zoological Journal of the Linnean Society*, **68**, 111–23.

Vadon, J. (1947) Les epilissiens de Madagascar (Coleoptera, Scarabaeidae, Canthoniini). II. Biologie. *Bulletin de l'Academie Malagache*, **26**, 173–4.

Walter, P. (1984) Contribution à la connaissance des Scarabéides coprophages du Gabon (Col.).2. Présence de populations dans la canopée de la forêt gabonaise. *Bulletin de la Société Entomologique de France*, **88**, 514–21.

20

Studies on the biology of a canopy-dwelling carabid beetle collected by canopy fogging in the rainforest of Sulawesi (Indonesia)

W. Paarmann and D. Paarmann

ABSTRACT

In an experiment lasting 180 weeks, *Colpodes buchanani* Hope (Coleoptera: Carabidae) was exposed to a distinct annual temperature cycle, more or less typical of warm temperate conditions. The beetles were the descendants of one pair collected by insecticide fogging of a tropical upland rainforest in North Sulawesi, Indonesia. Most of the females reacted to this temperature regime with a distinct annual reproductive period, passing the low winter temperatures and the highest summer temperatures in ovarian dormancy. Winter dormancy was more pronounced than summer dormancy. In two instances females only showed a reduced egg-laying rate, instead of a clear winter dormancy. *C. buchanani* showed great individual variation to the offered temperature regime. Typically, they had two reproductive periods during the warmer part of two successive years with retarded reproductive activity during the very warmest periods. Two females spent more than a year after hatching in ovarian dormancy. The longest life-span observed during the experiment was 1189 days. The possible reasons for the ability of this tropical ground beetle to adapt to warm temperate conditions are discussed. During periods of low reproduction the beetles tend to seek shelter in the canopy and stay higher above ground than during their main reproductive period.

Canopy Arthropods. Edited by N.E. Stork, J. Adis and R.K. Didham. Published in 1997 by Chapman & Hall, London. ISBN 0 412 74900 9

INTRODUCTION

Studying the biology of canopy-dwelling arthropods directly, by looking at them in the canopy, can be very difficult because of the typically low population densities of most rainforest species. Therefore, we made successful attempts to collect living insects from trees by canopy fogging (Paarmann and Stork, 1987a; Paarmann, 1994; Adis *et al.*, 1997, Chapter 4, this volume; Paarmann and Kerck, 1997, Chapter 3, this volume). During the first attempt in 1985 we collected a single pair of *Colpodes buchanani* Hope (Carabidae: Coleoptera) in an area of 'pole' forest in the Gunong Ambong Forestry Reserve (altitude approximately 1200 m), close to the western end of Danau (= Lake) Mooat in North Sulawesi (Indonesia). The offspring of this pair of beetles are still being reared in the laboratory in Göttingen, Germany.

We found that *C. buchanani* is a predator in both adult and larval stages. Females produce comparatively high numbers of small eggs. Thus, *C. buchanani* seems to be a r-strategist in its reproduction. Some of the females show a high rate of egg-laying combined with a short life-span, while females with a long life-span tend to have a low egg-laying rate or have a period of reproductive dormancy during their life. During the time of gonad dormancy, or low reproductivity, the beetles prefer to stay above ground, seeking shelter in the canopy (Paarmann and Bolte, 1990).

In this study we investigated the influence of a distinct annual temperature rhythm on life-span, egg-laying rate and vertical distribution on a 3-m tall 'artificial tree'.

MATERIALS AND METHODS

From October, 1989 to March, 1993, *C. buchanani* were kept in a 3-m tall cylinder of clear plastic (19 cm diameter) with a closed top. The bottom of the cylinder was covered with a 5-cm layer of moist peat. A wooden batten 3 m in length was fixed centrally inside the cylinder and four wooden tubes (7.5 × 2.3 cm) filled with folded cardboard were attached to the batten, with their open ends pointed downwards, at heights of 30, 105, 200 and 275 cm from the surface of the peat. These tubes served as shelter for the beetles. The cylinder was exposed to normal room temperatures in Göttingen, Germany: temperature was comparatively high during summer and low during winter. The temperature in the cylinder was recorded continuously. Similarly, day length varied according to the annual change in Göttingen, with long days in summer and short days in winter.

All beetles in the cylinder were marked individually with small punctures in their elytra and were fed twice a week with mealworms. Females belonged to one of two groups: (i) old females (numbers 1–4 in Figure 20.2) which were already egg-laying at the beginning of the experiment and had been kept under lowland rainforest temperature conditions; and (ii) young females (numbers 5–9 in Figure 20.2) which were newly hatched. Every 1 to 4 (usually 3) weeks we recorded the position of each beetle in the cylinder, and the egg-laying rate of each female by keeping the females in small Petri-dishes on wet sand for 24 hours. The eggs were then counted by washing the sand through a sieve.

RESULTS

The mean egg-laying rate throughout the 180-week experiment, calculated by summing the total number of eggs layed during each control day and dividing by the number of females, is shown in Figure 20.1. A long period of reduced egg-laying occurs during the winter and a shorter period of reduced egg-laying occurs during the warmest part of the summer.

The individual egg-laying rates of each of the nine females observed are shown in Figure 20.2. There was a tendency for those females that were egg-laying before the start of the experiment to have a longer period of ovarian dormancy. Female number 1 reduced, but did not halt, egg-laying during winter, thus showing that individuals of this tropical carabid species can still reproduce at low temperatures. Females 2–4 entered a dormancy which was shorter in female 2 than in the other two individuals.

Among the young newly hatched females, numbers 5–7 showed pronounced reproductive periods during their first (1990) and second (1991) summers, but the first reproductive cycle of female 6 started much earlier than the cycles of females 5 and 7. The reproductive rhythms of females 8 and 9 are different. Female 8 started egg-laying at an age of 21.5 months during summer 1991 and continued throughout the winter until its death in May 1992. Female 9 layed only a few eggs during summer 1991 after 18 months of ovarian dormancy; it then entered another 11 months of dormancy, its main reproductive phase starting in May 1992. This female reached an age of 1189 days (see Table 20.1). All five 'young' females reached comparatively high ages (Table 20.1). Thus, the reaction of these females to an annual temperature rhythm seems to be an extension of their life-spans through periods of reproductive dormancy.

From 23.12.1989 to 14.9.1990 we estimated the mean height above ground of the beetles of both sexes by multiplying the height of the shelters by the numbers of individuals found, and then dividing by the sum of the total number of individuals (separated for the sexes). Beetles on or in the ground were recorded as zero height. During the first part of

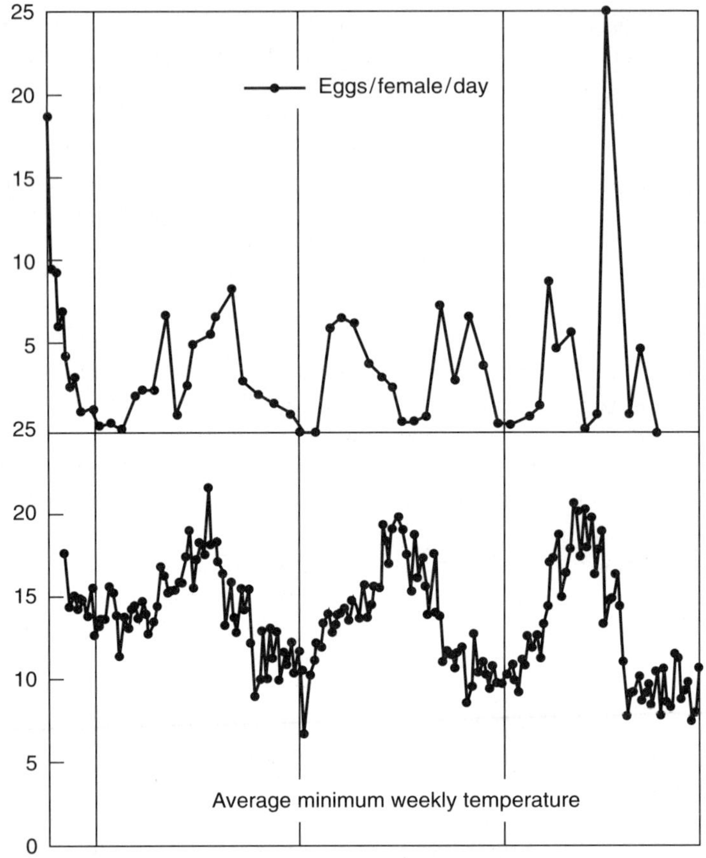

Figure 20.1 Average egg-laying rate of female *Colpodes buchanani* correlated with the average weekly minimum temperature(°C). Vertical lines in the graph from left to right indicate winters of 1989/1990, 1990/1991 and 1991/1992. For further information, see text.

Table 20.1 Life-span of females 5–9 (Figure 20.2), compared with the life-span of females kept under temperature conditions not influenced by an annual rhythm (light period 12 hours)

Female number	7	6	5	8	9
Life-span (days)	512	681	723	964	1189

Temperature and light conditions light times	*n*	*Average (range) life span (days)*
17/28°C (15/9 h)	10	342 (183–636)
21/27°C (15/9 h)	25	315 (141–520)

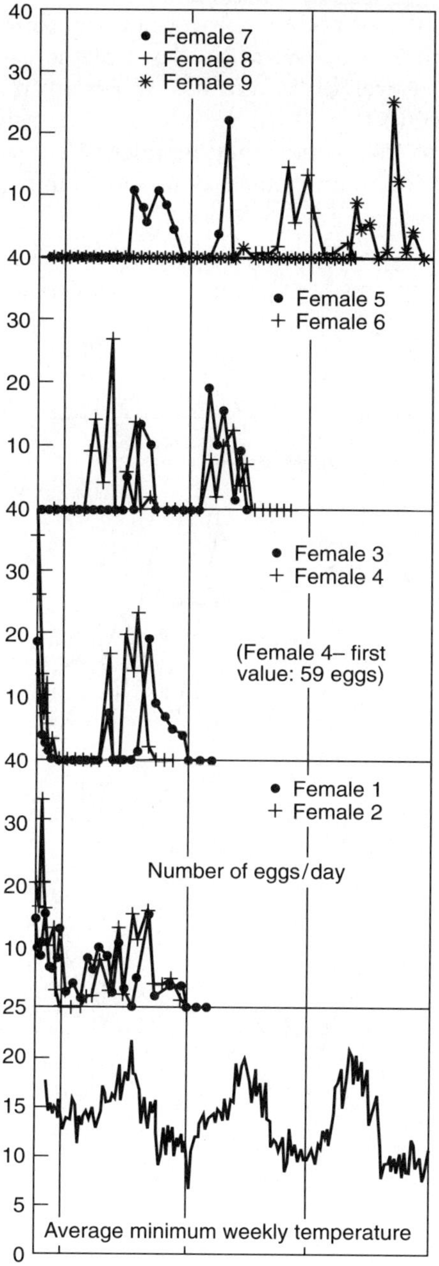

Figure 20.2 Egg-laying rates of females 1–9, correlated with the average weekly minimum temperature. Vertical lines from left to right indicate winters 1989/1990, 1990/1991 and 1991/1992. For further information, see text.

the experiment, beetles of both sexes were mostly found 150 cm or more above ground (Figure 20.3). With the onset of the reproductive period both sexes moved nearer to the ground: males from 13.4.90 to 11.5.90, females from 26.5.90 to 16.6.90. At the end of the experiment males were on average nearer to the ground than females. We tested the differences in vertical distribution with pooled data for certain periods using the Mann–Whitney *U*-test:

Males	23.12.89/13.04.90 to 11.05.90/14.09.90	$\alpha = 0.000$
Females	23.12.89/13.04.90 to 16.06.90/14.09.90	$\alpha = 0.350$
Females to males	11.05.95/14.09.95	$\alpha = 12.56$

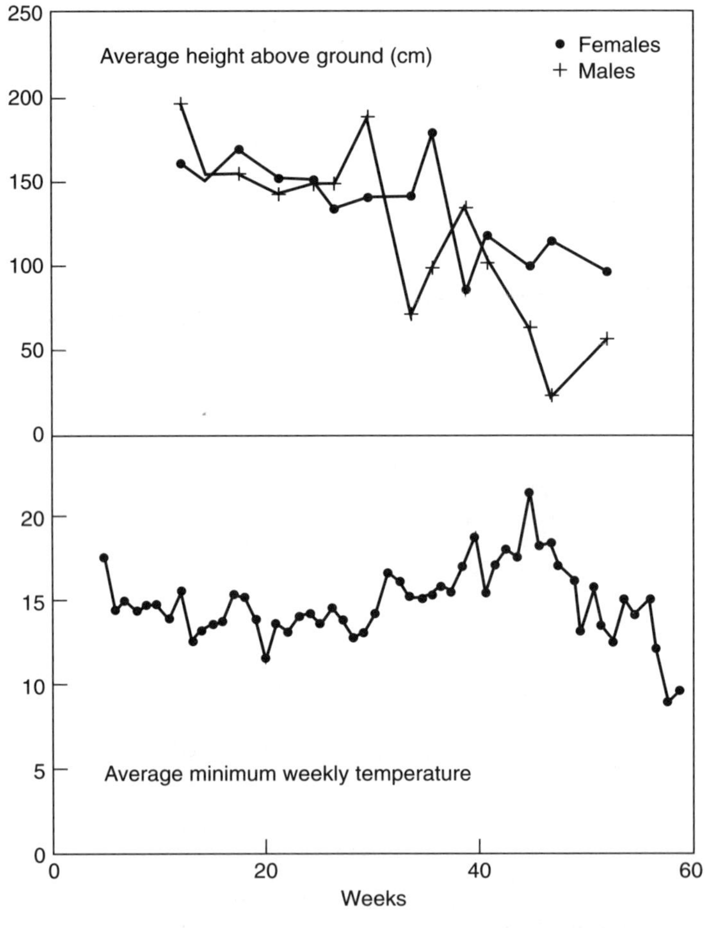

Figure 20.3 Average height above ground (cm) of both sexes, correlated with the average weekly minimum temperatures. For more information, see text.

DISCUSSION

The descendants of beetles collected in the upland rainforest of North Sulawesi are able to reproduce successfully under lowland rainforest conditions (Paarmann and Bolte, 1990) and also under more or less warm temperate conditions, as in this study. Females can enter long periods of ovarian dormancy during periods of unfavourable temperatures, thus extending their life-span compared with those females kept under lowland tropical conditions (Paarmann and Bolte, 1990; see also Table 20.1). Males also seem to react in a similar way because the deposited eggs were always fertilized. The question arises, why do members of a tropical population of *C. buchanani* have the ability to adapt to warm temperate conditions? Possibly there may be strong selective forces in the wild, such as large local variation in climatic conditions, which may lead to this wide tolerance. Previous studies of the seasonality of ground beetles in North Sulawesi showed 65% of 155 species to have at least one dormant female. These species showed a strong tendency to avoid reproduction in the cooler part of the year, although some species had the opposite strategy (Paarmann and Stork, 1987b; Stork and Paarmann, 1992). In *C. buchanani* both strategies seem to be available. Erwin (1979, 1981) and Erwin and Adis (1982) proposed the idea that the tropics, especially the lowland wet tropics, are the source of 'pulses' of new taxa (the 'taxon pulse' hypothesis). As new groups of Carabidae evolve there, they migrate along certain pathways, some leading to different climatic zones. One of us (W. Paarmann) came to similar conclusions through the study of the seasonal reproductive rhythms of Carabidae on a global scale (Paarmann, 1979).

The area of distribution of *C. buchanani* extends as far north as the south of Japan, according to specimens in the collection of The Natural History Museum, London. This may explain the ability of the tropical population originating from North Sulawesi to adapt to the warm temperate conditions of our experiment. Darlington (1971) suggested that many species of the tribe Platynini (= Agonini), to which *C. buchanani* belongs, migrated as island-hoppers from the Asian mainland to the south. In this case the reproductive ability of *C. buchanani* may have evolved from the warm temperate through to the tropics.

The high variability of individual responses to the temperature regime is surprising, particularly since all are descendants of one pair of beetles, inbred over 4 years and six or seven generations. We also found pronounced morphological differences among the beetles, such as shortened elytra of different types, different colour-morphs and individuals with elytra-like appendages on the pronotum. Furthermore, we found a variability in the hatching rate from the eggs of different females, often with a high percentage of larvae showing deformities. In our opinion, these only partly published results (Paarmann and Bolte, 1990) show that the ancestors of our laboratory stock carried a great genetic plasticity. The high egg-laying rate in fact may not represent a r-strategy, but rather it may be a way to increase the chance of survival of fit larvae, which use their less-fit relatives as a food source. High genetic variability in the gene-pool may be important if the population density becomes very low. This idea needs to be tested.

We confirmed earlier findings (Paarmann and Bolte, 1990) that the adults of *C. buchanani* prefer to rest above ground, especially when they are immature or in a state of low reproductivity. Egg deposition seems to take place mostly on the ground, but the females only stay there for a short period of time, possibly to avoid the risk of predation. Males may be more abundant than females at egg deposition sites because these sites provide the greatest opportunity for copulation. If this is so, then perhaps *C. buchanani* is not a true canopy-dwelling species but uses the trees mainly for shelter.

REFERENCES

Adis, J., Paarmann, W., da Fonseca, C.R.V. and Rafael, J.A. (1997) Knockdown efficiency of natural pyrethrum and survival rate of arthropods obtained by canopy fogging in Central Amazonia, in *Canopy Arthropods* (eds N.E. Stork, J. Adis and R.K. Didham), Chapman & Hall, London. pp. 67–79 (this volume).

Darlington, P.J. (1971) The carabid beetles of New Guinea. Part IV. General considerations; analysis and history of fauna; taxonomic supplement. *Bulletin of the Museum of Comparative Zoology*, **142**, 129–337.

Erwin, T.L. (1979) Thoughts on the evolutionary history of ground beetles: Hypotheses generated from comparative faunal analysis of lowland forest sites in temperate and tropical regions, in *Carabid Beetles: Their Evolution, Natural History, and Classification* (eds T.L. Erwin, G.E. Ball, D.R. Whitehead and A.L. Halpern), W. Junk Publishers, The Hague, pp. 539–92.

Erwin, T.L. (1981) Taxon pulses, vicariance, and dispersal: an evolutionary synthesis illustrated by carabid beetles, in *Vicariance Biogeography: A Critique* (eds G. Nelson and D. Rosen), Columbia University Press, New York, pp. 159–83.

Erwin, T.L. and Adis, J. (1982) Amazonian inundation forests, their role as a short term refugia and generators of species richness and taxon pulses, in *Proceedings of the Vth International Symposium of the Association of Tropical Biology 1979* (ed. G.T. Prance), Columbia University Press, New York. pp. 358–71.

Paarmann, W. (1979) Ideas about the evolution of the various annual reproduction rhythms in carabid beetles of different climatic zones. *Miscellaneous Papers of the Landbouhogeschool Wageningen*, **18**, 119–32.

Paarmann, W. (1994) Von der Tötung zum Lebendfang – Entwicklung der Insektizidnebeltechnik. *Allgemeine Forstzeitschrift*, **49**, 1457–9.

Paarmann, W. and Bolte, H. (1990) Studies on the biology of *Colpodes buchanani* (Hope) (Coleoptera, Carabidae), collected in Sulawesi rain forest (Indonesia) by application of knock-down insecticides with a canopy fogger, in *Insects and the Rain Forests of South East Asia (Wallacea)* (eds W.J. Knight and J.D. Holloway), The Royal Entomological Society of London, pp. 191–5.

Paarmann, W. and Kerck, K. (1997) Advances in using the canopy fogging technique to collect living arthropods from tree-crowns, in *Canopy Arthropods* (eds N.E. Stork, J. Adis and R.K. Didham), Chapman & Hall, London. pp. 53–66 (this volume).

Paarmann, W. and Stork, N.E. (1987a) Canopy fogging, a method of collecting living insects for investigations of life history strategies. *Journal of Natural History*, **21**, 563–6.

Paarmann W. and Stork, N.E. (1987b) Seasonality of ground beetles (Coleoptera: Carabidae) in the rain forest of N. Sulawesi (Indonesia). *Insect Science and its Application*, **8**, 483–7.

Stork, N.E. and Paarmann, W. (1992) Reproductive seasonality of the ground and tiger beetle (Coleoptera: Carabidae, Cicindelidae) fauna in N. Sulawesi, Indonesia. *Studies on Neotropical Fauna and Environment*, **27**, 101–15.

21

Bioacoustic monitoring of insect communities in a Bornean rainforest canopy

K. Riede

ABSTRACT

Animal calls were recorded within a mixed dipterocarp rainforest in Northeast Borneo (Mt Kinabalu, Sabah, Malaysia)[1]. Multiport recordings by four pairs of microphones distributed along a canopy walkway allowed an analysis of the spatial and temporal distribution of singing animals. High species diversity resulted in elevated background noise levels over a wide frequency range. This 'biotic noise' leads to a considerable reduction of communication range, which is ameliorated by a pronounced spatial and temporal segregation of songsters. Simultaneous recordings at canopy and ground level revealed that songs produced by ground-living mole crickets can be heard high up in the canopy, while songs from canopy-dwelling cicadas are damped considerably at ground level.

On the temporal scale, diurnal, nocturnal and dusk communities could be differentiated. Diurnal communities are dominated by birds and cicadas (Cicadidae), nocturnal communities by Orthoptera and frogs, while the dusk chorus contains species from all these groups. The latter is characterized by a well-defined set of six cicada species, mole crickets (Gryllotalpidae), crickets (Gryllidae, *Itara* spp.) and a few frog species which exhibit temporal fine tuning over a range of minutes at high song intensities.

The composition of the dusk community is similar at all study sites within the lowland mixed dipterocarp forest. In contrast, diurnal and nocturnal callers form species aggregations within forest patches of a few hectares in size (Cicadidae), or on certain tree crowns or understorey shrubs (Ensifera). Within such spatial assemblies, temporal synchronization is less precise than within the dusk chorus. In the understorey, pseudophylline Tettigoniidae (*Tympanophora* spp.) exhibit narrow low frequency calls between 1 and 2 kHz, probably exploiting optimal transmission channels on the forest floor.

[1] Some of the cicada songs described in the text can be heard via WWW under: http://www.biologie.uni-freiburg.de/data/zoology/riede/cicada.html

Canopy Arthropods. Edited by N.E. Stork, J. Adis and R.K. Didham. Published in 1997 by Chapman & Hall, London. ISBN 0 412 74900 9

INTRODUCTION

In tropical forests, a great number of animal species use acoustic signals for mate recognition, orientation and for the maintenance of permanent or temporary territories. Among the 3000 insect species found by Stork (1991) in the crowns of 10 trees from Borneo, around 2% were ensiferan Orthoptera, the most speciose group of sound-producing insects. A similar percentage was reported by Floren and Linsenmair (1994). A total of around 4000 insect species capable of sound production might be a realistic estimate for Bornean forests, plus quite a number of frog and bird species. Within such species-rich tropical communities, the probability of acoustic interference is high due to elevated species diversity. Elevated background noise levels due to 'biotic noise' might reduce communication distances considerably. Therefore, the ensemble of acoustically communicating animals can be considered as a community competing for acoustic broadcast channels (Riede, 1993a). As acoustic interference might occur between distinct taxa, the 'acoustic community' should not be defined by taxonomic criteria. In this study, sound recordings from Bornean lowland forest were analysed to determine possible assembly rules and mutual influences in a species-rich tropical acoustic community consisting of various higher taxa.

In addition, sound recording is a valuable tool for inventorying and monitoring singing animals, as exemplified by the dusk community of Bornean cicadas.

METHODS

Study sites and periods

Investigations were made at Kinabalu National Park, Sabah, Malaysia, in the canopy of a lowland mixed dipterocarp forest at Poring Hot Springs (500 m above sea level). A canopy walkway system of 900 m length was available. Observations were made from April to June 1991, March until May and August 1992, thereby covering different seasons and weather conditions.

Recording method and sound analysis

Recordings were made by four pairs of electret microphones (EM-3) installed at four stations about 30 m apart along the walkway. Microphones were fed into two, four-channel mixers and the output recorded by a stereo cassette recorder (Sony TCD5-PRO, frequency response: 40–16 000 Hz + 3 dB). Microphone stations could be selected by switching the respective mixer channels. Besides simple stereo recording at one site, multiport recording at two sites was possible by selecting microphones at different sites for the left and right channels, respectively (for further technical details, see Riede, in press). Overall sound intensity (dB[A]) was measured by a sound-level meter (Noris NM-3). For acoustic analysis, sound recordings were sampled with a computer (Tandon 486/33, Compaq) via an analog–digital converter (DT2821, Data Translation) at a sampling frequency of 30 kHz and analysed with signal-processing software (Hypersignal Acoustic, Hyperception Inc.).

RESULTS

Biotic noise and communication range

In temperate and tropical habitats, intensities of acoustic signals drop due to spherical spreading and frequency-dependent atmospheric attenuation losses (Figure 21.1, broken curves). In species-poor temperate habitats, a signal often competes with nothing other than 'abiotic noise' generated by air turbulence or running water, and octave-band noise levels scale with frequency^{-1} (Figure 21.1, broken lines, c.f. Fletcher, 1992). The signal becomes undetectable when its intensity equals noise, i.e. at the intersection of noise bands with the respective signal attenuation curves (filled circles, Figure 21.1). In contrast, communication range within tropical rainforests is reduced due to 'biotic noise' from other animals. The frequency bands of biotic noise do not drop with frequency^{-1} (Figure 21.1, broken lines at the left). Elevated noise levels within tropical forests reduce communication ranges considerably (Figure 21.1, open squares).

However, these theoretical considerations do not take into account the directional properties of the receivers, which may reduce ambient noise considerably. For example, most insect hearing systems are pressure gradient receivers and therefore inherently highly directional (Larsen *et al.*, 1989). The multiport recordings allow the calculation of pressure gradients which give a more realistic approximation of the acoustic situation as experienced by many insects (Figure 21.2). Higher frequencies are accentuated, and noise sources away from the axis of the receptor system have a lesser influence.

Temporal organization

Sound intensity increases considerably at dusk (see also Riede, 1996). This 'dusk chorus' consists of a well-defined set of cicada, cricket and frog species. Within this dusk chorus, song activities exhibit a clear, precise temporal segregation in the range of minutes (Figure 21.3). The

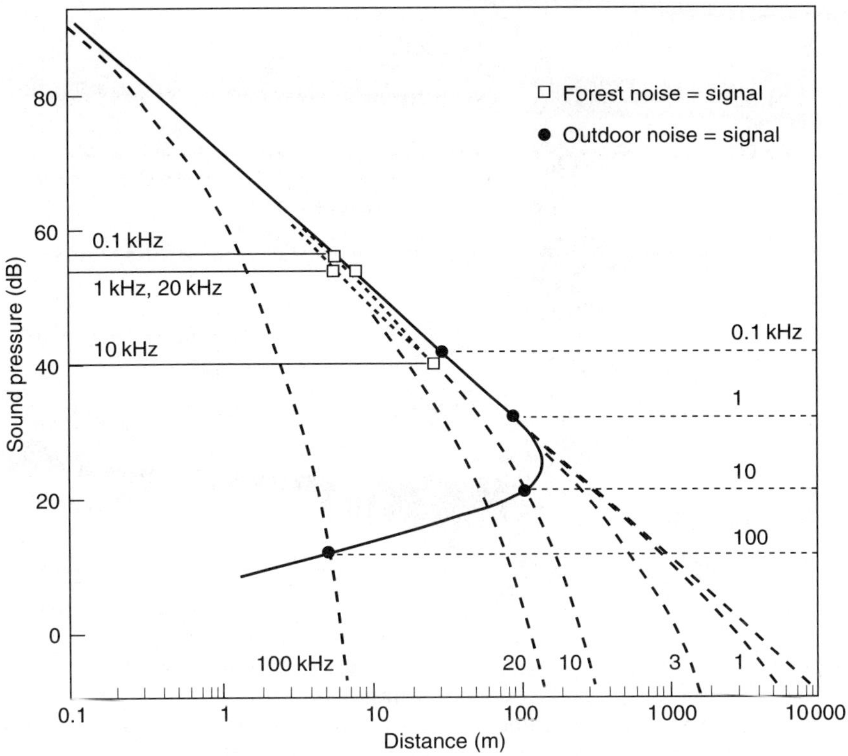

Figure 21.1 Communication distance for a sound source of 0.1 mW (equal to human speech) at various frequencies (broken curves: attenuation with frequency as a parameter) in the presence of typical outdoor octave-band noise levels at the same frequencies (about 30 dB(A) overall, broken lines at the right half, after Fletcher, 1992, p. 282), and in the presence of noise from Bornean rainforest (about 47 dB(A) overall, measured 20 m above ground in lower montane mixed dipterocarp forest, 11.4.1994, 17:45 h, Poring, Sabah). The intersections of signal and noise curves define the maximum distance beyond which the respective frequencies cannot be differentiated, i.e. the maximum communication distance. Note that the 20-kHz octave band coincides with the 1 kHz band at 53 dB. 'Biotic noise' frequency bands are generated by singing animals and possibly vegetation noise from dry rustling leaves, especially around 20 kHz.

first half-hour at dusk is dominated by cicadas (Cicadidae); the second half-hour by crickets (Grylloidea) and frogs (Anura).

The exact timing of song activity suggests that a precise trigger such as fading light is used; for example, the threshold of total light intensity (integral over some spectral sensitivity), the velocity of light intensity change, i.e. the first derivative of the light intensity versus time curve,

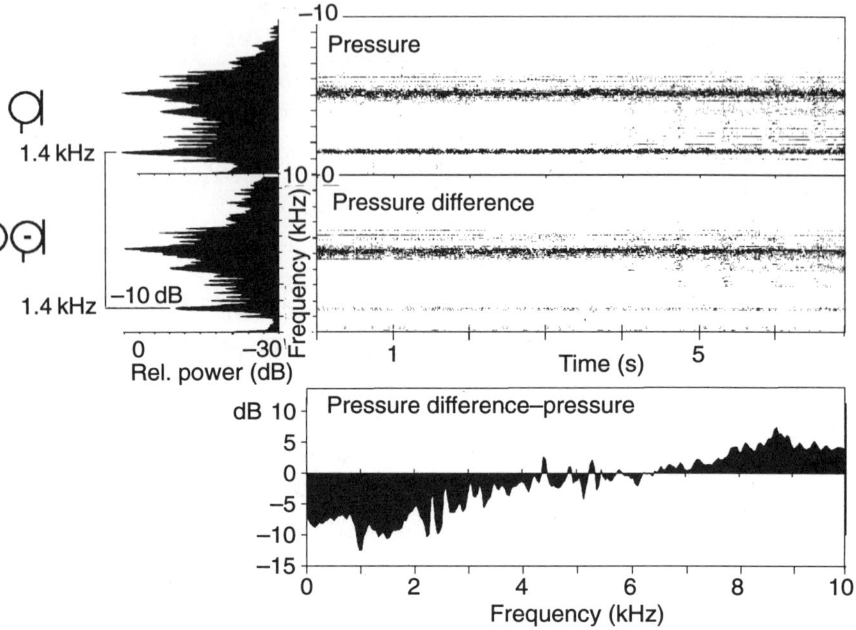

Figure 21.2 Calculation of pressure difference from multiport recordings. Upper half (pressure): power spectrum and spectrogram of sound pressure recording with one microphone at canopy level. Lower half (pressure difference): power spectrum and spectrogram of pressure difference calculated by subtraction of two sound pressure recordings from nearby microphones (distance = 1 cm). The comparison of power spectra is shown below (pressure difference – pressure): the pressure difference receiver system accentuates higher frequencies and damps lower components. Most insect ears work as pressure difference receivers and are inherently directional (see text).

or a measuring of the characteristic red shift at dusk. However, these hypotheses have not been investigated systematically.

Spatial organization

An important feature of acoustic communities is their spatial organization at various scales, from local to landscape level. Species participating in the dusk chorus are widely distributed. A similar 'dusk community' was observed at all lowland mixed dipterocarp forest sites visited in Sabah, Brunei and Sarawak (Poring, Sabah; Kuala Belalong Field Station, Brunei; Gunung Mulu National Park, Bako Park, Kuching, Sarawak). At the local scale, the spatial organization of this ubiquitous 'dusk community' is associated with certain forest strata. Most cicadas sing

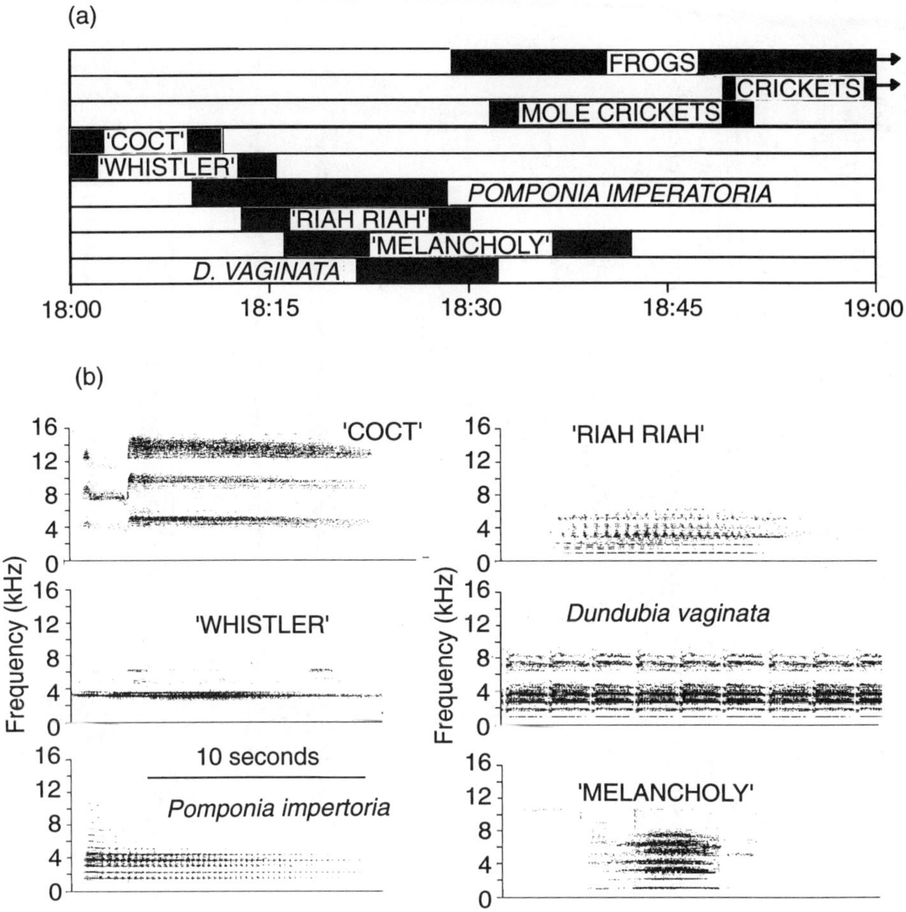

Figure 21.3 Temporal fine-tuning within the dusk chorus community. Unidentified cicada species are named mnemonically according to song characteristics (marked by inverted commas). (a) Activity patterns within the dusk chorus between 18:00 h and 19:00 h, local time. Song activity marked black. (b) Spectrograms of the cicada species participating in the dusk chorus (Fast Fourier Transform, 512 pt, Blackman window). Scale bar of 10 s applies to all spectrograms.

high up in the canopy, with only one species, *Dundubia vaginata*, singing in the lower forest strata (Riede and Kroker, 1995). Frogs, mole crickets and crickets (mainly *Itara* spp.) sing at ground level or on forest shrubs at a height of a few meters. Multiport recordings at various heights distinguished the degree of interference between ground and canopy singers (Figure 21.4). Sound produced by mole crickets reaches high up

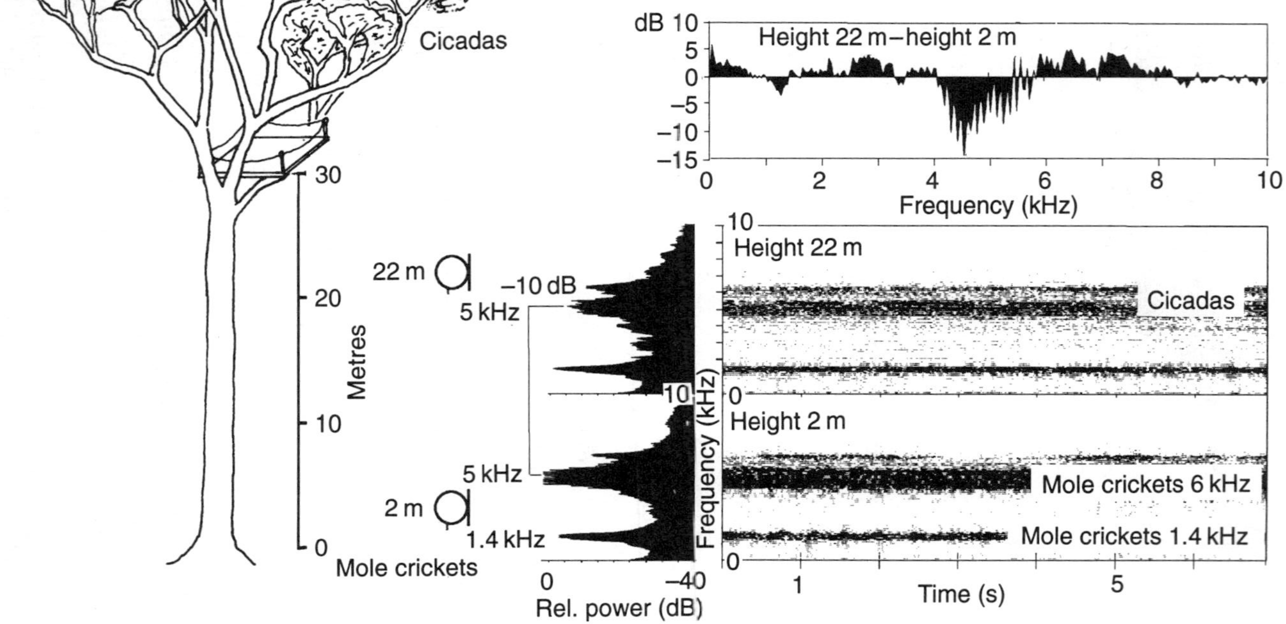

Figure 21.4 Recordings at different heights. A comparison of power spectra from recordings made simultaneously at different heights above the forest floor, with microphones hanging from the platform indicated at the left. Upper diagram (height 22 m–height 2 m): difference between power spectra at 22 m (middle canopy) and 2 m. Lower diagram: spectrograms and power spectra of the respective sound recordings at 22 m and 2 m. The difference spectrum reveals a 10-dB reduction at the canopy of the 5-kHz band produced by mole crickets at ground, while the sound of a second mole cricket species singing at 1.4 kHz reaches the canopy unattenuated. On the other hand, components of cicada songs heard at canopy level do not reach the ground.

into the canopy. Losses by spherical spread are compensated to a certain degree by a wider angle of interception for sound sources at ground level by a receiver located high up in the canopy. In contrast, the high frequency components of cicada sounds are highly damped at the ground. This has consequences for the optimal listening position of female conspecifics. For crickets, elevated listening positions would be advantageous. Cicada females should prefer the higher forest strata, where in fact most flight activity is observed.

Many species singing before or after dusk exhibit characteristic spatial organization at the landscape scale. A typical example is the 'waterfall community' of several ensiferan and anuran species singing at or near waterfalls. Their calls are adapted to overcome elevated background noise levels from running water (Riede, 1996). Other species form calling groups within the forest. These consist of small populations of a single species within a limited area, often defined by one bush or tree, or small forest plots. Single-species associations of diurnally singing cicadas concentrate within forest patches of a few hectares and occasionally move as a group.

Two species of the tettigoniid subfamily Pseudophyllinae (*Tympanophyllum* sp. and *Tympanophyllum* aff. *atroterminatum*) produce pure frequency songs of carrier frequencies between 1.4 and 2 kHz, which is the lowest carrier frequency known for Tettigoniidae. These calls strongly resemble vocalizations of the frog *Metaphrynella sundana* (Riede, 1993b). Both frogs and katydids aggregate within small calling groups in the lower forest strata, thereby alternating their calls.

DISCUSSION

In temperate or subtropical habitats, populations of certain species often outnumber those observed within species-rich tropical forests. In non-tropical oligospecies communities, singing animals often occur in considerable densities. Therefore, acoustic interference from conspecific signals may be greater than in tropical communities (c.f. Römer, 1993). As a result of intraspecific competition, a variety of well-described strategies and tactics such as chorusing and call alternation (c.f. Greenfield, 1994) or mute 'satellite male' crickets (Cade, 1981) have evolved. In contrast, tropical acoustic communities are dominated by interspecific competition which might culminate in genetically fixed patterns of acoustic resource partitioning among distinct species. Duellman and Pyles (1983) pioneered investigations on tropical acoustic guilds by analysing neotropical frog communities, Heller and von Helversen (1989) studied acoustic resource partitioning among bats and quite a number of studies deal with acoustic communication among tropical forest insects (Cicadidae: Pringle, 1955; Duffels, 1988. Grylloidea: Otte, 1992;

Riede, 1993a). All these studies concentrate on certain taxonomic groups. However, within tropical multispecies assemblages frequency range and even temporal structure of acoustic signals from distinct animal groups such as ensiferan Orthoptera and frogs overlap considerably, so that acoustic guilds cannot be delimited by taxonomic units.

Members of the Bornean lowland forest community avoid bioacoustic interference by spatial and temporal segregation. The temporal synchronization of the dusk community is very precise, and its composition is similar at all study sites throughout lowland mixed dipterocarp forest. In contrast, diurnal and nocturnal callers exhibit less precise temporal synchronization, but form spatial assemblies of species aggregations within forest patches a few hectares in size (Cicadidae), or on certain tree crowns or understorey shrubs (Ensifera). With the exception of one cicada species, *Dundubia vaginata* (Riede and Kroker, 1995), there is no overlap between the species composition of the dusk community and the diurnal/nocturnal species. This is in striking contrast to Neotropical cicada communities, where Young (1981) observed several species singing during the day and at dusk.

The precise temporal segregation within the Bornean dusk community could be a result of either proximate or ultimate mechanisms. Crickets and frogs start calling after termination of the acoustically dominant cicada community, which might be a result of 'direct' acoustic suppression as a proximate mechanism. Temporal fine-tuning within the dusk cicada community, however, might be an 'ultimate' evolutionary outcome of niche segregation on the temporal scale. Such temporal fine-tuning in the range of minutes is limited to the dusk community, indicating that fading light or changing spectral composition are used as precise temporal triggers.

Ensiferan Orthoptera species participating in the dusk chorus are widely distributed and common (Gryllidae, *Itara* sp.; Gryllotalpidae, *Gryllotalpa* sp.1, 2 and 3). In contrast to these well-known dominant species, our knowledge of the majority of orthopteran species singing at night is still incomplete, although preliminary data indicate a high beta-diversity of this fauna. In addition, song activity of some species seems to be weather-dependent. Inventorying of this fauna is still in progress, and an extension of recording duration and sites will probably reveal many more species.

Bioacoustic studies in tropical rainforests are of interest from various points of view. Sound transmission in complex habitats, such as forests, is an interesting bioacoustic problem which has been investigated in a number of studies (Marten and Marler, 1977; Richards and Wiley, 1980). The pioneer studies of Eyring (1946) in a Panamanian rainforest consider the forest as a room, with the canopy as a ceiling, leading to optimal sound transmission within a frequency window between 1 and 2 kHz.

The unusually low, euphonic calls between 1 and 2 kHz of Bornean *Tympanophyllum* katydids singing in the lower forest strata corroborate such a hypothesis.

Acknowledgements

I thank Sabah National Parks for generous support of field investigations, Dr N. Fletcher, Canberra, for permission to modify and reproduce Figure 21.1, and Dr N. Stork, London, for helpful comments. With support from the DFG (Theme programme: 'Mechanism of Maintenance of Tropical Diversity').

REFERENCES

Cade, W.H. (1981) Alternative male strategies: genetic differences in crickets. *Science*, **22**, 563–4.

Duellman, W.E. and Pyles, R.A. (1983) Acoustic resource partitioning in anuran communities. *Copeia*, **3**, 639–49.

Duffels, J.P. (1988) The cicadas of the Fiji, Samoa and Tonga Islands, their taxonomy and biogeography (Homoptera, Cicadoidea). *Entomonograph*, **10**, 1–108.

Eyring, C.F. (1946) Jungle acoustics. *Journal of the Acoustic Society of America*, **18**, 257–70.

Fletcher, N.H. (1992) *Acoustic Systems in Biology*, Oxford University Press, Oxford.

Floren, A and Linsenmair, E. (1994) Zur Diversität und Wiederbesiedelungsdynamik von Arthropoden auf drei Baumarten in einem Regenwald in Sabah, Malaysia. *Andrias*, **13**, 23–8.

Greenfield, M.D. (1994) Cooperation and conflict in the evolution of signal interactions. *Annual Review of Ecology and Systematics*, **25**, 97–126.

Heller, K.-G. and von Helversen, O. (1989) Resource partitioning of sonar frequency bands in rhinolophoid bats. *Oecologia*, **80**, 178–86.

Larsen, O.N., Kleindienst, H.-U. and Michelsen, A. (1989) Biophysical aspects of sound reception, in *Cricket Behavior and Neurobiology* (eds F. Huber, T.E. Moore and W. Loher), Cornell University Press, Ithaca, pp. 364–90.

Marten, K. and Marler, P. (1977) Sound transmission and its significance for animal vocalization. II. Tropical forest habitats. *Behavioural Ecology and Sociobiology*, **2**, 291–302.

Otte, D. (1992) Evolution of cricket song. *Journal of Orthoptera Research*, **1**, 25–49.

Pringle, W.S. (1955) The songs and habits of Ceylon cicadas, with a description of two new species. *Spolia Zeylandica*, **27**, 229–38.

Richards, D.G. and Wiley, R.H. (1980) Reverberations and amplitude fluctuations in the propagation of sound in a forest: Implications for animal communication. *The American Naturalist*, **115**, 381–99.

Riede, K. (1993a) Monitoring biodiversity: analysis of Amazonian rainforest sounds. *Ambio*, **22**, 546–8.

Riede, K. (1993b) Bioakustik der Rufgemeinschaften eines malaysischen Tieflandregenwaldes. *Verhandlungen der deutschen Zoologischen Gesellschaft*, **86**(1), 269.

Riede, K. (1996) Diversity of sound-producing insects of a Bornean lowland rain forest, in *Tropical Rainforest Research – Current Issues* (eds D.S. Edwards *et al.*), Kluwer Academic Publishers, Dordrecht, pp. 74–84.

Riede, K. and Kroker, A. (1995) Bioacoustics and niche differentiation in two cicada species from Bornean lowland forest. *Zoologischer Anzeiger*, **234**, 43–51.

Römer, H. (1993) Environmental and biological constraints for the evolution of long-range signalling and hearing in acoustic insects. *Transactions of the Royal Society of London, Series B*, **226**, 179–85.

Stork, N.E. (1991) The composition of the arthropod fauna of Bornean lowland rain forest trees. *Journal of Tropical Ecology*, **7**, 161–80.

Young, A.M. (1981) Temporal selection for communicatory optimization: the dawn–dusk chorus as an adaptation in tropical cicadas. *The American Naturalist*, **117**, 826–9.

22

Spatial and temporal use of microhabitats as a key strategy for the colonization of tree bark by *Entomobrya nivalis* L. (Collembola: Entomobryidae)

A. Prinzing

ABSTRACT

Investigations were carried out on the microhabitat utilization of tree bark by *Entomobrya nivalis* (Collembola: Entomobryidae). Population densities of *E. nivalis* were recorded from different microhabitats on the trunks of solitary oak, beech, lime and ash trees in northern Germany. The principal microhabitat types utilized by *E. nivalis* were horizontal and vertical bark crevices, a fruticose (= shrub-like) and a crust-like lichen species. The fruticose lichen species was found to be permanently used, while the crust-like lichen species and the horizontal bark crevices were densely populated under certain climatic conditions. Vertical crevices were particularly heavily used by juveniles. This complementary 'part-time use' of microhabitats was explained by food availability as well as by protection from very wet conditions offered by the crust-like lichen and wind- and sun-shelters in crevices. Preferred lichens were often scarce, isolated and small in size, but were still heavily populated.

A mosaic pattern of microhabitats enables *E. nivalis* to colonize bark despite the extreme climatic conditions that occur on exposed tree trunks throughout the year. This permitted: (i) the optimal use of the complete trunk when energetically favourable; (ii) the immigration of middle-aged individuals from adjacent population pools; and (iii) the coexistence of different age-classes. Minute size and eurytopy are considered to be prerequisites for the differentiated, dynamic and flexible use of small-sized microhabitats.

Canopy Arthropods. Edited by N.E. Stork, J. Adis and R.K. Didham. Published in 1997 by Chapman & Hall, London. ISBN 0 412 74900 9

INTRODUCTION

The edges of the canopy layer are extreme habitats. Abiotically, they are exposed to strong and changeable wind, sun and rain. Underneath the crown such extreme conditions can be found on the exposed upper parts of trunks (Braun, 1992). Moreover, lower trunk regions of isolated trees are also exposed to extreme conditions. This can be recognized by the absence of mosses at heights of only 0.5 m and the impact of wind on neighbouring shrubs and crowns (Nogushi, 1979). Biotically, bark crevices, algae and lichens are often the only microhabitats for arthropods in lower crown-regions and at trunks (at least in temperate latitudes). Each of these cryptogam species can occur as small, isolated patches of some square millimetres or may continuously cover several square metres of bark.

Generally, strategies for arthropods to cope with harsh conditions and patchy resources include: large size, strong sclerotization, shells, ability to dig into the substrate, large-scale wandering or dormancy interrupted by rapid development (Tischler, 1990). Microarthropods like Collembola lack such strategies and are therefore normally restricted to a sheltered, detritivorous life in crevices or below plant cushions (Travé, 1963; Joger, 1988; Bauer, 1993), except after precipitation (Hale, 1972; T. Bauer, 1979; R. Bauer, 1993).

The exposed and patchy lichens and algae on tree trunks, however, are mainly grazed on by Collembola, oribatid mites and small Psocoptera (Travé, 1963; André, 1975, 1979; Gjelstrup, 1979; Nicolai, 1985, 1986; Büchs, 1988, 1990; Prinzing and Wirtz, 1997, Chapter 23, this volume). None of these organisms is restricted to such extremely exposed, solitary tree trunks (Pschorn-Walcher and Gunhold, 1957; Travé, 1963). These species also occur under more sheltered conditions on the shaded base of tree trunks and in the forest litter-zone (Gisin, 1960; Günther, 1974; Gjelstrup, 1979; Weigmann and Kratz, 1981; Woltemade, 1982; Nicolai, 1985; Büchs, 1988).

How then can such sensitive and eurytopic (occurring in a wide range of habitats) microarthropod species utilize the exposed bark of tree trunks and canopy layers?

Microarthropods might respond to this problem by using *micro*habitats with exceptionally favourable conditions. To test this, information was needed on whether and how: (i) the type of microhabitat (cryptogam species or type of bark crevice) influences densities and phenologies of arthropod species; (ii) microhabitat use interacts with short-term climatic conditions and seasons; and (iii) how the use of the complete tree trunk 'macrohabitat' depends on climate, as reported by Bowden *et al.* (1976) and Bauer (1979).

In general, the spatiotemporal use of the trunk 'macrohabitat' can be restricted to certain ontogenetic stages, since many species need to transverse the complete trunk-layer in order to reach the food sources or hibernation habitats of the adult stage (Winter, 1972; Schauermann, 1973; Funke, 1979; Albert, 1982; von Allmen and Zettel, 1982; Nicolai, 1985, 1986; Büchs, 1988, 1990). Also, short-term migrations of Collembola from the surrounding soil lasting for only several hours were observed by Bowden *et al.* (1976) and Bauer (1979). They demonstrated strong increases in Collembola densities on sheltered trunks at night and after rain, by using light-, sticky- and pitfall-traps or by direct observation. Bauer (1979) was also able to enhance this vertical migration in the laboratory by using high air humidities. Under these conditions trunks were a suitable feeding habitat for Collembola and their risk of drowning was lower than in the soil.

Observed vertical and horizontal zonations of microarthropod distribution on tree trunks (Duffey, 1969; Niedballa, 1969; Gjelstrup, 1979; Woltemade, 1982; Nicolai, 1985, 1986; Büchs, 1988; Stubbs, 1989; Braun, 1992; Manhart, 1994) are correlated with climatic conditions. These are accompanied by changes in epiphyte cover, biomass and species composition (Rose, 1974 and authors mentioned above), in moss/lichen zonations (Travé, 1963; Gjelstrup, 1979; Woltemade, 1982; Manhart, 1994) and in lichen growth forms (André, 1979, 1983; Woltemade, 1982). The separate effect of such epiphytes has been demonstrated only twice. André (1979) found that bark covered with thalli of two selected, similar crust-like lichen species differed strongly in their oribatid mite faunas. Also, differentiation of living conditions was found within single thalli of a fruticose lichen species (Prinzing and Wirtz, 1997, Chapter 23, this volume).

No other investigations of the distribution of corticolous microarthropod populations have taken into consideration all common microhabitats and the different microclimatic zones on tree trunks. In this paper such results are presented for *Entomobrya nivalis* L. (Collembola: Entomobryidae), one of the most common Collembola species on trunks in northern Germany. This species often colonizes tree trunks despite its preference for a relative humidity above 80% (von Allmen and Zettel, 1982; André, 1983; Büchs, 1988; Müller-Kraenner, 1990; Prinzing and Wirtz, 1997, Chapter 23, this volume).

MATERIALS AND METHODS

Study sites

Research was conducted from August 1993 until April 1994 in five areas near Kiel (northern Germany, hilly landscape, oceanic–temperate climate, less than 50 m above sea level). Additionally, in August and September 1993 an area near Rostock was considered (flat and climatically more continental, north-east Germany). Trunks of large solitary trees (>2.5 m circumference at a height of 1.5 m) were searched for individuals of *E. nivalis* at heights between 0.8 and 1.8 m. Trees of *Quercus robur* L. (oak), *Fagus sylvatica* L. (beech), *Fraxinus excelsior* L. (ash) and, except in one area, *Tilia sp.* (mainly *T. platyphyllos* Scop.) (lime) were examined. Within each study area, trees were selected in order to minimize distances between the different tree species. Solitary trees are a common feature in these rural areas, mainly along paths and avenues and in meadows. Solitary beech trees mostly grow in small park-like forests or in clearings.

The common types of tree bark microhabitats investigated were:

Crevices

1. vertical crevices between bark scales
2. horizontal crevices, always starting at the bark surface inwardly

Epiphytes

3. Fruticose lichen species *Evernia prunastri* L.

4.–7. Crust-like lichen species *Lepraria incana* L., *Pertusaria albescens* Choisy and Werner, *Lecanora expallens* ACH., *L. conizaeoides* Nyl. ex Crombie,

8. Algae (*Pleurococcus* sp.),
9. Bark surface without epiphytes

Of these, only vertical crevices were absent on beech trunks. In addition, mosses and foliose lichens were found, but they never covered more than 5% of the bark surface – even when only a narrow zone of equal momentary exposures to wind, sun and precipitation was considered. On ash trees the lichens *Buellia punctata* Massal. and *B. griseovirens* Almb. sometimes covered up to 20% of the respective zone of exposure.

From August 1993 to January 1994, 45–50 trees were sampled per month (30% of them at night), in February and March 1994, 28 trees each, and in April, eight trees. Daytime visits started at 08:00 h or, in winter, from dawn and lasted until at least 2 hours before sunset, except when temperatures remained below –5°C (trunks were then sparsely populated). Night visits began at dusk and lasted for 6–9 hours. Different study areas and tree species were sampled in a random order. Nine trees were resampled, but only after a period of at least 3 months.

Sampling

Each trunk was divided, by eye, into faces according to their exposure to wind, sun and/or precipitation (= zones covered with a waterfilm) at the moment of investigation. As an example, during windy, sunny weather, mostly four trunk-faces could be differentiated: those exposed to wind and sun, only to wind, only to sun or to none of them. Each trunk-face was subdivided again according to different momentary exposures within its microrelief (due to positions on bark-ridges and in bark-valleys or in front of or behind fruticose thalli). 'Zones of exposure' were defined correspondingly by the combination of exposures of the respective trunk-face and microrelief-zone.

Within each zone, temperature (Gultan Tastotherm D 700), air velocity (thermic anemometer 641 N, Lambrecht Meßgeräte) and air humidity were measured (Valvo-hygrometer, sensor with detached cap and ventilator; after February 1994: Rotronic hygrometer A1), although the Valvo-sensor was mostly too large to measure within the 'valleys' of the bark-relief. All measurements were taken at a height approximately 2 mm above the bark. For each trunk-face five locations were tested for temperature, four for wind and one for air humidity measurements.

Entomobrya nivalis densities within different microhabitats were measured by direct counts of individuals with a hand-lens. For each single field of view the trunk was approached with careful movements and viewed from a distance of just 3 cm (× 10 magnification). In each zone 10 randomly distributed direct counts were made for each microhabitat type present. Thus, individual plots of 18 cm^2 of microhabitat surface were examined.

Inner surface examinations of cavities during the hand-lens-searches were as follows: (a) horizontal crevices were searched in a length that, multiplied by the approximate depth of the crevices, equalled approximately 18 cm^2 (this length was measured in field-of-view diameters); (b) thalli of the fruticose lichen, *Evernia prunastri*, were opened in several layers with a large needle; and (c) vertical crevices were broken up with a knife. The dark cavities of horizontal crevices and thalli of *Evernia prunastri* were illuminated with a small pen-light torch. At night, illumination was necessary for all types of microhabitats and was restricted to a single field of view for less than about 2 seconds at a time. Horizontal crevices were sometimes difficult to survey by eye. They were then additionally examined by scraping with a large needle. Investigation plots of crust-like thalli were completely scraped off on at least one in 10 trunks until mid-September. Later, this was done for all crust-like lichen plots.

Individuals of *Entomobrya nivalis* were mainly determined in the field. Animals were recorded as 'juvenile', 'middle-aged' or 'adult' according to the approximate lengths (<0.8 mm/0.8 – 1.7 mm/>1.7 mm). This field-classification was based on knowledge of the complete range of body-sizes of this species from laboratory-rearing. To ensure sufficient sample sizes, counts of *E. nivalis* were pooled across five subsequently investigated colonised trees. Relative proportions of the three age-classes were calculated for each sample.

In each exposure zone the amount of each of the epiphyte species was estimated according to their cover: <20%, >20 – <50%, >50%.

Statistical analysis

Tests were conducted on changes in age-class-compositions in different microhabitat types (χ^2-test) and on changes in relative densities of animals in a certain microhabitat when comparing different climatic environments (χ^2-tests of goodness of fit, see below). The non-parametric Kolmogorov–Smirnov test was applied when expected frequencies were very small (Lamprecht, 1992). Samples were pooled within each category of the independent variable (Fowler and Cohen, 1986; Mühlenberg, 1989).

The relative density on a given microhabitat '1', RD_1, expresses the ratio between the absolute density on this microhabitat, D_1, and the sum of absolute densities on all nine investigated microhabitats (indicated '1'–'9'). Thus, the relative density indicates the relative importance/suitability of '1' for the animals compared with the other microhabitats:

$$RD_1 = \frac{D_1}{\sum_{i=1}^{n} D_i} \qquad (n = 9) \tag{22.1}$$

If the microhabitat '1' was of equal relative importance/suitability during two contrasting climates I and II, the expected relative densities would be equal, to:

$$RD_{e1.I} = RD_{e1,II} \text{ (null hypothesis for the microhabitat '1')}$$

or:

$$\frac{D_{e1,I}}{\sum_{i=1}^{n} D_{i,I}} = \frac{D_{e1,II}}{\sum_{i=1}^{n} D_{i,II}} \tag{22.2}$$

Since the density in a microhabitat is calulated as the frequency F divided by the number of plots, in which this microhabitat was found and investigated (sampling effort P), Equation (22.2) can be rewritten as:

$$\frac{F_{e1,I} : P_{1,I}}{\sum_{i=1}^{n} (F_{i,I} : P_{i,I})} = \frac{F_{e1,II} : P_{1,II}}{\sum_{i=1}^{n} (F_{i,II} : P_{i,II})} \tag{22.3}$$

This equation allows the calculation of the null expected frequencies on '1' for the climates I and II, standardized for the differences in sampling efforts. Such standardized null expected frequencies were required for the subsequent χ^2 test of goodness. Four steps were required (E. Glück, H.-J. Krambeck and J. Schimmler, personal communication):

1. Rearrange Equation (22.3) to:

$$\frac{F_{e1,I} : P_{1,I}}{F_{e1,II} : P_{1,II}} = \frac{\sum_{i=1}^{n} (F_{ei,I} : P_{i,I})}{\sum_{i=1}^{n} (F_{ei,II} : P_{i,II})} = \frac{\sum_{i=1}^{n} D_{i,I}}{\sum_{i=1}^{n} D_{i,II}} = Z \qquad (22.4)$$

2. Replace the expected frequency for '1' during one of the climates by the difference between the other expected frequency and the sum of the two observed frequencies on '1'. This can be done, because the sum of expected frequencies must equal the sum of observed frequencies on a microhabitat:

$$\frac{F_{e1,I} : P_{1,I}}{((F_{1,I} + F_{1,II}) - F_{e1,I}) : P_{1,II}} = Z \qquad Z = \frac{\sum_{i=1}^{n} D_{i,I}}{\sum_{i=1}^{n} D_{i,II}} \qquad (22.5)$$

3. Solve Equation (22.5) for $F_{e1,I}$:

$$F_{e1,I} = \frac{Z^{\times} (F_{1,I} + F_{1,II})}{(Z + (P_{1,II} : P_{1,I})}$$

4. Calculate the second expected frequency as:

$$F_{e1,II} = (F_{1,I} + F_{1,II}) - F_{e1,I}$$

Observed frequencies were tested against the expected frequencies with a χ^2-test of goodness of fit applying Yates' correction (Welkowitz *et al.*, 1982).

Comparisons of frequencies during day and night were calculated only on counts for months and areas when night visits were also conducted; data for comparison of weather conditions are from day visits only.

Discriminant analysis was performed using the Systat statistical package (1992).

RESULTS

Abundance

The trunks were found to be a suitable macrohabitat for *E. nivalis* throughout the complete period of investigation, but with variable utilization of microhabitats. Densities fluctuated strongly, especially within thalli of *Evernia prunastri* (Figure 22.1). Fluctuations in overall densities on tree trunks was correlated with a higher trunk boundary layer temperature compared to the surrounding air. The only exception was in February when temperatures were extremely low (Figure 22.1). *E. nivalis* was also found on trunks during the rest of the year.

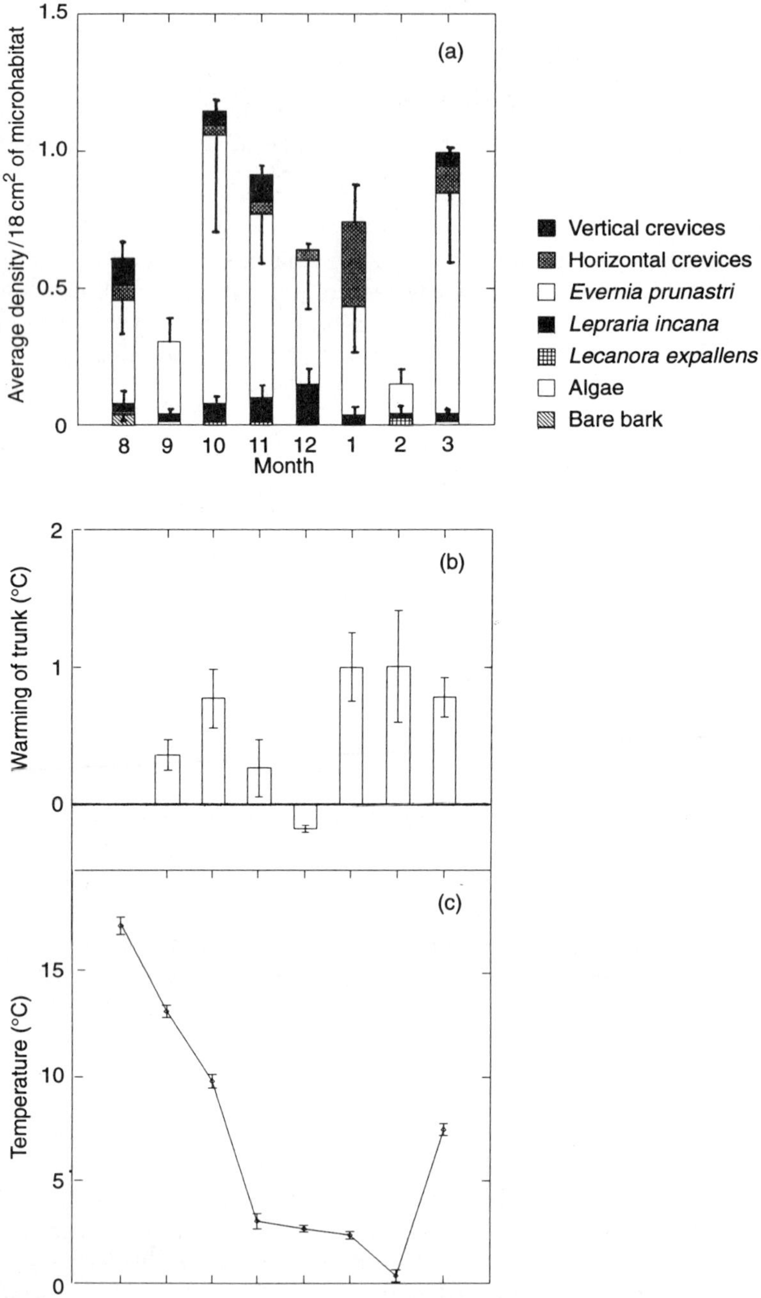

Figure 22.1 (a) Average population densities (± S.E.) of *Entomobrya nivalis* in different types of microhabitats during the investigation period. The area near Rostock (investigated only until September) is not included. (b) Average (± S.E.) differences between temperature of the boundary layer and the free surrounding atmosphere. (c) Average (± S.E.) temperature of the boundary layer.

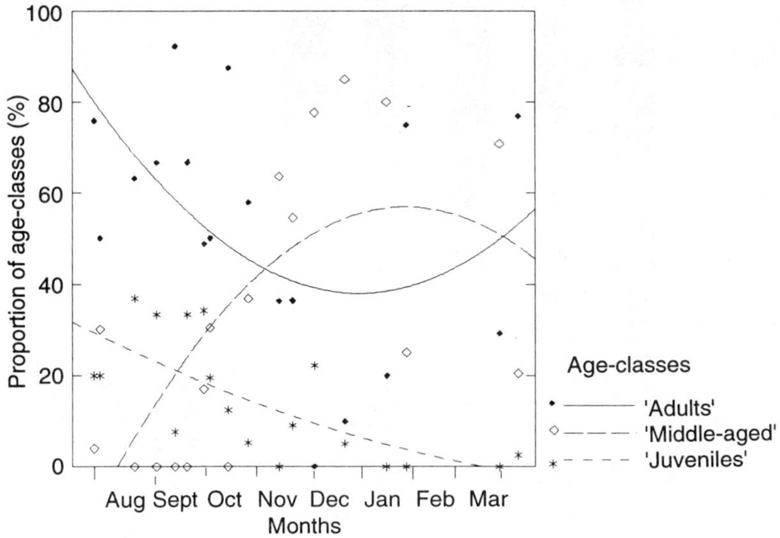

Figure 22.2 Phenology of *E. nivalis* during months of investigation (x-axis), plotted as percentages of different age-classes of *E. nivalis*, curves fitted by quadratic procedure. n = 25, 10, 19, 6, 13, 15, 41, 36, 8, 38, 33, 11, 9, 20, 20, 8, 24, 39 animals per sample, respectively.

Phenology and spatiotemporal distribution of age-classes

All age-classes of *E. nivalis* were found on trunks in high numbers and often simultaneously. Life-cycles seemed to be univoltine with hatching mainly in early summer (Figure 22.2). The simultaneously occurring age-classes were segregated in their temporal use of the tree trunks. The abundance of intermediate-aged animals increased significantly during rainy weather, whereas adults were slightly more abundant during sunny weather with <70% cover by clouds (χ^2 = 15.6 and 5.76, P <0.001 and <0.1, n = 77 and 129 respectively, d.f. = 2 each). Age-classes also varied in their microhabitat use (for all age classes: χ^2 = 46.11, 8 d.f., P <0.001, n = 331, Figure 22.3). Frequencies of juvenile animals were proportionally greater on *L. incana* and in vertical crevices (χ^2 = 7.74 and 27.04, P <0.05 and <0.001, n = 42 and 10, respectively, 2 d.f. each). In the latter, animals of intermediate age were completely absent. Adults were very uniformly distributed.

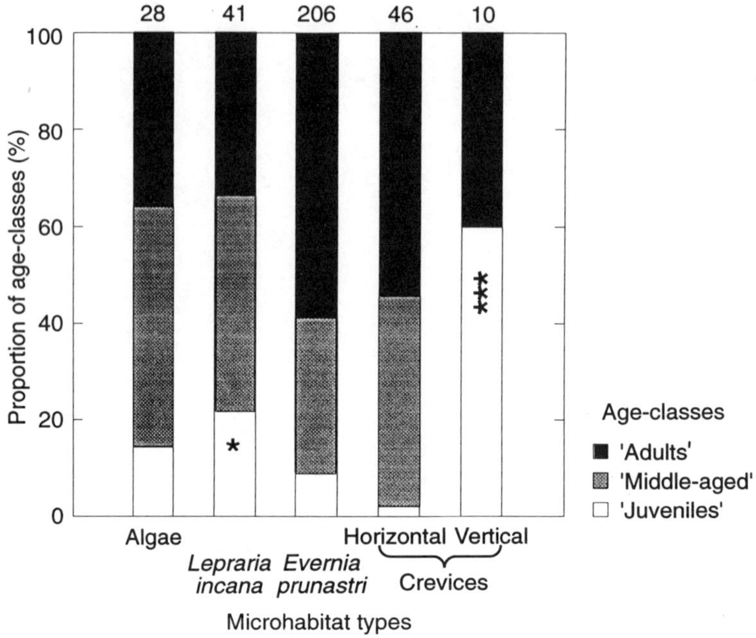

Figure 22.3 Percentage of age-classes in different microhabitats. Sum of animals per habitat-type given on top of each bar. Significantly higher proportional frequencies are stated as * or *** ($P < 0.05$ or $P < 0.001$ respectively, χ^2-test, 2 d.f.). Bare bark and *Lecanora* sp. have been omitted because of low numbers of individuals.

Spatial use of microhabitats

Out of nine microhabitat types, four were regularly used by *E. nivalis* (Figure 22.4). Densities were highest in *Evernia prunastri*, followed by the crust-like lichen *Lepraria incana*, horizontal crevices and vertical crevices. Microhabitat use was differentiated according to (i) the basic fruticose or crust-like growth-forms, and (ii) micromorphological characteristics. The avoidance of all crust-like lichens, except *L. incana*, is particularly notable since they are structurally very similar. Bare bark and crusts of algae were also avoided, although the latter was the most common type of microhabitat.

Effect of trunk-climate on temporal microhabitat use

Univariate analysis

Frequencies on *L. incana*, in *Evernia prunastri* and in horizontal crevices were sufficiently high for statistical analysis of their respective temporal

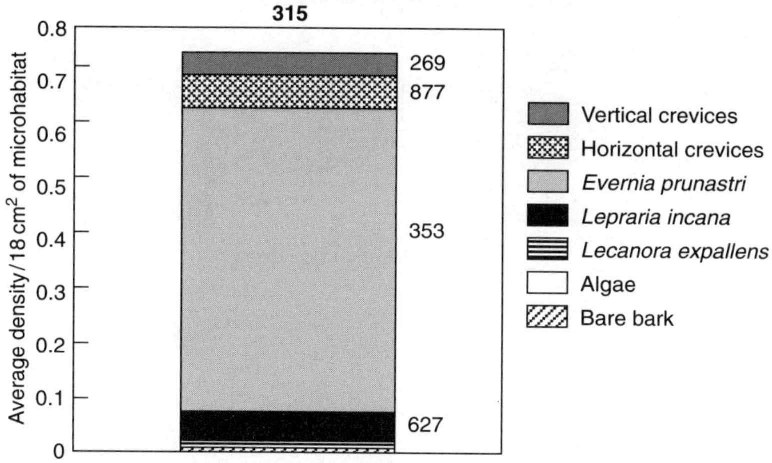

Figure 22.4 Average population densities of *Entomobrya nivalis* on different types of microhabitats. The total number of animals is given in boldface on top of the graph, and numbers of investigated plots of microhabitat types relevant to *E. nivalis* are given to the right of the respective bars. For other microhabitat types the following numbers of plots were investigated: algae, 1060; bare bark, 144; *Lecanora expallens*, 395; *L. conizaeoides*, 42; *Pertusaria albescens*, 138. Note the latter two were not populated at all.

use by *E. nivalis*. Under all climatic conditions considered *Evernia prunastri* was colonized by the largest proportions of *E. nivalis*. On the other hand, *L. incana* and horizontal crevices were used more under certain climatic conditions (Figure 22.5).

L. incana was preferred on trunk zones sheltered from wind (Figure 22.5(c)), from radiation (Figure 22.5(a), weather, daytime), from dry or medium humid air or from very high absolute temperatures (Figure 22.5(a)). Furthermore, shelter from high relative temperatures seemed to be required for the use of *L. incana* during strong winds (high convective desiccation), during the day (sun) and above 7°C (Figure 22.5(d)).

Horizontal crevices were preferred in the absence of a waterfilm (Figure 22.5(b)) and under conditions of very low *absolute* temperatures (Figure 22.5(a)). Furthermore, relatively low temperatures seemed to promote horizontal crevice use below 7°C or during low wind turbulence (high temperature gradient) (Figure 22.5(d)). High correlation of the use of *L. incana* and horizontal crevices with relative temperatures indicates an immense dynamic of *E. nivalis'* distribution, since the corresponding climatic zonation of a trunk quickly changes throughout each day (Nicolai, 1985). On sunny, calm days the respective temperature

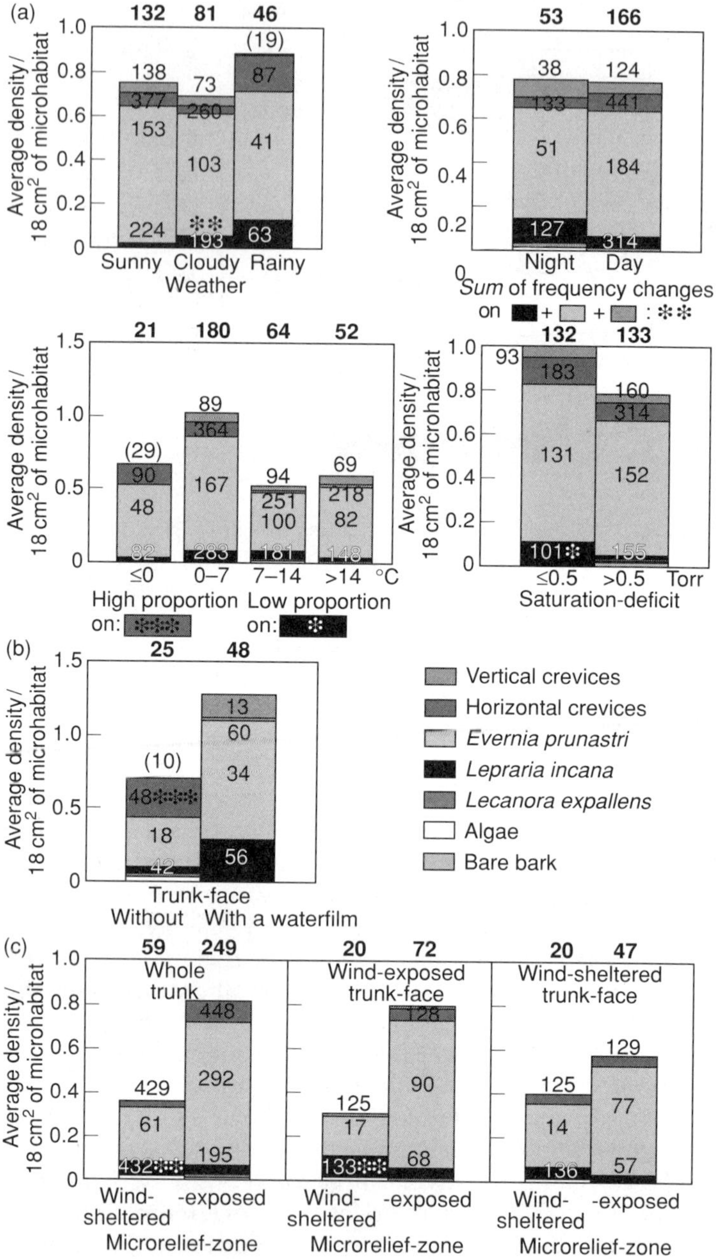

Figure 22.5 Changes in *proportions* of average densities of *Entomobyra nivalis* on microhabitat types under different climatic conditions: (a) absolute conditions; (b–d) zonations of waterfilm, wind and temperature. Significantly higher proportional frequencies on *L. incana*, in *Evernia prunastri* or in horizontal crevices compared with the neighbouring column are indicated by *

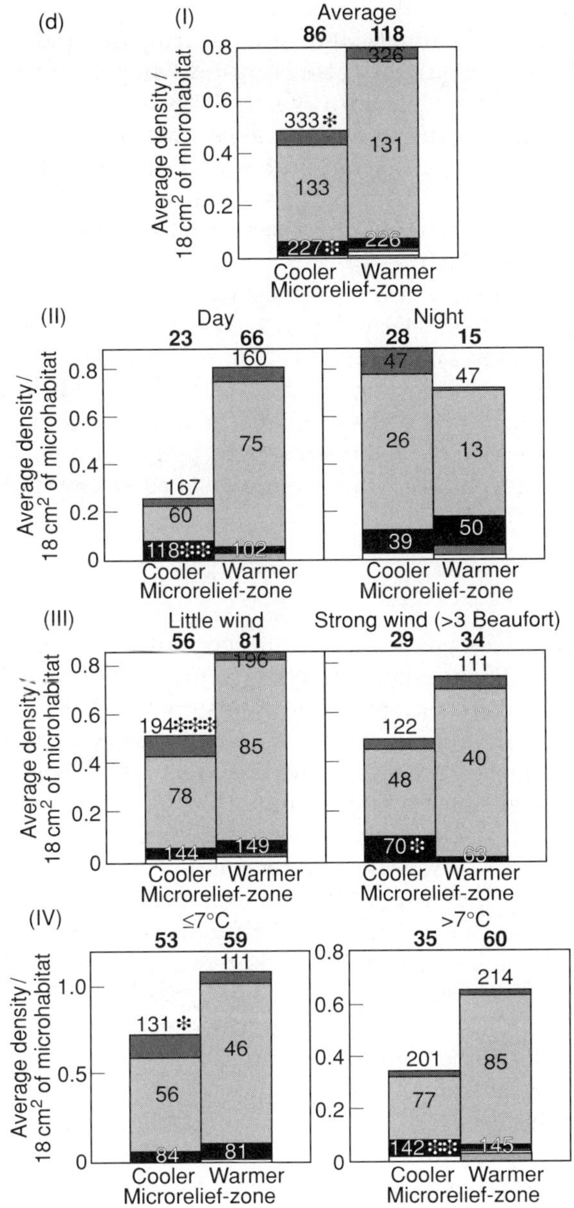

and ** (P <0.05 and <0.001 respectively, 1 d.f. χ^2-test of goodness, except for *L. incana* in Figure 22.5(d) II-IV, where the Kolmogorov-Smirnov test was applied. 'Cloudy' indicates >70% of sky covered by clouds. For details of sample sizes, see Figure 22.4

gradients could reach 20° C and 8°C respectively between opposed trunk faces and microrelief-zones.

In vertical crevices, animals were found during dry and warm weather conditions (Figure 22.5(a)), but also occurred while the bark was covered with a waterfilm (Figure 22.5(b)). Vertical crevices cannot be differentiated according to the microrelief because they are restricted to layers below bark-ridges. Algae seemed to be used only when sheltered from a waterfilm (Figure 22.5(b)) and from wind by trunk-face and microrelief (Figure 22.5(c)). At night the cooler zones of the microrelief were also used (Figure 22.5(d)).

No significant changes in frequencies were found when solely considering the effect of trunk-face climates. The statistical effects of several independent climatic parameters on *L. incana* use was similar: (i) wind-exposure on the scale of trunk-faces enhanced the effect of wind-exposure within the microrelief; (ii) high absolute temperatures enhanced the effect of relatively high temperatures; and (iii) the day/night changes showed the same effects as sunny/cloudy weather changes during the day.

Multivariate analysis

Microhabitat use was *predominantly* influenced by wind exposure within the microrelief (discriminant analysis factor 1 in Table 22.1) and by temperature and distribution of waterfilms (factor 2). These factors have high canonical correlation scores, indicating high explanatory value. The first factor segregated crust-like microhabitats from those with cavities (Figure 22.6, except of the single animal on *L. expallens* at 0.9,1.7). The latter were further divided by the second factor. Crusts of *L. incana* and algae were, therefore, only used under wind-sheltered conditions and at low or medium temperatures. Horizontal crevices were populated under cold, often wind-exposed conditions without waterfilm. In contrast, use of vertical crevices seemed to be positively correlated with heat and a waterfilm. Use of *Evernia prunastri* was not greatly dependent on any of the factors, as indicated by its position close to (0,0), although usage increased slightly under wind exposure.

The secondary effect of daytime and weather conditions on microhabitat use in multivariate analysis was probably due to the low number of visits at night or during sunny weather. The relative temperature within the microrelief only came into play in correlation with such day/night changes. This confirmed the strong interaction between these factors found in the univariate analysis. Measurements of saturation deficits were not included in the analysis because they were mostly only available for the exposed zones of the bark relief where the sensor could be fixed.

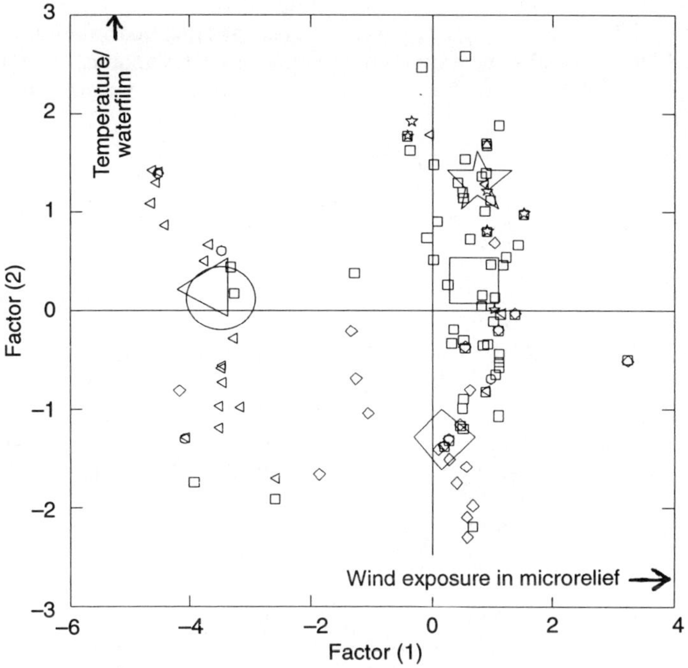

Figure 22.6 Use of different types of microhabitats according to the climatic variables summarized in the first and second canonical factors of discriminant analysis in Table 22.1. Small symbols indicate animals on algae (○), on *L. incana* (◁), in *Evernia prunastri* (□), in horizontal (◇) or in vertical crevices (☆). Large symbols indicate respective centroids.

The important first to third canonical factors were each strongly correlated with only one or two different microclimatic variables. This indicates the existence of clear-cut, well-separated climatic influences consisting of only one or a few parameters.

As discussed above, population densities on neighbouring *L. incana* and horizontal crevices were strongly correlated ($r_s = -0.897$, $P = 0.011$, $n = 17$ zones of exposure colonized by *E. nivalis*), but did not interact significantly with the use of *Evernia prunastri*.

Effect of patch size on microhabitat use

The preferred epiphytes *L. incana* and *Evernia prunastri* were highly patchily distributed and often of very low abundance. This did not influence the absolute or relative frequencies of *E. nivalis* on the respective thalli (Figure 22.7).

Table 22.1 Prevailing factors for microhabitat use according to discriminant analysis presented as significant and independent canonical factors. Numbers indicate correlations with original variables and canonical correlation (last row). The latter indicates the explanatory value of each canonical factor as a degree of its correlation with the original data. Bold numbers indicate correlations of > |0.33|

	Factors				
Original variables	*1*	*2*	*3*	*4*	*5*
a) tamabs	–0.02	**0.41**	–0.04	0.30	0.06
b) $dtemp_{tf}$	–0.06	–0.10	–0.20	–0.14	0.01
c) $dtemp_{mr}$	0.02	0.22	–0.10	**0.44**	0.22
d) I $dtemp_{tf}$I	0.14	0.28	–0.15	**–0.40**	–0.31
e) I $dtemp_{mr}$ I	0.05	0.08	0.07	**–0.34**	–0.32
f) $dtemp_{atm}$	0.09	0.32	–0.27	–0.12	**–0.41**
g) windsp	–0.03	–0.32	0.27	–0.10	0.33
h) $wind_{tf}$	–0.04	0.09	0.06	0.16	0.27
i) $wind_{mr}$	**0.77**	0.07	0.30	0.03	–0.16
j) sun_{tf}	0.02	–0.06	–0.21	–0.20	**0.42**
k sun_{mr}	0.05	0.11	0.05	0.21	0.08
l) wf_{tf}	–0.18	**0.40**	**0.55**	–0.12	–0.14
m) nd	0.20	–0.16	0.11	**0.39**	–0.26
n) wea	–0.20	0.26	0.03	–0.22	**0.40**
Canonical correlations:	0.81	0.51	0.31	0.25	0.22

a), absolute temperature; b)/c), difference in temperature between opposing trunk-faces/microrelief-zones; d)/e), strength of temperature gradient (= 0 or 1 when |b)|/|c)| < or >1); f), warming of boundary layer compared with surrounding atmosphere; g), wind speed (Beaufort scale); h)/i), wind-exposure of the trunk-face/microrelief-zone (+1/–1); j)/k), sun-exposure of the trunk-face/microrelief zone; l), trunk-face without/with waterfilm during precipitation (+1/–1); m), night/day (0/1); n), sunny/cloudy/rainy weather (1/2/3)

Effect of optical-searching method

Animals were not significantly chased away from exposed crusts into hide-outs by searching with a hand-lens. Altogether 36 animals were found on surfaces of crusts with the hand-lens (with 18-cm^2 plots). On plots where no animals were found in this way, the previous counts with a distance-lens (from a distance of 40 cm) on much larger plots of 31 cm^2 revealed only 11 additional animals. Moreover, animals that were spotted with the distance-lens were often 'tested' by approaching them with a hand-lens. This did not induce 'flight' behaviour.

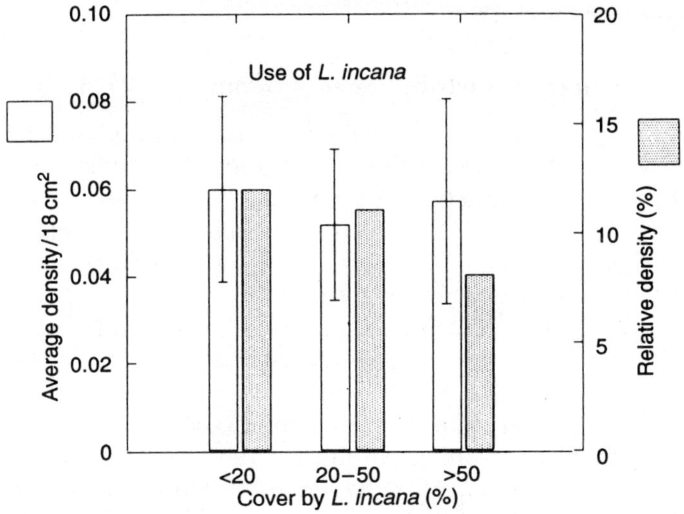

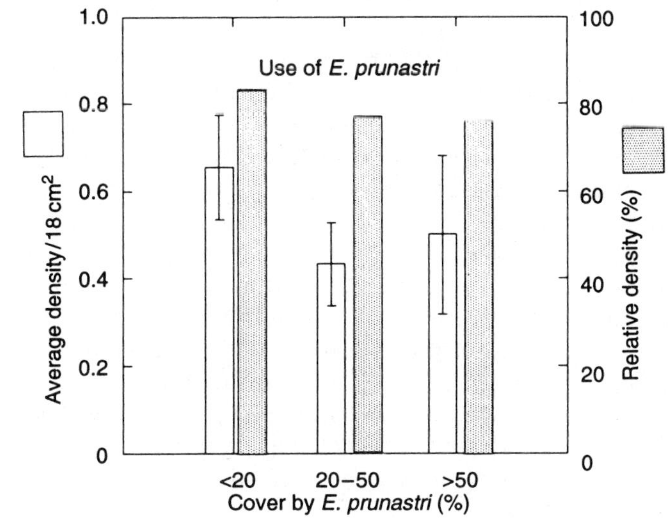

Figure 22.7 Densities of *Entomobyra nivalis* on thalli of *L. incana* and of *Evernia prunastri* covering small, medium or large parts of the trunk's respective zone of exposure. Absolute densities (± S.E.), relative density = (average density of *E. nivalis* on the respective lichen species) : (sum of average densities on all microhabitat types). Habitat use does not switch significantly with changing densities of the epiphyte.

DISCUSSION

Causes and functions of distributional patterns

Exposed trunks appear to be a habitat for *E. nivalis* throughout the year. Colonization of tree trunks may confer an energetic benefit for the animal as these vertical, dark structures can act as 'solar-panels', especially during winter (Nicolai, 1985). Indeed, densities on all microhabitats combined correlated strongly with the warming of the bark boundary layer, except in February when average temperatures were near 0°C. Such 'thermophilic' behaviour of colonization corresponds well to the preference of *E. nivalis* for high temperatures in the laboratory (17–27°C, depending on growth stage; Müller-Kraenner, 1990).

Alternatively, high air humidity and night conditions have been suggested as possible prevailing stimuli for the colonization of trunks by Collembola (Bowden *et al.*, 1976; Bauer, 1979). However, these factors appear to rarely correlate with the density of *E. nivalis* (Figure 22.5(a)). Furthermore, at night and at high air humidity the boundary layers of trunks have the same or even lower temperatures than the surrounding atmosphere (Nicolai, 1985; A. Prinzing, unpublished data). Night conditions and air humidity, therefore, do not explain the strong month-to-month fluctuations in abundance.

On solitary, exposed trunks this beneficial warming of the boundary layer is strongest due to the direct exposure to the sun (Nicolai, 1985). This might compensate for the higher risk of desiccation as well as drowning compared with much more sheltered trunks in forests. In fact, *E. nivalis* was found on the solitary trunks about as often as it was to be found on sheltered trees (von Allmen and Zettel, 1982; André, 1983; Büchs, 1988; Müller-Kraenner, 1990).

Tree trunks are utilized throughout the year, while positions within the mosaic of microhabitats changed. Where this microhabitat mosaic is absent, tree trunks are only colonized temporally after rain or at night: on the algae-covered trunks of ashes and legs of traps investigated by Bowden *et al.* (1976) as well as the smooth algae-covered beeches in Bauer (1979). Hence, trunks are energetically favourable macrohabitats throughout the year, as long as a variety of microhabitats is available.

In this study population density collapsed several times, followed by rapid recolonization. Recolonization in autumn and winter cannot be caused by new generations of individuals because development is univoltine. Instead, only the immigration of animals from the soil or from upper parts of the tree can account for population recovery. Such immigration would also explain the short-term increase in the density of adults during sunny weather and, to a greater extent, those of middle-aged individuals in rainy weather. Immigration of middle-aged individuals also explains the increase in total population density of *E. nivalis* under rainy as compared with cloudy weather. Therefore, middle-aged animals are capable of connecting corticolous populations to those in other habitats and of colonizing depleted trunk microhabitats.

The complementary increase in middle-aged and adult stages during rainy and sunny weather, respectively, corresponds well with the comparatively lower temperature- and higher air-humidity preferences of middle-aged animals in the laboratory (Müller-Kraenner, 1990). Juveniles, however – being much more sensitive to environmental conditions – tend to be constant in their use of the trunk. For juveniles, migration may be energetically too costly, or the risk of desiccation or drowning may be too great when crossing unsheltered microhabitats. Similarly, on subalpine spruce trees juvenile animals do not migrate onto trunks above 3–4 m (von Allmen and Zettel, 1982). Besides the preferences in the laboratory, the microhabitat use of middle-aged animals in the field also correspond to their large-scale migrations: these animals use horizontal crevices comparatively more than did other stages. Since these crevices are extremely common and interconnected to each other, this stage is able to rely on the dense network of sheltered microhabitats during migration.

All age-classes were found in large numbers in this study and often coexist on the same trunk, despite their very different sensitivities (Müller-Kraenner, 1990). Among these age-classes, middle-aged and adult animals segregate temporally on a macrohabitat scale, as discussed above. On the other hand, segregation of juveniles was detected on a microhabitat scale: the complete protection from sunny and windy weather conditions provided by vertical crevices is probably most important for juveniles, because juveniles (1) have the highest humidity- and lowest temperature-preferences (Müller-Kraenner, 1990), and (2) are not capable of compensatory large-scale migrations to adjacent macrohabitats as are other stages. Only middle-aged animals are found to be absent from vertical crevices, perhaps because these crevices are of no value during the preferred rainy weather conditions. High proportions of juveniles also are found on *L. incana*, followed by middle-aged animals. This corresponds to the high palatability of such extraordinarily soft thalli for animals with mouthparts as minute as those in young Collembola. These animals are often observed to graze upon *L. incana*.

Such spatiotemporal patterns permit a physiologically optimized use of microhabitats and avoid competition between age classes. They may also spread the risks of this unpredictable climatic environment among animals of very different physiological sensitivities and spatiotemporal habitat use. An alternative phenological strategy to cope with such extreme and unpredictable climates would be a very rapid life-cycle (Tischler, 1990).

The fruticose lichen *Evernia prunastri* is used ubiquitously and it seems that *E. nivalis* always find sufficient climatic shelter as well as food here. Within the thalli suitable shelter and food are always restricted to certain regions and these shift according to the climate, leading to microscale migrations of *E. nivalis* within thalli (Prinzing and Wirtz, 1997, Chapter 23, this volume). In contrast, microhabitats other than *Evernia prunastri* are used more under certain climatic conditions. This might be explained by the physiological suitability of the microhabitats for animals. Indications of such functions can be derived from the temporal use of microhabitats, taking into account: (i) their observed properties; and (ii) the insects' physiological needs (Prinzing and Wirtz, 1997, Chapter 23, this volume). *L. incana* has an unusually hydrophobic surface and this protects insects from drowning in a waterfilm. Indeed, *L. incana* was heavily utilized even on water-covered trunk-faces. In addition, *L. incana* has a soft, palatable thallus, even when dry and may be highly populated for nutritional reasons. *L. incana* is not used when exposed to dry air, direct radiation, high relative or absolute temperatures and convection and may not offer shelter from such risks of desiccation due to its crust-like growth-form.

The highly favourable conditions in *Evernia prunastri* and on *L. incana* probably also reduce the chance of insects switching to other cryptogam species when these lichens are very rare. This contrasts to the classic behaviour of herbivores on phanerogams (Crawley, 1983).

Crusts of *Pertusaria albescens* were not used at all, even though when the cortex layer was absent, they were hydrophobic and soft like *L. incana*. However, *P. albescens* has much higher concentrations of lichen acids in its medulla layer (Culberson, 1969), and these are known to reduce grazing (Stahl, 1904; Lawrey, 1980). Algal crusts only become soft, and thus readily palatable, above about 90% air humidity and during rain these algae quickly cover with a waterfilm. Correspondingly, *E. nivalis* was only found on algae when strongly sheltered from convective desiccation as well as from a waterfilm. If this characterization of algae as a microhabitat is true, Collembola would be forced to abandon tree trunks that are exclusively covered by algae during adverse weather conditions, as found by Bowden *et al.* (1976) and Bauer (1979).

In contrast to *L. incana* and algae, horizontal crevices seemed to offer sufficient shelter from desiccation due to wind and radiation. Correspondingly, in laboratory experiments cavity-microhabitats (dry thalli of *Evernia prunastri*) strongly increase the tolerance of corticolous microarthropods to desiccation as compared to a flat substrate (single branches of *Evernia prunastri*; Prinzing and Wirtz, 1997, Chapter 23, this volume). Horizontal crevices are also used more below 0°C. Here the crevices are free of hoarfrost or snow that otherwise often cover the trunk's surface, making food inaccessible in other microhabitats. These results confirm similar anecdotal winter observations made by Agrell (1941) and von Allmen and Zettel (1982). Horizontal crevices do not offer shelter from a waterfilm and are then avoided. They also offer very few algae or lichens as a food source and are correspondingly avoided when not needed for shelter. Thus the flexible and occasional use of horizontal crevices and *L. incana* enable *E. nivalis* to compensate for the extreme and changeable climates on exposed tree trunks.

Patterns of microhabitat use also may be determined at least in part by factors other than the physiological advantages presented above (e.g. natural enemies, competitors or different birthrates). However, the size of habitat patches of *L. incana* and horizontal crevices are small and their quality for the animals changes quickly compared with the larger range and longer lifespan of the individual animals. Thus, only an immediate response of each single animal to the momentary climatic and nutritional conditions in a habitat patch would result in the observed fine grained patterns of spatiotemporal habitat use (Schindler, 1988). On *Evernia prunastri* the spatiotemporal structure of the environment is different for *E. nivalis*: complete thalli do not change in suitability in the short term and most animals do not seem to leave the thalli when climatic conditions change. This is confirmed by the artificial wetting of bark and *Evernia prunastri* (Prinzing and Wirtz, 1997, Chapter 23, this volume). Animals stay in *Evernia prunastri* thalli but leave the bark crevices and move around for distances of tens of centimetres on the bark-surface. Furthermore, in *Evernia prunastri* the population densities are comparatively high. Under these conditions the effects of competition and natural enemies become more important (Schindler, 1988). *Evernia prunastri* is also preferred by most of *E. nivalis'* potential corticolous competitors (Entomobryidae, Psocoptera and Oribatei), predators (spiders, mainly *Lathys humilis* and *Entelecara penicillata*) and entomophagous fungi (mainly *Trichoderma viride* and *Beauveria bassiana*) (Prinzing and Wirtz, 1997, Chapter 23, this volume; A. Prinzing, unpublished data). In rare cases the thalli of *Evernia prunastri* are also obviously overgrazed (Laundon, 1971; Prinzing, 1996). Hence, long-term changes in *E. nivalis'* population density on *Evernia prunastri* may be due to competition and/or predation effects at the population and community levels. This evidence suggests that the spatiotemporal suitability of the microhabitats strongly influences the type of population regulation acting upon *E. nivalis* and the type of community interactions which subsequently develop.

Conclusions on microhabitat use: prerequisites and significance

Generally, the small patches of microhabitats present offer only enough food and favourable climatic conditions within the sheltered atmospheric boundary layer for minute organisms. Furthermore, only eurytopic species are able to show such a flexible use of microhabitats, including habitat changes on very different spatial scales: between different zones within a thallus of *Evernia prunastri*, between horizontal crevices and *L. incana*, and between the trunk and neighbouring habitats. Therefore, minute size and eurytopy might be prerequisites, rather than disadvantages for coping with extreme trunk environments. These characters permit the colonization of tree trunks through the extremely differentiated, flexible, accurate and adaptable use of microhabitats. Thus, sensitive Collembola can utilize the mosaic pattern of microhabitats on exposed tree trunks despite the generally extreme climate. The importance of this strategy is confirmed by the dominance of eurytopic microarthropods as grazers (Nicolai, 1985; Büchs, 1988; Prinzing and Wirtz, 1997, Chapter 23, this volume). The colonization of the canopy layer by non-flying arthropods might have been ontogenetically or phylogenetically preceeded by such eurytopic, small-scale stages on trunks.

Acknowledgements

'Generation' of this paper was strongly influenced by stimulating discussions with E. Glück, Th. Bauer, J. Adis, G. Weigmann, Y. Ayal, M. Badejo, W. Block, W. Kratz, M. Rosenzweig and the team of the Department of Ecology at the Zoological Institute in Kiel. Further ontogenesis was supported by E. Glück, H, P. Wirtz, R. Didham and N. Stork (critical revision) and C. Moore and V. Draack (correction of my English). Dr Nirenberg determined the fungus-mycelia. Thanks to all of them.

REFERENCES

Agrell, I. (1941) Zur Oekologie der Collembolen. Untersuchungen im Schwedischen Lappland. *Opuscula Entomologica Supplementum*, **3**, 1–236.

Albert, R. (1976) Zusammensetzung und Vertikalverteilung der Spinnenfauna in Buchenwäldern des Solling. *Faunistisch-ökologische Mitteilungen*, **5**, 65–80.

André, H.M. (1975) Observations sur les Acariens corticoles de Belgique. *Fondation Universitaire Luxemburgoise, Série 'notes de recherche'*, **4**, 5–31.

André, H.M. (1979) Notes on the ecology of corticolous epiphyte dwellers: 1. the mite fauna of fruticose lichens. *Recent Advances in Acarology*, **1**, 551–7.

André, H.M. (1983) Notes on the ecology of corticolous epiphyte dwellers: 2. Collembola. *Pedobiologia*, **25**, 271–8.

Bauer, R. (1993) Zur Zönologie und Phänologie granitblockbesiedelnder Collembolen. *Zoologischer Anzeiger*, **230**, 237–48.

Bauer, T. (1979) Die Feuchtigkeit als steuernder Faktor für das Kletterverhalten von Collembolen. *Pedobiologia*, **19**, 165–75.

Bowden, J., Haines, I. H. and Mercer, D. (1976) Climbing collembola. *Pedobiologia*, **16**, 298–312.

Braun, D. (1992) Aspekte der Vertikalverteilung von Spinnen (Araneae) an Kiefernstämmen. *Arachnologische Mitteilungen*, **4**, 1–20.

Büchs, W. (1988) *Stamm- und Rindenzoozönosen verschiedener Baumarten des Hartholzauenwaldes und ihr Indikatorwert für die Früherkennung von Baumschäden.* PhD Thesis, University of Bonn, Germany.

Büchs, W. (1990) Zur Bedeutung der Stammregion von Bäumen als Lebensraum von Arthropoden und anderen Evertebraten. *Zeitschrift für angewandte Zoologie*, **77**, 453–77.

Crawley, M.J. (1983) *Herbivory*, Blackwell Scientific Publications, Oxford.

Culberson, C.F. (1969) *Chemical and Botanical Guide to Lichen Products*, University of North Carolina Press, Chapel Hill.

Duffey, E. (1969) The seasonal movements of *Clubonia brevipes* Blackwall and *Clubonia compta* C.L. Koch on oak trees in Monks Wood, Huntingdonshire. *British Arachnology Society Bulletin*, **1**, 29–32.

Fowler, J. and Cohen, L. (1986) *Statistics for Ornithologists*, BTO Guide 22.

Funke, W. (1971) Food and energy turnover of leaf-eating insects and their influence on primary production, in *Integrated Experimental Ecology* (ed. H. Ellenberg), Ecological Studies 2, Springer-Verlag, pp. 81–93.

Gisin, H. (1960) *Collembolenfauna Europas*, Museum d'Histoire Naturelle, Genéve.

Gjelstrup, P. (1979) Epiphytic cryptostigmatic mites on some beech- and birch-trees in Denmark. *Pedobiologia*, **19**, 1–8.

Günther, K.K. (1974) Staubläuse, Psocoptera, in *Die Tierwelt Deutschlands 61* (eds K. Senglaub, H.-J. Hannemann and H. Schumann), VEB Gustav Fischer Verlag, Jena.

Hale, M.E. (1972) Natural History of Plummers Island, Maryland. XXI. Infestation of the lichen *Parmelia baltimorensis* Gyel. and For. by *Hypogastura packardi* Folsom (Collembola). *Proceedings of the Biological Society of Washington*, **23**, 287–96.

Joger, H.G. (1988) Untersuchungen über die Tierwelt einer Stadtmauer. *Zoologische Jahrbücher (Systamatik, Ökologie und Geographie)*, **115**, 69–91.

Lamprecht J. (1992) *Biologische Forschung: Von der Planung bis zur Publikation*, Verlag Paul Parey, Berlin.

Laundon, J.R. (1971) Lichen communities destroyed by psocids. *Lichenologist*, **5**, 177.

Lawerey, J.D. (1980) Correlation betwen lichen secondary chemistry and grazing activity of *Pallifera varia. The Bryologist*, **83**, 328–34.

Manhart, C. (1994) Spiders on bark in a tropical rainforest (Panguana, Peru). *Studies on Neotropical Fauna and Environment*, **29**, 49–53.

Mühlenberg, M. (1989) *Freilandökolgie*, UTB, Quelle and Meyer, Heidelberg.

Müller-Kraenner, S. (1990) *Entwicklung, Ökologie und Verhalten von Entomobrya corticalis (Nicolet 1841), im Vergleich mit drei anderen epigäischen Collembolen-Arten*, MSc Thesis, University of Bayreuth, Germany.

Nicolai, V. (1985) *Die ökologische Bedeutung verschiedener Rindentypen bei Bäumen*, PhD Thesis, University of Marburg, Germany.

Nicolai, V. (1986) The bark of trees: thermal properties, microclimate and fauna. *Oecologia*, **69**, 148–60.

Niedballa, W. (1969) Fauna mechowców (*Acari, Oribatei*) nadrzewnych w okolicach Poznania (Arboreal moss-mites fauna (*Acari, Oribatei*) in Poznan environs). *Polskie Pismo Entomologiczne*, **39**, 83–94.

Nogushi, Y. (1979) Deformation of trees in Hawaii and its relation to wind. *Journal of Ecology*, **67**, 611–28.

Prinzing, A. and Wirtz, H.-P. (1997) The epiphytic lichen, *Evernia prunastri* L., as a habitat for arthropods: shelter from desiccation, food-limitation and indirect mutualism, in *Canopy Arthropods* (eds N.E. Stork, J. Adis and R.K. Didham), Chapman & Hall, London, pp. 477–94 (this volume).

Pschorn-Walcher, H. and Gunhold, P. (1957) Zur Kenntnis der Tiergemeinschaften in Moos- und Flechtenrasen an Park- und Waldbäumen. *Zeitschrift für Morphologie und Ökologie der Tiere*, **46**, 342–54.

Rose, F. (1974) The epiphytes of oak, in *The British Oak* (eds M.G. Morris and F.H. Perring), E. W. Classey, Ltd., Berkshire, pp. 274–97.

Schauermann, J. (1973) Zum Energieumsatz phytophager Insekten im Buchenwald: II. Die produktionsbiologische Stellung der Rüsselkäfer (Curculionida) mit rhizophagen Larvenstadien. *Oecologia*, **13**, 313–50.

Schindler J.E. (1988) Freshwater ecosystems: a perspective, in *Concepts of Ecosystem Ecology* (eds L.R. Pomeroy and J.J. Alberts), Ecological Studies 67, Springer-Verlag, pp. 57–74.

Stahl, E. (1904) *Die Schutzmittel der Flechten gegen Tierfraß*, Festschrift zum siebziegsten Geburtstage von Ernst Haeckel, Jena, pp. 355–75.

Stubbs, C.S. (1989) Patterns of distribution and abundance of corticolous lichens and their invertebrate associates on *Quercus rubra* in Maine. *The Bryologist*, **92**, 453–60.

Systat for Windows (1992) *Statistics and graphics*, Version 5 Edition, Evanston, IL: Systat Inc.

Tischler, W. (1990) *Ökolgie der Lebensräume*, G. Fischer Verlag, Stuttgart.

Travé, J. (1963) Écologie et biologie des Oribates (Acariens) saxicoles et arboricoles. *Vie et Milieu (Supplement)*, **14**, 1–267

von Allmen, H. and Zettel, J. (1982) Populationsbiologische Untersuchungen zur Art *Entomobrya nivalis* (Collembola). *Revue Suisse de Zoologie*, **89**, 919–26.

Weigmann, G. and Kratz, W. (1981) Die deutschen Hornmilbenarten und ihre ökologische Charakteristik. *Zoologische Beiträge*, **27**, 459–89.

Welkowitz, J., Robert, B.E. and Cohen, J (1982) *Introductory Statistics for the Behavioural Sciences*, HBJ Publishers, San Diego.

Winter, K. (1972) *Zum Energieumsatz phytophager Insekten im Buchenwald: Untersuchungen an Lepidopterenpopulationen*, PhD Thesis, University of Göttingen, Germany.

Woltemade, H. (1982) Zur Ökologie baumrindenbewohnender Hornmilben (Acari, Oribatei). *Sitzungsberichte der Gesellschaft naturforschender Freunde zu Berlin*, **22**, 118–39.

23

The epiphytic lichen, *Evernia prunastri* L., as a habitat for arthropods: shelter from desiccation, food-limitation and indirect mutualism

A. Prinzing and H.-P. Wirtz

ABSTRACT

On exposed surfaces of tree trunks, the minute structures of the fruticose lichen thalli of *Evernia prunastri* positively influence abundance and living conditions of various corticolous arthropods. The densities of Collembola, Psocoptera and oribatid mites that feed on the lichen's phycobiots are much higher than on the neighbouring bark where algal cover represents the only food resource. The preference for *E. prunastri* is explained by a unique combination of food availability and protection from desiccation in the lichen thalli. Such benefits outweigh the various restrictions placed on grazers by the thalli of *E. prunastri*: solid cortex layers, densely packed and inaccessible, or loose and wind-exposed, lichen branches, cover with sand and detritus, waterfilm after precipitation and the reduced edibility of phycobiots. The balance between such restrictions and the protection afforded to arthropods depends on the thallial growth form. This is a new perspective on primary producers' defence mechanisms. Moreover, for the lichen itself, thallus peripheries of dense growth provide protection from desiccation and wind-adapted growth forms offer optimal shelter from damage by wind. This adaptation is shown to depend on variation in the growth of lichen branches, induced largely by arthropod grazing. Thus, indirect mutualism seems to exist between *E. prunastri* and its grazers under exposed microclimatic conditions.

Canopy Arthropods. Edited by N.E. Stork, J. Adis and R.K. Didham. Published in 1997 by Chapman & Hall, London. ISBN 0 412 74900 9

INTRODUCTION

The trunks of trees are exposed to variation in climatic conditions such as wind, solar radiation and precipitation (Haarlov and Petersen, 1952; Nicolai, 1985, 1986, 1989). The microclimate of epiphytes on the tree trunk, therefore, may be very important for grazing microarthropods. In fact, a correlation has been demonstrated between the degree of epiphyte cover on different orientations of trunk-faces and the species composition and abundance of the corticolous fauna (Pschorn-Walcher and Gunhold, 1957; Travé, 1963; Gjelstrup, 1979; Nicolai, 1985; Büchs, 1988, 1990; Stubbs, 1989; Prinzing, 1997, Chapter 22, this volume). Similar differences might exist between lichens of crust-like and fruticose (shrub-like) growth forms with their correspondingly different microclimates (André, 1983). Other observations of animals that use tree trunks for passage, or for roosting also suggest that orientation/epiphyte cover and weather conditions may be important for arthropod species composition (Bowden *et al.*, 1976; Bauer, 1979; Funke, 1979; J. Adis, personal communication). Until now this complex of biotic and abiotic conditions has been little analysed with regard to arthropod distributional patterns (but see André, 1975; Bauer, 1979).

Solitary tree trunks prove particularly interesting for investigating such epiphytic cryptogams as a habitat for arthropods because of their climatic exposure and species-poor, eurytopic fauna (Pschorn-Walcher and Gunhold, 1957; Travé, 1963). The fruticose lichen *Evernia prunastri* (L.) is a common and conspicuous epiphytic structure found on exposed tree trunks all over Europe, especially in northern Germany (Poelt, 1969; Jacobsen, 1992). This lichen displays an extremely variable growth form and growth density. Morphological observations, comparison of growth forms in different locations and experimental simulation of grazing have shown that such polymorphism is the result of the combined impact of arthropod grazing and microclimate (Prinzing, 1992 and in preparation). The resulting growth forms obviously influence the habitat of corticolous arthropods, sheltering them from solar radiation, wind and rainfall. However, the lichen also displaces *Pleurococcus* algae which are known to be a food source for corticolous arthropods (Jentsch, 1940; Bowden *et al.*, 1976; Bauer, 1979).

In order to study such interactions, the food and climatic requirements of corticolous arthropods were investigated in the laboratory and behavioural reactions to climate were observed in the field. The results were then related to the structural and climatic properties of *E. prunastri* thalli in comparison to bark covered by algae (which is the most common type of epiphyte cover on tree trunks), in order to distinguish which of these microhabitat types better suited microarthropods under different conditions. Suitability for grazers might be important to the lichen itself, as grazers can induce strong morphological variability in the thalli (for *E. prunastri*, see Prinzing, 1992, also in preparation; for similar fruticose lichen species, see Zopf, 1907; Bachmann, 1929). This feedback effect was also investigated by relating thallial morphology to thallial tolerance to windfall and desiccation.

MATERIALS AND METHODS

Investigations were conducted from August 1991 to February 1992 and in April and May 1992 in the area surrounding Kiel (northern Germany, mostly hilly landscape below 50 m above sea level, oceanic climate). The thalli of *E. prunastri* were investigated two to four times each week at 24 sites on solitary trees or groups of trees at heights from 0.5–2.5 m above the ground. The trees studied were mostly oak (*Quercus robur* L.), as well as lime (*Tilia platyphyllos* Scop., *T. cordata* Mill), ash (*Fraxinus excelsior* L.) and horse-chestnut (*Aesculus hippocastanum* L.). *E. prunastri* thalli were found mainly on north-west- to west-exposed trunk-faces, often in dense stocks. Thallial development and grazing patterns were documented photographically at 3-month intervals at six sites from August 1991 to April 1992.

Distribution and requirements of the arthropods

Arthropod abundance was determined 10 times a week, mainly by 5 minutes of hand-searching of both thalli and bark covered with algae (including bark crevices) using a hand-lens (i.e. time-sampling; Southwood, 1966; Dunger, 1989). Numbers of animals were also determined in 20 thalli of fixed volume (approximately 3 cm^3) each week by hand-searching. The drought-tolerance and food requirements of arthropods were determined in the laboratory. The animals were kept at room temperature in plexiglas containers (inner radius 3 cm, height 4 cm). Humidity was supplied from a Petri dish with water put below the gauze floor of the container, avoiding direct water contact on the gauze (Figure 23.1). The containers could be illuminated from below by cold light in order to chase the animals to the shaded upper-side of the substrate (lichen or bark). Feeding behaviour of the animals could then be easily observed from above without further disturbance. To investigate drought tolerance, animals of different species were kept together in the same container without humidification from below.

Structure and climate of the thalli

To characterize the major structural variations of *E. prunastri* under different living conditions, 43 lichen specimens were collected from

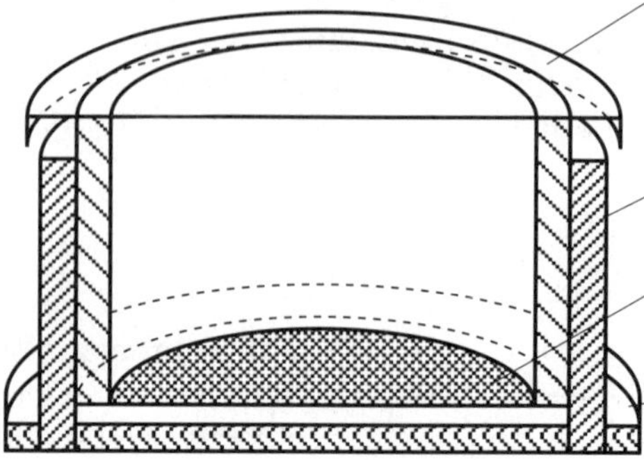

Figure 23.1 Containers for cultivation, observation and experimentation of arthropods, made from two tubes of plexiglass covered by a transparent Petri dish. Base made of gauze and placed in a Petri dish.

different sites on a single oak tree. The following variables were considered: (i) height on the trunk (0.5 and 1.7 m); (ii) exposure to main wind direction (frontal and both lateral trunk faces; no thalli were found on the sheltered face); and (iii) solitary thalli and thalli surrounded by neighbouring *E. prunastri*. At each of the sites differentiated by combinations of these factors, three to four thalli of obviously different growth form were collected. The thalli were cut longitudinally with a razor blade and the proportions of thallus surface with (a) feeding traces, (b) detritus/sand-cover, (c) shelter from wind and sun by outer lichen branches, (d) without dense cover by neighbouring lichen branches and (e) a combination of (c) and (d), were estimated at 10% intervals. The thalli were then sprayed with water three times. After some minutes, when the soaked thalli reached their final form, the proportions of surface were noted with: (i) a cover of waterfilm; (ii) shelter from wind and sun by outer lichen branches; (iii) no dense cover by neighbouring lichen branches; and (iv) both (i) and (ii).

For measurements of climatic properties of the thalli, 15 intact specimens were each weighed before and after spraying (giving a measure of water uptake capacity), then blow-dried with a hair-dryer from a distance of 30 cm for 1 minute and re-weighed. The ratio of water loss due to drying to the initial water uptake was used as a measure of the wind accessibility of the thalli. A photograph was taken of the front sides of each thallus and used to estimate lichen branch percentage cover (i.e. growth density of frontal thalli) by overlaying transparent graph paper (equivalent to standard quantification of grazing on leaves, for example).

Identifications of lichens and arthropods were made with the aid of reported studies by Wirth (1980) (lichens), Günther (1974) (Psocoptera),

Gisin (1960) (Collembola), Gruner (1976) (Isopoda), Sellnick (1960), Willmann (1931) (Oribatei) and Heimer and Nentwig (1991) (Araneae).

RESULTS

Climatic properties of thalli

Generally, the inner parts of the thalli were sheltered from air flow by the fruticose growth form. This was demonstrated in two different ways. First, the inner, thallial cavities between the branches were hardly penetrated by cigarette smoke blown onto the lichen. Second, after precipitation the central lichen branches stayed moist for hours. This could be easily seen with the naked eye as the thalli of *E. prunastri* and many other lichen species are darker green in colour when wet.

Accessibility to wind (simulated with a hair-dryer) significantly decreased with increasing growth density of the thallus' outer face (Spearman's R = –0.963, P <0.001, see Figure 23.2). Optical estimation of the proportion of thalli surfaces that were wind-sheltered confirmed the enormous inter-thallial variability (Figure 23.3(a)). Moreover, movements of lichen branches during soaking by artificial precipitation led to a significant reduction in sheltered areas (Figure 23.3(a)). Both water-uptake during soaking and the extent of cover with a waterfilm also varied markedly between thalli (Figure 23.3(a)).

Arthropods in *E. prunastri* and their requirements

Ten microphyte-grazing arthropod species were regularly found (Table 23.1). Although none of these was restricted to *E. prunastri* alone, large species (such as *Cerobasis guestifalica* (Psocoptera) and species of Entomobryidae) were significantly more common on *E. prunastri* than on bark with algal growth (Table 23.1). Spiders (predominantly juvenile Dictynidae, Linyphiidae and Theridiidae) were also found on *E. prunastri*, mainly in late summer and in autumn.

Artificial wetting of bark induced both Psocoptera and entomobryids to move to the outer surface of the bark, where almost all the algae were growing, and to migrate for long distances. In contrast, Psocoptera and entomobryids living in dry thalli of *E. prunastri* became much less mobile after artificial spraying (Figure 23.4). This basic difference between bark- and thallus-dwelling animals was found in all 15 replicates (P <0.001, sign-test). Similarly, in the laboratory, arthropods were more drought-tolerant when kept in dry, intact *E. prunastri* thalli than on single lichen branches (Table 23.2). Generally, entomobryids were the most sensitive to desiccation and to drowning in a waterfilm, followed by *C. guestifalica* (Table 23.2).

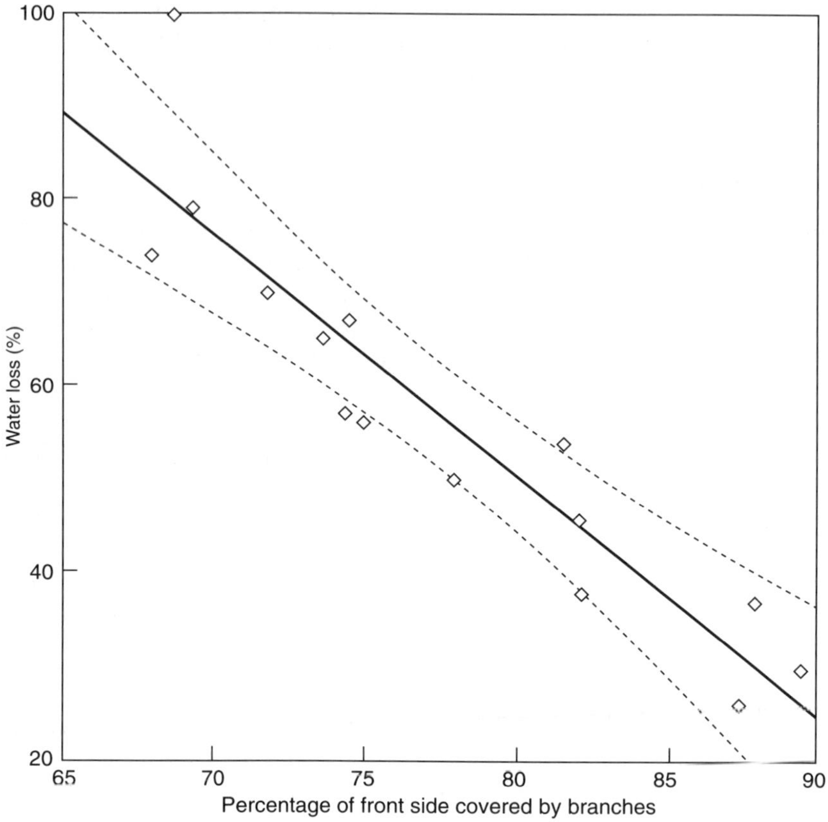

Figure 23.2 Linear correlation between growth density of frontal faces of thalli (percentage of area covered by lichen branches) and accessibility to wind (water loss of soaked thalli due to exposure to a hair-dryer), including 95% confidence limits (Spearman's R = –0.963, n = 15, y = –2.58x + 256).

None of the arthropods was able to feed on all layers of the lichen thallus. The phycobiot layer and algal epibionts were the most frequently grazed. Only *C. guestifalica*, *Orchesella cincta* (Entomobryidae) and oribatids seemed capable of cracking the cortical layer and thus also feeding on the phycobiots of intact lichen branches (Figure 23.5). *Entomobrya* spp. especially could only graze upon phycobiots which were opened up along the edges by feeding traces of larger grazers or at mechanical ruptures of the cortex layer. Otherwise these species were restricted to grazing on the algal cover of bark and lichens.

Entomobryids, oribatids and *C. guestifalica* were kept alive in the laboratory for up to 5 months when supplied with *Pleurococcus* algae or thalli

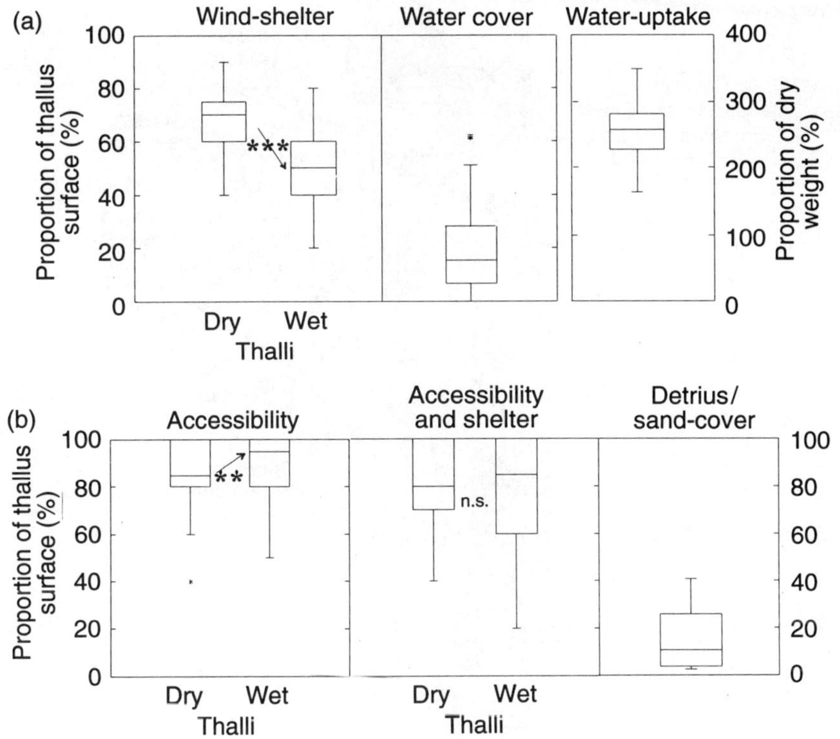

Figure 23.3 Variation of (a) climatic and (b) relevant structural properties on thalli of *E. prunastri* (box–whisker plots with median, outer limits as first and third quartile and exceptional outlying values). Temporal changes due to soaking by artificial precipitation are tested by the Wilcoxon rank test (** and ***, $P<0.01$ and $P<0.001$, respectively, n.s. = not significant). For detailed description see text. $n = 43$ thalli, except for water-uptake ($n = 15$).

of *E. prunastri*. Generally, no thallial grazing was found in the field or observed in the laboratory on lichen branches covered by detritus or sand or by the granular lichen, soredia. Furthermore, feeding traces in the thallus' phycobiot layer were rarely found in thalli densely covered by algal epibionts. Correspondingly, in the laboratory, the animals always fed strictly on covers of algal epibionts when presented with a choice. While doing this, the animals were never observed to push and pull with their whole body as they had to when grazing on the thallus itself, especially on its stiff cortex.

Grazing on the cortex and phycobiot layers seemed to be most effective in slightly moistened thalli, whereas in lichen branches with a higher

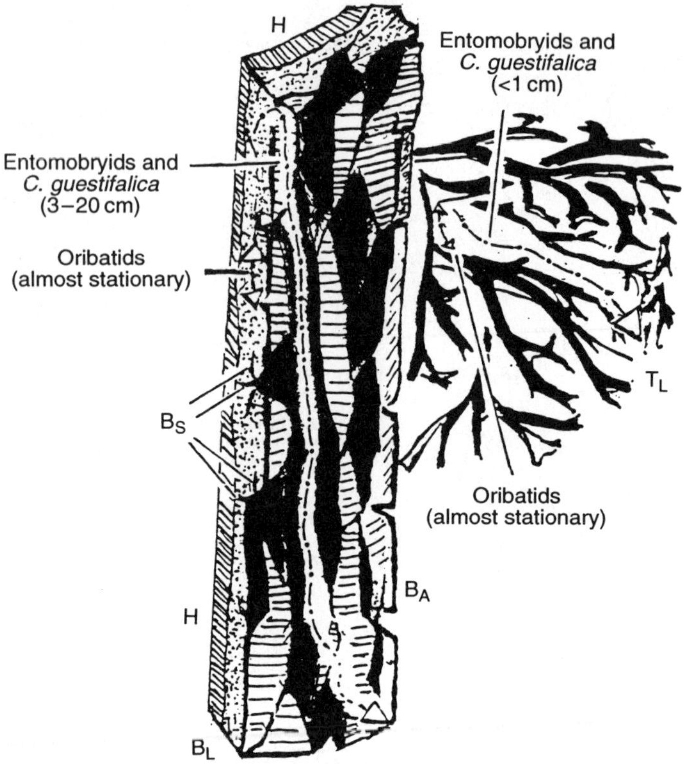

Figure 23.4 Migration distances of arthropods within 10 minutes after moistening thalli and surrounding bark by spraying (during lull conditions, animals stayed stationary with wind intensity of three degrees or more on the Beaufort scale). B_A, upperside of bark covered by algae; B_L, bark in longitudinal section; B_S, bark crevice, mostly free of algae; H, wood; T_L, *E. prunastri* in longitudinal section. Experiments were repeated 15 times (with more than three individuals of entomobryid Collembola, of *Cerobasis guestifalica* (Psocoptera) and of oribatids).

water content the inner cells of the phycobiot layer were avoided (Table 23.3). Correspondingly, in the field after extreme precipitation at the end of September, the majority of feeding traces in all of the 43 collected thalli were covered with such inner phycobiot cells rejected by grazers. During September such restriction of the food resource was probably significant for *C. guestifalica*. Population densities strongly decreased and animals often showed a shortened abdomen (Table 23.1). In the laboratory the only alternative cause of a shortened abdomen was desiccation, which was certainly irrelevant to the above-mentioned field observations. Laboratory observations also showed that oviposition did not result in a shortened abdomen.

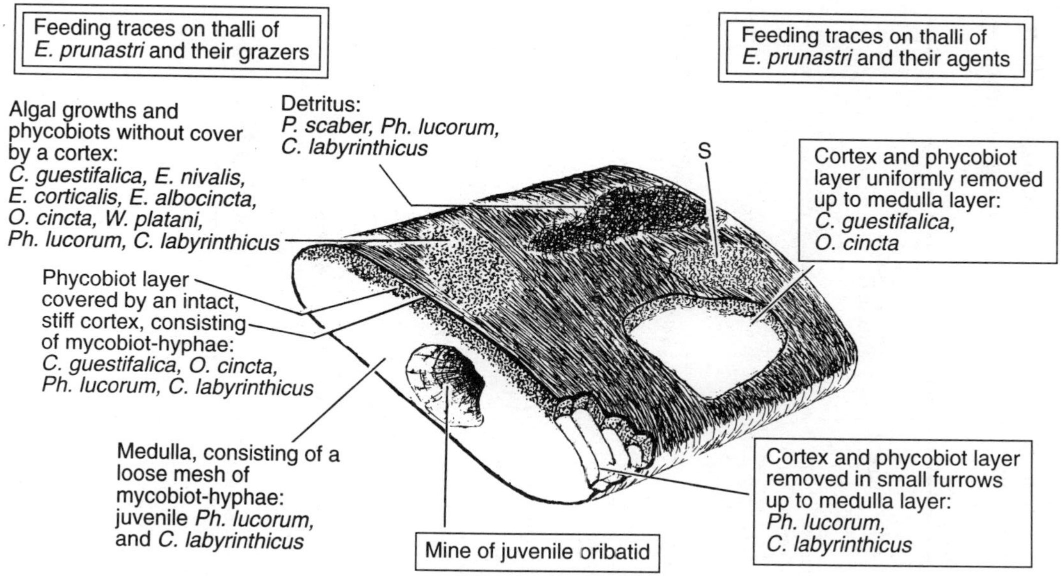

Figure 23.5 Lichen branch of *E. prunastri* as a food source of grazers and the corresponding feeding traces, cross-section from the front, slightly schematic. S, soredia, which are not grazed upon. Full species names are given in Table 23.1.

Table 23.1 Abundance and phenology (except for March) of arthropods found on *Evernia prunastri* and on adjacent bark (including crevices) covered by algae, sampled for fixed time-spans or fixed thallial volumes. When animals occurred extremely patchily, the latter number is not given. *Entomobryidae are mainly *Entomobrya nivalis* L., *E. albocincta* Templeton and *Orchesella cincta* L. as well as some *E. corticalis* Nicolet and *Willowsia platani* Nicolet

taxa	abundance: per 5 min search time, on thalli	abundance: per 5 min search time, on bark	abundance: per app. $3cm^3$ thallus volume	phenology on *E. prunastri* (months: June–May)
Insecta: Psocoptera *Cerobasis guestifalica* Kolbe	10–50	max. 4	≤ 7 adults ≤ 15 juveniles	J J A S O N D J F (M) A M eggs; adults; juveniles
	comment: found in all areas of research. Drastic decrease during end of September/ beginning of October 1991 after the first rainfall that lasted for several days. >50% of the surviving individuals with extremely shortened abdomen.			
Collembola: *Entomobryidae**	2–10	max. 2	≤ 3 adults ≤ 5 juveniles	J J A S O N D J F (M) A M juveniles and adults of *Entomobrya spec.*; *Orchesella cincta*
	comment: found on 30 to 80% of investigated trunks, mainly on oaks			
Isotoma viridis Bourlet	strongly clustered ≤ 20	> 20	(distribution too patchy)	J J A S O N D J F (M) A M juveniles
	comment: found in large numbers below foliose lichens and bark of oaks and limes throughout the whole year, also as adults			
Isopoda: *Porcellio scaber* Latreille	strongly clustered ≤ 20	>> 20	"	J J A S O N D J F (M) A M
	comment: not found on exposed trees on farmland			
Acari: Oribatei: *Carabodes labyrinthicus Michael, also Phauloppia lucorum C.L. Koch*	strongly clustered with no obvious differences between bark and lichens (< 40 ind. per thallus)		"	J J A S O N D J F (M) A M

Table 23.2 Tolerance to desiccation and to a waterfilm of common arthropod species on *E. prunastri*. Desiccation experiments were conducted in five containers with four dry lichen branches of *E. prunastri* each, or in five containers with one complete dry thallus of 3 cm^3 each, both without water supply. Tolerance is given as survival time in days (d) or hours (h), respectively. Number of animals investigated is given in parentheses. Tolerance to water-cover is measured as the parts of the body that the animals were able to detach from a waterfilm on a thallus (Wp) and as survival time under complete cover of water in hours (Ws, n >10 individuals per species)

	Desiccation		*Tolerance to water-cover*	
	Dry branches	*Dry thallus*	*Wp*	*Ws (h)*
C. guestifalica	5–7 d (*n*=20)	10–15 d (n= 24) abdomen shrinks to 2/3 of normal size	Antenna, bristle, often also leg	0.5
O. cincta	4–5 d (*n*=10)	–	Antenna, bristle	0.5
E. nivalis	2–3 d (*n*=15)	–	Antenna, bristle	0.5
Ph. lucorum/ *C. labyrinthicus*	5–10 d (*n*=25)	9–17 d (*n*=32)	Bristle	>72
P. scaber	3–4 h (*n*=10)	1 d (*n*=5)	Whole body	–

Table 23.3 Grazing pattern of phycobiot-feeding arthropods on *E. prunastri* branches of different moisture content (three branches per container and greater than three adults of *Orchesella cincta* (Entomobryidae), four *Cerobasis guestifalica* (Psocoptera), seven *Phauloppia lucorum* (Orabatei), or three *Carabodes labyrinthicus* (Orabatei), respectively

	Humidity of lichen branches			
Fraβ	*Dry*	*Slightly*	*Mediumly moistened*	*Soaked*
Extent	Small patches	Large areas		Patches, by oribatids only
Depth	Cortex + complete phycobiot layer		Cortex + upper phycobiot layer only	
Distribution	Irregular, including edges and tips of branches			

When estimating the extent of feeding traces in the cortex and the phycobiot layer, grazing of >80% of the surface was only found in the peripheries of one out of 43 thalli and in only four of the thalli centres. The centre of the thallus is shaded and therefore less important for the lichen's photosynthesis.

Figure 23.6 Fate of *E. prunastri* thalli on an oak, reflecting the impact of windfall on thalli of different growth patterns between (a) 15 August, 1991 and (b) 22 April, 1992. Lichen branches marked (z) in Figure 23.6(b) were windswept, those marked (a) were even torn off. The latter was also preceded by a windswept stage, as intermediate photographs showed. Such branches were comparatively constant in the growth layer and regular in branching pattern, but they were not more strongly grazed than surviving lichen branches. Feeding traces of *Cerobasis guestifalica* (Psocoptera), *Orchesella cincta* (Entomobryidae) or oribatids were located in comparatively wind-sheltered microsites, recognizable as white patches (marked F) in Figure 23.6(a).

Effect of spatial structure of the thalli on arthropods

Traces of feeding were only found on lichen branches accessible to grazers, but not where there was dense contact with other branches (as found sometimes on up to 60% of the thallus surface, especially when thalli were dry) (Figure 23.3(b)). Proportions of areas both accessible to grazers and sheltered by outer lichen branches varied more between thalli, but were not affected by artificial soaking (P = 0.860, Wilcoxon rank test; Figure 23.3(b)).

The grazed portions of the thallus surface were 2.1 times more extensive in central parts of thalli sheltered by outer lichen branches than in the more exposed branches at the periphery of the same thalli (P <0.001, Wilcoxon rank test; e.g. Figure 23.6(a)). In contrast, wind exposure due to the height of thalli on the tree trunks (1.7 m compared with 0.5 m) correlated with a significant increase in the percentage of surface grazed at thallus peripheries (P = 0.012, Mann–Whitney U-statistic = 143), whereas grazing was not affected at the thallus centres (P = 0.417).

Most feeding traces with intact inner cells of the phycobiot layer (see Table 23.3) were found on the sheltered, central lichen branches, corresponding to the constant, high moisture content. In the thallus peripheries incomplete grazing occurred only after extreme precipitation at the end of September and was then even found in all of the 43 thalli.

Rarely, in very dense thalli, large parts of the surface were covered by detritus and sand which made phycobiots inaccessible for grazing by arthropods, recognizable by the lack of feeding traces (Figure 23.3(b)).

Effect of thallus growth-form on survival of thalli

During rain, and in particular during stormy weather, thalli from exposed parts of the bark relief were often torn from the trunk. Destruction by other agents such as birds or squirrels was not observed. The destructive impact of wind was also reflected by the windswept shape of thalli before detachment and was recognized in all photographically documented cases (Figure 23.6). In contrast, intact thalli on wind-exposed zones of the trunk showed an aerodynamic dense mesh of lichen branches, some of which even grow back into the bark.

At seven sites near Kiel, 15 to 25 thalli or lichen branches that had fallen from the trunk were collected. All were predominantly constant in growth layers and isotomous–dichotomous in branching pattern in all investigated areas. Moreover, undamaged lichen branches that remained on the trunks were more irregular in growth pattern than the neighbouring detached thalli or branches (observed photographically in 47 out of 52 thalli).

DISCUSSION

E. prunastri as a habitat for grazers: advantages and disadvantages

The results of this study show that thalli of *E. prunastri* are not only an adequate food source for arthropods (even when dry) but also retain humidity and provide shelter from convective desiccation. This protection increases on strongly wind-exposed sites due to an increasing growth density of the lichen (Prinzing, 1992, also in preparation; Zimmer, 1994). Therefore, grazing on the central parts of thalli does not become less intense at the upper, wind-exposed heights of a trunk.

It is possible that grazers also create a humid environment by oxidative synthesis of water from ingested lichen particles. However, in such a case food should eventually become limiting – at least with regard to certain optimally utilized parts of the lichen branches. This was not found to be the case.

Food source and shelter from desiccation within *E. prunastri* could be segregated spatially by distances of not more than several millimetres and, therefore, resources were faster and more flexibly accessible than on bark with algal cover. Effects of the exposure of the whole trunk to wind, sun and rain on the accessibility of different corticolous microhabitats are currently being investigated (Prinzing, 1997, Chapter 22, this volume).

Thalli might also protect grazers from predatory ants, which were observed to be largely unable to climb densely growing thalli of *E. prunastri*. The importance of the above-mentioned advantages of *E. prunastri* as a habitat, compared with bark with algae, is reflected by the greater arthropod density on *E. prunastri*. Nevertheless, even in thalli of *E. prunastri* access to food (mainly consisting of phycobiots) is restricted to some degree, for example by densely packed lichen branches. After soaking, the lichen is less dense, but at the same time the animal's exposure to wind and rain is increased. Moreover, loosely packed, dry lichen branches are especially exposed to desiccation and are correspondingly less grazed. Oribatei are to some extent independent of such restrictions within a thallus, but they are least capable of moving quickly to a favourable microsite.

Even when lichen branches are easily accessible, access to their phycobiot layer is often prevented or complicated by an intact, stiff cortical layer, by a detritus cover, or by inappropriate inner phycobiot cells. The latter two factors depend on the density of thallial growth. The quality of phycobiot cells is also influenced by the actual weather conditions. Overall, these factors help to avoid overgrazing in all regions of a thallus, despite its very low regenerative capacity. Such indirect defence mechanisms are very different from most defence strategies described for

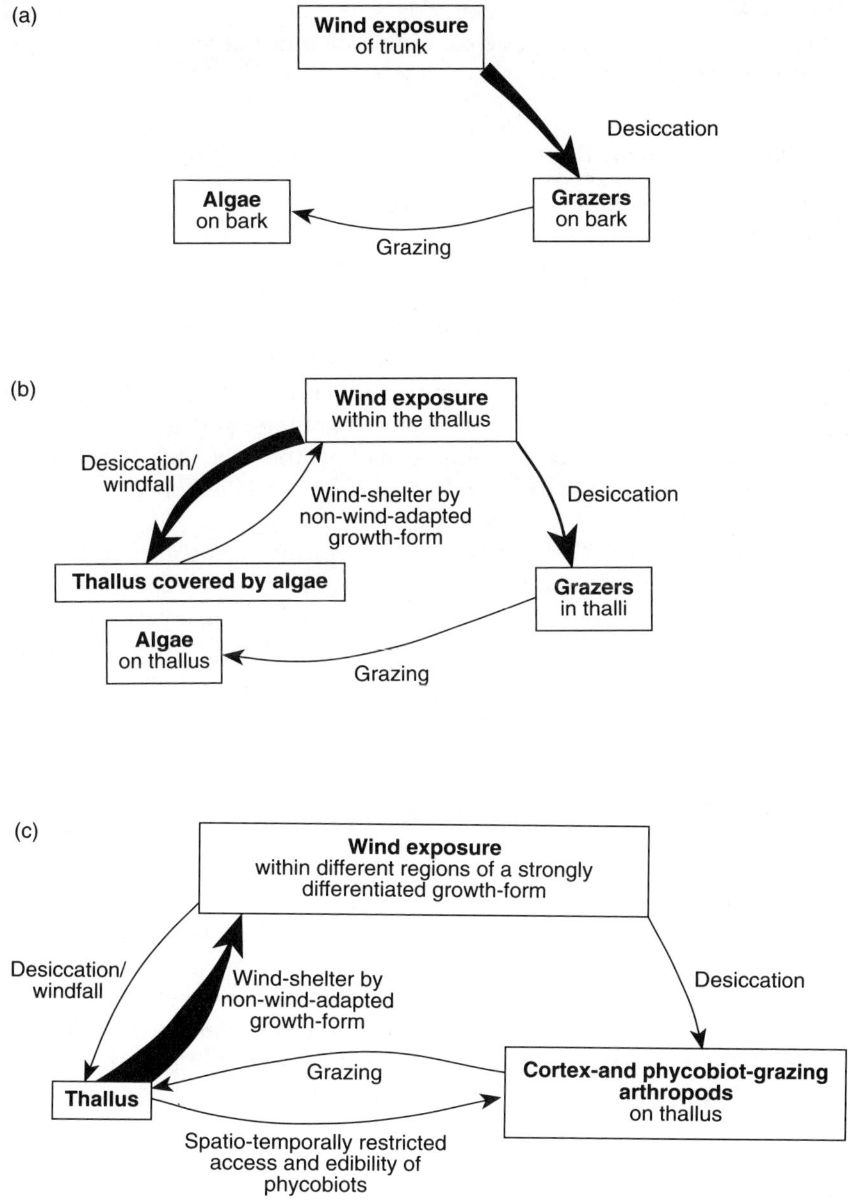

Figure 23.7 Simplified system of interactions between wind-exposure, grazers and (a) algal cover on bark, (b) algal cover on thalli of *E. prunastri* and (c) thalli of *E. prunastri* without algal cover. Only direct effects are presented. Width of arrows represents the probable strength of the effect.

higher plants (Crawley, 1983; Strong *et al.*, 1984; Howe and Westley, 1988). In *E. prunastri*, defence depends strongly on the spatial structure of the polymorphic lichen species, which changes rapidly according to moisture content.

Interactions between *E. prunastri* and grazing arthropods

Grazing by Psocoptera, oribatid mites and *Orchesella cincta* (Entomobryidae) induces variability in the growth of lichen branches. Variable growth of lichen branches is correlated with a low risk of damage by wind, due to the wind-adapted shape of thalli. This shape also restricts desiccation of *E. prunastri*. Therefore, *E. prunastri* does benefit from the activity of the above-mentioned grazers, especially on wind-exposed sites where arthropods most strongly depend on lichen habitat. The direct, detrimental, interactions between grazers and lichen become relatively unimportant. This is supported by the fact that overgrazing of *E. prunastri* was never reported from exposed sites, only from trees in a climatically sheltered forest (Laundon, 1971).

Arthropods and thalli not only inhibit or promote each other, but can also differentiate each others' living conditions (Figure 23.7(a,c)). This effect reaches down to the cellular level of phycobiot layers as a food source and to growth directions of lichen branches following grazing. Strong algal epibiosis on thalli, however, changes the interaction between lichens and grazers (Figure 23.7(b)), as the grazers confine themselves to the more easily accessible algae, thus interrupting most of the mutualistic and regulating feedback mechanisms. Wind damage to such algae-covered thalli is correspondingly high (Prinzing, 1992, also in preparation).

Grazing by *Entomobrya* species does not affect the differentiation and adaptability of thallus growth forms. On the other hand, the means by which the animals feed at the edges of existing feeding traces may hinder the thallus' regeneration, which was often observed to start at that place.

The results presented here might help us to understand the colonization of the canopy layer from arthropod communities on tree trunks. Microarthropod species colonize the canopy layer in a similar fashion to tree trunks and are probably engaged in the same kind of mutualism. Before canopy colonization, however, epiphyte grazers must cross the trunk and use its cryptogam flora.

Acknowledgements

We wish to thank N.E. Stork, M. Williams and K. Dierßen for very helpful comments on earlier versions of the manuscript, Th. Bauer, K.-K. Günther, W. Seeger and G. Weigmann for detailed discussions,

G. Weigmann and J. Grabo for determining oribatids and spiders, respectively, and C. Moore and I. Draack for correction of the English manuscript.

REFERENCES

André, H.M. (1975) Observations sur les Acariens corticoles de Belgique. *Fondation Universitaire Luxemburgoise, Série 'notes de recherche'*, **4**, 5–31.

André, H.M. (1983) Notes on the ecology of corticolous epiphyte dwellers: 2. Collembola. *Pedobiologia*, **25**, 271–8.

Bachmann, E. (1929) Tiergallen auf Flechten. *Archiv für Protistenkunde*, **66**, 61–103.

Bauer, T. (1979) Die Feuchtigkeit als steuernder Faktor für das Kletterverhalten von Collembolen. *Pedobiologia*, **19**, 165–75.

Bowden, J., Haines, I. H. and Mercer, D. (1976) Climbing collembola. *Pedobiologia*, **16**, 298–312.

Büchs, W. (1988) *Stamm- und Rindenzoozönosen verschiedener Baumarten des Hartholzauenwaldes und ihr Indikatorwert für die Früherkennung von Baumschäden.* PhD Thesis, University of Bonn, Germany.

Büchs, W. (1990) Zur Bedeutung der Stammregion von Bäumen als Lebensraum von Arthropoden und anderen Evertebraten. *Zeitschrift für angewandte Zoologie*, **77**, 453–77.

Crawley, M.J. (1983) *Herbivory. The Dynamics of Animal–Plant Interactions*, Blackwell Scientific Publications, Oxford.

Dunger, W. (1989) Quantitative Erfassung der Bodenfauna im Gelände, in *Methoden der Bodenbiologie* (eds W. Dunger and H.J. Fiedler), Gustav Fischer Verlag, Stuttgart, pp. 229–40.

Funke, W. (1979) Wälder, Objekte der Ökosystemforschung. Die Stammregion – Lebensraum und Durchgangszone von Arthropoden. *Jahresberichte des Naturwissenschaftlichen Vereins zu Wuppertal*, **32**, 45–50.

Gisin, H. (1960) *Collembolenfauna Europas*, Museum d'histoire naturelle, Genéve.

Gjelstrup, P. (1979) Epiphytic cryptostigmatic mites on some beech- and birch-trees in Denmark. *Pedobiologia*, **19**, 1–8.

Gruner, H.-E. (1976) Crustacea – Krebse, in *Exkursionsfauna für die Gebiete der BRD und der DDR* (ed. E. Stresemann), Volk und Wissen volkseigener Verlag, Berlin, pp. 394–450

Günther, K. (1974) Staubläuse, Psocoptera, in *Die Tierwelt Deutschlands* (eds F. Dahl, K. Senglaub and H. Schumann), VEB Gustav Fischer Verlag, Jena, vol. 61.

Haarlov, N. and Petersen, B.B. (1952) Temperaturmalinger i bark og ved af sitkagran (*Picea sitchensis*). *Der. forst. Fors. Danm.*, **21**, 43–91.

Heimer, S. and Nentwig, W. (1991) *Spinnen Mittelaeuropas*, Paul Parey, Berlin.

Howe, H.F. and Westley, L.C. (1988) *Ecological Relationships of Plants and Animals*, Oxford University Press, Oxford.

Jacobsen, P. (1992) Flechten in Schleswig-Holstein: Bestand, Gefährdung und Bedeutung als Bioindikatoren. *Mitteilungen der Arbeitsgemeinschaft Geobotanik in Schleswig-Holstein und Hamburg*, Vol. 42, pp. 1–234.

Jentsch, S. (1940) Vergleichend entwicklungsbiologische und ökologische Untersuchungen an einheimischen Psocopteren unter besonderer Berücksichtigung der Art *Hyperetes guestifalicus* Kolbe 1880. *Zoologische Jahrbücher Abteilung für Systematik, Ökologie und Geographie der Tiere*, **73**, 1–46.

Laundon, J.R. (1971) Lichen communities destroyed by psocids. *Lichenologist*, **5**, 177.

Nicolai, V. (1985) *Die ökologische Bedeutung verschiedener Rindentypen bei Bäumen*, PhD Thesis, University of Marburg, Germany.

Nicolai, V. (1986) The bark of trees: thermal properties, microclimate and fauna. *Oecologia*, **69**, 148–60.

Nicolai, V. (1989) Thermal properties and fauna on the bark of trees in two different African ecosystems. *Oecologia*, **80**, 421–30.

Poelt, J. (1969) *Bestimmungsschlüssel europäischer Flechten*, Verlag J. Cramer, Lehre.

Prinzing, A. (1992) *Epiphytische Flechten als Lebensraum für Arthropoden: Wechselwirkungen zwischen Tierverhalten und Thalluswuchs am Beispiel der Flechte Evernia prunastri (L. 1753) in der Umgebung von Kiel (Schleswig-Holstein)*, Unpublished MSc Thesis, University of Kiel, Germany.

Prinzing, A. (1997) Spatial and temporal use of microhabitats as a key strategy for the colonization of tree bark by *Entomobrya nivalis*. (Collembola: Entomobryidae), in *Canopy Arthropods* (eds N.E. Stork, J. Adis and R.K. Didham), Chapman & Hall, London, pp. 453–76 (this volume).

Pschorn-Walcher, H. and Gunhold, P. (1957) Zur Kenntnis der Tiergemeinschaften in Moos- und Flechtenrasen an Park- und Waldbäumen. *Zeitschrift für Morphologie und Ökologie der Tiere*, **46**, 342–54.

Sellnick, M. (1960) Formenkreis: Hornmilben, Orabatei, in *Die Tierwelt Mitteleuropas, vol. 3, Ergänzung* (eds P. Brohmer, P. Ehrmann and G. Ulmer), Quelle and Meyer, Leipzig, pp. 45–134.

Southwood, T.R.E. (1966) *Ecological Methods*, Chapman & Hall, London.

Strong, D.R., Lawton, J.H. and Southwood, T.R.E. (1984) *Insects on Plants. Community Patterns and Mechanisms*, Harvard University Press, Cambridge, Massachusetts.

Stubbs, C.S. (1989) Patterns of distribution and abundance of corticolous lichens and their invertebrate associates on *Quercus rubra* in Maine. *The Bryologist*, **92**, 453–60.

Travé, J. (1963) Écologie et biologie des Oribates (Acariens) saxicoles et arboricoles. *Vie et Milieu (Supplement)*, **14**, 1–267

Willmann, C. (1931) Moosmilben oder Oribatiden (Oribatei), in *Die Tierwelt Deutschlands und der angrenzenden Meeresteile* (eds D. Dahl, M. Dahl and H. Bischoff), Fischer Verlag, Jena, Vol. 22.

Wirth, V. (1980) *Flechtenflora*, UTB, Stuttgart.

Zimmer, D. (1994) Erfassung der Flechtenflora der Stadt Kiel sowie ihre Auswertung unter immissionsökologischen Aspekten. *Mitteilungen der Arbeitsgemeinschaft Geobotanik in Schleswig-Holstein und Hamburg*, Vol. 46, pp. 1–140.

Zopf, W. (1907) Biologische und morphologische Beobachtungen an Flechten: III Durch tierische Eingriffe hervorgerufene Gallenbildungen an Vertretern der Gattung *Ramalina*. *Berichte der Deutschen Botanischen Gesellschaft*, **25**, 233–7.

Part Five

The Management and Conservation of Canopy Arthropods

24

Distinguishing the woods from the trees

B.J. Richardson, S. Burgin, F.F. Azarbayjani and S. Lutubula

ABSTRACT

It is a truism of experimental design that the set of sampling units chosen must accurately represent the experimental unit of interest. Policy makers and managers of biodiversity will properly ask biologists questions about the diversity present in a particular area or habitat. In collecting data to answer such questions it is essential to develop a sampling regime that accurately reflects the nature of the experimental unit. In developing a sound sampling protocol for collecting arboreal arthropods many questions need to be answered. Are different parts of a host tree equivalent? Are trees of the same host species growing side-by-side equivalent? Are trees of the same species growing in different parts of the same woodland equivalent? Data are presented that show that the answer to each of these questions is **no**, they are not. A great deal more work is required in developing effective sampling protocols if biologists are to adequately advise managers and policy makers.

INTRODUCTION

Biodiversity has received steadily increasing international attention since the World Commission on Environment and Development met in 1987 (Mittermeier and Bowles, 1993). Attention reached a peak at the United Nations Conference on Environment and Development (UNCED) in 1992, with the negotiation of the Convention on Biological Diversity and Agenda 21, along with the development of a global fund for biodiversity conservation and associated environmental priorities (GEF). The recommendations emanating from UNCED identified inventory, survey and

Canopy Arthropods. Edited by N.E. Stork, J. Adis and R.K. Didham. Published in 1997 by Chapman & Hall, London. ISBN 0 412 74900 9

monitoring of basic biological resources as necessary to quantify biodiversity in all nations (Lovejoy, 1994).

These decisions raise serious questions for policy makers and managers over what this biological diversity is that they are responsible for, and how they might identify, map and monitor it in a meaningful way. They also need to know what factors affect the quality and interpretation of the available data. In this situation managers are likely to ask biologists questions of the kind: 'What can you measure? How robust is the estimate and what factors would affect the estimate? Over what geographical range does the estimate have relevance? How can changes in estimates over time be interpreted?' The work summarized here addresses these issues as they relate to the organismal level of biodiversity (c.f. Harper and Hawksworth, 1994).

Although various groups of organisms have been studied, for example birds and higher plants, in the attempt to measure and monitor organismal biodiversity, the use of arthropod species, especially insects, as indicators has increased in popularity over the last decade (Pyle *et al.*, 1981; Collins and Morris, 1985; Rosenberg *et al.*, 1986; Morris and Rispin, 1988; Murphy and Weiss, 1988; Samways, 1988, 1989, 1990a,b; Viejo *et al.*, 1989; Webb, 1989; den Boer, 1990; Rushton *et al.*, 1990; Thomas, 1991). Arthropods are chosen because they dominate terrestrial ecosystems in terms of species, numbers and biomass (Erwin, 1982, 1988; Wilson, 1985, 1988; Stork, 1988; Gaston, 1991a,b) and, consequently, their diversity is a rich potential source of information for conservation planning and management (Kremen *et al.*, 1993).

The number of arthropods and their distribution provide an easy, cost-effective and sensitive means to measure the effects of anthropogenic stress on biodiversity and the environment in general (Kim, 1993). Kremen *et al.*, (1993) point out the usefulness of insects in detecting environmental impacts, such as fragmentation, disturbance, habitat modification, ecological disturbance, climate change and chemical pollution, thereby making them potentially useful to scientifically based management programs (Kremen *et al.*, 1993). Despite this, little is known about factors influencing population persistence of invertebrates in ecosystems that are most at risk (Stork, 1991; Western, 1992).

The level and dynamics of biological diversity are the result of the interplay of physical and biological forces acting at a wide range of spatial and temporal scales (cf. Huston, 1994). These range from the long-term effects of evolution and continental drift, through to the effects of fine-scale habitat structure and seasonality. It is unclear at present which factors control the level of biological diversity at a site and their relative importance, what interpretation is to be placed on the estimates obtained, and the effects of remedial conservation options on biological diversity (Western, 1992). The purpose of the work summarized here is to examine issues related to the measurement of arboreal arthropod diversity, particularly those that are likely to be raised by managers.

The aim of this study was to collect the practical information needed to answer the questions: What is the effect of sampling different parts of the tree? What is the effect of sampling at different times of the year? Does a single tree provide an adequate sample of insect diversity on the host species in the collection area? How much variation is there in biodiversity found on examples of the same and of different tree species? How robust and informative are the present descriptive statistics?

METHODS

Many factors affect the values obtained when estimating the level of organismal diversity and it is essential to identify them and determine their effects if managers are to have any confidence in the estimates given to them. Care must therefore be taken to identify relatively simple systems in which to examine the issues if one is to avoid getting lost in the practical problems of analysing the abundant and diverse faunas of large trees. It is important to be able to replicate experiments so that confidence in the estimates obtained can be quantified, if such estimates are to be used in policy decisions. Adequate replication is often impractical if the number of specimens in a single sample is very large.

To address these issues, work has commenced sampling the arthropod faunas found on woodland tree species present on the 1500-ha campus of the University of Western Sydney, Hawkesbury at Richmond, New South Wales, Australia (33°38'S, 150°45'W). The level of arboreal arthropod diversity on small trees (<7 m in height), collected by knock-down insecticide fogging, is being measured both in terms of numbers of specimens and numbers of recognizable taxonomic units (RTUs). RTUs were defined by Rees (1983) 'as taxa that differ from each other clearly' and consist of units that can be distinguished on the basis of easily observable morphological criteria. Many studies since have been undertaken which utilize the technique (Oliver and Beattie, 1993). At Hawkesbury, typically 200–2000 specimens representing 35–100 RTUs were obtained per tree-crown fogged. In replicated experiments, mature trees were selected, matched with respect to location, height and health. Trees were sprayed from the ground or a stepladder using standard horticultural spraying equipment and a spray consisting of 0.02% (v/v) Permethrin and 0.05% x-77 surfactant. Trees were sprayed from four sides using a total of 5 l of insecticide on mornings of calm, dry days. The standard technique used consisted of placing six circular trays (total collection area 3.8 m^2) under each tree. Trays were left in position for 90 minutes. A range of tree species have been sampled, depending on the experiment, as specified where relevant in the results. Each tree sample consisted of the specimens from 30–100% of the crown area of the tree. Specimens from each tray were sorted separately and then cross-correlated, as well as combined to provide a single sample for the tree, depending on the questions being asked.

In an on-going series of studies, seven tree species have been studied and over 100 trees sampled so far.

RESULTS AND DISCUSSION

Effect of sampling different parts of the tree

The effect of collecting from different sides of each of six trees of the species *Melaleuca linariifolia* Smith (Myrtaceae) is presented in Figure 24.1. The trees were sampled on different days over a five week period in Autumn. Trays were distributed around each tree and classed by compass bearing: north (13 trays), south (11 trays), east (six trays) and west (six trays). There were 36% more species and 120% more specimens on the southern side of the trees than on the northern (sunny) side of the trees (Wilcoxon signed rank test, $P < 0.005$ in each case). Clearly, estimates of organismal biodiversity taken without regard to such differences will be misleading. For example, sample sets taken along the north side of a patch of woodland, rather than the south side, will differ significantly even though the same species of tree was sampled in the same general locality.

Comparing the sample sets obtained in trays from the same tree showed that there were still large differences in the number of speci-

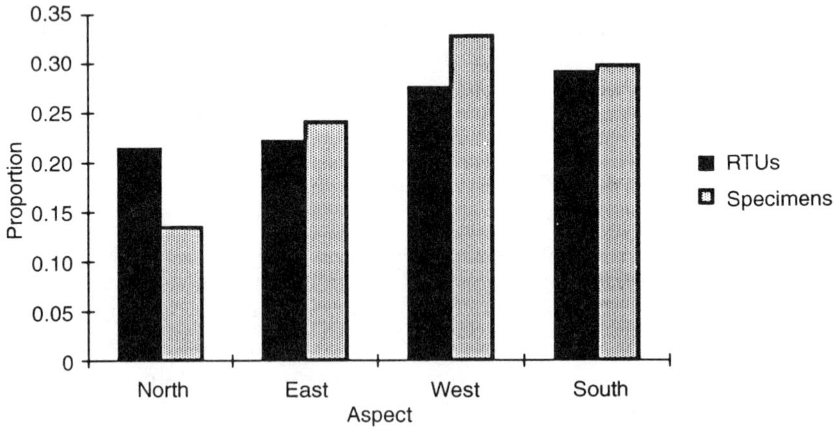

Figure 24.1 The effect of the side of the tree sampled (i.e. aspect) on the number of specimens and RTUs obtained.

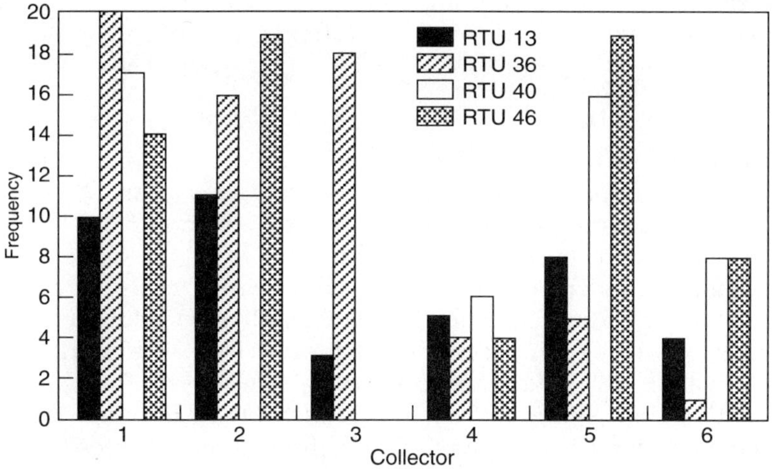

Figure 24.2 An example of correlated changes in the number of specimens of different species found in different trays under the same tree.

mens collected in different trays after aspect was discounted. Much of this variation is due to correlated differences in the number of specimens of several RTUs. An example of a series of trays under one *M. linariifolia* tree is shown in Figure 24.2. Sample sets 1 and 2 had high numbers of specimens for these and other RTUs, while other trays (e.g. 4 and 6) had low numbers of all species. Some trays (e.g. 3) had markedly different numbers of specimens for all species. That is, the number of specimens of different species often varied together from sample set to sample set. While the nature of this phenomenon needs further study, great care must be taken in comparative studies to ensure that the area sampled under a tree is sufficiently well distributed to represent the diversity present on the tree.

Effect of sampling at different times of the year

Studies of seasonal changes in biodiversity using nine trees of the species *M. linariifolia* showed a 70% reduction in the number of specimens between autumn and late winter. This is matched by a 50% change in the number of RTUs present. This result is unsurprising and reflects changes found in other studies over many years. From the management perspective, however, it is worth reiterating that in describing the organismal diversity of an area, or comparing diversity in different areas, the season of collection must be taken into consideration. As altitude, latitude and year-to-year differences can affect the timing of seasonal cycles, this factor can be very important in interpreting the significance of differences between data sets from different locations.

Arthropod diversity on host-trees

The question can be asked, 'Does a single tree provide an adequate sample of arthropod diversity on the host-tree in that area?' A cross-comparison of sample sets obtained from different trays from the same or different trees sampled at the same time from the same locality showed that complementarity values, which are a measure of biotic distinctness (see statistical section below; Colwell and Coddington, 1994) calculated between sample sets from the same tree were more similar than the values obtained when comparing sample sets from different trees of the same species. Each tree is distinct in its fauna (average complementarity values of pairs of trays from different trees 73% and from the same tree 60%). Whether these differences are historical and due to different colonization histories, due to the presence of particular keystone species, or due to differences between host trees, for example in the type or concentration of secondary plant compounds, needs further investigation. Unfortunately, complementarity is counter-intuitive and when providing information to managers it is better to translate the value. One minus complementarity is the widely used Jaccard index; however, for analytical reasons (see below), complementarity is a more useful statistic.

Insect biodiversity between tree species

Likewise, how much variation is there in insect biodiversity of the same and of different tree species? Comparison of species accumulation curves from sample sets from single trees of different, but related, species sampled at the same time in the same location (Figure 24.3), showed that the shape of the curves was markedly different and that the same size and distribution of trays under trees of the same size sampled a different proportion of the available diversity. Evidently, the number of RTUs found on each tree species is different, with most of the species on *M. linariifolia* trees being sampled, while a much smaller proportion of the species pool was sampled on *Melaleuca decora* (Salisb.) Britten (Myrtaceae) trees. It cannot be assumed, therefore, that the sampling regime for one host species is suitable for another host species or that the estimates obtained are comparable.

An experiment aimed at comparing matched trees of the same size, but different species, growing in the same woodland was carried out. Three trees of each of five species (*Eucalyptus amplifolia* Naudin (Myrtaceae), *Acacia parramattensis* Tind. (Mimosaceae), *Bursaria spinosa*

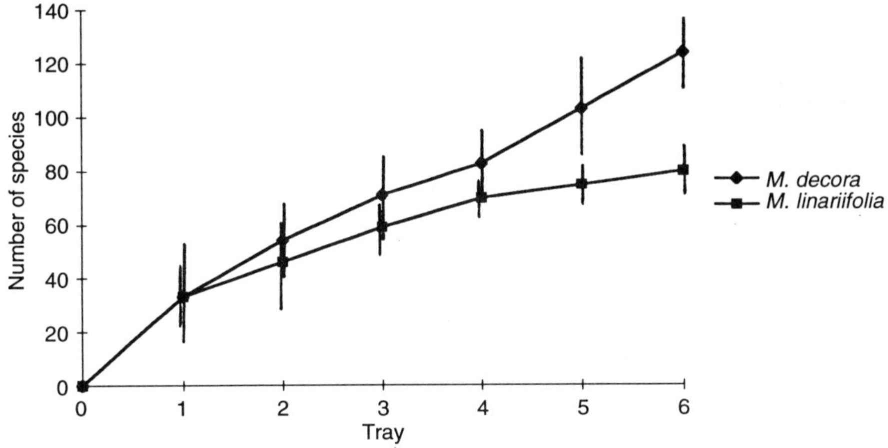

Figure 24.3 Average species accumulation curves for trays from two host species, *Melaleuca linariifolia* and *M. decora*.

Cav. (Pittosporaceae), *Jacksonia scoparia* Smith (Fabaceae) and *Angophora sublevutina* F. Muell. (Myrtaceae)) were sampled at the same time. The combined sample from each tree was cross-compared with all other trees. Similar ranges of diversity, both in the number of specimens and the number of RTUs, were observed on each tree (Table 24.1), irrespective of species. The between-tree complementarity values were also calculated. Frequency histograms of the within host-tree species comparisons and between host-tree species comparisons are shown in Figure 24.4. The values shown are not independent and normal statistical procedures cannot be used. Clearly, some arthropod species are found on more than one of the tree species studied, a series of arthropod species are found on particular tree species and a further set are found only on single trees. The relationship between tree morphology and other variables on the observed distribution of RTUs will be considered elsewhere.

The distribution of RTUs and specimens between major taxonomic groups can also be compared (Table 24.1). It can be seen that the distribution of the proportion of different taxonomic groups varies with tree species and that sampling a single species of tree does not provide a taxonomically representative sample of the arboreal fauna of the woodland.

Quality of the descriptive statistics

How robust and informative are the present descriptive statistics? Several parameters are of interest to managers; these include the total

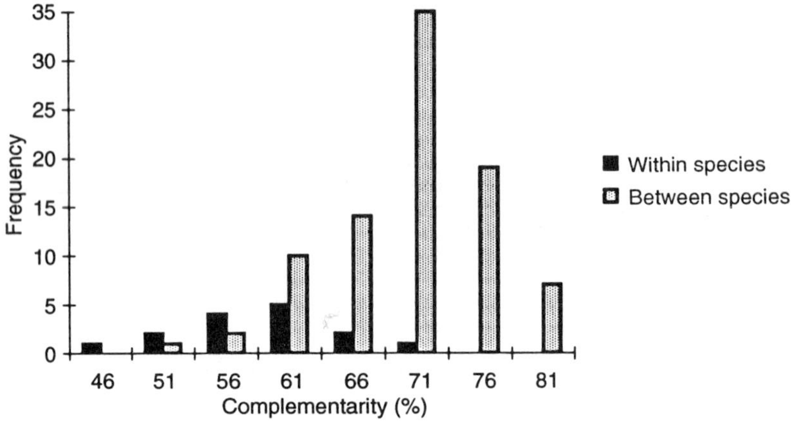

Figure 24.4 Frequency histogram showing the distribution of complementarity values for comparisons within and between host species.

number of species present in an area and the similarity of diversity at one location relative to other locations.

The work reported here shows that it is exceedingly difficult to estimate the number of species present in an area. From species accumulation curves it is possible to estimate the likely number of species present in the sampled population (Colwell and Coddington, 1994). However, the species accumulation curves obtained during these studies demonstrate that single trees, much less parts thereof, are not representative of the faunas of surrounding similar trees. Further, examination of the shape of the accumulation curve shows significant differences when the host species changes (Figure 24.3), even when sampled at the same locality at the same time. As the shape of the curve is a key factor in the use of various mathematical methods of estimating the number of species in the sampled population, their adequacy must be tested. The robustness of these methods is presently being examined using the data collected at Hawkesbury.

The most urgent need of managers is to be able to ask questions related to the level and significance of differences between locations and at different times at the same location. That is, a measure is needed to allow comparison of the diversity present in sample sets. Complementarity has been proposed in this context because of its known statistical attributes as an example of a Marczewski–Steinhaus distance (specifically it obeys the 'triangle of inequality', thereby offering the opportunity to make integrated analyses of data sets) (Colwell and Coddington, 1994). In a replicate study using *M. linariifolia*, sample sets

Table 24.1 The distribution of RTUs (%) and number of specimens between major arthropod taxa collected from different tree species

Taxon	*Host species**						
	M. l	*M. d*	*E. a*	*B. s*	*J. s*	*A. p*	*A. s*
RTUs							
Hymenoptera	19–26	6–28	19–25	20–28	19–31	17–23	13–24
Diptera	6–15	0–4	7–11	10–16	10–18	5–14	11–15
Coleoptera	6–20	9–15	13–18	8–14	4–10	14–21	14–22
Hemiptera	5–9	5–14	14–20	13–17	14–21	16–20	15–20
Araneae	5–18	14–31	10–13	9–19	6–9	9–12	11–12
Thysanoptera	1–4	2–4	1–2	2	2–4	2–3	1–3
Psocoptera	2–5	0	4–6	5–10	4–6	2–6	2–7
Total	68–99	34–137	54–66	90–99	45–63	58–92	46–92
Specimens							
Hymenoptera	4–10	5–28	14–37	7–17	9–28	8–25	11–15
Diptera	6–21	0–1	3–4	3–6	3–7	2–6	1–4
Coleoptera	1–7	2–22	4–11	4	1–3	9–24	1–8
Psocoptera	9–31	0	1–14	1–6	1–2	2–3	1–2
Hemiptera	3–18	3–13	3–24	5–10	4–12	24–62	10–23
Araneae	2–7	3–12	4–5	10–23	4	5–7	4–10
Thysanoptera	1–15	13–55	2–8	4–17	1–6	1–4	1
Total	394–1810	239–1593	288–664	624–1126	282–611	219–898	689–1026
No. trees	6	8	3	3	3	3	3

* Abbreviations: *M. l, Melaleuca linariifolia; M. d, Melaleuca decora; E. a, Eucalyptus amplifolia; B. s, Bursaria spinosa* Cav. (Pittosporaceae); *J. s, Jacksonia scoparia* Smith (Fabaceae); *A. p, Acacia parramattensis* Tind. (Mimosaceae); *A. s, Angophora sublevutine* F. Muell. (Myrtaceae)

from six trays from each of two trees were compared (average 73% complementarity, range 67–79%) and contrasted with the results obtained by comparing the combined sample sets from the trees (71% complementarity). In each situation similar complementarity values were obtained. If this preliminary result is upheld in further large-scale studies, then it seems clear that this would be a robust and highly informative statistic that can be measured with relatively small sample sets.

Ecological geneticists face problems not unlike those faced by biologists studying organismal biodiversity. Different forms are scattered across an area and it is essential to describe – in convenient statistical form – the size and geographical area covered by a unit of population.

For geneticists, this value is the genetic neighbourhood size. A genetic neighbourhood is defined as the area 'in continuum from which the parents of individuals born near the centre may be treated as if drawn at random'. This area can be measured in several ways. For example, the radius of the neighbourhood is about twice the distance including 40% of the parents and the estimate is little affected by the kurtosis of the distribution of distances moved by individuals during their lifetime (Wright, 1969). If the distribution of distances is normal, then a neighbourhood is twice the variance of the distance moved and contains 86.5% of the parents.

A similar statistic, a biodiversity neighbourhood, would have the radius of the area containing 86.5% of the species, i.e. 13% of the species remaining in common, or a complementarity value of 91%. If complementarity is as robust as predicted, then it will be relatively easy to estimate the size of a biodiversity neighbourhood. Such a statistic would allow the comparison of biodiversity neighbourhood size for different tree species in the same or different habitats. Present statistics are unsatisfactory for describing diversity parameters, even in the oversimplified situations described here, and a great deal more work is needed if meaningful descriptions, taking account of different habitats and ecosystems within the one landscape, are to be developed.

While the work at Hawkesbury to date has concentrated on the issues affecting estimates at a single location, before moving on to the complexities of distance and other ecological variables, two examples show the value of the approach. Trees of the species *Melaleuca decora* were sampled at logarithmically increasing distances up to 20 km. The effect of distance on complementarity values found in the pilot study is presented in Figure 24.5. By the time 20 km is reached the fauna has changed significantly, but has not reached the level of divergence suggested above that would be necessary for them to be considered to belong to different biodiversity neighbourhoods. Clearly this preliminary result will need further testing.

Geographers and geneticists have also developed other methods of analysing spatial structures (c.f. Sokal and Oden, 1978; Epperson, 1993). For example, analysis using Moran's I^2 allows the level of structuring in a population to be compared over different geographic distances. When the value of I^2 is reduced to zero, then similarity has been reduced to the point where only random associations remain and the resultant geographical distance may be considered the size of a neighbourhood. A similar technique is presently being evaluated for use in biodiversity studies.

In summary, there is hope that the estimation of complementarity, in combination with other statistics, will allow many of the questions asked by managers to be answered. However, the robustness of the statistics

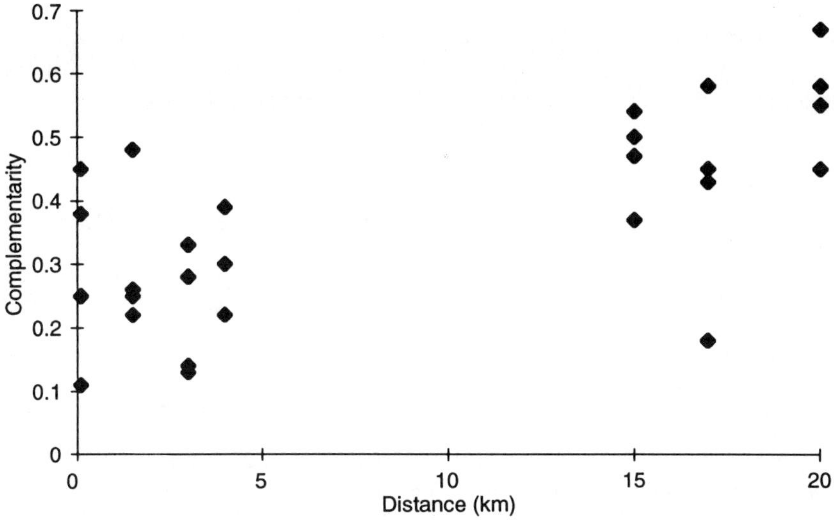

Figure 24.5 Scattergram of distances between *Melaleuca decora* host trees and complementarity values. As the points are not independent, a regression line cannot be fitted.

needs much more testing. It also remains to be seen whether it is possible to estimate total diversity and whether this statistic is meaningful and useful to anyone interested in more than impressive numbers.

It is a truism of experimental design that the set of sampling units chosen must accurately represent the experimental unit of interest. In the present case the questions asked are often about the diversity present in a particular area or habitat. Under such circumstances it is essential to develop a sampling regime that accurately reflects the nature of the experimental unit. For example, if a particular habitat is the issue of concern, then the different parts of the habitat must be sampled. In developing a sound sampling protocol many questions need to be answered. Are different parts of a host tree equivalent? Are trees of the same host species equivalent? Are trees of the same species growing in different parts of the same woodland equivalent? We have identified that the answer to each of these questions is no, they are not. A great deal more ingenuity and work are required in developing effective sampling protocols and statistical tools if biologists are to adequately advise managers and policy makers.

Acknowledgements

This project was supported by a grant from the Australian Research Council to B.J.R. and S.B. F.F.A. would like to acknowledge the Ministry of Science of the Islamic Republic of Iran for support during the project.

REFERENCES

Collins, N.M. and Morris, M.G. (1985) *Threatened Swallowtail Butterflies of the World*. IUCN Red Data Book IUCN, Gland.

Colwell, R.K. and Coddington, J.A. (1994) Estimating terrestrial biodiversity through extrapolation. *Philosophical Transactions of the Royal Society of London, Series B*, **345**, 101–18.

den Boer, P.J. (1990) The survival value of dispersal in terrestrial arthropods. *Biological Conservation*, **54**, 175–92.

Erwin, T.L. (1982) Tropical forests: Their richness in Coleoptera and other arthropod species. *The Coleopterist's Bulletin*, **36**, 74–5.

Erwin, T.L. (1988) The tropical forest canopy: the heart of biotic diversity, in *Biodiversity* (ed. E.O. Wilson), National Academy Press, Washington, pp. 123–9

Epperson, B.K. (1993) Recent advances in correlation studies of spatial patterns of genetic variation. *Evolutionary Biology*, **27**, 95–156.

Gaston, K.J. (1991a) The magnitude of global insect species richness. *Conservation Biology*, **5**, 283–96.

Gaston, K.J. (1991b) Estimates of the near-imponderable: a reply to Erwin. *Conservation Biology*, **5**, 564–6.

Harper, J.L. and Hawksworth, D.L. (1994) Biodiversity: measurement and estimation. Preface. *Philosophical Transactions of the Royal Society of London Series, Series B*, **345**, 5–12.

Huston, M.A. (1994) *Biological Diversity: The Coexistence of Species in Changing Landscapes*. Cambridge University Press, Cambridge.

Kim, K.C. (1993) Biodiversity, conservation and inventory: why insects matter. *Biodiversity and Conservation*, **2**, 191–214

Kremen, C., Colwell, R.K., Erwin, T.L., Murphy, D.D., Noss, R.F. and Sanjayan, M.A. (1993) Terrestrial arthropod assemblages: their use in conservation planning. *Conservation Biology*, **7**, 796–808.

Lovejoy, T.E. (1994) The quantification of biodiversity: an esoteric quest or a vital component of sustainable development? *Philosophical Transactions of the Royal Society of London, Series B*, **345**, 81–7.

Mittermeier, R.A. and Bowles, I.A. (1993) The global environment facility and biodiversity conservation: lessons to date and suggestions for future action. *Biodiversity and Conservation*, **2**, 637–55.

Morris, M.G. and Rispin, W.E. (1988) A beetle fauna of Oolitic limestone grassland, and responses of species to conservation management by different cutting regimes. *Biological Conservation*, **43**, 87–105.

Murphy, D.D. and Weiss, S.B. (1988) Ecological studies and the conservation of the Bay checkerspot butterfly, *Euphydryas editha bayensis*. *Biological Conservation*, **46**, 183–200.

Oliver, I. and Beattie, A.J. (1993) A possible method for the rapid assessment of biodiversity. *Conservation Biology*, **7**, 562–8.

Pyle, R., Bentzien, M. and Opler, P. (1981) Insect conservation. *Annual Review of Entomology*, **26**, 233–58.

Rees, C.J.C. (1983) Microclimate and the flying Hemipteran fauna of a primary lowland rainforest in Sulawesi, in *Tropical Rain Forest: Ecology and Management* (eds S.L. Sutton, T.C. Whitmore and A.C. Chadwick), Blackwell Scientific Publications, Oxford, pp. 121–36.

Rosenberg, D.M., Danks, H.V. and Lehmkuhl, D.M. (1986) Importance of insects in environmental impact assessment. *Environmental Management*, **10**, 773–83.

Rushton, S.P., Eyre, M.D. and Luff, M.D. (1990) The effects of scrub management on the ground beetles of Oolitic limestone grassland at Castor Hanglands National Nature Reserve, Cambridgeshire, U.K. *Biological Conservation*, **51**, 97–111.

Samways, M.J. (1988) Classical biological control and insect conservation: are they compatible? *Environmental Conservation*, **15**, 97–111.

Samways, M.J. (1989) Insect conservation and landscape ecology: a case history of bush cricket (Tettigoniidae) in southern France. *Environmental Conservation*, **16**, 217–26.

Samways, M.J. (1990a) Land forms and winter habitat refugia in the conservation of montane grasshoppers in Southern Africa. *Conservation Biology*, **4**, 375–82.

Samways, M.J. (1990b) Insect conservation ethics. *Environmental Conservation*, **17**, 7–8.

Sokal, R.R. and Oden, N.L. (1978) Spatial autocorrelation in biology 1. Methodology. *Biological Journal of the Linnean Society*, **10**, 199–228.

Stork, N.E. (1988) Insect diversity: facts, fiction and speculation. *Biological Journal of the Linnean Society*, **35**, 321–7.

Stork, N.E. (1991) The composition of the arthropod fauna of Bornean lowland rain forest trees. *Journal of Tropical Ecology*, **7**, 161–80.

Thomas, C.D. (1991) Habitat use and geographic ranges of butterflies from the wet lowlands of Costa Rica. *Biological Conservation*, **55**, 269–81.

Viejo, J.L., Deviedma, M.G. and Martinez, F. (1989) The importance of woodlands in the conservation of butterflies (Lep. Papilionidae and Hesperioidae) in the centre of the Iberian Peninsula. *Biological Conservation*, **48**, 101–14.

Webb, N.R. (1989) Studies on the invertebrate fauna of fragmented heathland in Dorset, UK, and the implications for conservation. *Biological Conservation*, **47**, 153–65.

Western, D. (1992) The biodiversity crisis: a challenge for biology. *Oikos*, **63**, 29–38.

Wilson, E.O. (1985) The biological diversity crisis: a challenge to science. *Issues in Science and Technology*, **2**, 20–9.

Wilson, E.O. (1988) The current state of biological diversity, in *Biodiversity* (eds E.O. Wilson and F.M. Peter), National Academy Press, Washington, pp. 3–17.

Wright, S. (1969) *Evolution and Genetics of Populations*, Vol. 2, *The Theory of Gene Frequencies*, University of Chicago Press, Chicago.

25

Insect biomass in Amazonian forest fragments

J.R. Malcolm

ABSTRACT

At four sites approximately 80 km north of Manaus, Brazil, arthropod (primarily insect) biomass and vegetation structure were measured in five major habitat types: (i) continuous forest (CF); (ii) the edge of CF; (iii) a 10-ha forest fragment; (iv) a 1-ha forest fragment; and (v) the matrix of secondary vegetation surrounding the forest fragments (including pasture and young secondary forest). Total biomass of arthropods was calculated for two types of traps (sticky and baited pitfall) at two heights (understorey and overstorey), and the biomass of commonly-caught arthropod groups was also calculated for pitfall traps. The hypothesis was tested that variation in arthropod biomass among primary forest habitats was entirely attributable to edge-induced habitat changes and was independent of insularization *per se*. Total arthropod biomass did not vary significantly among habitat types for overstorey sticky traps and understorey pitfall traps, but did for understorey sticky traps and overstorey pitfall traps. As predicted from the edge hypothesis, habitat-specific means from these latter two traps could be ranked in the sequence: 1-ha fragment, 10-ha fragment, CF edge, CF. Variation among habitats was correlated with understorey and/or overstorey vegetation structure (which also varied with distance from an edge). Also, when the relationship between arthropod biomass and foliage thickness was compared between isolated (1-ha fragments and 10-ha fragments) and non-isolated (CF edge and CF) habitats by use of analysis of covariance, little evidence of differences in regression lines was obtained. Thus, the increase in insect biomass in fragment understories and the decrease in the overstorey of fragments could be attributed to edge effects. This change in the vertical distribution of insect biomass has important implications for vertebrate insectivores.

Canopy Arthropods. Edited by N.E. Stork, J. Adis and R.K. Didham. Published in 1997 by Chapman & Hall, London. ISBN 0 412 74900 9

INTRODUCTION

Insects are important members of tropical forest ecosystems, and play key roles as pollinators, herbivores and detritivores. They act as a food source for numerous other organisms; for example, more than 50% of tropical Asian and Neotropical bird and bat species (Harrison, 1962; Lein, 1972; Wilson, 1973) and 30–40% of non-volant mammal species (Harrison, 1962; Robinson and Redford, 1986) are insectivores. Insects are also an enormously diverse group in the tropics, perhaps numbering in the millions of species (Erwin, 1982). This unparalleled diversity and overall ecological importance makes them particularly valuable to conservation. Much can be gained by studying and understanding their diverse responses to habitat modification. Also, changes in insect populations may directly or indirectly affect other ecosystem components, and hence ripple through the ecosystem.

Increasingly, large stands of continuous forest in the tropics are being reduced to isolated patches in a sea of pasture and secondary vegetation. The response of insect populations to this habitat modification will be determined in part by the pattern of landscape changes. According to the theory of island biogeography (MacArthur and Wilson, 1967), population densities in fragments will depend on both the rate at which new individuals colonize fragments (which will depend on the spatial configuration of the remaining forest and the role of the intervening secondary habitat as a barrier to movements) and the likelihood of stochastic population extinctions (a function of the absolute size of the populations). It is becoming increasingly clear, however, that the forest habitat itself is modified to a large extent by the proximity of the adjoining secondary habitats (Lovejoy *et al.*, 1984, 1986; Kapos, 1989; Malcolm, 1994; Camargo and Kapos, 1995), and hence that the carrying capacities of the fragment will be determined not just by their size, but also by the extent of edge-induced habitat changes. Thus, quite independently of insularization, insect density in fragments may vary with fragment size (Murcia, 1995), just as expected from island theory.

In the present study, I attempt to answer two questions: does fragmentation in the central Amazon influence insect biomass? and does insularization play a role in determining habitat-specific variation in insect biomass? Concerning the latter question, two hypotheses were tested of habitat-specific variation in insect biomass that were independent of insularization *per se*: (i) insect biomass varied solely as a function of proximity to edges; and (ii) variation in insect biomass was solely attributable to variation in vegetation structure.

Table 25.1 Characteristics of the four field sites approximately 80 km north of Manaus, Brazil

	Site			
Site characteristic	*1*	*2*	*3*	*4*
Ranch name	Esteio	Esteio	Porto Alegre	Dimona
Type of matrix habitat surrounding fragments and abutting CF edge	Pasture	Secondary forest	Secondary forest	Pasture
History of matrix habitat	Forest cut and burned in 1980; grass planted, some areas maintained as pasture, others abandoned; all secondary vegetation cleared in dry season 1987	Forest cut in 1983, never burned or recut; secondary forest approx. 12 m high at time of study	Forest cut in 1983, never burned or recut; secondary forest approx. 12 m high at time of study	Forest cut and burned in 1984; grass planted, maintained as pasture, secondary vegetation most recently cleared in dry season 1987
Number of 1-ha sub-sampling-units	15 (4 in CF, 3 along CF edge, 4 in 10-ha fragment, 1 in 1-ha fragment, 3 in pasture)	14 (4 in CF, 2 along CF edge, 4 in 10-ha fragment, 1 in 1-ha fragment, 3 in secondary forest)	14 (4 in CF, 3 along CF edge, 4 in 10-ha fragment, 1 in 1-ha fragment, 2 in secondary forest)	15 (4 in CF, 3 along CF edge, 4 in 10-ha fragment, 1 in 1-ha fragment, 3 in pasture)
Forest fragment identification codes*	1104 (1–ha); 1202 (10-ha)	1112 (1-ha); 1207 (10-ha)	3114 (1-ha); 3209 (10-ha)	2107 (1-ha); 2206(10-ha)

* See Lovejoy *et al.* (1986) for general descriptions of the fragments.
CF, continuous forest.

MATERIALS AND METHODS

Study area

The study, part of the Biological Dynamics of Forest Fragments Project, was carried out on three cattle ranches 80 km north of Manaus, Brazil. Primary forest in the area is on moderately rugged terrain and is dissected by small creeks that form the headwaters of the tributaries of three small rivers: the Cuieiras, the Preto da Eva, and the Urubú. Soils are nutrient-poor latosols and annual rainfall averages about 2200 mm, with a pronounced dry season of <100 mm per month from July to September. Further details of the study site can be found in Lovejoy *et al.*, (1984, 1986) and Lovejoy and Bierregaard (1990).

Experimental design

At each of four sites, arthropod (primarily insect) biomass and vegetation structure were measured in five major habitat types: (i) continuous forest (CF); (ii) the edge of CF; (iii) a 10-ha forest fragment; (iv) a 1-ha forest fragment; and (v) the matrix of secondary vegetation surrounding the forest fragments (at two sites it was pasture, at the other two it was young secondary forest). Two to four 1-ha (100 × 100 m) sub-sampling units were established in each habitat at a site, for a total of 58 units (Table 25.1). The average proximity of a unit to forest/clear-cut edges (and hence the importance of edge effects) varied with habitat type: units in CF had no edges and were at least 400 m from the nearest edge, units in CF edge had an edge along one side, units in 10-ha fragments had an edge along one or two sides (in a few cases, units were in the fragment interior), and units in 1-ha fragments had edges along all four sides (Malcolm, 1994). The distance from a fragment to the nearest continuous forest varied from 100 to 800 m. Sub-sampling units in the matrix were at least 150 m from the forest.

Insect sampling techniques

Each site was sampled for arthropods once during each of three censuses: (i) September 1987 to February 1988; (ii) March 1988 to September 1988; and (iii) October 1988 to March 1989. The five habitats at a site were sampled simultaneously and/or sequentially during each census, so that all habitats were sampled within, on average, a 6-week period (range 3–9 weeks). Exceptions were the first census of site 1 (15 weeks), the first census of site 4 (17 weeks) and the second census of site 3 (15 weeks).

Sticky traps

Each 1-ha unit was divided into two 50 × 100 m halves, and in the centre of each half was established a 100-m long transect marked at 20-m intervals. Transects in units on the edge of continuous forest and on the edge of 10-ha fragments were perpendicular to the edge. Sticky traps were set at three randomly selected points per transect during census one, and at three of the remaining six points in the unit during census two. Sticky traps were not set during census three. Each point had two traps (except in matrix habitat, where only understorey traps were set): one approximately 14 m high (overstorey), suspended 20 cm below a 'pulley' small mammal trap (Malcolm, 1991a), and another 0.5 m high (understorey).

Traps were a 20 × 20 cm piece of glass covered on both sides with an approximately 2-mm thick layer of 'tangle-trap'. Tape along two edges reduced the trap surface to 20 × 17 cm per side. After eight consecutive trap-nights, captures were washed in gasoline and stored in a solution of alcohol, acetic acid and formalin (FAA). A few months later, samples were washed in gasoline four more times, dried to constant weight, and weighed to the nearest mg. Arthropod taxa were not identified, hence analyses were performed using total dry weight (= biomass). Because of vagaries of sampling, a few traps were set for less than eight nights; effort was therefore standardized by calculating biomass per trap-night.

Terrestrial baited pitfall traps

Two 100-m long parallel transects were established in each 1-ha unit, one transect being located 20 m from each unit border. Again, transects were marked at 20-m intervals and were situated perpendicular to the forest edge along continuous forest edge and along the edge of 10-ha fragments. Traps were set at three randomly selected points per transect during census two and at the three remaining points during census three. Terrestrial pitfall traps were not set during census one. As bait, a piece of banana was suspended over the cup (Figure 25.1). After three trap-nights, captures were removed and the bait and soap solution replaced; after a further three nights additional captures and the trap were removed and the captures stored in FAA.

Arachnids and insects (other than larvae) were identified to order (CSIRO, 1973), except for spiders and harvestmen (which were lumped), whip-scorpions (which were lumped), and larvae (not identified). Other invertebrates were identified to class. In addition, individuals were assigned to one of five size classes: (i) <5 mm; (ii) 5–10 mm; (iii) 10–20 mm; (iv) 20–35 mm; and (v) 35–55 mm. Individuals >55 mm in length

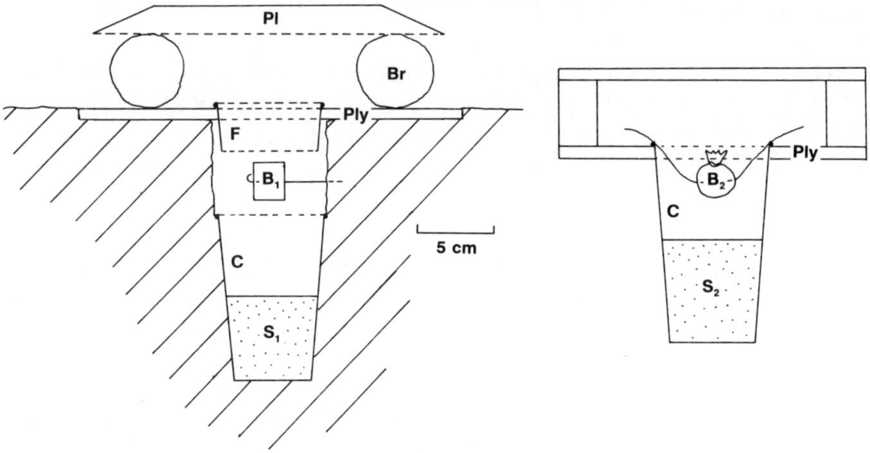

Figure 25.1 Scale drawings of terrestrial (left) and arboreal (right) baited pitfall traps. Pl, plastic plate; Br, sticks to support plate; Ply, plywood; F, funnel; B_1, banana; C, plastic cup; S_1, soap and water solution; B_2, banana in cloth sac; S_2, soap, 5% ethanol and water solution.

were measured to the nearest mm. Biomass was estimated from the relationship:

$$W = 0.0305\ L^{2.62}, \tag{25.1}$$

where W is the dry weight in mg and L the length in mm (Rogers *et al.*, 1976; Winnett-Murray, 1986). The midpoint of each size class was used for L (except for arthropods >55 mm in length, where the actual length was used). Total biomass was the summed biomass of all taxa captured. Again, trapping effort was standardized by calculating biomass per trap-night. To verify the use of Equation (25.1) as a method to estimate biomass, estimated dry biomass was regressed against wet biomass for 24 samples (three samples from each of eight units; see the randomized block experiment described below). Correlation was very high (model with no intercept: $R^2 = 0.98$, $P < 0.01$).

Arboreal baited pitfall traps

During census three, arboreal pitfall traps were set at three randomly selected points on the same transects established for sticky traps, to provide six traps per 1-ha unit. Traps were suspended immediately below small mammal traps at approximately 14 m in height in primary forest and 2 m in height in secondary forest. Traps were not set in pasture. As bait, a small cloth sac containing banana was suspended over the cup (Figure 25.1).

Traps at site 2 were set for eight nights, whereas at sites 3 and 4, and at four units in site 1, they were set for six nights. To compare results given this unequal effort, it was assumed that capture rate did not vary with the number of nights a trap was set and unequal effort was controlled for by calculating biomass per trap-night. To test the validity of this assumption, a randomized block experiment was conducted at site 1. Within each of eight 1-ha units, two randomly selected traps were set for four nights, two for six nights, and two for eight nights (= treatment). The arthropod samples were drained for 5 minutes, and wet biomass measured to the nearest g. Mean biomass per trap-night did not differ significantly among the three treatments ($P = 0.14$). Arthropods were stored in FAA, identified, measured and their biomass estimated as described for terrestrial pitfall traps.

Vegetation sampling techniques

In each 1-ha unit, vertical sightings were made into the canopy and, with the aid of a range-finder, foliage density estimated in six height intervals: 0–2, 2–5, 5–10, 10–20, 20–30 and 30–40 m. From these measurements estimates of understorey (0–5 m) and overstorey (10–30 m) vegetation density (hectare-specific means), variability (residuals of the hectare-specific variances regressed against the means) and vegetation 'surface area' (an inverse measure of vegetation 'grain' size, estimated from residuals of hectare-specific surface areas regressed on variances and log-transformed means) were derived (see Malcolm, 1995 for details).

Data analysis

Variance in arthropod biomass among habitats

As 1-ha units were sub-samples, not replicates, data from the units were combined in each habitat-by-census-by-site combination and the mean calculated. Because the means were calculated from different numbers of 1-ha units, weighted analyses were used where possible. A three-factor ANOVA (matrix type, habitat and census) with two repeated measures (habitat and census) was used to analyse the sticky trap and terrestrial pitfall data, and a two-factor ANOVA (matrix type and habitat) with one repeated measure (habitat) was used to analyse the arboreal pitfall data [see Cody and Smith (1987) for examples of these designs]. Matrix type at two of the sites was pasture and at the other two it was secondary forest; hence site was 'nested' within matrix (Cody and Smith, 1987). Two analyses were conducted: one with all five habitats, and, because matrix type seemed unlikely to influence samples in continuous forest, one in which continuous forest was excluded. Total biomass was not transformed, whereas biomasses of frequently captured taxa (≥200 individuals) were square-root-transformed. Biomasses of infrequently captured taxa (<200 individuals) were not analysed.

Tests of hypotheses: proximity to edges and vegetation structure

According to the 'edge' hypothesis, variation in arthropod biomass among primary forest habitats was solely attributable to proximity to the forest/matrix edge. The average proportion of edge-modified forest in a 1-ha sampling unit varied in the sequence: CF, CF edge, 10-ha fragment, and 1-ha fragment; hence, a simple prediction from the hypothesis was that it would be possible to rank habitat-specific biomass in the same sequence. Isotonic regression (Gaines and Rice, 1990) was used to test for the predicted ordering. The test is similar to ANOVA, except that the alternative hypothesis is directional. Therefore, for taxa that exhibited significant habitat effects (as tested by the ANOVAs described above), site effects were removed (by subtracting the grand site mean) and isotonic regression used to test the null hypothesis of no habitat effect against one of two alternatives: (i) $\mu_{CF} \leq \mu_{CF\ Edge} \leq \mu_{10\ ha} \leq \mu_{1\ ha}$ (with at least one strict inequality); or (ii) $\mu_{CF} \geq \mu_{CF\ Edge} \geq \mu_{10\ ha} \geq \mu_{1\ ha}$ (with at least one strict inequality). The choice of alternative hypothesis was made *a posteriori*, so the test was a liberal one.

According to the 'vegetation' hypothesis, variation in arthropod biomass among habitats was attributable to variation in vegetation structure. Note that if variation in vegetation structure among habitats is a function of proximity to edge, this hypothesis and the previous one are equivalent. Under this hypothesis, it was predicted that when arthropod biomass varied significantly among primary forest habitats (as tested by the ANOVAs described above), biomass would also be significantly correlated with variation in vegetation structure among the habitats. The predicted relationship between arthropod biomass and vegetation structure was tested using simple correlation. The first two sets of variables from a canonical correlation of arthropod biomass versus vegetation structure were also examined, but because of the restrictive assumption of multivariate normality, the analyses were not used for statistical testing.

More specifically, under this second hypothesis any significant variation in arthropod biomass among habitats would be expected to disappear given vegetation structure as a covariate. Therefore, two treatment groups (non-isolated and isolated sites) were defined and the hypothesis of equal arthropod biomass given equal vegetation structure

Table 25.2 Mean biomass (dry weight in mg/trap-night)* from understorey and overstorey sticky traps during two censuses of five habitats at four sites. The matrix at two of the sites was pasture; at the other two it was secondary forest. Before calculating means, site effects were removed by computing $x_{ijk} - x_{i.k} + x_{....}$, where x_{ijk} is the biomass per trapnight during the ith census of the jth habitat at the kth site

		Matrix type								
		Pasture (n = 2)			Secondary forest (n = 2)			Combined (n = 4)		
		Census			Census			Census		
Trap height	Habitat	1	2	$\bar{X}$	1	2	$\bar{X}$	1	2	$\bar{X}$
Understorey	Matrix	9.5	24.3	16.9	4.7	7.2	5.9	7.1	15.8	11.4
		(0.2)	(3.3)	(1.7)	(3.6)	(0.1)	(1.7)	(3.4)	(10.1)	(6.5)
	1-ha fragment	11.4	5.5	8.5	19.3	14.5	16.9	15.4	10.0	12.7
		(2.2)	(1.1)	(1.7)	(1.4)	(3.4)	(1.0)	(4.8)	(5.5)	(5.0)
	10-ha fragment	7.0	2.2	4.6	10.3	11.2	10.8	8.7	6.7	7.7
		(1.2)	(1.8)	(1.5)	(2.2)	(3.7)	(0.7)	(2.4)	(5.7)	(3.7)
	CF edge	9.8	6.9	8.3	5.3	10.7	8.0	7.5	8.8	8.1
		(0.7)	(5.5)	(2.4)	(0.3)	(5.7)	(3.0)	(2.6)	(5.1)	(2.2)
	CF	4.1	2.8	3.4	2.1	−1.6	0.2	3.1	0.6	1.8
		(0.1)	(1.5)	(0.8)	(0.3)	(12.7)	(6.5)	(1.1)	(7.8)	(4.2)
Overstorey	1-ha fragment	1.9	3.3	2.6	9.5	3.5	6.5	5.7	3.4	4.5
		(2.5)	(1.3)	(0.6)	(1.7)	(1.9)	(0.1)	(4.7)	(1.3)	(2.3)
	10-ha fragment	3.0	4.0	3.5	8.3	9.5	8.9	5.6	6.7	6.2
		(2.6)	(0.3)	(1.4)	(4.6)	(1.2)	(2.9)	(4.3)	(3.2)	(3.6)
	CF edge	9.2	8.0	8.6	0.7	6.6	3.6	4.9	7.3	6.1
		(2.7)	(2.3)	(0.2)	(4.6)	(8.5)	(6.6)	(5.8)	(5.2)	(4.8)
	CF	5.8	4.5	5.2	1.4	0.2	0.8	3.6	2.4	3.0
		(2.5)	(1.2)	(1.8)	(1.7)	(9.3)	(3.8)	(3.1)	(5.9)	(3.5)

* Values in parentheses are standard deviations.
CF, continuous forest.

tested by use of analysis of covariance (ANCOVA). In the special case of a linear relationship, the hypothesis of equal arthropod biomass given equal vegetation structure was equal to the null hypothesis of no treatment effect in ANCOVA. Significant interaction (i.e. a slope effect) could indicate a non-linear, but continuous relationship, or a discontinuity between the two treatment groups.

Isotonic regression was performed using a program supplied by Gaines and Rice (1990). Other analyses were performed using SAS (SAS, 1985). Statistical tests were judged significant at $P < 0.05$.

RESULTS

Sticky traps

The 1-ha fragments had the greatest understorey arthropod biomass, 10-ha fragments and CF edge were intermediate, and CF had the lowest biomass (habitat effect $F_{4,8}$ = 6.74; P = 0.01) (Table 25.2). Biomass in pasture was greater than in all other habitats, whereas biomass in secondary forest only exceeded that in CF (habitat-by-matrix interaction $F_{4,8}$ = 4.96, P = 0.03).

In the analysis of all habitats except CF, the habitat effect and habitat-by-matrix interaction described above were again significant ($F_{3,6}$ = 6.53, P = 0.03 and $F_{3,6}$ = 23.59, P <0.01, respectively). In addition, biomass in the matrix increased from census one to census two (habitat-by-census interaction $F_{3,6}$ = 6.17, P = 0.03), especially in pasture (habitat-by-matrix-by-census interaction $F_{3,6}$ = 4.69, P = 0.05).

Significant interaction terms were attributable to differences in arthropod biomass in the matrix, as none was significant when matrix habitat was excluded from the analyses. In an analysis of the four primary forest habitats for example, habitat-by-matrix interaction was no longer significant ($F_{3,6}$ = 2.23, P = 0.19) and habitat remained significant ($F_{4,8}$ = 5.29, P = 0.04, isotonic regression P <0.01). The remaining weak interaction was due to higher biomass in fragments surrounded by secondary forest than in those surrounded by pasture.

In the three-way, weighted ANOVA of overstorey biomass per trap-night, no main effect or interaction terms were significant, either when all habitats were included, or when CF was excluded. As in the analysis of understorey biomass excluding matrix habitat, weak habitat-by-matrix interaction ($F_{3,6}$ = 2.65, P = 0.14 for all habitats; $F_{2,4}$ = 2.58, P = 0.19 with CF excluded) was due to higher biomass in fragments surrounded by secondary forest than in those surrounded by pasture (Table 25.2).

To compare differences in the ratio of overstorey to understorey biomass, overstorey biomass was regressed against understorey biomass, and deviations from the line Y = X analysed (Figure 25.2). A three-way, weighted ANOVA was close to significant for habitat ($F_{3,6}$ = 4.53, P = 0.06), but not for other main effects or for interaction terms. A one-way ANOVA comparing mean deviations among habitats (censuses combined) was significant ($F_{3,12}$ = 6.73, P <0.01, isotonic regression P <0.01). In continuous forest, biomass in the overstorey was greater than in the understorey, whereas the converse was true in 1-ha fragments. These results thus confirmed the separate analyses on the understorey and overstorey data which showed that understorey biomass increased with increased edge in a habitat, whereas overstorey biomass changed little and hence the ratio overstorey : understorey biomass decreased

Table 25.3 Weighted correlation between insect biomass per trap-night and vegetation structure, and for pitfall traps, between each set of variables and their first and second canonical variables (CV). See text for definitions of vegetation variables

Data set	*Taxon*	*Understorey thickness*	*Overstorey thickness*	*Understorey residual variance*	*Overstorey residual variance*	*Understorey residual surface area*	*Overstorey residual surface area*	*Insect CV 1*	*Insect CV 2*
Understorey sticky	Total bimass	0.46	–0.69**	–0.18	–0.06	–0.09	–0.22	–	–
Overstorey sticky	Total biomass	0.29	–0.29	0.33	–0.02	0.06	–0.44	–	–
Transformed O:U	Total biomass	0.43	–0.64**	–0.49	0.01	–0.14	0.03	–	–
Terrestrial pitfall	Total biomass	0.36	0.21	0.68**	0.38	0.14	–0.54*	0.72	–0.04
	Blattodea	0.31	0.35	0.73**	0.38	0.06	–0.45	0.78	–0.13
	Coleoptera	0.30	–0.11	0.14	0.02	0.18	–0.59*	0.07	0.10
	Dermaptera	0.62*	–0.60*	0.08	0.32	0.01	–0.30	0.30	0.73
	Diptera	0.13	–0.19	0.23	–0.01	0.31	–0.32	0.13	0.13
	Hymenoptera	–0.05	0.11	–0.02	0.10	0.06	–0.02	–0.06	–0.11
	Isoptera	0.11	0.23	0.34	0.14	0.17	–0.20	0.28	–0.07
	Orthoptera	–0.38	0.16	<0.01	–0.16	0.22	0.06	–0.19	–0.36
	Arachnida	0.13	0.12	0.65**	0.45	0.15	–0.18	0.73	–0.05
	Vegetation CV 1	0.34	0.29	0.80	0.55	–0.07	–0.23	>0.99	–
	Vegetation CV 2	0.82	–0.78	–0.35	0.24	–0.27	–0.17	–	0.94
Arboreal pitfall	Total biomass	–0.12	0.31	0.32	0.25	0.14	–0.24	0.21	0.64
	Blattodea	–0.22	0.41	0.23	0.09	–0.06	–0.02	0.25	0.48
	Coleoptera	0.08	0.26	0.68**	0.12	0.26	–0.55*	0.67	0.61
	Diptera	–0.39	0.37	0.11	–0.05	0.19	0.06	–0.13	0.59
	Hymenoptera	0.58**	–0.30	–0.03	0.50*	–0.35	–0.08	0.28	–0.63
	Lepidoptera	0.05	0.13	0.20	0.38	0.01	–0.23	0.19	0.40
	Vegetation CV 1	0.41	0.18	0.68	0.31	–0.21	–0.41	0.96	–
	Vegetation CV 2	–0.59	0.53	0.42	–0.21	0.58	–0.34	–	0.76

O:U overstorey : Understorey biomass.
$*P < 0.05$; $**P < 0.01$

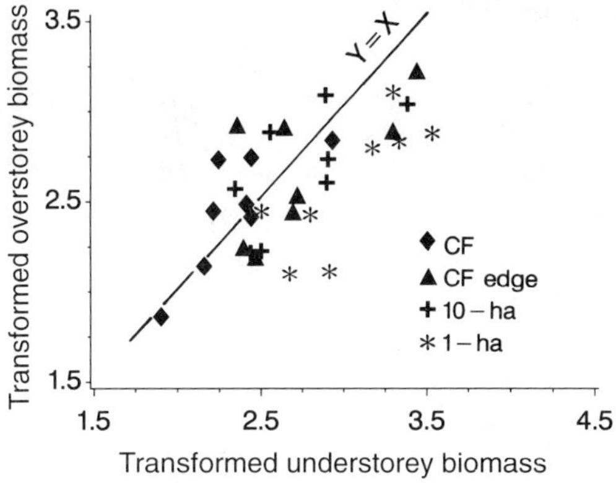

Figure 25.2 Log-transformed overstorey biomass versus log-transformed understorey biomass as determined from sticky traps (five was added to each biomass, and the sums were log transformed). Data are from two censuses.

with increasing edge in a habitat. Because matrix habitat was excluded, no interaction terms were significant. Finally, the slightly higher biomass in fragments surrounded by secondary forest than in those surrounded by pasture was true in both the overstorey and understorey samples, hence habitat-by-matrix interaction was not close to significant ($F_{3,6}$ = 0.78, *P* = 0.55).

Weighted correlations between sticky trap biomass in non-matrix habitats and the six vegetation variables are shown in Table 25.3. The relationship between understorey arthropod biomass and overstorey thickness was strong (*P* <0.01) and negative; habitats with more open overstories had higher understorey arthropod biomass (Figure 25.3(a)). Not surprisingly (given the negative correlation between overstorey and understorey thickness), the correlation between understorey thickness and understorey arthropod biomass was positive, although not quite significant (*P* = 0.07). The regression between arthropod biomass and overstorey thickness among isolated habitats (1- and 10-ha fragments) was similar to that among non-isolated habitats (CF edge and CF) (ANCOVA $F_{1,13}$ = 1.37, *P* = 0.26). None of the correlations between overstorey arthropod biomass and the vegetation variables was significant. Analysis of the overstorey : understorey biomass ratios gave corroborative results. The ratio decreased with increasing overstorey thickness (*P* <0.01, Figure 25.3(b)) and there was also some evidence of a decrease in the ratio with decreasing understorey thickness and increasing understorey variance (p = 0.10 and 0.06 respectively). These three correlations reflected correlations among the vegetation variables. In addition to the

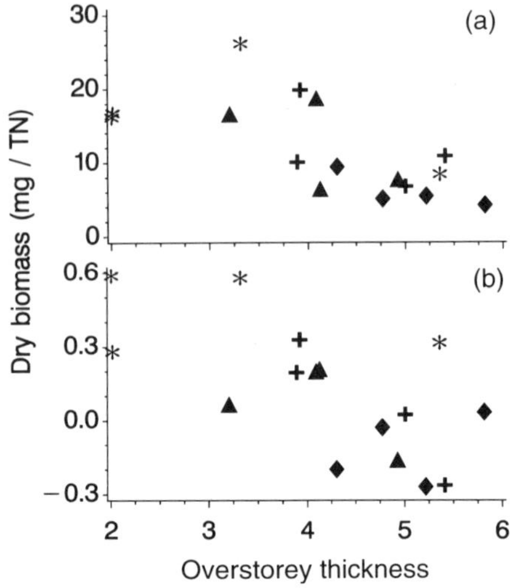

Figure 25.3 (a) Mean dry biomass per trap-night (TN) of insects from understorey sticky traps and (b) understorey : overstorey biomass from sticky traps versus overstorey thickness. Understorey : overstorey biomasses are mean deviations (per habitat/site combination) from the line Y = X in Figure 25.2. Symbols are as in Figure 25.2.

negative correlation between overstorey and understorey thickness, high overstorey thickness was correlated with high understorey variance. Again, with overstorey thickness as a covariate, arthropod biomass did not appear to differ between isolated and non-isolated habitats (ANCOVA $F_{1,13} = 2.06$, $P = 0.17$).

Terrestrial pitfall traps

Averaged across the 58 1-ha units sampled, the rank order of taxon biomass (mg/trap-night and total number of individuals in parentheses) was Blattodea (64; 4556), Orthoptera (29; 2681), Diplopoda (28; 172), Hymenoptera (20; 63 932), Coleoptera (11; 6559), Dermaptera (9; 1468), Chilopoda (3; 19), Lepidoptera (2; 41), Arachnida (1; 355), Scorpiones (1; 23), Isoptera (1; 9168), Annelida (1; 1), Diptera (1; 5611), Gastropoda (1; 7), whip scorpions (<1; 10), mites (<1; 69), Homoptera (<1; 28), Malacostraca (<1; 14), Hemiptera (<1; 2), Diplura (<1; 6), Neuroptera (<1; 2), Archaeognatha (<1; 1), Collembola (<1; 4) and Pseuodoscorpiones (<1; 3) (see Table 25.4 for biomass in the various habitats).

A weighted, three-way ANOVA of total biomass was not significant for main effects or interaction terms, either when all five habitats were tested ($F_{1,2}$ = 2.42, P = 0.26 for matrix, otherwise P (≥0.40), or when CF was excluded ($F_{1,2}$ = 3.78, P = 0.19 for matrix, otherwise P (≥0.30). Nor was multivariate analysis of variance on the eight frequently caught taxa (>200 individuals) significant for main effects or interaction terms (all five habitats, Wilk's Lambda, P always >0.30). Degrees of freedom were insufficient for a multivariate test of a matrix effect, or for a test excluding CF. Univariate, three-way ANOVAs on each of the eight taxa yielded the following significant results: matrix and matrix-by-habitat-by-census for Diptera (all habitats $F_{1,2}$ = 25.43, P = 0.04 and $F_{4,8}$ = 4.12, P = 0.04, respectively), matrix for Hymenoptera (all habitats $F_{1,2}$ = 22.49, P = 0.04), habitat (CF excluded $F_{3,6}$ = 5.41, P = 0.04, isotonic regression P = 0.02) and matrix-by-habitat (all habitats $F_{4,8}$ = 4.81, P = 0.03, CF excluded $F_{3,6}$ = 12.72, P <0.01) for Dermaptera, and matrix-by-habitat for Coleoptera (all habitats $F_{4,8}$ = 4.12, P = 0.04; CF excluded $F_{3,6}$ = 5.01, P = 0.04). Of the 112 F-statistics (two tests with seven statistics for eight taxa), only 7% were significant, and probably represented type I errors.

The following correlations between terrestrial arthropod biomass and vegetation structure were significant: (i) total, Blattodea, and Arachnida biomass increased with increasing variance in understorey foliage thickness (corroborated by the first 'arthropod' and 'vegetation' canonical variables); (ii) total and Coleoptera biomass decreased with increasing overstorey grain; and (iii) Dermaptera biomass increased with increasing understorey thickness and decreased with increasing overstorey thickness (corroborated by the second canonical variables) (Table 25.3). Recall that among frequently captured taxa, only Dermaptera showed evidence of variation in biomass with habitat. Thus, as predicted, Dermaptera biomass varied with understorey and overstorey thickness. Analysis of covariance comparing isolated (1- and 10-ha fragments) and non-isolated (CF edge and CF) habitats was not significant for this taxon ($F_{1,13}$ = 2.20, P = 0.16, Figure 25.4(a)).

Arboreal pitfall traps

Averaged across the 52 1-ha units sampled, the rank order of estimated taxon biomass (milligrams per trap-night and total number of individuals in parentheses) was Lepidoptera (192; 7797), Diptera (106; 41 803), Coleoptera (43; 6277), Blattodea (19; 844), Hymenoptera (11; 2870), Orthoptera (6; 62), Neuroptera (3; 114), Mantodea (<1; 1), Arachnida (<1; 18), Dermaptera (<1; 2), Isoptera (<1; 4), Pseudoscorpiones (<1; 2) and Homoptera (<1; 1).

A two-way, weighted ANOVA on total biomass (secondary forest excluded) was significant for habitat ($F_{3,6}$ = 4.89, P <0.05, isotonic regression P <0.01), and close to significant when CF was excluded ($F_{2,4}$ = 6.37, P = 0.06), but was not significant for matrix effects or habitat–matrix

Table 25.4 Estimated biomass (dry weight in mg per terrestrial pitfall trap-night) (± S.D.) of all captures and taxa with ≥200 individuals. Before calculating means for the total and for each taxon, site effects were removed by computing $x_{ijk} - x_{i.k} + x_{...}$, where x_{ijk} is the biomass per trapnight during the ith census in the jth habitat at the kth site. Data from the two censuses are combined

		Habitat				
Matrix type	*Taxon*	*Matrix*	*1-ha fragment*	*10-ha fragment*	*CF edge*	*CF*
Pasture	Total	179.4 ± 54.2	173.0 ± 30.9	139.2 ± 41.7	237.4 ± 24.6	140.1 ± 40.4
(n = 2)	Blattodea	26.7 ± 28.1	78.0 ± 13.9	46.8 ± 3.4	115.0 ± 44.4	46.4 ± 5.8
	Coleoptera	10.4 ± 6.8	10.3 ± 5.3	7.3 ± 2.7	14.7 ± 2.7	15.9 ± 6.9
	Dermaptera	0.2 ± 3.6	17.2 ± 2.3	7.4 ± 2.7	19.8 ± 4.6	6.0 ± 6.1
	Diptera	0.8 ± 0.8	0.6 ± 0.9	0.8 ± 0.1	0.8 ± 0.5	0.7 ± 0.3
	Hymenoptera	31.7 ± 3.6	14.1 ± 1.1	16.8 ± 1.1	20.5 ± 3.5	17.0 ± 7.0
	Isoptera	<0.1 ± 0.7	2.8 ± 3.4	0.2 ± 1.5	1.4 ± 0.2	1.0 ± 1.3
	Orthoptera	19.9 ± 3.3	16.8 ± 2.5	26.8 ± 18.0	35.3 ± 9.6	38.8 ± 14.3
	Arachnida	2.1 ± 0.8	0.5 ± 0.1	1.1 ± 0.4	2.7 ± 0.8	0.4 ± 0.5
Secondary Forest	Totai	193.8 ± 75.1	142.7 ± 24.5	256.8 ± 142.8	156.9 ± 94.5	118.7 ± 2.3
(n = 2)	Blattodea	27.5 ± 27.0	44.1 ± 38.4	146.8 ± 129.0	55.6 ± 33.4	38.9 ± 30.2
	Coleoptera	3.8 ± 7.6	19.2 ± 17.2	23.4 ± 0.1	10.8 ± 2.5	1.3 ± 7.0
	Dermaptera	14.0 ± 4.7	12.7 ± 4.2	11.7 ± 0.2	8.3 ± 4.4	4.0 ± 4.8
	Diptera	0.6 ± 0.6	0.9 ± 0.4	1.0 ± 0.3	0.9 ± 0.1	0.4 ± 0.0
	Hymenoptera	20.8 ± 26.0	23.1 ± 4.9	17.4 ± 3.8	9.3 ± 7.9	29.4 ± 32.8
	Isoptera	0.5 ± 0.3	0.8 ± 0.3	2.7 ± 0.5	0.5 ± 0.4	0.8 ± 0.5
	Orthoptera	50.5 ± 13.3	13.6 ± 16.7	22.2 ± 7.1	21.8 ± 0.9	29.6 ± 9.6
	Arachnida	1.2 ± 0.3	0.7 ± 0.6	2.0 ± 1.1	1.6 ± 0.6	1.3 ± 0.5

Table 25.4 continued

Matrix type	*Taxon*	*Habitat*				
		Matrix	*1-ha fragment*	*10-ha fragment*	*CF edge*	*CF*
Combined	Total	186.6 ± 54.1	157.8 ± 28.7	198.0 ± 109.5	197.2 ± 73.1	129.4 ± 26.4
(*n* = 4)	Blattodea	27.1 ± 22.5	61.0 ± 30.6	96.8 ± 94.2	85.3 ± 46.9	42.6 ± 18.3
	Coleoptera	7.1 ± 7.0	14.8 ± 11.6	15.4 ± 9.5	12.7 ± 3.1	8.6 ± 10.2
	Dermaptera	7.1 ± 8.7	15.0 ± 3.8	9.5 ± 2.9	14.1 ± 7.6	5.0 ± 4.6
	Diptera	0.7 ± 0.6	0.8 ± 0.6	0.9 ± 0.2	0.9 ± 0.3	0.6 ± 0.2
	Hymenoptera	26.2 ± 16.4	18.6 ± 6.0	17.1 ± 2.3	14.9 ± 8.2	23.2 ± 20.7
	Isoptera	0.2 ± 0.5	1.8 ± 2.3	1.5 ± 1.7	1.0 ± 0.6	0.9 ± 0.8
	Orthoptera	35.2 ± 19.4	15.2 ± 9.9	24.5 ± 11.5	28.5 ± 9.6	34.2 ± 11.3
	Arachnida	1.7 ± 0.7	0.6 ± 0.4	1.6 ± 0.9	2.1 ± 0.9	0.9 ± 0.7

CF, continuous forest.

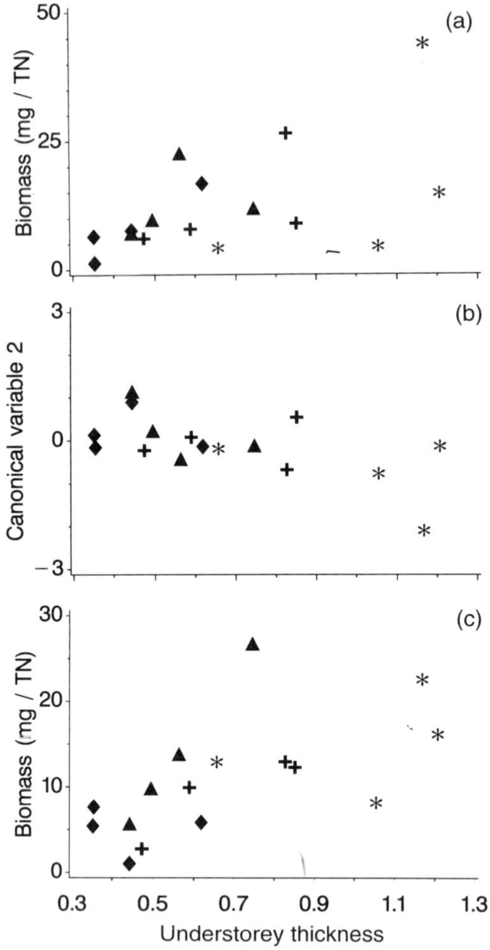

Figure 25.4 (a) Dermaptera biomass per trap-night (TN) from terrestrial pitfall traps, (b) canonical variable 2 and (c) Hymenoptera biomass per trap-night from arboreal pitfall traps versus understorey thickness. Symbols are as in Figure 25.2. See text for biomass estimation and derivation of the canonical variable.

interaction. Highest biomass per trap-night was obtained in CF edge, followed by CF, 10-ha fragments and 1-ha fragments (Table 25.5). Matrix and habitat–matrix interaction were not significant for any of the frequently caught taxa, and habitat was significant only for Diptera ($F_{3,6}$ = 10.16, P <0.01 all primary forest habitats, and $F_{2,4}$ = 8.35, P = 0.04 CF excluded, isotonic regression P <0.01). However, among 1-ha fragments, 10-ha fragments and CF edge, all frequently caught taxa but

Hymenoptera were least abundant in 1-ha fragments and most abundant in CF edge (Table 25.5).

To compare arthropod biomass in secondary forest with that in the other habitats a randomized block ANOVA was used and data were included only from the two sites where the matrix was secondary forest. Habitat was close to significant for total biomass ($F_{4,4}$ = 5.87, P = 0.06) and significant for Diptera biomass ($F_{4,4}$ = 25.95, P <0.01). In general, biomass in secondary forest was similar to that in 1-ha fragments and lower than that in CF edge and CF (Table 25.6).

Weighted simple correlation between arthropod biomass and vegetation variables yielded the following significant relationships: (i) Coleoptera biomass increased with increasing understorey variance and overstorey grain (corroborated by the first canonical variables); and (ii) Hymenoptera biomass increased with increasing understorey thickness and increasing variance in overstorey foliage thickness (Table 25.3). The second arthropod canonical variable contrasted Hymenoptera biomass with total biomass and the biomass of the other taxa. Hymenoptera biomass increased with increasing understorey thickness, decreasing overstorey thickness, and increasing understorey grain, whereas for other taxa and for total arthropod biomass, the opposite was true. The second canonical correlation thus described differences in biomass observed among the various habitats; biomass of Hymenoptera increased with increased proportions of edge in a habitat, whereas the opposite was true for other taxa and for total biomass. With understorey thickness as a covariate, there was little evidence to suggest that isolated and non-isolated habitats differed, either for Hymenoptera biomass (Figure 25.4(c), $F_{1,13}$ = 0.09, P = 0.76) or for the canonical variable (Figure 25.4(b), $F_{1,13}$ = 1.09, P = 0.32).

DISCUSSION

Fragmentation of once continuous forest had a profound effect on the spatial variability of arthropod biomass. Significant differences among habitats were obtained for: (i) total biomass from understorey sticky traps; (ii) Dermaptera biomass from terrestrial pitfall traps; and (iii) total biomass from arboreal pitfall traps. Also, the abundance of most taxa captured in arboreal pitfall traps seemed to vary systematically with habitat type, but significant differences among habitats were only obtained for Diptera.

Three lines of evidence suggest that this variation among primary forest habitats is due to edge-induced habitat changes. First, just as the proportion of edge-modified forest in a habitat followed the sequence: CF, CF edge, 10-ha fragment and 1-ha fragment, so did insect biomass. In all cases, when significant variation in arthropod biomass was found

Table 25.5 Estimated dry biomass (mg per arboreal pitfall trap-night) (± S.D.) of all captures and taxa with ≥200 individuals. Before calculating means for the total and for each taxon, site effects were removed by computing $x_{jk} - x_{k} + x_{..}$, where x_{jk} is the biomass per trapnight in the jth habitat at the kth site

		Habitat			
Matrix type	*Taxon*	*1-ha fragment*	*10-ha fragment*	*CF edge*	*CF*
Pasture	Total	325.0 ± 50.5	398.6 ± 59.0	468.3 ± 114.8	328.4 ± 5.2
(n = 2)	Lepidoptera	192.1 ± 40.8	223.2 ± 8.5	266.6 ± 61.5	116.9 ± 29.2
	Diptera	56.8 ± 4.3	92.7 ± 53.3	98.6 ± 16.7	139.9 ± 40.9
	Coleoptera	39.0 ± 1.8	43.6 ± 0.2	53.7 ± 13.4	37.8 ± 15.1
	Blattodea	10.0 ± 7.5	12.5 ± 5.0	25.5 ± 5.6	24.4 ± 8.1
	Hymenoptera	21.1 ± 18.4	13.4 ± 4.8	14.4 ± 17.2	1.0 ± 5.9
Secondary forest	Total	143.3 ± 61.9	240.9 ± 33.4	635.8 ± 192.3	501.2 ± 97.0
(n = 2)	Lepidoptera	154.8 ± 13.9	126.8 ± 48.3	326.2 ± 201.6	191.1 ± 139.4
	Diptera	–30.0 ± 40.8	8.0 ± 35.7	171.2 ± 10.5	238.7 ± 66.0
	Coleoptera	–7.3 ± 33.5	63.7 ± 58.0	105.7 ± 1.7	12.1 ± 26.2
	Blattodea	11.8 ± 11.1	18.1 ± 2.9	12.3 ± 17.1	30.1 ± 3.1
	Hymenoptera	12.4 ± 7.4	11.2 ± 0.8	11.3 ± 1.8	15.0 ± 6.4
Combined	Total	234.6 ± 115.1	319.7 ± 99.1	552.0 ± 161.5	414.8 ± 114.4
(n = 4)	Lepidoptera	173.5 ± 32.9	175.0 ± 62.5	296.4 ± 126.4	154.0 ± 92.7
	Diptera	13.4 ± 55.4	50.3 ± 61.4	134.9 ± 43.5	189.3 ± 72.6
	Coleoptera	15.9 ± 33.0	53.6 ± 35.4	79.7 ± 31.0	25.9 ± 22.9
	Blattodea	10.9 ± 7.8	15.3 ± 4.6	18.9 ± 12.9	27.2 ± 6.0
	Hymenoptera	16.7 ± 12.5	12.3 ± 3.1	12.9 ± 10.2	8.0 ± 9.5

CF, continuous forest.

Table 25.6 Estimated dry biomass (mg per arboreal pitfall trap-night) (± S.D.) of all captures and taxa with a total of ≥200 individuals. Before calculating means for the total and for each taxon, site effects were removed by computing $x_{jk} - x_{.k} + x_{..}$, where x_{jk} is the biomass per trapnight in the jth habitat at the kth site. Data are from the two sites where the matrix habitat was secondary forest

Taxon	*Habitat*				
	Secondary forest	*1-ha fragment*	*10-ha fragment*	*CF edge*	*CF*
Total	269.8 ± 35.2	232.2 ± 70.7	329.8 ± 42.1	724.7 ± 183.5	590.1 ± 105.8
Lepidoptera	185.2 ± 40.3	179.4 ± 24.0	151.4 ± 58.3	350.8 ± 191.5	215.7 ± 149.4
Diptera	43.5 ± 3.6	9.7 ± 41.7	47.6 ± 36.6	210.9 ± 9.6	278.4 ± 65.1
Coleoptera	23.4 ± 11.3	17.3 ± 30.6	88.3 ± 60.8	130.3 ± 4.5	36.7 ± 23.4
Blattodea	8.6 ± 1.6	13.9 ± 11.5	20.1 ± 3.3	14.4 ± 16.7	32.2 ± 3.5
Hymenoptera	7.4 ± 0.6	9.6 ± 7.2	8.4 ± 0.6	8.4 ± 1.9	12.2 ± 6.5

CF, continuous forest.

among habitats, habitat-specific means could also be ranked in this sequence (as tested by isotonic regression).

Secondly, in almost all cases, variation in biomass among habitats was correlated with understorey and/or overstorey thickness, vegetation variables that appeared to vary according to an edge model (Malcolm, 1994). The exception was Diptera biomass from arboreal pitfalls; however, in the canonical correlation analysis total biomass and the biomass of Diptera and all other taxa (except Hymenoptera) correlated highly and positively with understorey thickness and negatively with overstorey thickness.

Finally, with understorey or overstorey thickness as a covariate, little evidence was found of differences in biomass between 'isolated' primary forest (1- and 10-ha fragments) and 'non-isolated' primary forest (CF edge and CF), suggesting that processes based on insularization *per se* were unimportant in determining arthropod biomass in forest fragments. The measurement of arthropod biomass in CF edge provided information critical to this test of a 'null' model of edge effects, i.e. a model of edge effects independent of island effects. CF edge and fragment edges share many environmental features in common; however, the possibilities for movement of individuals to and from adjacent CF may differ radically between the two. CF edge, therefore, provided a control for changes in community structure resulting solely from edge-driven environmental changes. Because arthropod biomass in fragments could be predicted from vegetation structure in continuous forest and its edge and vice versa, little reason was found to invoke island processes such as density compensation (MacArthur *et al.*, 1972; Case, 1975; Case *et al.*, 1979), differential immigration/extinction (MacArthur and Wilson, 1967), or 'fence' effects (Krebs *et al.*, 1969).

Communities in habitat islands may in some cases be influenced by the action (and perhaps interaction) of both edge effects and insularization-based (immigration/extinction) processes. As a possible example, Webb and Hopkins (1984; see also Hopkins and Webb, 1984) found that as the area of heathland patches decreased, beetle diversity and abundance increased, a result contrary to predictions from island biogeographic theory (MacArthur and Wilson, 1967). They attributed the increase to edge effects from a richer and more abundant beetle community in the surrounding matrix. However, for species 'typical' of heathland, diversity decreased with island area, as predicted by island theory. Two approaches can be used to test whether this decrease was due to a negative edge effect rather than any island effect:

1. Localities with the same multivariate edge response should have the same beetle diversity, regardless of the isolation of the localities or the size of the patch they are located on (the approach used here).
2. Given a suitable edge model, it should be possible to predict beetle diversity based solely on distance to matrix habitat (Malcolm, 1994).

The few data available indicate that insect biomass is greater in the overstorey of tropical forests than in the understorey (Wolda, 1982), a result corroborated by the present data from overstorey and understorey sticky traps. The net effect of fragmentation appeared to be an increase in the proportion of insect biomass close to the ground. Biomass from understorey sticky traps was greater in fragments than in CF, whereas the opposite was true for arboreal pitfall traps. This change in the spatial distribution of insect prey may have important consequences for vertebrate predators, and at least in part may account for more frequent observations of canopy bird and mammal species close to the ground in forest fragments (Bierregaard and Lovejoy, 1989; Malcolm, 1988, 1991b). If insectivorous species are resource-limited, then the shift in prey distributions in fragments may eventually lead to decreases in the densities of canopy predators and to a superabundant understorey insectivore fauna.

Differences in the distribution of prey biomass among habitats may also correlate with other characteristics of the insect community relevant to an insect predator. For example, Winnett-Murray (1986) conducted visual sampling of understorey insects in several habitats in Costa Rica, including pasture and early successional scrub, woodland edge and woodland. As in the present study, she found that biomass was highest in open habitats, intermediate in woodland edges and least in woodland. In addition, she found that several estimates of the temporal and spatial variability of insect populations were higher in forest than in more open habitats, and that insects in the forest were more likely to be in concealed microhabitats. Therefore, she reasoned that predators in more open habitats would be more likely to find not only more prey, but more prey of the same types from place to place and from month to month.

Edge-correlated increases in the biomass of understorey arthropod taxa are probably due in part to increased vegetation volume and productivity in the forest close to the edge. Because of an increased proportion of actively growing tissues, palatability of the vegetation may also be higher along the edge (Janzen, 1973). Alternately, insects may 'overflow' from the adjacent matrix habitat. The importance of the two, i.e. *in situ* versus external increases, probably varies from taxon to taxon. Data from understorey sticky traps in general, suggested that increases along the edge resulted in large part from *in situ* increases: despite large differences in arthropod biomass in the two types of matrix habitat, forests abutting them showed similar arthropod biomasses (matrix effect not significant). In contrast, the matrix effect was significant for Diptera and Hymenoptera from terrestrial pitfall traps, but matrix-by-habitat interaction was not, indicating greater biomass in primary forest abutting pasture sites and in the pastures themselves than in primary forest abutting secondary forest and in the secondary forests themselves. More mobile taxa will presumably overflow more than less mobile taxa. Also, one might expect the relative importance of *in situ* production to decrease as taxon size increases. In the same vein, a less abundant overstorey insect fauna along edges may result in part from decreased production of insects at the edge and in part from fewer arboreal insects in the adjacent matrix. Decreased production at the edge is expected from a decreased resource base: wind damage resulted in less foliage at the edge. If overflow from adjacent, intact canopy is important, then one might expect relatively greater edge populations in CF edge than in fragment edge, a result not predicted by simple edge models (Malcolm, 1991b).

In conclusion, fragmentation of tropical rainforest had a profound effect on arthropod biomass. The increase in understorey and decrease in overstorey biomass along primary forest edges could be predicted from measurements of forest structure and appeared to be independent of any island effects. Extensive habitat and resource changes along edges will likely have important consequences for insect predators and for ecosystem function, and will complicate attempts to apply island biogeography theory to the study of tropical forest fragments.

Acknowledgements

Helpful comments on an earlier draft of the manuscript were made by R. Didham, J. Eisenberg, P. Feinsinger, L. Harris, R. Kiltie, J. Putz, J. Ray, and N. Stork. I am indebted to R. Cardoso, A. Cardoso, J. Santos, C. Martins, D. Oliveira, and J. Voltolini for assistance in the field. R. Bierregaard, J. Eisenberg, and T. Lovejoy provided me with the opportunity to conduct the research. Funding was provided by World Wildlife Fund – US, the Insituto Nacional de Pesquisas da Amazônia, the National Geographic Society, the Tinker Foundation, and Sigma-Xi, and by a postgraduate scholarship from the Natural Sciences and Engineering Research Council of Canada and graduate assistantships from the Department of Wildlife and Range Sciences and the Katharine Ordway Chair of Ecosystem Conservation of the University of Florida. This is publication number 146 in the Biological Dynamics of Forest Fragments Project Technical Series.

REFERENCES

Bierregaard, R.O., Jr and Lovejoy, T.E. (1989) Effects of forest fragmentation on Amazonian understorey bird communities. *Acta Amazônica*, **19**, 215–41.

Camargo, J.L.C. and Kapos, V. (1995) Complex edge effects on soil-moisture and microclimate in central Amazonian forest. *Journal of Tropical Ecology*, **11**, 205–21.

Case, T.J. (1975) Species numbers, density compensation, and colonizing ability of lizards on islands in the Gulf of California. *Ecology*, **56**, 3–18.

Case, T.J., Gilpin, M.E. and Diamond, J.M. (1979) Overexploitation, interference competition, and excess density compensation in insular faunas. *The American Naturalist*, **113**, 843–54.

Cody, R.P. and Smith, J.K. (1987) *Applied Statistics and the SAS Programming Language*, North-Holland, New York.

CSIRO (Commonwealth Scientific and Industrial Research Organization) (1973) *The Insects of Australia; a Textbook for Students and Research Workers*, Melbourne University Press, Melbourne.

Erwin, T.L. (1982) Tropical forests: their richness in Coleoptera and other arthropod species. *The Coleopterist's Bulletin*, **36**, 74–5.

Gaines, S.D. and Rice, W.R. (1990) Analysis of biological data when there are ordered expectations. *The American Naturalist*, **135**, 310–17.

Harrison, J.L. (1962) The distribution of feeding habits among animals in a tropical rain forest. *Journal of Animal Ecology*, **31**, 53–63.

Hopkins, P.J. and Webb, N.R. (1984) The composition of the beetle and spider faunas on fragmented heathlands. *Journal of Applied Ecology*, **21**, 935–46.

Janzen, D.H. (1973) Sweep samples of tropical foliage insects: effects of seasons, vegetation types, elevation, time of day, and insularity. *Ecology*, **54**, 687–708.

Kapos, V. (1989) Effects of isolation on the water status of forest patches in the Brazilian Amazon. *Journal of Tropical Ecology*, **5**, 173–85.

Krebs, C.J., Keller, B.L. and Tamarin, R.H. (1969) *Microtus* population biology: demographic changes in fluctuating populations of *M. ochrogaster* and *M. pennsylvanicus* in southern Indiana. *Ecology*, **50**, 587–607.

Lein, M.R. (1972) A trophic comparison of avifaunas. *Systematic Zoology*, **21**, 135–50.

Lovejoy, T.E. and Bierregaard, R.O., Jr (1990) Central Amazonian forests and the minimum critical size of ecosystems project, in *Four Neotropical Rainforests* (ed. A.H. Gentry), Yale University Press, New Haven, pp. 60–71.

Lovejoy, T.E., Rankin, J.M., Bierregaard, R.O. Jr, Brown, K.S., Emmons, L.H. and Van der Voort, M.E. (1984) Ecosystem decay of Amazon forest fragments, in *Extinctions* (ed. M.N. Nitecki), University of Chicago Press, Chicago, pp. 295–325.

Lovejoy, T.E., Bierregaard, R.O. Jr, Rylands, A.B., Malcolm, J.R., Quintela, C.E., Harper, L.H., Brown, K.S. Jr, Powell, A.H., Powell, G.V.N., Schubart, H.O.R. and Hays, M.B. (1986) Edge and other effects of isolation on Amazon forest fragments, in *Conservation Biology: the Science of Scarcity and Diversity* (ed. M.E. Soulé), Sinauer, Sunderland, pp. 257–85.

MacArthur, R.H. and Wilson, E.O. (1967) *The Theory of Island Biogeography*, Princeton University Press, Princeton.

MacArthur, R.H., Diamond, J.M. and Karr, J.R. (1972) Density compensation in island faunas. *Ecology*, **53**, 330–42.

Malcolm, J.R. (1988) Small mammal abundances in isolated and non-isolated primary forest reserves near Manaus, Brazil. *Acta Amazônica*, **18**, 67–83.

Malcolm, J.R. (1991a) Comparative abundances of Neotropical small mammals by trap height. *Journal of Mammalogy*, **72**, 188–92.

Malcolm, J.R. (1991b) *The Small Mammals of Amazonian Forest Fragments: Pattern and Process*, Unpublished Ph.D Thesis, University of Florida, USA.

Malcolm, J.R. (1994) Edge effects in Amazonian forest fragments. *Ecology*, **75**, 2438–45.

Malcolm, J.R. (1995) Forest structure and the abundance and diversity of neotropical small mammals, in *Forest Canopies* (eds M.D. Lowman and N.M. Nadkarni), Academic Press, San Diego, pp. 179–97.

Murcia, C. (1995) Edge effects in fragmented forests: implications for conservation. *Trends in Ecology and Evolution*, **10**, 58–62.

Robinson, J.G. and Redford, K.H. (1986) Body size, diet, and population density of Neotropical forest mammals. *The American Naturalist*, **128**, 665–80.

Rogers, L.E., Hinds, W.T. and Buschbom, R.L. (1976) A general weight vs length relationship for insects. *Annals of the Entomological Society of America*, **69**, 387–9.

SAS (1985) *SAS User's Guide: Statistics, Version 5 Edition*, SAS Institute Inc., Cary, USA.

Webb, N.R. and Hopkins, P.J. (1984) Invertebrate diversity on fragmented *Calluna* heathland. *Journal of Applied Ecology*, **21**, 921–33.

Wilson, D.E. (1973) Bat faunas: a trophic comparison. *Systematic Zoology*, **22**, 14–29.

Winnett-Murray, K. (1986) *Variation in the Behavior and Food Supply of Four Neotropical Wrens*, Unpublished PhD Thesis, University of Florida, USA.

Wolda, H. (1982) Seasonality of homoptera on Barro Colorado Island, in *The Ecology of a Tropical Forest: Seasonal Rhythms and Long-term Changes* (eds E.G. Leigh, Jr, A.S. Rand and D.M. Windsor), Smithsonian Institution Press, Washington, pp. 319–30.

26

The significance of edge effects in the management of forests for invertebrate biodiversity

C.M.P. Ozanne, C. Hambler, A. Foggo and M.R. Speight

ABSTRACT

The influence of edge effects on forest canopy invertebrates was studied in a model system, coniferous plantations, using pyrethrin knockdown. Conifer forests in Britain support an abundant and species-rich canopy community, with an edge assemblage differing from that deeper into the forest. Overall abundance dropped significantly near the edge, with small organisms being particularly affected. Taxa showing lower abundance at the edge included Psocoptera, Lepidoptera, Coleoptera, Hymenoptera, Collembola and Araneae, whereas Thysanoptera and Homoptera showed higher abundances at the edge. Edge-functions varied with taxon.

Species richness was examined using spiders (Araneae) as indicators. In Norway spruce forest species richness was 57% lower at the edge. Notably, the abundance, proportion and species richness of woodland specialist spiders (Araneae) was lower near the edge. The trophic structure of the arthropod community differed at the edge, which supported proportionately more generalist predators and fewer specialist predators than the core. Adult spiders were more common at the edge. Edge effects were most pronounced within approximately 10 m of the edge, but significant differences between the edge and interior were still evident for up to 25 m in some taxa. Arthropod abundance was inversely correlated with light levels, supporting a role for microclimate in determining edge effects. Small forests, irregularly shaped stands and narrow forest corridors will be less likely to sustain assemblages of forest-core arthropods. Fragments or corridors narrower than 50 m across will be of limited value for the most sensitive specialist arthropods.

Canopy Arthropods. Edited by N.E. Stork, J. Adis and R.K. Didham. Published in 1997 by Chapman & Hall, London. ISBN 0 412 74900 9

INTRODUCTION

Forest canopies support abundant and species rich assemblages of arthropods (Ozanne, 1991; Kitching *et al.*, 1993; Stork and Blackburn, 1993). These communities are increasingly subject to the effects of fragmentation, since forests are a prime example of habitats which have been markedly disturbed. One of the consequences of fragmentation is an increase in habitat affected by proximity to an edge (Rolstad, 1991).

Forests in Britain display the characteristics of severe fragmentation (Rackham, 1990). The current forest cover, standing at 9–10% of the land area, represents the remnants of a previously extensive and near continuous native forest and also the results of 20th century afforestation (Kreysa, 1987; Forestry Commission, 1992). The average size of current forest areas in Britain is small, with much new planting in compartments of less than 25 ha. This study examines patterns on scales relevant to these forests and to corridors between woodland patches.

In order to manage fragmented habitats for high biodiversity, or for diversity of selected taxa, an understanding of patch dynamics is required. Ecologists must take into account not only the interactions between populations in separated patches, such as migration between the elements of metapopulations (Hanski and Gilpin, 1991; Nee and May, 1992), but also within-patch dynamics. One of the primary effects of fragmentation is to disrupt the integrity of remaining patches by reducing the proportion of 'core' habitat and increasing the proportion affected by proximity to the 'edge' (Schonewald-Cox and Bayless, 1986; Laurance and Yensen, 1991).

Evidence gathered from studies of vertebrates suggests that edges influence several aspects of population and community structure and that the penetration of edge effects varies with taxon and region. For example, a recent review suggests that the impact of edge effects on the success of bird nests is most apparent within 50 m of the edge (Paton, 1994), although the woodland avifauna of Britain is not always strongly influenced by edges (Fuller and Whittington, 1987). Medley (1993) demonstrated that both abundance and group size of primates were negatively affected by reduction in forest area or increase in forest edge habitat, and Crome and Richards (1988) found assemblages of bats to be influenced by the disturbance caused by gap formation.

To date, studies of the influence of edge effects on forest arthropods have mainly been confined to the ground-dwelling fauna, in particular Coleoptera. Helle and Muona (1985) found edge effects in several taxa. More recently, coleopteran forest core communities have been identified which differ in composition from those near edges (Halffter *et al.*, 1992; Buse and Good, 1993; Halme and Niemalä, 1993). Lovejoy *et al.* (1986)

demonstrated that the composition of a tropical forest butterfly fauna is influenced over distances of hundreds of metres by those species associated with the forest edge. Although canopy invertebrate faunas have been shown to be remarkably rich (Southwood *et al.*, 1982; Stork, 1987; Ozanne, 1991), no studies of the responses of canopy invertebrates to edges, other than that of Malcolm (1997, Chapter 25, this volume), have previously been carried out.

Data from empirical studies, such as those described above, have been utilized in the development of models predicting the impact of patch geometry on forest fauna and flora (Laurance, 1991; Laurance and Yensen, 1991). Use of core-area models may assist in determining critical patch sizes for sensitive species and may be applicable to the design of nature reserves (Blouin and Connor, 1985; Schonewald-Cox and Bayless, 1986), to the planting of new forests (Watkins, 1993) and to the management of existing areas. However, like other models, improvements will depend on an understanding of the population and community dynamics occurring within the patch. Thus, the management of woodlands for optimal diversity, whether this means maximum diversity or the sustainability of characteristic communities, requires further empirical evidence on the penetration of edge effects for as wide a range of taxa as possible (Hambler and Speight, 1995).

This study makes a preliminary investigation of the pattern and penetration of edge effects for canopy arthropods in coniferous forests in lowland Britain and draws implications from the data for the management of established and newly created woodlands.

METHODS

Studies were carried out in Norway spruce (*Picea abies*) and Scots pine (*Pinus sylvestris*) plantations. These species were selected because there are background data for the canopy faunas (Ozanne, 1991; also unpublished data) and because they have a significant role in the British landscape. Scots pine is native to parts of Scotland and Norway spruce was native to Britain in the last interglacial. Norway spruce is widely used in high-density plantations across Europe where it supports a rich and abundant community of arthropods, many of which are associated with the tree species elsewhere in its natural range. For Norway spruce, the distinct, south-facing edge between a 2-ha stand and a short rye grassland (*Lolium* sp.) at Bernwood Forest, Buckinghamshire, was examined. The edge was characterized by herbaceous vegetation. The three stands of Scots pine, located in the New Forest, averaged 13 ha, typical of plantation compartments and farm-forests in Britain. All the trees were approximately 14 m in height, of mid-rotation age (at 35 years since planting) and were planted at 1.5–2.0 m spacings.

Samples were collected from the canopy of Scots pine in August 1988 and from Norway spruce in August 1994, using pyrethrin knockdown (Ozanne *et al.*, 1988). Samples were taken from 'tree units' (the canopy between adjacent trees). For Scots pine, three samples were taken at the edge and three at 6 m into each stand. For Norway spruce, samples were collected at 3-m intervals along three transects, spaced 10 m apart, running perpendicularly from the edge to 45 m into the interior of the patch. Each canopy unit was sprayed for 30 seconds using Pyrethrin 2/16 (Roussel Uclaf) and knocked-down arthropods were collected on rigid plastic trays of 0.57 m^2. After 1 hour invertebrates were removed from the trays and placed in alcohol for storage.

To provide an indication of canopy density, light levels were recorded at 12:00 h on an overcast day using a photographer's light-meter with a diffusing cover. Three measurements were taken at each sample position on the transects and light levels were regressed against abundance data to investigate a possible causal relationship between the two variables.

Invertebrates were initially sorted to order and the abundances recorded. Subsequently, specimens were size-classed and assigned to feeding guilds according to Moran and Southwood (1982). Where orders showed multiple feeding strategies, the arthropods were identified to family or species in order to facilitate assignment. Subsamples from each order were dried at 40°C for 72 hours and used to calculate typical biomass values for organisms within each of several size classes.

Spiders (Araneae) were identified to species, genus or family, depending on maturity, by C. Hambler. Minimum species richness was calculated by assuming that any immature spiders which could not be identified to species or genus belonged to the same species as mature individuals of the same genus or family within a sample (Gibson *et al.*, 1992). To examine patterns in species richness, spiders (Araneae) were used as indicators. The Araneae includes many specialists and previously has been employed with some success as an indicator group (Duffey, 1978; Di Castri *et al.*, 1992; Gibson *et al.*, 1992). Woodland specialists (Hambler and Speight, 1995) were defined as those generally recorded in woodland and grassland spiders as those generally recorded in grassland (Ratcliffe, 1977). The proportion of adult versus immature spiders was also investigated since it is plausible that edges and interiors, with different degrees of predation and different microclimates, might differ in the maturation rate of canopy invertebrates and their survivorship to maturity.

Abundance data were analysed using parametric ANOVAR (SPSS) after testing for homogeneity of variances using Bartlett's test. Duncan's multiple range test was used to determine significant differences between means for the distances along the transects (edge penetration distance). *t*-Tests were used to compare abundances 6 m into Scots pine patches with those at the extreme edge (Siegel, 1956; Sokal and Rohlf, 1981).

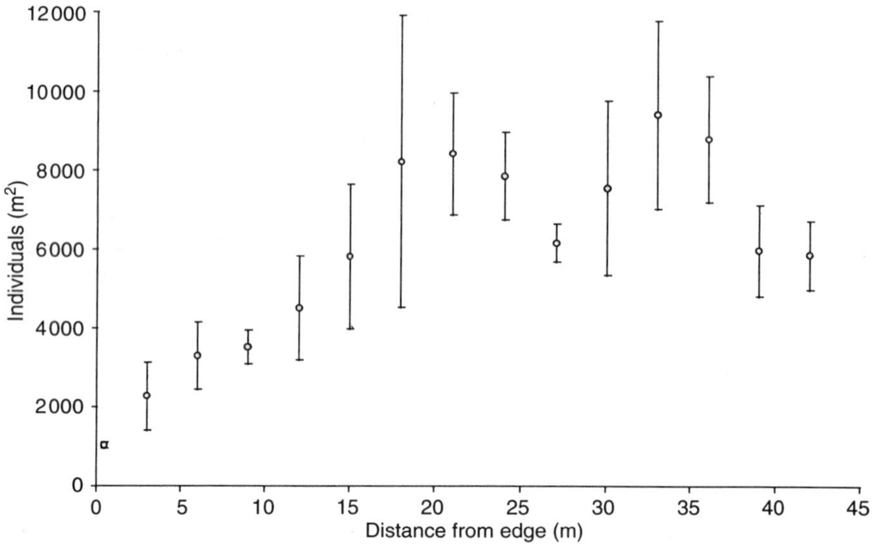

Figure 26.1 Mean abundance/m^2 (± S.E.) of all arthropods in the canopy of Norway spruce. Samples were taken at 3-m intervals from the edge to 45 m into the stand. (For all points, n = 3 trees.)

RESULTS

Overall abundance and biomass

The abundance of arthropods in the canopy of Norway spruce was very high; at up to 13 000/m^2 it was among the highest recorded for any tree species in Britain. Abundance declined very rapidly at the extreme edge (Figure 26.1). A similar trend was observed for biomass (Figure 26.2).

Differences in responses among taxa

Several taxa were found to increase significantly in population densities towards the core. These are summarized in Table 26.1, which also indicates the maximum distances into the forest to which samples differ significantly from those at the edge (a measure of the edge effect). Psocoptera and Collembola show the most marked variations. Data for Psocoptera, Collembola and Araneae presented in Table 26.1 support the results from Scots pine which demonstrate significantly higher densities of these three groups at 6 m into several 13-ha patches (Ozanne, 1991; C.M. Ozanne *et al.*, unpublished results).

By contrast, abundance of Thysanoptera decreased strongly towards the core in Norway spruce (Figure 26.3) and Homoptera showed the

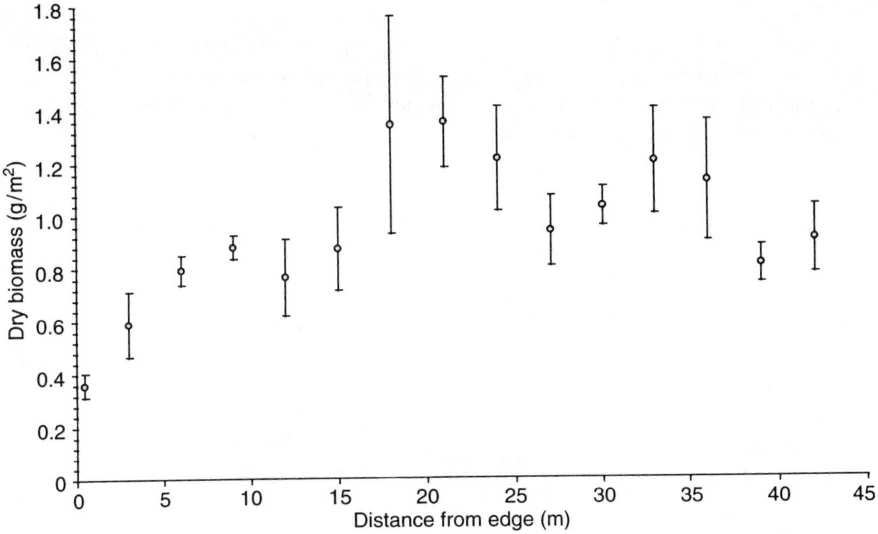

Figure 26.2 Mean dry biomass/m^2 (± S.E.) of all arthropods in the canopy of Norway spruce. (For all points, n = 3 trees.)

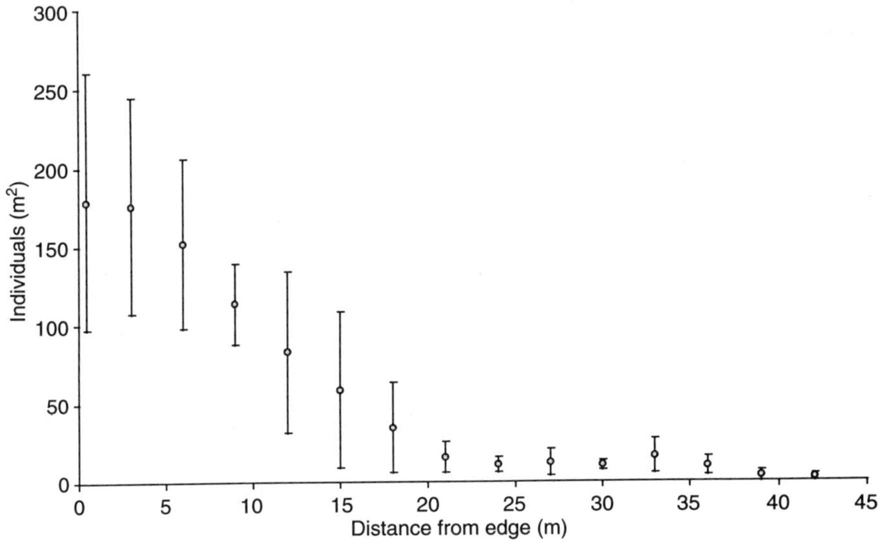

Figure 26.3 Mean abundance (± S.E.) of Thysanoptera/m^2 in the canopy of Norway spruce. (For all points, n = 3 trees.)

Table 26.1 Taxa showing significant variation in abundance with depth into the forest for two tree species. Data from Norway spruce were analysed using one-way ANOVAR with *a posteriori* comparisons by Duncan's multiple range test. Data from Scots pine were analysed using *t*-tests to compare means. Edge penetration is defined here as the depths to which significant differences from the interior to the edge occur

Norway spruce

Taxon	*Edge penetration (m)*	*Significance (P)*
All taxa	<15	<0.001
Araneae	<24	<0.001
Lepidoptera	<15	<0.05
Thysanoptera	<18	<0.001
Psocoptera	<15	<0.001
Coleoptera	<9	<0.005
Collembola	<9	<0.001

Scots pine

Taxon	*Mean density* m^{-2}		*Significance (P)*
	0 m	*6 m*	
All taxa	285.9	674.2	0.004
Araneae	4.9	11.7	0.004
Psocoptera	60.7	274.3	0.002
Homoptera	57.9	33.5	0.016
Collembola	15.9	49.3	0.005

same pattern on Scots pine (Table 26.1). Thysanoptera, which are generally tolerant of dry, open environments, appear in this site to be edge species, either filtered out by the foliage as they disperse through the air, or actually exploiting the shelter of the forest edge. Homoptera are likely to be responding to increased foliage on edge trees (Ranney *et al.*, 1981; Denno and Roderick, 1991).

Variations in community structure

The proportion of small size classes (individuals less than 2 mm) increased with depth into the Norway spruce forest (Figure 26.4). The absolute biomass and abundance of predators declined near the edge (see Table 26.1) and significant differences were found up to 15 m into the patch (P <0.001, d.f. = 14). The relative abundance of herbivores and

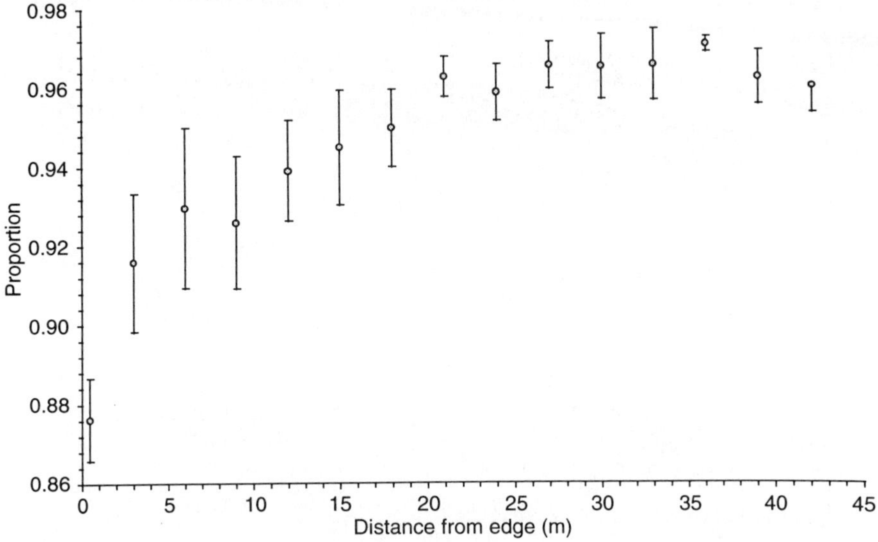

Figure 26.4 Mean proportion (± S.E.) of total numbers of arthropods less than 2 mm in body length in the canopy of Norway spruce.

predators (all predatory arthropods) increased, while the proportion of detritivores declined towards the edge (Figure 26.5). Relatively, specialized predators such as parasitoids were more abundant toward the core. Predator–prey ratios (Figure 26.6) showed significant differences up to 21 m into the patch (P <0.001, d.f. = 14).

Richness and specialisms of assemblages

The mean spider species richness at the extreme edge of Norway spruce was 6.0 species (per 0.57 m^2) and for interior samples was 10.6 species (Figure 26.7). These measures are relatively high for British forests (C. Hambler and S. Roberts, unpublished data). Species richness is significantly lower at the edge than in the interior (P = 0.016, d.f. = 14). Woodland specialist species were significantly more abundant in the forest interior than at the edge (P = 0.025, d.f. = 14) and their species richness was also higher (P = 0.025, d.f. = 14). No significant differences were detected further along the transects. Grassland spiders did not show any such significant changes in abundance or richness.

Proportions of adults and phenology

For spiders, the proportion of adults was significantly lower towards the edge on Norway spruce (P = 0.021, d.f. = 14) and the densities lower

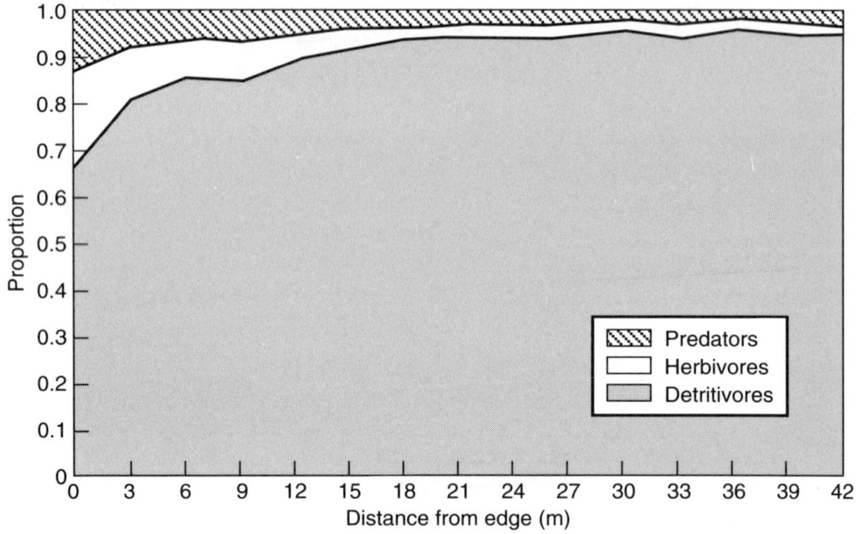

Figure 26.5 Mean proportion of individuals from different trophic guilds (predators, herbivores, detritivores) in the community.

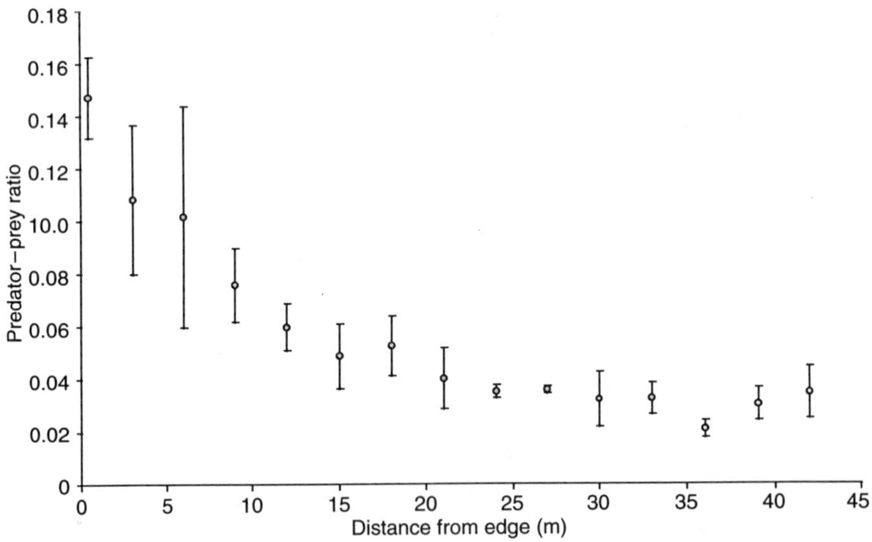

Figure 26.6 Mean (± S.E.) predator–prey ratios for total arthropod abundance.

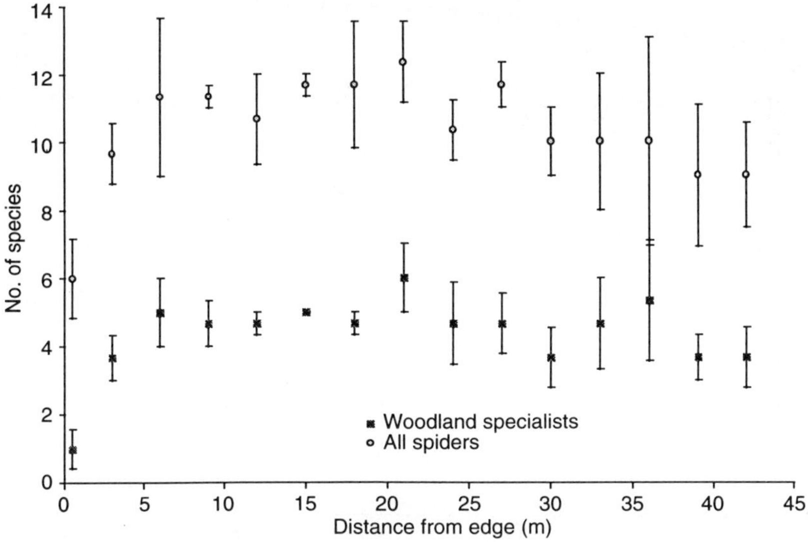

Figure 26.7 Mean species richness of all spiders, and woodland specialist spiders, in the canopy of Norway spruce.

on Scots pine ($P = 0.032$, d.f. = 14), suggesting more rapid maturation deeper in the woodland and/or higher survival rates.

Light levels

Light levels at different distances from the edge are presented in Figure 26.8. These were inversely correlated with the overall abundance of invertebrates ($r^2 = 0.7$, d.f. = 13, $P < 0.001$) and with the abundance of many of the individual taxa.

DISCUSSION

The results of these studies, based in coniferous forests in southern Britain, demonstrate that canopy arthropod populations and communities are subject to significant edge effects. This concurs with past studies of avifauna, primates and vascular plants which have been carried out in temperate and tropical forests. Our studies were concerned with established, or inherent, edges (Yahner, 1988), indicating that even in the relatively uncomplicated environment of a coniferous plantation, proximity to an edge, and not simply edge creation, affects the canopy fauna.

The penetration of edge effects, described by Laurance and Yensen (1991) as the edge-function, *d*, is difficult to determine from the data gathered to date and will depend upon studies of a wider range of forest

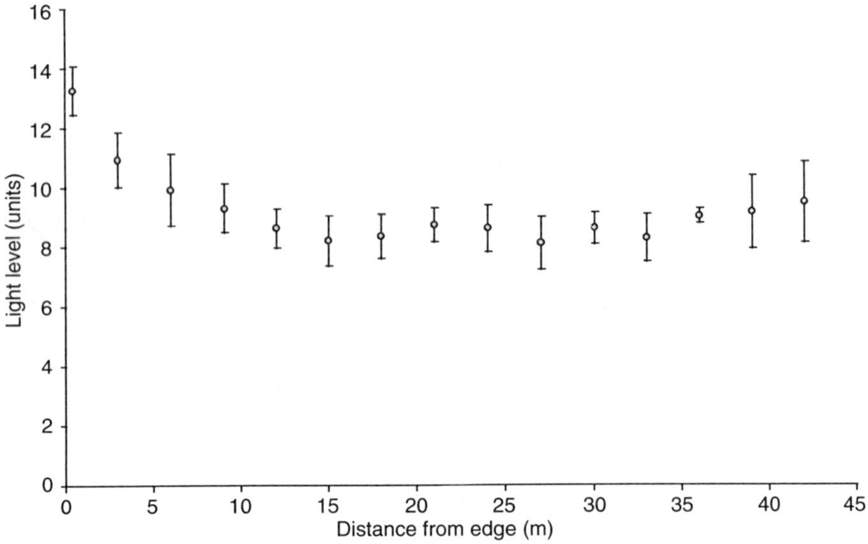

Figure 26.8 Mean light levels (arbitrary units) (± S.E.) at 3-m distance intervals into a Norway spruce plantation.

edges and patch sizes. However, this study suggests that edge penetration distance varies with taxon and that for some invertebrates effects are ameliorated rapidly within 3–6 m of the edge, whereas for other taxa an effect can still be observed 25 m into the stand.

Extensive work on avian populations in woodlands has identified effects on a similar scale (Lovejoy *et al.*, 1986; Burkey, 1993; Paton, 1994), but the penetration of edge effects for other taxa seems variable. The data for canopy arthropods confirm that edge-functions may be taxon-specific and support the cautionary note expressed by Murphy and Wilcox (1986) that management strategies employed for vertebrate conservation in fragmented habitats are unlikely to be automatically suitable for invertebrates.

The sharp variations found between samples taken on the outer edge of the canopy and those within the stand, and the penetration of the edge effect within the stand, indicate a reduction in the effective forest patch area. The outer zone does not support a community typical of the forest interior – the forest canopy is evidently a heterogeneous environment. Thus, models of patch dynamics assuming homogeneity (Hanski and Gilpin, 1991; Nee and May, 1992) may be too simplistic and some models of reserve design misleading (Simberloff, 1986). The results of this study strongly support the use of models in which edge to core ratios are incorporated, such as those proposed by Laurance and Yensen (1991). Diamond's (1975) influential recommendations for greater reserve size, circularity and low fragmentation may be appropriate – but for the wrong reasons.

Community composition

The conventional view of edges is that they are rich in species and individuals (Greatorex-Davies, 1991). The data from this study indicate that within the canopy itself sharp edges, such as those observed in plantations or in tree-fall gaps, river-sides or cliffs, may be relatively impoverished in terms of invertebrates.

Woodland specialists decline to very low levels at the edge, but, as in the case of spiders, some can recover rapidly and reach normal forest-interior levels within 10 m of a sharp edge. It is possible that more sensitive species are now rare in the heavily fragmented British landscape. More gentle edges, with scrubby transitional zones, might support more species, but might also support species inimical to the core (Whitcomb *et al.*, 1981; Brash, 1987). In Bernwood Forest, grassland spiders (dispersing by ballooning) apparently rain down into the woodland at random depth, rather than from one side. Although such species were rare in the samples, they illustrate that some edge effects may be dorsal, rather than horizontal (Soulé, 1986).

Underlying parameters

A number of underlying parameters have been proposed as affecting the distribution and abundance of flora and fauna at edges. These include the penetration of disturbance (Laurance, 1991; Medley, 1993), predation pressure (Andrén and Angelstam, 1988) and microclimatic gradients (Kapos *et al.*, 1993). Meteorological studies in temperate and tropical forests have clearly demonstrated an edge-affected zone, which differs significantly in humidity, temperature, light quality, wind speed and turbulence from that of the core (Williams-Linera, 1990; Kapos *et al.*, 1993; Matlack, 1993; Young and Mitchell, 1994). These studies have indicated that such edge effects may penetrate up to 50 m into forest stands, perhaps with a more severely affected zone within the first 15 m (Young and Mitchell, 1994).

This study investigated a model system in which the potential impact of disturbance was reduced by the use of established edges and plantations in which disturbance caused by management activities was evenly distributed with depth. The patterns in abundance and biomass detected here support a central role for microclimate in determining edge effects in canopy arthropods. The response of small organisms, which decline in proportion at the edge, suggests that the dense forest structure provides more favourable conditions for many invertebrates than the exposed edge. Small organisms, with high surface area-to-volume ratios, are more likely to be sensitive to microclimatic stresses such as low humidity and high wind speeds at the edge. Psocoptera and Collembola, which feed mainly on fungi and algae on needle and bark surfaces, are known to require high humidity (often exceeding 65–70%; Broadhead and Thornton, 1955; Bowden *et al.*, 1976). The negative correlation between high light intensity and invertebrate abundance and biomass supports the importance of the stable, humid microclimate of the forest interior for these invertebrates and stands in contrast to conservation dogmas suggesting that open glades and rides are necessarily advantageous (Hambler and Speight, 1995).

The results also suggest that predation may affect invertebrate abundance and distribution. Predation pressure from invertebrate predators may indeed be greater at edges where predator–prey ratios were higher. However, the densities of predators were not significantly greater at the edge as has been found for studies of nest predation in birds (Wilcove *et al.*, 1986). It is thus difficult to ascertain if predator–prey ratios are simply lower in the interior because the habitat at the edge is not ideal for the detritivore component of the community.

The results presented indicate the likely order of magnitude and direction of trends in the density and species richness of canopy invertebrates in dense temperate coniferous forests. However, much more work involving different orientations of forest plots, tree densities, patch sizes and seasons will be required before detailed statements as to edge effects can be supported.

Implications for management

Edge effects of the magnitude and width detected here are particularly relevant to small forests. Edge effects of some 25 m, as detected, suggest that circular or square forest patches or fragments of 1 ha can include some forest-core conditions, with aspects of community structure at the centre typical of that in larger blocks. Many of Britain's forests have tracks and clearings, or are shaped in other ways which may render them devoid of such core conditions.

Further, if edge effects are similar on all compass-orientations, corridors may need to be over 50 m wide to include a central area suitable for the most sensitive species. This is much wider than hedgerows, the main landscape elements assumed to act as woodland corridors.

CONCLUSIONS

This preliminary examination of canopy invertebrates extends the range of taxa known to be responsive to forest edges. Edge effects were detected on scales similar to those for better-known, and more popular, taxa such as birds and mammals. Inverse relationships between canopy density (measured by light levels) and invertebrate biodiversity may surprise conservationists who have hitherto focused on light-loving taxa or forest-floor communities.

The scale of edge effects is likely to be site- as well as taxon-specific. Nevertheless, there are indications of the order of magnitude of forests and corridors which may be required to sustain forest specialists – inevitably relatively rare species in the highly fragmented British forest canopy.

Acknowledgements

We thank the Forest Enterprise and the J.N.C.C. for permission to work in the Bernwood S.S.S.I. The work was supported by the Oxford University pump-priming grant (A. Foggo and M. Speight), the Forestry Commission and the Roehampton Institute London (C. Ozanne), and Oxford Environmental Consultancy (C. Hambler). Thanks also go to those who helped in the field, in particular P.D.B. Embden and S.P. Roberts.

REFERENCES

Andrén, H. and Angelstam, P. (1988) Elevated predation rates as an edge effect in habitat islands: experimental evidence. *Ecology*, **69**, 544–7.

Blouin, M.S. and Connor, E.F. (1985) Is there a best shape for nature reserves? *Biological Conservation*, **32**, 277–88.

Bowden, J., Haines, I.H. and Mercer, D. (1976) Climbing Collembola. *Pedobiologia*, **165**, 298–312.

Brash, A.R. (1987) The history of avian extinction and forest conversion on Puerto Rico. *Biological Conservation*, **39**, 97–111.

Broadhead, E. and Thornton, I.W.B. (1955) An ecological study of three closely related psocid species. *Oikos*, **6**, 1–50.

Burkey, T.V. (1993) Edge effects in seed and egg predation at two neotropical rainforest sites. *Biological Conservation*, **66**, 139–43.

Buse, A. and Good, J.E.G. (1993) The effects of conifer forest design and management on abundance and diversity of rove beetles (Coleoptera: Staphylinidae): Implications for conservation. *Biological Conservation*, **65**, 67–76.

Crome, F.H.J. and Richards, G.C. (1988) Bats and gaps: microchiropteran community structure in a Queensland rain forest. *Ecology*, **69**, 1960–9.

Denno, R.F. and Roderick, G.K. (1991) Influence of patch size, vegetation texture, and host plant architecture on the diversity, abundance, and life history styles of sap-feeding herbivores, in *Habitat Structure: The Physical Arrangement of Objects in Space* (eds S.S. Bell, E.D. McCoy and H.R. Mushinsky), Chapman & Hall, London, pp. 169–96.

Diamond, J.M. (1975) The island dilemma – lessons of modern biogeographic studies for the design of natural reserves. *Biological Conservation*, **7**, 129–46.

Di Castri, F., Vernhes, J.R. and Younes, T. (1992) Inventorying and monitoring

biodiversity. A proposal for an international network. *Biology International*, Special Issue 27.
Duffey, E. (1978) Ecological strategies in spiders, including some characteristics of species in pioneer and mature habitats. *Symposia of the Zoological Society of London*, **42**, 109–23.
Forestry Commission (1992) *FC Facts and Figures*, HMSO, London.
Fuller, R.J. and Whittington, P.A. (1987) Comparison of neotropical winter bird populations in isolated patches vs. extensive forest. *Acta Oecologia*, **6**, 285–92.
Gibson, C.W.D., Hambler, C. and Brown, V.K. (1992) Changes in spider (Araneae) assemblages in relation to succession and grazing management. *Journal of Applied Ecology*, **29**, 132–42.
Greatorex-Davies, J.N. (1991) Woodland edge management for invertebrates, in *Edge Management in Woodlands* (ed. R. Ferris-Khan), Forestry Commission Occasional Paper 28, HMSO, pp. 25–30.
Halffter, G., Favila, M.E. and Halffter, V. (1992) A comparative study of the structure of the scarab guild in Mexican tropical rain forests and derived ecosystems. *Folia Entomológica Mexicana*, **84**, 131–56.
Halme, E. and Niemalä, J. (1993) Carabid beetles in fragments of coniferous forest. *Annales Zoologici Fennici*, **30**, 17–30.
Hambler, C. and Speight, M.R. (1995) Biodiversity conservation in Britain: science replacing tradition. *British Wildlife*, **6**, 137–47.
Hanski, I. and Gilpin, M.E. (eds) (1991) *Metapopulation Dynamics: Empirical and Theoretical Investigations*, Linnean Society of London, Academic Press, London.
Helle, P. and Muona, J. (1985) Invertebrate numbers in edges between clear-fellings and mature forest in northern Finland. *Silva Fennica*, **19**, 281–94.
Kapos, V., Ganade, G., Matsui, E. and Victoria R.L. (1993) $\delta^{13}C$ as an indicator of edge effects in tropical rainforest reserves. *Journal of Ecology*, **81**, 425–32.
Kitching, R.L., Bergelson, J.M., Lowman, M.D., McIntyre, S. and Carruthers, G. (1993) The biodiversity of arthropods from Australian rainforest canopies: general introduction, methods, sites and ordinal results. *Australian Journal of Ecology*, **18**, 181–91.
Kreysa, J. (1987) *The Forest Resource of the EEC-12: A Statistical Analysis*. EEC-FAST Occasional Paper, no. 162.
Laurance, W.F. (1991) Edge effects in tropical forest fragments: Application of model for the design of nature reserves. *Biological Conservation*, **57**, 205–19.
Laurance, W.F. and Yensen, E. (1991) Predicting the impacts of edges in fragmented habitats. *Biological Conservation*, **55**, 77–92.
Lovejoy, T.E., Bierregaard, R.O., Rylands, A.B., Malcolm, J.R., Quintela, C.E., Harper, L.H., Brown, K.S. Jr, Powell, A.H., Powell, G.V.N., Schubart, H.O.R. and Hays, M.B. (1986) Edge and other effects of isolation on Amazon Forest fragments, in *Conservation Biology: The Science of Scarcity and Diversity* (ed. M.E. Soulé), Sinauer Associates Inc., Sunderland, Mass., pp. 257–85.
Malcolm, J.R. (1997) Insect biomass in Amazonian forest fragments, in *Canopy Arthropods* (eds N.E. Stork, J. Adis and R.K. Didham), Chapman & Hall, London, pp. 510–33 (this volume).
Matlack, G.R. (1993) Microenvironment variation within and among forest edge sites in the Eastern United States. *Biological Conservation*, **66**, 185–94.
Medley, K.E. (1993) Primate conservation along the Tana river, Kenya: an examination of the forest habitat. *Conservation Biology*, **7**, 109–21.
Moran, V.C. and Southwood, T.R.E. (1982) The guild composition of arthropod communities in trees. *Journal of Animal Ecology*, **51**, 289–306.

Murphy, D.D. and Wilcox, B.A. (1986) Butterfly diversity in natural habitat fragments: a test of the validity of vertebrate based management, in *Wildlife 2000: Modelling Habitat Relationships of Terrestrial Vertebrates* (eds J. Verner, M.L. Morrison and C.J. Ralph), Wisconsin Press, Madison, pp. 287–92.

Nee, S. and May, R.M. (1992) Patch removal favours inferior competitors. *Journal of Animal Ecology*, **61**, 37–40.

Ozanne, C.M.P. (1991) *The Arthropod Fauna of Coniferous Plantations*, Unpublished DPhil Thesis, University of Oxford, UK.

Ozanne, C.M.P., Speight, M.R. and Evans, H.F. (1988) Spray deposition and retention in the canopies of five forest tree species. *Aspects of Applied Biology*, **17**, 245–9.

Paton, P.W.C. (1994) The effect of edge on avian nest success: how strong is the evidence? *Conservation Biology*, **8**, 17–26.

Rackham, O. (1990) *Trees and Woodland in the British Landscape*, Dent, London.

Ranney, J.W., Brunner, M.C. and Levenson, J.B. (1981) The importance of edge in the structure and dynamics of forest islands, in *Forest Island Dynamics in Man Dominated Landscapes* (eds R.L. Burgess and D.M. Sharpe), Ecological Studies 42, Springer, New York, pp. 67–95.

Ratcliffe, D.A. (1977) *A Nature Conservation Review*, Vol. 1, Cambridge University Press, Cambridge, UK.

Rolstad, J. (1991) Consequences of forest fragmentation for the dynamics of bird populations: conceptual issues and the evidence. *Biological Journal of the Linnean Society*, **42**, 149–63.

Schonewald-Cox, C.M. and Bayless, J.W. (1986) The Boundary model: a geographical analysis of design and conservation of nature reserves. *Biological Conservation*, **38**, 305–22.

Simberloff, D. (1986) Design of nature reserves, in *Wildlife Conservation Evaluation* (ed. M.E. Usher), Chapman & Hall, London, pp. 315–37.

Siegel, S. (1956) *Nonparametric Statistics for the Behavioural Sciences*, McGraw-Hill, New York.

Sokal, R.R. and Rohlf, F.J. (1981) *Biometry*, 2nd edn, Freeman & Co., New York.

Soulé, M.E. (1986) *Conservation Biology: The Science of Scarcity and Diversity*, Sinauer Associates Inc., Sunderland, Massachusetts.

Southwood, T.R.E., Moran, V.S. and Kennedy, C.E.J. (1982) The assessment of arboreal insect fauna: comparisons of knockdown sampling and faunal lists. *Ecological Entomology*, **7**, 331–40.

Stork, N. (1987) Guild structure of arthropods from Bornean rain forest trees. *Ecological Entomology*, **12**, 69–80.

Stork, N.E. and Blackburn, T.M. (1993) Abundance, body size and biomass of arthropods in tropical forest. *Oikos*, **67**, 483–9.

Watkins, C. (1993) Forest expansion and nature conservation, in *Ecological Effects of Afforestation* (ed. C. Watkins), CAB International, Wallingford, pp. 1–13.

Whitcomb, R.F., Robins, C.S., Lynch, J.F., Whitcomb, B.L. Klimkiewicz, M.K. and Bystrak, D. (1981) Effects of forest fragmentation on avifauna of the eastern deciduous forest, in *Wildlife 2000: Modelling Habitat Relationships of Terrestrial Vertebrates* (eds J. Verner, M.L. Morrison and C.J. Ralph), Wisconsin Press, Madison, pp. 125–205.

Wilcove, D.S., McLellan, C.H. and Dobson, A.P. (1986) Habitat fragmentation in the temperate zone, in *Conservation Biology: The Science of Scarcity and Diversity* (ed. M.E. Soulé), Sinauer Associates Inc., Sunderland, Massachusetts, pp. 237–56.

Williams-Linera, G. (1990) Vegetation structure and environmental conditions of forest edges in Panama. *Journal of Ecology*, **78**, 356–73.

Yahner, R.H. (1988) Changes in wildlife communities near edges. *Conservation Biology*, **2**, 333–9.

Young, A. and Mitchell, N. (1994) Microclimate and vegetation edge effects in a fragmented podocarp-broadleaf forest in New Zealand. *Biological Conservation*, **67**, 63–72.

27

Canopy arthropod studies for the future

N.E. Stork, R.K. Didham and J. Adis

INTRODUCTION

Early canopy research developed as a natural extension of ground-based studies (Haddow *et al.*, 1961), but biologists quickly met the limitations imposed by canopy access. In the early 1970s, however, canopy science developed as a field in its own right, spurred on by new tree-climbing techniques. Gradually over the years, the logistical constraints on working in the canopy have been removed and more intensive, long-term studies have been initiated. As a consequence, we are now moving away from the sensationalism that surrounded pioneering efforts to open up the so-called 'last biotic frontier', toward more ecologically directed questions of greater scientific import. This book has brought together a broad cross-section of canopy arthropod research carried out around the world over the past 10 years. In this concluding chapter we review some of the rapidly expanding areas of canopy arthropod research and highlight other areas which have so far been neglected. We also suggest some future priorities for canopy arthropod studies.

For this chapter we draw upon a number of sources. First, the postal survey of European scientists carried out by Stork and Best (1994) for the European Science Foundation's (ESF) Tropical Canopy Research Programme in 1994. This survey was carried out over a two-month period and hence the questions asked were limited in scope. Two of the questions asked are relevant here: (i) which fields of tropical canopy research are you particularly interested in? and (ii) what do you think

Canopy Arthropods. Edited by N.E. Stork, J. Adis and R.K. Didham. Published in 1997 by Chapman & Hall, London. ISBN 0 412 74900 9

are the most important questions to be answered or hypotheses to be tested in canopy research? This questionnaire was circulated to about 150 European canopy researchers including those on an existing ESF database for tropical research and those on the North American-based canopy e-mail network. Some 39 of the 70+ respondents (56%) were studying canopy arthropods. A parallel, but more extensive, survey of North American canopy researchers was carried out by Nadkarni and Parker (1994). Their questions related more to operational aspects of research rather than the hypotheses being tested.

Second, this chapter draws upon numerous discussions with colleagues working in forest canopies world-wide. Most recently, some particularly constructive comments and ideas for future canopy arthropod studies came from discussions at the First International Canopy Arthropod Symposium in Manchester, UK, in August 1994 and the 2nd European Science Foundation's Tropical Canopy Research Workshop in Ulm, Germany, in July 1995. A number of general considerations concerning the scientific issues being addressed by canopy arthropod studies have emerged from these discussions, and we first discuss these.

GENERAL CONSIDERATIONS

Hypothesis testing or descriptive research

A number of general comments preface our thoughts on future canopy research. Perhaps the most important of these is whether the questions being asked relate solely to the canopy, or could equally well be answered through studies of the ground layer, or indeed through investigations of other terrestrial or aquatic ecosystems? We can say unreservedly that as one of the main life zones in forests, the canopy must hold crucial answers to the way in which forest ecosystems function. As such, some questions can only be fully answered through the study of the forest canopy; questions such as 'how does herbivory affect carbon fluxes between the forest canopy and the atmosphere?', 'what factors account for high biodiversity in tropical forests?' and 'what is the degree of host-specificity of forest arthropods?'. It is important to distinguish those studies that test hypotheses and general ecological theory from those that are purely descriptive. Usually studies of the latter kind provide less valuable information that sometimes borders on the anecdotal. It would not be an understatement to say that many canopy studies to date have been descriptive, in one form or another, and have not tested strict hypotheses. We believe that this situation arises from a perception of the canopy as different from other biotopes, rather than as an integral part of the forest ecosystem. Ecological theory may be tested with as much rigour in the canopy as in any other biotope. It is pleasing to see recognition of this fact in a number of contributions in this volume. On the other hand, canopy arthropod samples have been well utilized in the development of some fields of ecological theory, such as body size : species abundance relationships and host specificity.

In order to move future canopy arthropod studies away from the anecdotal end of the scale, it is important that data be collected and analysed in a comparable manner between studies. There is still a need for good comparative studies of arthropod communities in different types of forests and in different continents, latitudes and climates. Although there have been many individual studies of tree-crown arthropod communities around the world, it remains difficult to draw many general conclusions about arthropod community structure. The lack of replication and the variety of methodologies used have made it impossible to distinguish local variation from variation between sites and between studies. Sadly, previous calls for a standardization of sampling methods have not led to any appreciable change in entomologists' attitudes to sample collection. With sound reason, different trap types are employed to tackle different questions under differing field conditions (Basset *et al.*, 1997, Chapter 2, this volume), but new canopy studies should follow a standard methodology for their selected trap type, as outlined by various chapters in this volume.

We also suggest that small- and large-scale entomological studies of the canopy should be integrated with studies of other groups of organisms and studies of ecosystem processes (e.g. Didham *et al.*, 1996). In this regard the collaborative research programme of Professor Morawetz and co-workers presently underway in Venezuela (Figure 27.1) shows how the sum is greater than the individual parts, and as such can lead to a better understanding of the diversity of organisms and processes in tropical canopies.

The scale of the task

Studies of single species of canopy arthropods are extremely difficult, except where the species concerned reaches pest proportions. Also, because of the large number of arthropod species found in trees it is impossible to extrapolate the findings from single species studies up to whole communities. As a result, most canopy biologists have taken a community approach to the study of arthropods. Perhaps the greatest impact in this respect has been made by knockdown insecticide sampling. However, as Erwin (1995) pointed out, it requires an enormous effort for canopy arthropod samples to be sorted, the resulting data compiled and analysed, and the papers written. For example, in 1981 it took one of us (N.E.S.) just 12 days to sample 10 Bornean lowland

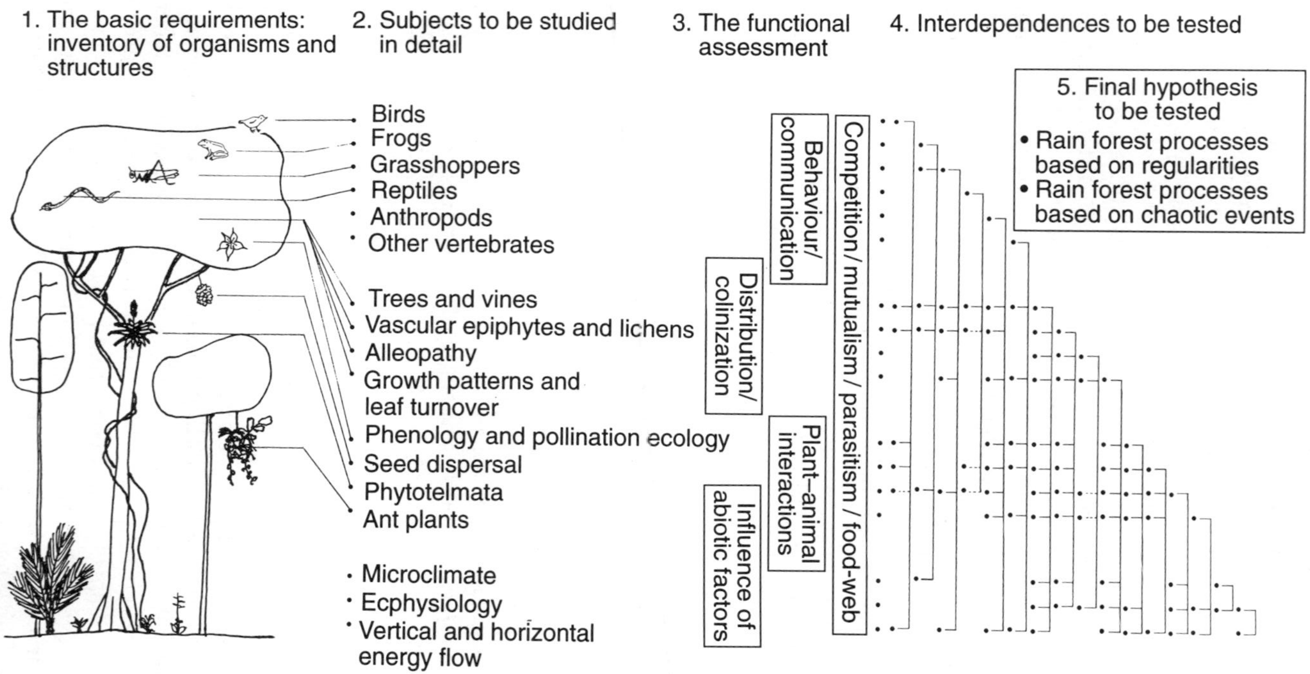

Figure 27.1 Schematic drawing of the basic requirements, investigation topics and interdisciplinary connections of a collaborative research programme to investigate canopy ecosystems in Venezuela. (Courtesy of Professor Morawetz, University of Leipzig, Germany.)

forest trees using a fogging machine pulled into the canopy, 2 years' effort by more than 20 taxonomists at The Natural History Museum, in London, to sort the 24 000 arthropods to orders, families and almost 4000 species, and yet a further 4 years for the resulting publications to start to appear (Stork, 1987a,b)! As a word of warning to those about to embark on studies of arthropods in trees using mass-sampling techniques, it is extremely important to evaluate carefully just how much arthropod material the proposed sampling programme will produce, how long it will take to sort to morphospecies, whether the human resources are available to carry out this task, and whether the resulting data will answer the questions posed. The research programme carried out by Southwood and colleagues looking at arthropod community structure on native and introduced species of trees in Britain and South Africa (Southwood *et al.*, 1982a,b; Moran and Southwood, 1982) is a model for how to manage such a canopy arthropod programme. The results of this study are still relevant today. For every completed Southwood-style study there are dozens of others that failed to be completed or successfully make their mark. These comments are made to forewarn, rather than to discourage prospective canopy arthropod researchers.

FUTURE RESEARCH AREAS

Generation and maintenance of arthropod diversity and community structure in tree canopies

Two of the most critical questions for canopy researchers are 'why is the canopy arthropod fauna so rich?' and 'how is this diversity maintained?'. In answering these, and related, questions, we need to be aware of the spatial and temporal variation in patterns of arthropod diversity and of the hierarchies of ecological and evolutionary processes operating at different scales (di Castri, 1991). Another related question is 'what is the contribution of canopy arthropods to total arthropod diversity in forests?'. For Indonesian forests, Hammond *et al.* (1997, Chapter 10, this volume) tested the widely held belief that the canopy beetle fauna is more diverse than the ground fauna, and found that this is not the case. If this finding is applicable to all forests, then it will drastically change our current perceptions of global biodiversity and arthropod community interactions in forest ecosystems. The 'last biotic frontier' may not be the canopy as some have suggested, but rather the soil. This said, it does not detract from the importance and value of future studies of canopy arthropods.

Many comparisons have been made between different arthropod communities, both in this volume and elsewhere, and in almost all cases we find that the faunal similarity of different trees, even of trees of the same species, is low. A high proportion of the species found in trees are singletons, in many cases presumably 'tourists' (*sensu* Moran and Southwood, 1982). Their presence in particular trees would seem to be dictated in large part by chance. One area for future research is to examine the relative contributions of deterministic (predetermined) and stochastic (chance) processes in defining the arthropod assemblages of individual trees or groups of trees. Floren and Linsenmair (1997, Chapter 16, this volume) provide some insights into this problem.

Canopy ecology and structure

One important area of research that has been little studied is the role of canopy architecture in shaping the structure of insect assemblages. A number of eminent botanists have produced models which describe the architectural complexity of trees (Hallé, 1995) and yet few entomologists have related arthropod diversity and distribution to plant architectural attributes (Lawton, 1978; Strong *et al.*, 1984; Didham, 1997, Chapter 15, this volume). Arthropods are not evenly distributed from the top to the bottom of the canopy, nor are they evenly distributed between the canopies of different trees. The structure of leaves, flowers, plants and branches of different trees must influence the composition of the associated arthropod communities in many ways. Morse *et al.*, (1985) showed how the fractal structure of plant surfaces determined, at least in part, the body size : abundance distribution of the associated arthropod communities. This, and the importance of microclimate in determining the composition of arthropod assemblages, needs further study.

The arthropod assemblage in a tree

Several chapters in this book have addressed, directly or indirectly, the question of the specificity of arthropod species to their host-trees. There is a surprising lack of data on how many species of arthropods are specific to one or a few tree species. This is largely because few researchers have attempted to look at the autecology of canopy arthropod species, leading to the collection of 'point' samples in time and space which tell us nothing about a species' distribution. Feeding trials are one solution to the problem (Basset, 1997, Chapter 12, this volume), albeit an imperfect one in some cases. The question of host-specificity is further complicated by problems in the interpretation of sample data. All sampling methods collect only a portion of the fauna that inhabits a tree, and differences between samples may be due to pseudo-turnover, rather than real differences in the faunas. Much more work is required on arthropod species turnover between tree-crowns, using approaches such as the calculation of 'effective specialization' (Mawdsley and Stork, 1997, Chapter 6, this volume).

It is often very difficult to tell whether many of the arthropods found in canopy samples are associated with the host tree itself or rather with the many epiphytes and vines on the tree. We know very little about the arthropod communities associated with these plants, or with canopy fungi, bryophytes, pteridophytes or lichens.

SYSTEMATICS OF CANOPY ARTHROPODS

The first well-publicized knockdown insecticide samples of tropical canopy arthropods created considerable excitement among the taxonomic community in the 1970s and 1980s. Groups previously thought to be rare were found to be abundant and rich in species in the forest canopy (Barnard *et al.*, 1986). However, in general it is difficult to assess how much impact canopy studies have had on arthropod systematics because of the long lag-time between the discovery and subsequent taxonomic description of a new species. Many individuals have provided off-the-cuff estimates of the proportion of a particular sample that may be new to science, but no-one has taken a complete set of samples and provided published figures. Nor do we know how the proportions of described to undescribed species change with increasing sample size.

FOREST FRAGMENTATION, SPECIES LOSS AND ENVIRONMENTAL CHANGE

The world's forests are rapidly diminishing in the face of ever-increasing human population pressure. It has been suggested that the widespread destruction of tropical forests is causing a modern 'mass extinction' event. While we have some certainty about the scale of current vertebrate extinction rates, there are still no firm data to support speculation that vast numbers of arthropod species are becoming extinct. If we had a more complete inventory of forest arthropods would the actual loss of biodiversity caused by habitat reduction be more obvious? The answer is 'probably not', because of the natural rarity of many species and the practicalities of sampling diverse assemblages. The problem revolves around relating habitat loss to species extinction rates, but there are two crucial pieces of missing data: species tolerances to different degrees of deforestation and forest fragmentation, and the size of species' geographic ranges. Both are major unknown variables in the measurement of species extinction rates.

Forest fragmentation is a rapidly expanding field in conservation biology. The number of studies of invertebrate responses to forest fragmentation is also increasing dramatically (Didham, in press), but has, as yet, largely involved ground-dwelling arthropod assemblages. As far as we are aware only Malcolm (1997, Chapter 25, this volume) and Ozanne *et al.* (1997, Chapter 26, this volume) have looked at the effects of forest fragmentation on canopy arthropods. Clearly, we are very interested in how forest disturbance affects canopy arthropods, but are we able to infer much about canopy responses to disturbance from ground-based studies? Ozanne *et al.* found a decrease in the abundance and diversity of canopy arthropods at the edge of forest fragments, a result contrary to the findings of most ground studies. However, Malcolm showed a more typical increase in arthropod abundance at the forest edge, but also a relative shift in arthropod biomass from the canopy to the ground at the edge. It is not clear if there are differences in the responses of ground versus canopy arthropod communities to forest fragmentation, or if the variation in response is simply a function of different forest types and different locations.

Deforestation and fragmentation are widely implicated in current global environmental change. At least for the foreseeable future, habitat loss will outstrip global climate change (among other factors) as the single most important component of global change (Vitousek, 1994). Thus, one of the most urgent concerns is how forest ecosystems themselves are affected by fragmentation and loss of biodiversity. How important are arthropods in the maintenance of ecosystem function? How does fragmentation alter trophic interactions and energy flow in the canopy? What are the consequences of altered ecosystem processes for the maintenance of global biogeochemical cycles? The sheer complexity of biotic interactions in the canopy makes these questions almost intractable at our current state of knowledge, but they are nevertheless important for a complete understanding of fragmentation-induced changes in forest ecosystems. For example, fragmentation and associated anthropogenic influences increase the susceptibility of the forest canopy to pest outbreaks, causing defoliation and, in some cases, death of the trees (Bellinger *et al.*, 1989; Landsberg, 1990; Roland, 1993; Roland and Taylor, in press). The loss of insect predators or parasitoids in fragmented habitats has also been shown to release herbivores from natural control, leading to population outbreaks and habitat damage (Kareiva, 1987; Lasalle and Gauld, 1991; Kruess and Tscharnkte, 1994). Such pervasive, multitrophic level interactions are the most difficult to identify and predict in disturbed forests, but are potentially the most damaging to ecosystem function. This is likely to be a productive area of study for canopy arthropod researchers.

In summary, there are innumerable avenues of research open to biologists in the new age of scientific exploration of the canopy. We have inevitably left some stones unturned, yet touched on other problems which may eventually prove to be blind alleys. Hopefully, though, we have highlighted some of the more important questions of theoretical and practical importance in the study of canopy arthropods. There is certainly fertile ground for future canopy arthropod studies.

Acknowledgements

We thank P. M. Hammond and P. Eggleton for valuable help, comment and discussion, and The Natural History Museum, London, for support.

REFERENCES

Barnard, P.C., Brooks, S.J. and Stork, N.E. (1986) The seasonality and distribution of Neuroptera, Raphidioptera and Mecoptera on oaks in Richmond Park, Surrey, as revealed by insecticide knock-down sampling. *Journal of Natural History*, **20**, 1321–31.

Basset, Y. (1997) Species abundance and body-size relationships in insect herbivores associated with New Guinea forest trees, with particular reference to insect host-specificity, in *Canopy Arthropods* (eds N.E. Stork, J. Adis and R.K. Didham), Chapman & Hall, London, pp. 237–64 (this volume).

Basset, Y., Springate, N.D., Aberlenc, H.P. and Delvare, G. (1997) A review of methods for sampling arthropods in tree canopies, in *Canopy Arthropods* (eds N.E. Stork, J. Adis and R.K. Didham), Chapman & Hall, London, pp. 27–52 (this volume).

Bellinger, R.G., Ravlin, F.W. and McManus, M.L. (1989) Forest edge effects and their influence on Gypsy Moth (Lepidoptera: Lymantriidae) egg mass distribution. *Environmental Entomology*, **18**(5), 840–3.

di Castri, F. (1991) Ecosystem evolution and global change, in *Perspectives on Biological Complexity* (eds O.T. Solbing and G. Nicolis), IUBS Monograph Series, No. 6, Paris, pp. 189–217.

Didham, R.K. (1997) Dipteran tree crown assemblages in a diverse southern temperate rainforest, in *Canopy Arthropods* (eds N.E. Stork, J. Adis and R.K. Didham), Chapman & Hall, London, pp. 320–43 (this volume).

Didham, R.K. An overview of invertebrate responses to forest fragmentation, in *Forests and Insects* (eds A.D. Watt, N.E. Stork and M. Hunter), The 18th Symposium of the Royal Entomological Society of London, Chapman & Hall, London (in press).

Didham, R.K., Ghazoul, J., Stork, N.E., and Davis, A.J. (1996) Insects in fragmented forests: a functional approach. *Trends in Ecology and Evolution* **11**, 255–60.

Erwin, T.L. (1995) Measuring arthropod biodiversity in the tropical forest canopy, in *Forest Canopies* (eds M.D. Lowman and N.M. Nadkarni), Academic Press, London, pp. 109–27.

Floren, A. and Linsenmair, K.E. (1997) Diversity and recolonization dynamics of selected arthropod groups on different tree species in a lowland rainforest in Sabah, Malaysia, in *Canopy Arthropods* (eds N.E. Stork, J. Adis and R.K. Didham), Chapman & Hall, London, pp. 344–81 (this volume).

Haddow, A.J., Corbet, P.S. and Gillett, J.D. (1961) Entomological studies from a high tower in Mpanga forest, Uganda. I. Introduction. *Transactions of the Royal Entomological Society London*, **113**, 249–56.

Hallé, F. (1995) Canopy architecture in tropical trees: a pictorial approach, in *Forest Canopies* (eds M.D. Lowman and N.M. Nadkarni), Academic Press, London, pp. 27–44.

Hammond, P.M., Stork, N.E. and Brendell, M.J.D. (1997) Tree-crown beetles in context: A comparison of canopy and other ecotone assemblages in a lowland tropical forest in Sulawesi, in *Canopy Arthropods* (eds N.E. Stork, J. Adis and R.K. Didham), Chapman & Hall, London, pp. 184–223 (this volume).

Kareiva, P. (1987) Habitat fragmentation and the stability of predator–prey interactions. *Nature*, **326**, 388–90.

Kruess, A. and Tscharntke, T. (1994) Habitat fragmentation, species loss, and biological control. *Science*, **264**, 1581–4.

Landsberg, J. (1990) Dieback of rural eucalypts: the effect of stress on the nutritional quality of foliage. *Australian Journal of Ecology*, **15**, 97–107.

Lasalle, J. and Gauld, I.D. (1991) Parasitic Hymenoptera and the biodiversity crisis. *Redia*, **74**, 315–34.

Lawton, J.H. (1978) Host-plant influences on insect diversity: the effects of space and time, in *Diversity of Insect Faunas* (eds L.A. Mound and W.A. Waloff), Symposium of the Royal Entomological Society of London, Volume 9, Blackwell Scientific Publications, Oxford, pp. 105–25.

Malcolm, J.R. (1997) Insect biomass in Amazonian forest fragments, in *Canopy Arthropods* (eds N.E. Stork, J. Adis and R.K. Didham), Chapman & Hall, London, pp. 510–33 (this volume).

Mawdsley, N.A. and Stork, N.E. (1997) Host-specificity and the effective specialization of tropical canopy beetles, in *Canopy Arthropods* (eds N.E. Stork, J. Adis and R.K. Didham), Chapman & Hall, London, pp. 104–30 (this volume).

Moran, V.C. and Southwood, T.R.E. (1982) The guild composition of arthropod communities in trees. *Journal of Animal Ecology*, **51**, 289–306.

Morse, D.R., Lawton, J.H., Dodson, M.M. and Williamson, M.H. (1985) Fractal dimension of vegetation and the distribution of arthropod body lengths. *Nature*, **314**, 731–3.

Nadkarni, N.M. and Parker, G.G. (1994) A profile of forest canopy science and scientists – who we are, what we want to know, and obstacles we face: results of an international survey. *Selbyana*, **15**(2), 38–50.

Ozanne, C.M., Hambler, C., Foggo, A. and Speight, M.R. (1997) The significance of edge effects in the management of forests for invertebrate biodiversity, in *Canopy Arthropods* (eds N.E. Stork, J. Adis and R.K. Didham), Chapman & Hall, London, pp. 534–50 (this volume).

Roland, J. (1993) Large-scale forest fragmentation increases the duration of tent caterpillar outbreaks. *Oecologia*, **93**(1), 25–30.

Roland, J. and Taylor, P. Forest insect dynamics in fragmented and continuous forests, in *Forest and Insects* (eds A.D. Watt, N.E. Stork and M.D. Hunter), 18th Symposium of the Royal Entomological Society of London, Chapman & Hall, London (in press).

Southwood, T.R.E., Moran, V.C. and Kennedy, C.E.J. (1982a) The assessment of arboreal insect fauna – comparisons of knockdown sampling and faunal lists. *Ecological Entomology*, **7**, 331–40.

Southwood, T.R.E., Moran, V.C. and Kennedy, C.E.J. (1982b) The richness, abundance and biomass of the arthropod communities on trees. *Journal of Animal Ecology*, **51**, 635–49.

Stork, N.E. (1987a) Guild structure of arthropods from Bornean rain forest trees. *Ecological Entomology*, **12**, 69–80.

Stork, N.E. (1987b) Arthropod faunal similarity of Bornean rain forest trees. *Ecological Entomology*, **12**, 219–26.

Stork, N.E. and Best, V. (1994) European Science Foundation – results of a survey of European canopy research in the tropics. *Selbyana*, **15**, 51–62.

Strong, D.R., Lawton, J.H. and Southwood, T.R.E. (1984) *Insects on Plants: Community Patterns and Mechanisms*, Blackwell Scientific Publications, Oxford.
Vitousek, P.M. (1994) Beyond global warming: ecology and global change. *Ecology*, **75**(7), 1861–76.

Index